Experimental Biochemistry

EXPERIMENTAL BIOCHEMISTRY

by

Robert L. Dryer
Late of The University of Iowa

and

Gene F. Lata
The University of Iowa

New York Oxford
OXFORD UNIVERSITY PRESS
1989

OXFORD UNIVERSITY PRESS

Oxford New York Toronto
Delhi Bombay Calcutta Madras Karachi
Petaling Jaya Singapore Hong Kong Tokyo
Nairobi Dar es Salaam Cape Town
Melbourne Auckland

and associated companies in
Berlin Ibadan

Published by Oxford University Press, Inc.,
200 Madison Avenue, New York, New York 10016

Library of Congress Cataloging-in-Publication Data
Dryer, Robert L.
Experimental biochemistry/Robert L. Dryer and Gene F. Lata.
p. cm. Includes index. ISBN 0-19-505083-5
1. Biochemistry—Laboratory manuals. 2. Biochemistry—Research—Methodology.
I. Lata, Gene F. II. Title. QP519.D79 1988 574.19′2′078—dc19
88-11754 CIP AC

ISBN 0-19-505083-5

9 8 7 6 5 4 3 2 1
Printed in the United States of America
on acid-free paper

Dedicated to the Memory of
Robert L. Dryer
1921–1984

This book is a resultant of two separate volumes used together in our course for several years. The original theoretical component was written as a syllabus by Dr. Dryer shortly before his untimely death; the experimental section was in large part also assembled by him. Both portions have been revised during the preparation of this volume, but the genesis of the idea for this product must properly be attributed to Dr. Dryer.

ACKNOWLEDGMENTS

I would like to acknowledge the good-natured patience and efficiency of Brenda Kunkel and Cay Wieland in typing the manuscript, as well as the competence of Todd Erickson in drawing the illustrations.

Contents

Section I Theory

Chapter 3. Nuclear Instability and Radioactive Transformations

Chapter 4. Buffers and pH Regulation

Chapter 11. Introductory Aspects of Molecular Biology

Section III Appendix

Introduction

From its beginning, biochemistry has been a largely experimental science. Early biochemists were forced, by the crudity of their tools and methods, to work with whole plants or animals. As the field progressed, successive generations of biochemists were enabled to study tissue slices, tissue homogenates, isolated organelles, and even what are now popularly known as "molecular" systems. This search for ever simpler systems arose from the belief that the simpler the system the more precisely it could be defined and studied.

Recently, as part of the generalized fusion of the biological sciences, we have seen a return to more complex systems such as tissue culture, in which intact cells may be isolated, grown, harvested, and then marvelously manipulated. Even more complex systems, such as isolated pancreatic islets, are now considered to be effective biochemical tools for certain studies. Bacteria are being "engineered" into producing materials that are rare and costly when produced by more traditional methods. With all of this, we now have analytical methods capable of dealing with nanogram (and frequently with subnanogram) quantities of materials.

Unfortunately, however, baccalaureate students of today get an all too limited exposure to basic laboratory science. It is time-consuming and very expensive to mount a good experimental course using modern tools and techniques. Student time is at a premium, given the pressures of other courses and legitimate interests. In spite of these difficulties, we have a strong belief in the value of laboratory training in baccalaureate as well as in graduate programs. We hope to introduce young students to what is new and exciting, but at the same time we want to be sure they are reasonably grounded in what is not so new, perhaps not so exciting, but still very important. In this book and in the course from which it was derived, emphasis is placed not just on the operational mechanics of the experiments, but also on an understanding of the rationale behind each protocol and of the theoretical bases for the individual methods and procedures used. The book is divided into two parts. In section I are presented fundamental treatments of physical, chemical, and mathematical topics to serve as a background for the experiments in section II. At the beginning of each experiment, reference is made to those chapters in section I that most directly pertain to the experiment; a student should study them together with any other assigned readings before coming to class.

The experiments in section II constitute a tested, one-semester course, which is required of our undergraduate majors usually in the first semester of their junior year. Most of the students have completed at least one semester of didactic biochemistry and are co-registered in another. This class involves one weekly lecture and two laboratory periods of four to eight hours each.

The experiments we have selected emphasize three major points: first, all systems that obey the laws of chemistry are "molecular"; second, it is necessary to be as quantitative as possible in all experimental work; third, basic principles and fundamental knowledge form the basis of all later professional work. Only a few of the later exercises deal with what might be called molecular biology; its full treatment requires an additional course dedicated to that purpose. We make no claim to originality in these experiments. Indeed, they all had their genesis in previously published research. Neither is this collection the work of any single individual. Since its inception, a number of our biochemistry faculty have made significant contributions. In particular, we want to recognize the efforts of Bryce Plapp, Peter Rubenstein, Earle Stellwagen, and Joseph Walder. Additionally, at least several dozens of graduate student assistants have helped us refine these protocols to their present state.

This program is not static; it will continue to evolve, reflecting technical developments in the tools of biochemistry as well as its expanding theoretical horizons.

SECTION I

Theory

CHAPTER ONE

Manipulation of Volumes, Masses, and Concentrations

A fundamental chemical principle, first clearly enunciated by John Dalton, states that in any chemical reaction materials combine or are transformed in proportion to their equivalent weights, or some small multiple of those weights. As early as 1805, some sixty years before Mendeleyev, Dalton had formulated a prototypic table of atomic weights based on what we now know as Dalton's law. The history of quantitative chemistry goes back at least this far.

Most biochemical reactions take place in solution, so that when one writes a chemical reaction it indicates not only how reactants are converted to products, but also the stoichiometry (the quantitative relationships among them). For example,

$$\begin{array}{c} COO^- \\ | \\ H_2NCH \\ | \\ CH_2 \\ | \\ CH_2 \\ | \\ COO^- \end{array} + NAD(P)^+ + H_2O \underset{\text{dehydrogenase}}{\overset{\text{glutamate}}{\rightleftharpoons}} \begin{array}{c} COO^- \\ | \\ O{=}C \\ | \\ CH_2 \\ | \\ CH_2 \\ | \\ COO^- \end{array}$$

$$+ NAD(P)H + H^+ + NH_4^+$$

specifies that glutamate can be reversibly converted to α-ketoglutarate by the catalytic action of the enzyme glutamate dehydrogenase in the presence of $NAD(P)^+$; it also specifies that for each equivalent (or microequivalent, μeq) of glutamate oxidized, the sample equivalent quantity of $NAD(P)^+$ must be reduced. To apply this stoichiometric principle in quantitative analysis one must employ solutions of known concentration. In turn, this requires precise measurement of volumes and masses of materials, and gives to such measurements a primacy among all the unit operations of the chemistry laboratory.

DETERMINATION OF VOLUME

Except for approximations, volumes should be measured in volumetric flasks and *not* in graduated cylinders for two reasons: First, manufacturers use much closer tolerances for the calibration of volumetric flasks than for graduated cylinders. Second, the shape of graduated cylinders, which ordinarily are right cylinders, makes for a large relative error in volume for a given error in adjusting the height of the contained fluid column. When the use of a cylinder is tolerable, the size of the cylinder should be selected so that the measured volume is at least one-third to one-half of the cylinder capacity.

Volumetric flasks are available in sizes ranging from 3 ml to more than 1 L. They are built with bulbous portions that contain the greater part of the nominal volume, surmounted by a long, narrow neck on which is engraved a ring or mark indicating the correct filling point. For use with aqueous solutions, the bottom of the fluid meniscus is set just at this mark. The narrow neck minimizes filling errors. Most volumetric flasks are fitted with stoppers, the tightness of which should always be checked before use.

Good practice dictates the following usage. After some solute is transferred to the flask, the bulb of the flask is filled approximately half-way with water or other diluent. The flask is then stoppered and shaken well. More diluent is added, the flask is stoppered and shaken again, and the process is repeated until the diluent is up to the mark. In this way, one is not taken by surprise should the solution change in volume as the solute is dissolved. If there is a temperature change on solution, the height of the fluid column should be left below the mark until the solution has come to room temperature; then the volume should be finally adjusted.

Dispensing Fluid Volumes: Use of Pipets

Having prepared an appropriate solution, one must be able to dispense known portions, or aliquots, of the total volume. In most instances, the aliquot will be fixed in size and relatively small. The classic tool for this purpose is the pipet. Pipets are small glass tubes with drawn-out tips. Virtually all pipets are filled by suction applied to the open end of the tube; however, in a few instances involving pipets of very small volume, they may be filled by capillary action.

Pipets take a number of forms. *Volumetric pipets* (Fig. 1–1A) usually bear a single calibration mark; depending on the size, there may or may not be a bulb blown between the tip and the calibration mark. When suction is released, the contents drain down until only a small volume, held by capillarity, remains just above the tip. This should *not* be blown out, since the calibration is *to deliver.* Volumetric pipets are therefore marked *TD. Serologic pipets* and *Mohr pipets* (Fig. 1–1B,C) are commonly calibrated in partial volumes, so that a 5-ml pipet may have a calibration mark indicative of 0.5 ml along its entire working length. These are customarily calibrated right to the tip. After free drainage has ceased, the

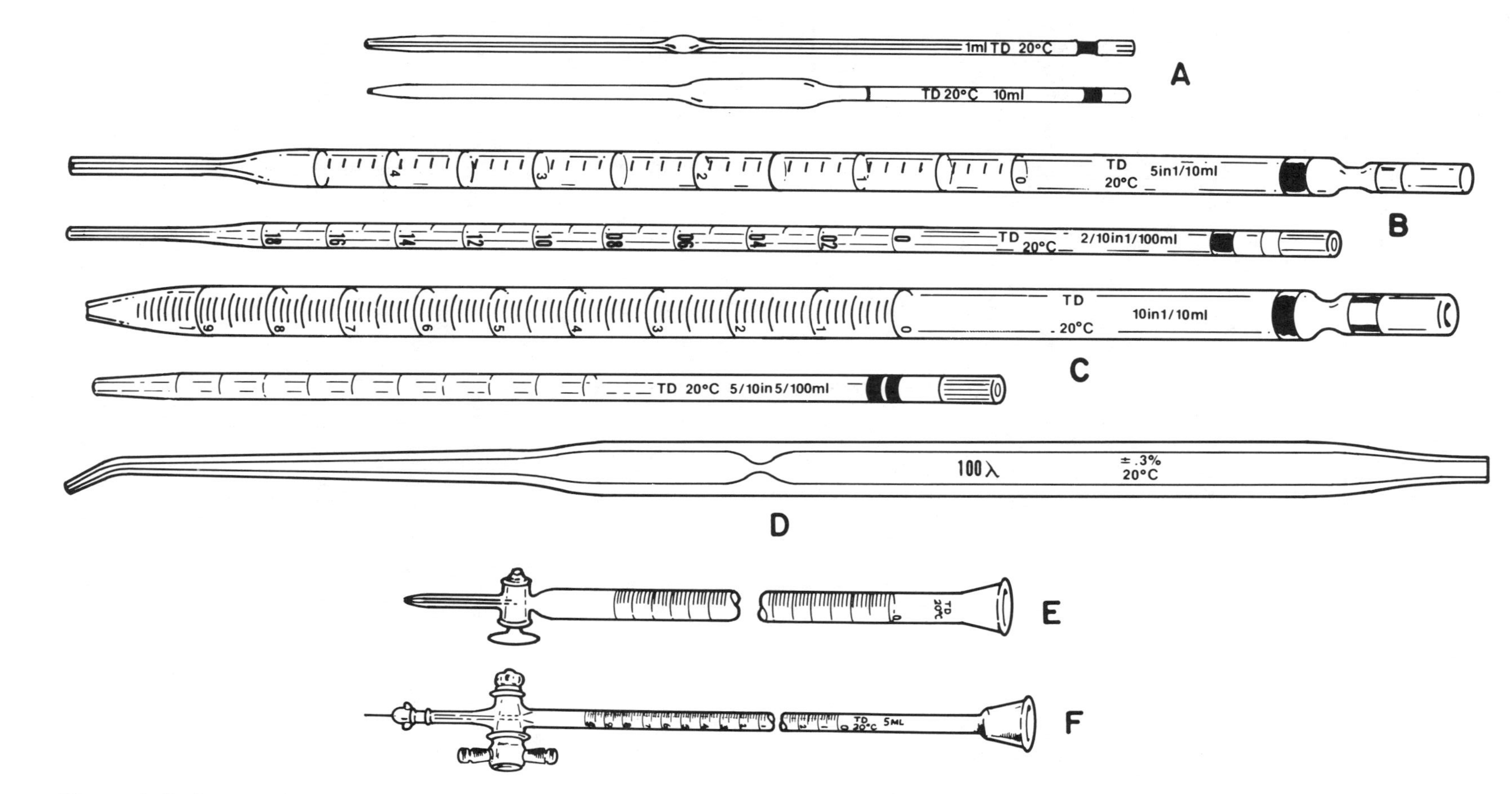

Figure 1–1. Some typical glass pipets and classical burets. Volumetric pipets (A) of several styles; two forms of transfer, or Mohr, pipets (B); two sizes of serologic pipets (C); and a Linderström–Lang micropipet (D). Also shown are burets, one with a fixed tip (E), and one with a replaceable tip made of hypodermic needle tubing (F). Note that these illustrations are not to scale. See the text for further details.

residual volume *must* be blown out since these are calibrated *to contain*. They are therefore marked *TC*. Some manufacturers, however, have introduced serologic-type pipets where the calibration ends just above the tip; these pipets are delivering between two marks, so they should not be blown out. Generally pipettes that must be blown out are marked with a ground-glass ring near the suction end. The practice is confusing, and one should check the nature of every pipet before use. Pasteur pipets are uncalibrated tubes, drawn out to fine tips, and are used for transfer of fluid without regard to precise volume. They generally hold about 1.5–2.0 ml, and are most conveniently employed with a small rubber suction bulb.

For volumes of 100 μl or less, pipets of the sorts described above (e.g., Fig. 1–1A–C) are not practical. As chemistry progressed to the microscale, new pipet designs were developed. Among the very best is the Linderström–Lang pipet (Fig. 1–1D). In appearance it is similar to a volumetric pipet, but the internal bore is much smaller. Instead of a filling mark, the best pipets of this variety have a narrow constriction above the bulb. This constriction offers great resistance to further filling. Loading the Linderström–Lang pipets is begun by gentle suction, and allowed to come to the constriction by capillarity. They must be blown out by air pressure applied through a rubber mouthpiece. Good-quality pipets of this form are surprisingly accurate, but they are also expensive and fragile. Fragility is due to the fact that the wall thickness at the tip is made as thin as possible to minimize the surface on which fluid may accumulate.

Cheaper and less accurate micropipets are made of thin-wall tubing of constant cross section and are usually disposable. These may have one or more calibration marks, but the marks cannot be engraved. They are painted on the tubes as heavy bands or stripes of enamel. One is never sure of where to set the meniscus, or if the bore is constant along the length of the tube. Other versions bear no calibration marks whatever; instead, the tube is filled from end to end, and the total fluid column is taken to be the contained volume. Again, the possibility that the bore is not constant makes precision work with such pipets impossible, as may errors in the length. At best, they must be assumed to contain approximate volumes, perhaps within ±10% of the nominal volume.

Pipet Errors. All of the above pipets are subject to serious errors when used with viscous aqueous solutions or with nonaqueous solutions. Viscous solutions do not drain readily, and the retained film of fluid may constitute an appreciable part of the total volume. Nonaqueous solutions have surface tension characteristics that are very different from those of aqueous solutions; the drainage may be greater than anticipated by the calibrator, and an error will result.

All glass pipets are intrinsically fragile. Properly drawn-out tips tend to chip or break easily, destroying the initial drainage rate and perhaps affecting the original volume calibration. If the glass at the tip is made thicker in order to withstand hard usage, too much fluid may be retained over the added surface; conversely, if the glass at the tip is too thin, breakage will surely result. In any event, no pipet with a chipped or broken tip should ever be employed in a serious measurement.

Pipets are also hard to clean properly, especially when used with protein- or lipid-containing solutions. A properly cleaned pipet should drain freely, leaving a virtually invisible liquid film over its inner surface. If droplets can be seen on the inside of a pipet, that is evidence of improper cleaning. Volumes measured under these conditions are clearly in error. Linderström–Lang pipets are particularly hard to clean; they also have a tendency to plug at the tip if the solutions contain traces of suspended dust or other particulate matter. A comprehensive review of volumetric and pipet errors in particular is given in the references.

Piston-Displacement Pipets. Chemists have long sought an alternative to the classic pipets and the problems they pose. A number of instruments are now available which differ radically from those discussed above. The body of the instrument consists of a spring-loaded piston moving inside a carefully machined cylinder. The external end of the piston is fitted with a knob by which the piston may be pressed into the cylinder for some predetermined distance that is a measure of the volume. The other end of the cylinder is shaped to form an airtight seal with a disposable plastic tip. When the piston is depressed, the measured volume of air is expelled through the plastic tip; release of the piston draws an equal volume of liquid into the tip. Models of fixed and variable volume are manufactured. Because fluid never enters the working part of the cylinder and never contacts the piston (except in cases of misuse), the integrity of the calibrated volume can be maintained over long periods of time. Because the fluid is drawn into the instrument by piston-generated suction, mouth pipeting is eliminated. The thickness of the tip wall is very small so that drainage errors are reduced and held closely at a constant value. Very viscous and nonaqueous solutions are still a problem, but less so than with glass pipets. The advantages of piston-displacement devices are such that they are rapidly taking the place of the glass pipets in most biochemistry laboratories (see Section II, Expt. 1, Fig. E1–1).

Titrations: Use of Burets

In experiments such as acidimetry or ligand-binding studies, one wishes to determine the quantity of a molecular species by adding, in the presence of an indicator, as much of a reagent as is needed to reach an equivalence point. Such a procedure is known as a *titration,* and it involves the repeated addition of small, varying aliquots of the reagent. It would be quite tedious and inconvenient to do this accurately with a pipet. A more suitable tool is known as a *buret.* A buret is a long tube, calibrated for volume, bearing at its lower end a stopcock terminating in a drawn-out tip (Fig. 1–1E,F). Depending on the cross section of the tube and the structure of the tip, it is possible to dispense drops of about 10–20 μl. As is true of pipets, burets are difficult to clean properly, and the fixed tips are easily damaged or chipped. Once the tip is chipped, it is impossible to deliver as small a drop as before. To avoid the problem of chipping, some microburets are built to accept tips constructed of fine hypodermic needles, which are replaceable.

Syringe burets are far more practical than open-tube burets of the sort just described. Syringe burets consist of syringes designed to have a constant cross-sectional area, such that advance of the piston into the barrel by a standard distance expels a known fluid volume. Syringes provided for use as burets come in a variety of sizes ranging from 0.5 to 5.0 ml per inch of piston displacement. They may be recognized easily since they bear no volume information other than a statement of milliliters expressed per inch. The syringe is clamped on a stand that also carries a micrometer screw with a scale of 1000 divisions. With the smallest syringe installed, each scale unit therefore represents expulsion of 0.5×10^{-3} ml or 0.5 μl. The surface tension of most aqueous solutions does not permit the formation of so small a drop, and delivery from the syringe outlet is normally made through a length of very fine plastic tubing that dips into the solution being titrated.

Syringe burets are a great convenience because the syringe can be demounted and taken apart for thorough cleaning, eliminating a major source of error encountered with tube burets. The plastic delivery tip does not chip or break, and it can be replaced easily and cheaply if it gets contaminated. Because no stopcocks are involved, there is no stopcock grease to foul the bore of the delivery tube; neither are there leaks.

The use of a buret in titrations is a direct application of Dalton's law. Suppose one has a volume of solution A, of unknown concentration. If A reacts with B in a stoichiometric manner, and if the concentration of solution B is known, then Dalton's law states that:

$$E_a = E_b \tag{1-1}$$

where E represents the equivalents of the subscripted material. Since $E = NV$, where N is the normality (i.e., equivalents per liter) and V is the volume, Eq. 1–1 can be stated as:

$$N_a V_a = N_b V_b \tag{1-2}$$

$$N_a = \frac{N_b V_b}{V_a} \tag{1-3}$$

and thus the concentration of A is determined. Equation 1–3 is the fundamental equation of volumetric analysis.

DETERMINATION OF MASS

From a chemist's point of view, when one "weighs" an object, one is actually making an estimate of its mass, and the tool used is known as a balance. The unknown mass is balanced against a number of precalibrated

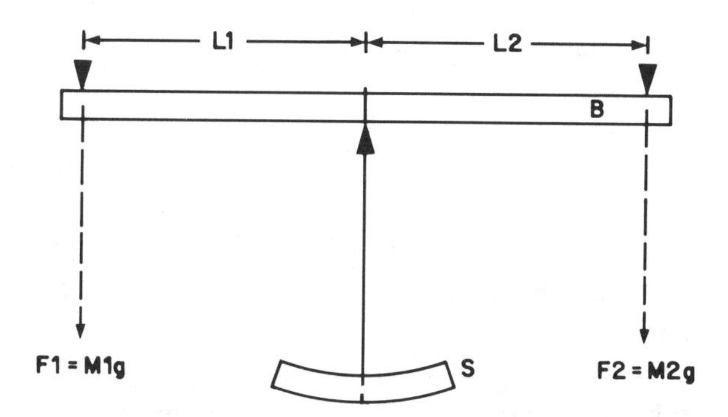

Figure 1–2. Principle of equal-armed analytical balance. A beam of light alloy rests on a central fulcrum. At either end, two pans (not shown for clarity) are suspended from other fulcri (shown by the filled triangles). The center of the beam also carries a pointer, the exact angle of which can be read against markings on the scale, S. At equilibrium, $L_1 \times F_1 = L_2 \times F_2$. The "weight" of an object is its mass times the force of gravity. In common practice, the force of gravity is ignored, and the weight is taken to be the mass. In biochemical usage, the buoyancy error is similarly ignored.

bits of metal, certified by a manufacturer as having masses of 1.0000 ± 0.0001, or 5.0000 ± 0.005, or whatever. Balances differ in their capacity, sensitivity, and precision according to their design, intended purpose, and the quality of their construction, but in all cases the fundamental principle is the same. It is important to understand the limitations of balances generally available, and to know how to use them correctly.

Analytical Balances

In its simplest form, a balance is a lever system of the first class, which means that $L_1 = L_2$ (see Fig. 1–2). At equilibrium, $M_1L_1 = M_2L_2$, where M_1 and M_2 are the known and unknown masses, and the indicating pointer will lie just over the central index mark of the fixed scale. When an object of unknown mass is placed on a pan, $M_1L_1 > M_2L_2$, and the pointer will swing away from the index mark until some collection of items of known mass is placed on the other pan to restore equilibrium.

This version of the equal-arm, double-pan balance is still very widely used. Analytical balances of this kind are enclosed in a case to protect the moving parts from drafts of air, the beam is of a more complex pattern than that shown in Fig. 1–2, and both the beam and the pans are fitted with arresting mechanisms to protect the finely cut and polished agate knife edges and to facilitate loading or unloading of the pans. In some models the indicating pointer is replaced by a light source, a system of mirrors, and an engraved reticle onto which a shadow is cast to determine the rest point of the beam. Most balances are also equipped with a level and leveling feet to ensure that at equilibrium the beam is truly horizontal. With a balance of moderate quality, it is not difficult to determine masses of as little as 10^{-4} g; and with balances of highest quality, it is possible to do somewhat better, perhaps 10^{-5} g. The capacity of analytical balances is usually 100–200 g, and this limit should never be exceeded, to avoid beam damage.

Single-Pan Balances. No matter how careful the design and no matter how well-engineered the construction, the beam of a balance will always flex to some extent under load, affecting the sensitivity of a balance. To offset this problem, newer balances operate under full load at all times, and weighings are actually made by a process of substitution. This is the operational basis of the so-called single-pan balances. In most single-pan balances, all one sees through the doors of the balance case is the single pan. Perhaps suspended well above it is a system of closed, lightweight cylinders which serve as an air damper to slow the moving parts. It is important to realize that, hidden behind an opaque shield, there is a beam and a countermass, just as in the classic double-pan instruments. When one removes the case top from a single-pan balance, the following components are revealed: Just behind the opaque shield, and suspended from its own knife edge, is a mass equivalent to 200 g, if that is the capacity of the balance. At the far end of the beam is a mass equal to the mass of the single pan plus the air damper (if that is installed). In addition, the far end of the beam carries a hanger from which are suspended a number of mass pieces, 1.0000, 2.0000, 5.0000, 10.0000, and so forth, the sum of which is also 200 g. A system of hooks, arranged to be operated by controls on the exterior of the case, can lift one or more of the fixed mass pieces from the hanger or replace them into the hanger. With all of the mass pieces on the hanger, the balance would be at equilibrium and the indicator should rest at the zero point. When an unknown object is placed on the pan, the equilibrium is disturbed and can be restored only by removing an equivalent mass from the hanger. Because the weights are never touched, they are more likely to maintain their original calibration. With care, single-pan balances will perform at specified limits for several years without any service whatever.

Mechanical Top-Loading Balances. These balances are modifications of the single-pan balance and frequently have a considerably larger capacity. The case encloses only the mechanism; the pan is exposed for ease in weighing dry reagents, flasks, etc. Typical top-loading balances have unit division equal to 100 mg and are of far greater convenience and reliability than direct weighing machines of comparable capacity. As noted above, the principle of weighing by substitution means that they will retain that degree of accuracy and reliability for long periods of time with only modest care and cleaning. Some models are equipped with digital readout in place of analog devices, although this offers little or nothing by way of performance.

Electromagnetic Balances. In all of the models discussed thus far, balance is accomplished by the physical act of adding (or removing) finite mass pieces from one end of the beam. Another way of accomplishing balance is by the application of electromagnetic force. Two of the major types of electromagnetically based balances are the top-loading models and the Cahn balance. The physical basis of the top-loading balance (see

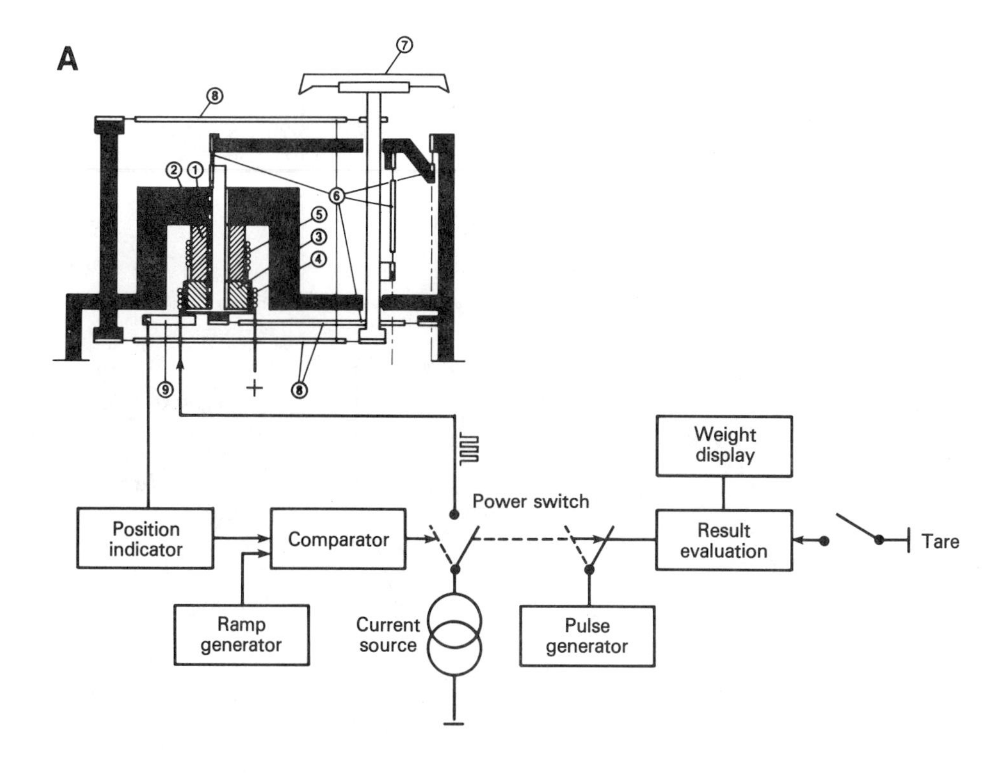

Figure 1-3. Top-loading electromagnetically compensated balance. (A) Schematic of essential components: ① and ② magnets; ③ pole shoe; ④ compensation coil; ⑤ temperature compensation; ⑥ flexible bearings; ⑦ weighing pan; ⑧ guides; ⑨ position indicator. The position indicator ⑨ determines the vertical position changes of the pan support ⑧ when the pan ⑦ is loaded. These changes are used to generate the compensation current which then zeroes the compensation coil ④ of the wiring system. This current is digitized and translated into a corresponding weight display. (B) External view showing digital display. (Courtesy of Mettler Instrument Corporation.)

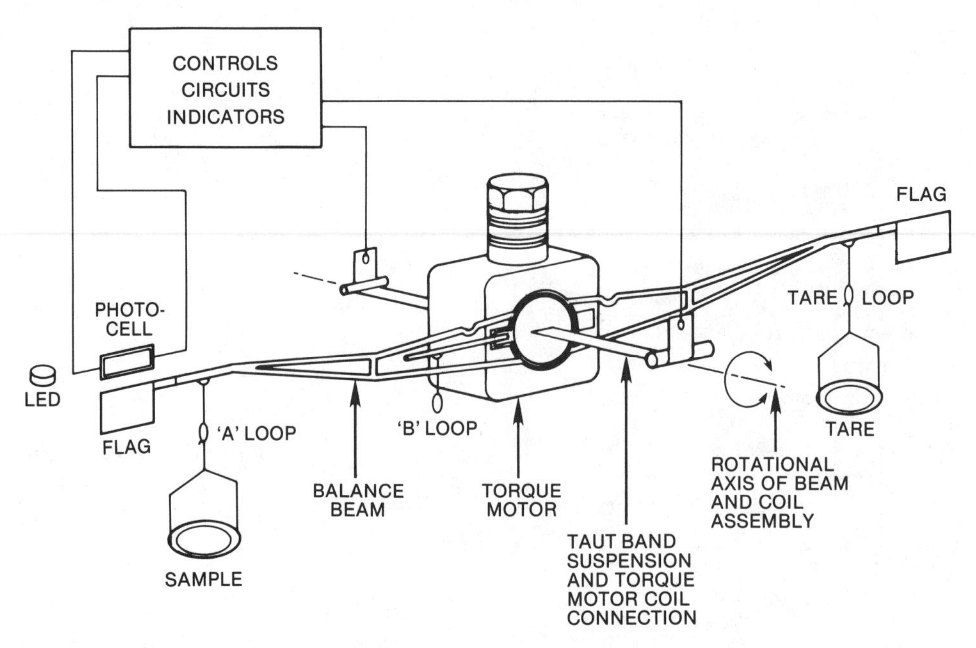

Figure 1–4. Principle of an electromagnetic balance. Loop A and tare loop C are equidistant from the ribbon suspension. Loop B is used only for larger masses, according to the principle of lever moments. No knife edges are involved since friction at the suspension points of the sample pan or the calibrating pan is very slight. The circuitry may be powered by a solid-state power supply. (Modified from data supplied by Cahn Instruments, Inc., Cerritos, Calif.)

Fig. 1–3) is briefly described as follows: If a wire is placed between the poles of an electromagnet and a current passed through the wire, the interaction between the electromagnetic field generated around the wire and the existing electromagnetic field results in a force which moves the wire out of the field. In a balance there is an arrangement whereby a weighing pan is connected to the wire. As weight is added to the pan, the movement of the wire out of the region between the poles of the electromagnet is opposed.

The current flow through the wire is automatically increased until the original "no-weight" position of the wire is regained. The difference in wire current between the empty-pan condition and the weighted-pan condition is proportional to the weight. This difference, appearing as a weight, is displayed on a calibrated scale. The unit division for such balances may be 1, 10, or 100 mg.

The Cahn balance is somewhat similar to the above. A coil, delicately suspended in a magnetic field, experiences a torque when current passes through the coil. This is the principle common to galvanometers and electric motors. Figure 1–4 shows how this principle can also be extended to quite rugged but very sensitive balances. A light-weight beam is mounted at right angles to the axis of a coil in a magnetic field. At loop C, one can place a calibrating weight of as little as 5 mg. This causes the left-hand end of the beam to rise, moving the shadow of the beam pointer above the zero rest point etched on a translucent screen. The beam is restored to its equilibrium position by means of the resistor, R_1, which regulates current passing through the coil and so applies a restoring torque. When a sample is added to loop A, the left end of the beam drops below the rest point and equilibrium is again restored by adjustment of the resistor (unmarked in Fig. 1–4). Finally, the voltage across the coil is measured by a one-turn precision potentiometer, R_2, and the galvanometer, G. The potentiometer is connected to a dial scale marked off in 1000 divisions. Since the calibration mass in this example was 5×10^{-3} g, and since the scale is marked off in 10^{-3} divisions, each scale unit is equal to $(5 \times 10^{-3})(10^{-3}) = 5 \times 10^{-6}$ g, or 5 μg. The typical range of this type of electromagnetic balance is 1–5 g, and it can easily weigh samples of a few micrograms. Frequently, the sample container will weigh more than the sample itself. Electromagnetic balances are ruggedly built, but especial care must be taken not to damage the very lightly constructed beam. Within their limited range, they obviously qualify for use where analytical accuracy is required.

Care of Analytical Balances. Analytical balances of any type are expensive, precision instruments and should be treated with care and respect. They provide after all, one of the two primary standards of any laboratory. The rules for balance care are few and simple, but they should not be neglected.

1. *Always* check, before use, that the balance is properly leveled.
2. *Never* weigh materials by adding them to the balance pan directly. *Always* use weighing paper, plastic, or aluminum weighing dishes or tared watch glasses.
3. *Always* arrest the beam when adding or removing material from the balance, or when changing the mass pieces in steps of more than 100 mg. (**Note:** In the case of top loading balances, there is no beam arrest; thus this rule does not apply.)
4. *Carefully* remove any spilled material from the balance pan or the balance case to avoid corrosion of the glass or metal parts.
5. When weighing is completed, *always* return the controls to zero readings.

Secondary Balances

For bulk work, or for work of less than analytical accuracy, balances of secondary quality are important and useful. Secondary balances have sensitivities in the range of ± 10 to ± 100 mg, depending on their capacities, which range from 100 g to 10 kg. These balances come in a wide variety of styles. Some are double-pan units, of 1- to 5-g capacity, and depend on a torsion suspension instead of knife edges. They were once quite popular for pharmaceutical purposes. Some are single-pan, unequal-arm balances, with a series of counterweights that slide along one or more calibrated beams. In many of these, one counterweight measures tens of grams, another measures grams, and still another measures hundredths of grams.

Still others are patterned after grocery store balances and have capacities large enough to weigh small animals. Even though secondary balances are less precise than analytical balances, it is best to follow the rules outlined above.

EXPRESSIONS OF CONCENTRATION, OR MASS PER UNIT VOLUME

As was stated at the beginning of this chapter, a major need for measurement of volume and mass is to define the concentration of a solution. Concentration can be expressed in many ways, and one expression can be converted to another. Certain ways of expressing concentration are more common than others, only because they are particularly suited for a wide range of applications.

Weight/Weight: A 10% solution (wt./wt.) of NaCl would be made by taking 10 g of the salt in 90 g of water; the result would be expressed as 10% (w/w).

Weight/Volume: The weight of solute per deciliter of solution is an example. A 10% solution (w/v) is made by adding 10 g of the salt to a vessel calibrated at 100 ml and then adding water to the mark.

Volume/Volume, or Volume %: This expression is used primarily for solutions of one liquid in another. A 70% ethanol solution would be made by taking 70 ml of absolute ethanol and adding water sufficient (after thorough mixing) to make 100 ml. The result would be expressed as 70% (v/v) ethanol. The usual laboratory grade of ethanol is only 95% by volume, however, so the solution made as above but using 95% ethanol would be only 66.5% (v/v). Common convention ignores the slight difference unless absolute (100%) ethanol is clearly indicated.

The latter example points out the ambiguity in statements of concentration expressed on a percentage basis. Matters become worse when one is dealing with reagents such as concentrated hydrochloric acid, where the usual solution is only 37% HCl and the density of the reagent is quite different from 1.

Molal solutions: One gram molecular weight dissolved in 1000 g solvent gives a 1 molal solution. Although this measure has considerable utility in some physicochemical studies, it is not widely used because of the inconvenience in weighing large volumes.

Molar solutions: One gram molecular weight dissolved in solvent sufficient to make 1 L of solution is expressed as 1 M or 1 mol/L. This is the most widely used expression of concentrations.

Normal solutions: One gram equivalent weight dissolved in solvent to make 1 L gives a 1 N solution. This may or may not be identical with a 1 M solution. Thus, a solution of H_2SO_4 containing 1 mol/L would be 2 N, since each mole of H_2SO_4 contains two replaceable, or dissociable, protons.

Care must be taken here of the use to which the solution is put. For example, a 1 mol/L of $CuSO_4$ might be 1 N or 2 N, depending on whether a reaction resulted in change of oxidation state, as

$$Cu^{2+} \rightarrow Cu^{1+}, \quad \text{or } Cu^{2+} \rightarrow Cu^{0}$$

One must also distinguish between the chemical and physiologic meaning of the term "normal." Physiologic saline, or "normal" saline, is 0.154 N and is isosmotic with body fluids.

REFERENCES

Christian, G. *Analytical Chemistry.* 2nd ed. John Wiley and Sons, New York (1977).

Conway, E. J. *Microdiffusion Analysis and Volumetric Error.* Crosby Lockwood and Son, London (1950).

Kolthoff, I. M., and Sandell, E. B. *Textbook of Analytical Chemistry,* 3rd ed. Macmillan Publishing, New York (1952).

CHAPTER TWO

The Interaction of Radiant Energy with Matter

NATURE AND PROPERTIES OF ELECTROMAGNETIC RADIATION

Radiant energy is composed of electromagnetic waves. These are best described as the resultant of an *electric vector* and a *magnetic vector,* at right angles to each other and to the direction of wave propagation (see Fig. 2–1). All electromagnetic waves travel at the speed of light, which in a vacuum is about 3×10^{10} cm/sec ($= c$). In other media, the speed of electromagnetic waves is slightly retarded by a factor known as the *refractive index, n.* This informs us that while in a vacuum $n = 1.0000$, for all other media $n > 1.0000$. Because the component electric and magnetic vectors vary in a periodic fashion (usually in a sinusoidal fashion), electromagnetic waves are commonly described in terms of their wavelengths (λ) and frequency (ν), which are related by the expression

$$\lambda = \frac{c}{n\nu} \tag{2–1}$$

The wavelength of light is that property which we perceive as its color.

An alternative description of electromagnetic radiation is provided by quantum theory, which indicates that radiant energy travels in small packets, or quanta, known as *photons.* The laws of quantum physics require that interactions between radiant energy with any other form of matter involve only integral numbers of photons. The energy content of a photon is given by the expression

$$E = h\nu \tag{2–2}$$

where h is defined as Planck's action constant. This constant has a value of 6.6×10^{-27} erg·sec, which is a very small quantity. It is frequently more useful to deal with the energy content of 1 "mol" of photons or *1 Einstein,* obtained by multiplying by Avogadro's number. Such calculations indicate that for red light, 1 Einstein of photons corresponds to approximately 50 kcal/mol. Equations 2–1 and 2–2 indicate that the energy per photon increases directly with its frequency and inversely with its wavelength.

The Electromagnetic Spectrum

The useful range of the electromagnetic spectrum is shown in Table 2–1. The division into identifiable regions is entirely arbitrary; in several

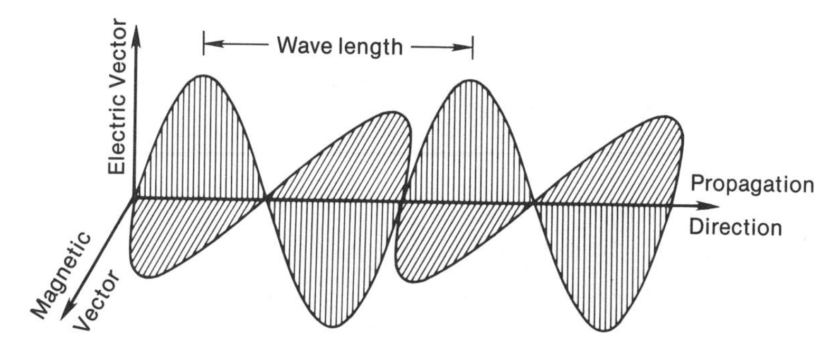

Figure 2–1. *An electromagnetic wave. The electric vector and the magnetic vector are at right angles to each other.*

regions the useful boundaries may overlap one another. Virtually all of the identified spectral regions are now used in technology and research, frequently as aids in probing the structure or the quantity of some kind of matter. Consequently, we say that interaction of radiant energy with matter is the basis of spectroscopy. Chemists and biologists can command the use of x-ray spectrometers, infrared (IR) and nuclear magnetic resonance (nmr) spectrometers, as well as visible and ultraviolet spectrophotometers. These instruments tend to be complicated and expensive, but their use is justified by the speed and quality of the data they furnish. A second important advantage is that they do not, as a rule, consume or damage the material sample.

Wave Phase and the Interaction of Two Waves

A sine curve can be generated by rotating a unit vector within a circle of unit radius, as shown in the diagram below. By plotting h as a function of

Table 2–1. The Electromagnetic Spectrum

Wavelength (nm)[a]	Frequency (Hz)[b]	Spectral Region	Interactions Between Energy and Matter	
10^{10}	3×10^{7}			
10^{9}	3×10^{8}			
10^{8}	3×10^{9}	Radio waves	Spin orientations (e.g., nmr, esr[c], astrospectroscopy)	
10^{7}	3×10^{10}			
10^{6}	3×10^{11}			
10^{5}	3×10^{12}			
10^{4}	3×10^{13}	Infrared	Vibrations and/or rotations, bond stretching or bending	
10^{3}	3×10^{14}			
10^{2}	3×10^{15}	Visible light	Valence shells	electronic transitions
10^{1}	3×10^{16}	Ultraviolet		
10^{0}	3×10^{17}			
10^{-1}	3×10^{18}	X-rays	Inner shells	
10^{-2}	3×10^{19}			
10^{-3}	3×10^{20}	Gamma rays	Nuclear transitions	
10^{-4}	3×10^{21}			

[a] The unit is the nanometer, 10^{-9} meter; in the older literature this is sometimes identified by the term, millimicron, represented by mμ. By more recent conventions, mμ should no longer be employed.

[b] The unit is the Hertz, or cycles per second.

[c] esr, electron spin resonance.

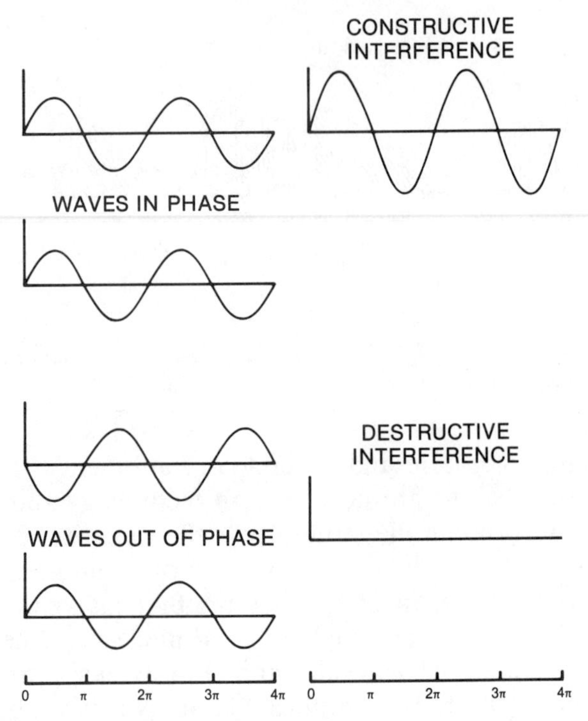

Figure 2–2. Phase relationships of two waves. Note the reinforcement of wave amplitude when waves are in phase (column 1) and the "canceling out" effect when they are out of phase.

θ, the sine θ curve results. This convention is employed in analysis of wave motion in general, and we shall use it here in dealing with electromagnetic radiation.

It is necessary to understand what is meant by the term *wave phase.* Two waves are said to be in phase if comparable points on both waves are displaced from the origin in the same direction by exactly the same magnitude. Two waves are said to be of opposite phase if the comparable points are displaced from the origin in opposite directions by exactly the same magnitude. A diagram of these two statements is given in the left-hand column of Fig. 2–2. If two in-phase waves impinge upon a suitable detector simultaneously, the result is a wave of exactly the same frequency as the original waves but of twice the *amplitude,* or maximum displacement from the origin. This is shown at the top of the right-hand column of Fig. 2–2. Such a result is termed *constructive interference.* It is constructive because the two waves reinforce each other. If, instead, the two waves are of exactly opposite phase, the result is termed *destructive interference.* It is destructive because two equal but opposite quantities sum to zero and thus cancel each other. A third possibility is easily recognized in which two waves are out of phase by some quantity other than π radians. The waves again suffer destruc-

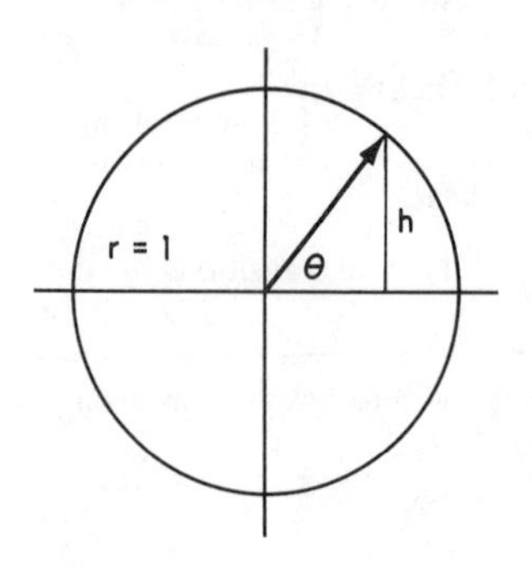

tive interference, but the resultant will not be zero, and the amplitude will be diminished. Furthermore, the form of the resultant wave will no longer be a pure sine curve. It must be emphasized, in this part of our discussion, that the above considerations apply only if the two waves are passing through a medium of constant refractive index. Later on, we shall indicate that phase changes may be caused by passage of waves through media of differing refractive indices.

PROCESSES CHARACTERIZING INTERACTION OF RADIANT ENERGY WITH MATTER

Just how radiant energy interacts with a particular form of matter depends on the wavelength range of the radiant energy and on the order (structure) intrinsic in the matter. Thus, an aluminum plate some millimeters thick will not transmit visible or ultraviolet light; these will be reflected almost totally. On the other hand, such a plate will transmit typical x-ray, as well as radio or heat waves. Very thin films of aluminum, as well as similarly thin films of gold, silver, and some other metals, will even transmit light waves. Some practical uses of thin-film transmission will be indicated below.

One or more of the following processes must be involved in any real interaction between radiant energy and some form of matter. The six most important processes are described below.

Reflection

Reflection is the bending back of all or part of a ray or wave as it passes from one medium to another of different density. Reflection may therefore be total or partial, depending on the media in question and on the wavelength of the impinging radiation. According to Snell's law, the angle of reflection is always equal to the angle of incidence. Mirrors are designed to be efficient reflectors, but some reflection will always occur when a beam of radiant energy passes through a discontinuity of refractive index.

Diffraction

Diffraction is the breaking up of a ray into alternate "light" and "dark" bands or (in the case of visible light) into the colors of the spectrum. This is caused by interference of one part of a beam with another when a ray is deflected at the edge of an opaque object or when it passes through a narrow slit or series of slits.

Diffraction gratings may be prepared by ruling many fine lines on the surface of a glass or metal plate, as will be discussed below. Such devices are important dispersing elements in visible and ultraviolet monochromators. In the x-ray region, the ordered array of atoms in crystals of various salts acts as a diffraction grating. Because of the very short wavelengths of x-radiation, it would not be possible to "rule" suitable gratings.

Refraction

Refraction is the bending of a ray or wave as it passes obliquely through one medium to another of different density (refractive index), or through layers of different density in the same medium.

The classic refractive element is the lens, which may be made of any material transparent to the radiation in question. For light waves, the traditional material is glass, but a lens may also be made of plastic, diamond, or even ice. The purpose of a lens is to form an image, and its substance is unimportant. A second class of refractive element is the aplanar mirror, which can also focus radiant energy to form an image, as in reflecting telescopes and radio telescopes.

Scattering

Scattering is the reflection, diffraction, or refraction of radiant energy from the surface of matter in an irregular or diffuse manner. Scattering frequency involves more than one of these modalities.

Considerable pain is taken to avoid energy scattering in most instrumentation, since it degrades performance. The measurement of light *scattered* by suspended or dissolved particles is called *nephelometry,* whereas measurement of light *not* scattered in such a system is called *turbidometry.* The extent of scattering depends on the size, number, and concentration of particles as well as the wavelength of radiation.

Absorption

Absorption is the partial loss of intensity of a ray or wave in passing through a medium that can take up some portion of the impinging energy. The captured photon energy is ultimately released as heat. The loss of intensity may be observed as a decrease in wave amplitude. The wavelength of the rays is not affected.

The inverse of absorption is called *transmission.* A medium is said to be transparent to the extent that it allows transmission. The transmissivity of a medium depends on the wavelength of the radiant energy. This is the basis of *absorption spectroscopy.*

Fluorescence

Fluorescence is the ability of certain molecules, when exposed to radiant energy of particular wavelengths, to absorb photons and thereby to pass to an excited state. (The half-life of the excited state is quite short, typically 10^{-8} to 10^{-5} sec.) Part of the energy is then released as photons of lower energy (longer wavelength) than that of the exciting radiation. The remainder is released as heat or transfered internally via various quenching modes.

FOCUSING, COLLIMATION, AND DISPERSION OF RADIANT ENERGY

Most sources of radiation have fairly large emitting surfaces. They emit a broad band of wavelengths, and the radiation spreads through large spatial angles; they act neither as point sources nor as monochromatic sources.

In modern instruments, some combination of slits or similar apertures, together with lenses and/or aplanar mirrors, simulates a point source. Effectively, the lens–mirror combination provides a focused image of the slit(s). This image, being uniformly filled with radiant energy, functions as the apparent source. Additional lenses and/or mirrors are employed to *collimate* the entrance beam, that is, to ensure that all rays in the beam travel in essentially parallel paths. Collimation minimizes errors due to scattering and reflections in the system.

Lastly, it is necessary to *disperse* the broad band of emitted energy ("white" radiation, or "white" light in the case of optical spectrophotometers) into its component wavelengths. An exit slit then passes a beam containing energy of a narrow-wavelength band, or as nearly as possible, monochromatic energy. A variety of dispersing elements are employed in modern instruments. The combination of (1) source, (2) entrance slit and focusing lenses, and (3) dispersing elements and exit slit constitutes a monochromator. The quality of the monochromator frequently limits the quality of the total instrument. Typical monochromators have output bandwidths in the range of 2–20 nm.

One pays a significant price for this kind of energy purity. The output bandwidth contains only a tiny fraction of the total physical source output. Consequently, the lamps employed must be quite intense; they commonly generate a good deal of heat which must be carefully kept from the remainder of the instrument. The entire system must be allowed to come to thermal equilibrium in order to ensure optical and electrical equilibrium. Because some sources emit large amounts of ultraviolet radiation along with other wavelengths, it is a good idea *never* to operate such instruments while one's eyes are directly exposed to the lamp output. The intense ultraviolet energy can be damaging to one's eyes.

DISPERSION DEVICES (MONOCHROMATORS): THEORY AND PROPERTIES

The oldest and simplest dispersive element, suitable for use in the visible and ultraviolet (UV) regions, is the prism. Prisms are made of various glasses ($n \leqslant 1.89$) or of fused, ground, and polished quartz ($n = 1.553$). Quartz prisms are preferred for instruments in the UV region because of their greater transparency to such radiation.

Figure 2–3 shows a ray from a beam of collimated, "white" light striking the surface of a prism cut with a fairly acute angle, θ. The continuing dotted line through the prism and beyond shows the path that a ray would have taken if the ray had not been deviated by the prism (i.e., if the refrac-

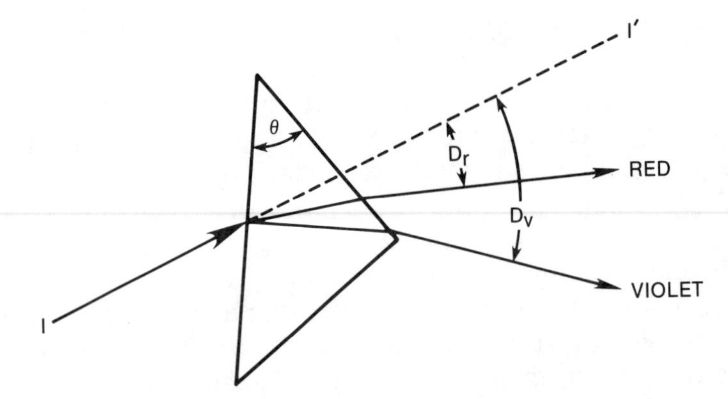

Figure 2–3. Dispersion by a prism: refraction optics. "I" is the incident ray, θ is the prism angle, D_v is the extreme angular deviation, and D_r is the angular deviation for any given wavelength.

tive index of the prism had been 1.000). For any real prism, where $n >$ 1.000, the component radiation is refracted (deviated) to a degree that varies with the wavelength. Longer wavelengths are deviated less, and shorter wavelengths more, from the original path. The relative angular spread between the undeviated beam and any of the emergent wavelengths is known as the dispersion for that particular wavelength. Because the dispersion varies with wavelength, it is the usual convention to make such measurements at the wavelength of a particularly intense line in the emission spectrum of sodium vapor. This is known as the *D line;* it has a wavelength of 586.9 nm, roughly in the middle of the visible region.

When the source and the prism are so arranged that the exiting beam is minimally deviated, it is possible to show that

$$n = \frac{\sin\,[(D_m + \theta)/2]}{\sin\,(\theta/2)} \tag{2–3}$$

where n is the refractive index, D_m is the angle of minimal deviation, and θ is the angle at which the prism is cut. Since for small values of θ (up to about $\pi/6$) little error is introduced in the value for n by the substitution, $\sin\theta = \theta$, we can thus rewrite Eq. 2–3 to read

$$n = \frac{(D_m + \theta)/2}{\theta/2} \tag{2–4}$$

Solution of Eq. 2–4 for D_m gives Eq. 2–5:

$$D_m = \theta(n - 1) \tag{2–5}$$

Equation 2–5 demonstrates that the dispersing power of a prism, at any wavelength, depends on its included angle and on its refractive index. In an ideal world, prisms might be made of diamond, for which $n = 2.417$. This explains why faceted diamonds sparkle with such brilliance. It also provides us with a means of distinguishing diamond from nondiamond gems.

In many applications, one prism surface is silvered. Instead of exiting from that surface, dispersed rays are reflected back through the body of the prism. This doubles the dispersion but introduces some complications into the geometry of the monochromator. Because n is a function of wavelength, Eq. 2–5 tells us that prisms are not constant-dispersion devices. The angular displacement of a prism required to select between two output wavelengths such as 650 nm and 675 nm is very different from the angular displacement required to select between 225 nm and 250 nm. Instead of mounting a prism on a simple circle, it must be mounted on a cam with a very complicated shape. This is a source of considerable expense to the designer, and it imposes some limitations on the ultimate performance of a monochromator. On the other hand, prisms introduce very little light loss; over their useful range they absorb very little energy from the incident beam. Although prisms are still employed by a number of manufacturers of quality monochromators, the popularity of prisms has been seriously challenged by development of dispersing elements based on diffraction as opposed to refraction.

Thin-Film Optics: Interference Filters and Interference Wedges

Certain natural phenomena depend on interference of light rays. These include the colors seen in soap bubbles or in oil slicks on wet pavement and the colors of some bird feathers and some insect wings. The colors observed are clearly not a bulk property of these materials; rather they are a property of thin films or lamellae. The mechanism of light dispersion by thin films is described by the diagram shown in Fig. 2–4.

Imagine two infinitely thin, parallel glass layers, L_1 and L_2, the opposing surfaces of which are coated with a very thin film of silver. The silver layers will reflect part of the incident beam (ray ABE_1) and will transmit part of the beam (ray ABC). If the space between the films contains a medium

Figure 2–4. An interference filter: thin-film optics. A is the incident beam; t is the film thickness. See text for detailed description.

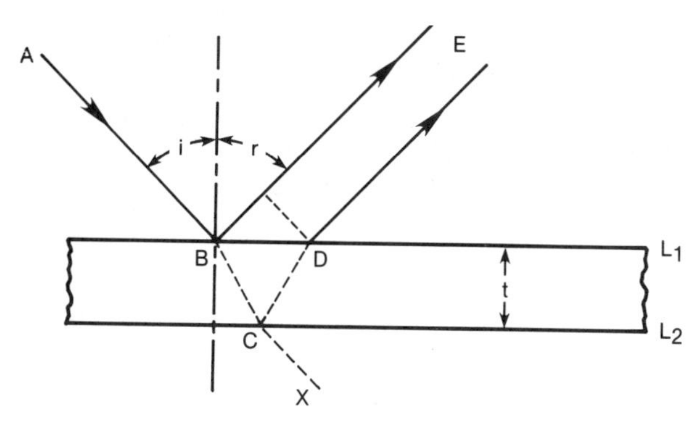

other than air, the ray ABC will be refracted as shown. At point C, a portion of the beam will again be reflected (ray CD). For the moment, we shall ignore the ray, BCX, transmitted through L_2. At point D, some transmission and refraction will again occur, generating ray DE_2. The reflected rays, BE_1 and DE_2 will be parallel, as required by Snell's law governing angles of incidence and reflection.

To reach an observer stationed at E, the ray $ABCDE_2$ had to travel farther than the ray, $ABFE_1$. The difference between these optical paths is given by the expression $2a - b$, where $BC = CD = a$ and $BF = b$. However, because the interspace between L_1 and L_2 has the higher refractive index, n, the equivalent optical distance in air is $2na$, so the actual retardation of ray 2 with respect to ray 1 (at point E) is $2na - b$. This retardation is due solely to passage of a part of the beam through the thin film.

The distance, a, clearly depends on the separation between the silver films, described as t; on the refractive index of the medium between them; and on the angle of incidence, i, of the light beam. The retardation, R, due to path length difference can be expressed as:

$$R_{\text{path length}} = 2t\sqrt{n^2 - \sin^2 i} \qquad (2\text{–}6)$$

If the light beam strikes the thin film normal to its surface, this reduces to

$$R_{\text{path length}} = 2nt = R_1 \qquad (2\text{–}7)$$

There is a second retardation component due to the reflections themselves, one going from a medium of lower refractive index to a medium of higher index, and the second in the reverse direction. In such dual reflections, there is always a phase shift, or retardation, equal to exactly one-half the wavelength, or

$$R_{\text{reflection}} = \lambda/2 = R_2 \qquad (2\text{–}8)$$

The total retardation, R_T, is the sum of R_1 and R_2. If destructive interference is to be avoided, it follows that

$$R_T = R_1 + R_2 = 2nt + \frac{\lambda}{2} = N\lambda \qquad (2\text{–}9)$$

In other words, the path length difference must be an integral number of wavelengths, N. If an interference filter is to pass light of a certain wavelength, λ, and is to contain a medium of known refractive index, n, its thickness can be determined by solving Eq. 2–9 to give

$$t = \frac{\lambda(N - 1/2)}{2n} = \frac{\lambda(2N - 1)}{4n} \qquad (2\text{–}10)$$

Equation 2–10 is the fundamental equation of interference filters.

Interference filters are very "sharp cutoff" devices; that is, they can be made with bandpasses as narrow as 5–10 nm. This is probably an order of magnitude better than most other available kinds of optical filters. They have some serious drawbacks as well. First, recall that a certain portion of the incident beam is transmitted without alteration (giving rise to the ray *CX* shown in Fig. 2–4). In a precision instrument this would amount to stray light and would degrade performance if it were not removed. A second disadvantage is that the filter depends essentially on some reflection. This means that a very significant portion of the incident energy never passes through the filter at all. Interference filters are therefore inefficient devices, in terms of radiant energy flux.

Several extensions of this theory of thin-film interference optics are interesting and have proven useful. First, one may well ask why thick films show no color. Equation 2–10 indicates that many conditions satisfy the requirement for constructive interference. For thicker films, many values of t, for example:

$$\frac{\lambda}{4n}, \frac{3\lambda}{4n}, \frac{4\lambda}{4n}, \frac{5\lambda}{4n}, \ldots$$

would constructively interfere. There would be so many of these satisfactory conditions that the combination of these rays would appear as white light.

Extremely thin films, on the other hand, appear black, because such ultrathin films have very small values of t, reducing the actual interference due to path length differences to the vanishing point. What is left is interference due to reflections; reflections always produce destructive interference because the waves are always of opposite phase. So-called black membranes are usually only a few molecules thick, and are of great current interest as models of membrane structure. Their appearance follows from the principles outlined above.

Another application of thin-film optics is the coating of lens and mirrors in optical equipment such as cameras, microscopes, etc., to reduce surface reflections. These coatings consist of very thin layers of compounds such as MgF_2. These are much more transparent than silver films, so they do not exact such a cost in terms of optical efficiency. They act essentially as "one-sided" interference filters, designed to pass more or less white light.

Dispersion by Diffraction Gratings

Diffraction gratings suitable for the near infrared, visible, and ultraviolet regions can be prepared by ruling a series of parallel grooves in the surface of a glass or metal plate. A typical grating might have 5000 grooves/cm. To prepare such a surface requires elaborate machines, called ruling engines, and elaborate precautions must be taken in their use. Ruling engines are commonly installed in specially designed underground vaults to avoid mechanical vibrations and are usually operated by remote control lest the body heat of a nearby operator cause unwanted expansion of key components. The actual cutting tool is a precisely shaped diamond edge. It may take up to one full working week to cut a single grating 10 cm long. Clearly, the cost of such an original grating is likely to be quite high. For-

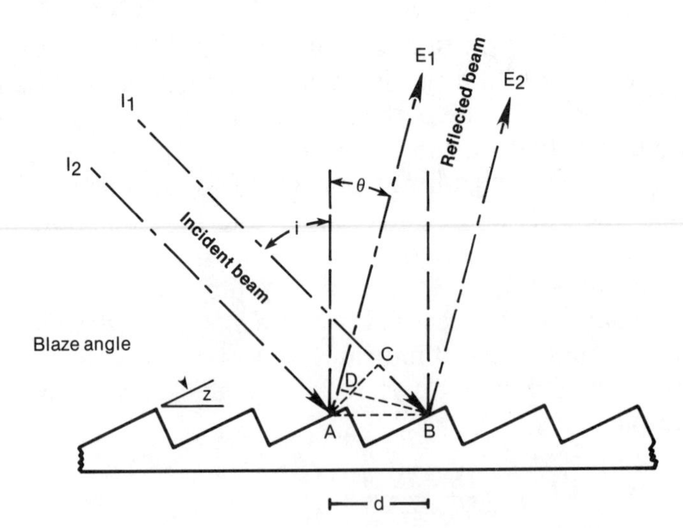

Figure 2–5. Dispersion by a diffraction grating. The incident beams I_1 and I_2 strike the grating and are reflected as beams E_1 and E_2. The relationships of the angles of diffraction are described in the text.

tunately, once an original grating is available, it is possible to cast plastic replicas at little expense and with faithful reproduction of the grooved surface. These may then be mounted on a clear reflecting support surface (substrate).

Diffraction gratings depend on interference optics. The basic theory is substantially the same for transmission and reflection gratings. We shall examine the behavior of such gratings in terms of a reflection device with the aid of Fig. 2–5, in which the size of the ruled grooves is grossly exaggerated for clarity.

Two parallel rays of the incident and reflected beams are called out for discussion. We define the path lengths, P_1 and P_2, of the identified rays as they travel from the source, located at I, to an observer located at E. Suitable expressions for the path lengths are:

$$P_1 = I_1BE_2 \quad \text{and} \quad P_2 = I_2AE_1$$

In order to have constructive interference at E, the two waves must be in phase at E. In other words,

$$\Delta P = P_2 - P_1 = N\lambda = CB - AD \tag{2–11}$$

where N is some integral number of wavelengths (the order of the spectrum), and CB and AD are the distances shown in Fig. 2–5. CB and AD can be expressed in terms of the angles of incidence and reflectance, i and θ, respectively. From the geometry of the figure, we can write the pair of equations

$$CB = d(\sin i) \quad \text{and} \quad AD = d(\sin \theta) \tag{2–12}$$

Note, however, that the directions of i and θ must be regarded as opposite to each other. We can make use of the substitution, $\sin(-\theta) = -\sin \theta$. If the values for CB and AD from Eq. 2–12 are substituted into Eq. 2–11, the result is

$$N\lambda = d(\sin i + \sin \theta) \tag{2-13}$$

Equation 2–13 is the fundamental equation of a reflection diffraction grating. It states that the dispersion is strictly a function of the angles of incidence and reflection for any given grating. Note also that if energy of wavelength λ is reflected at a certain angle, then energy of wavelengths $\lambda/2$, $\lambda/3$, $\lambda/4$, . . . etc. will also be reflected at the same angle. These secondary and tertiary reflections, known as spectra of higher order, are interpolated in the primary spectrum. Higher-order spectra are usually much less intense than primary spectra, and ordinarily can be removed by suitable filters. Nevertheless, overlap of the primary spectrum by spectra of higher order does lower the efficiency of diffraction gratings. The performance of a grating can be shown to depend on the shape of the grooves with respect to the normal surface. By careful attention to the way grooves are cut, which is commonly known as the *blaze angle,* it is possible to concentrate as much as 75% of the incident energy into the primary spectrum.

Diffraction gratings have become increasingly popular as dispersion elements for several reasons. First, they can be made with higher resolving power simply by increasing the number of lines per centimeter. Recall that resolving power of a prism is limited by the refractive index of the glass. Second, at least in the visible and ultraviolet regions, the dispersion of a grating is very nearly a linear function of the angles of incidence and reflection. A prism is not a linear device, as noted earlier. Third, diffraction gratings are less massive than prisms of equal power. These facts make the mechanical problems of grating mountings less severe and make automated wavelength scanning easier to accomplish. Holographic ruling and laser engraving promise to make still better gratings.

SOURCES OF RADIANT ENERGY

The source of radiant energy clearly must be appropriate for the spectral region under investigation. We shall limit the present discussion to sources useful in the infrared, visible, and ultraviolet regions. In these regions, radiant energy may be generated either by heating a solid source or by exciting molecules of a vapor to emit a characteristic line spectrum. In the case of line spectra, one seeks a spectrum that is rich in closely spaced lines of as nearly as possible uniform intensity. One also seeks a source that can be excited under relatively moderate conditions of power input. The two elements that have been most widely adopted are *deuterium* and *xenon.* The deuterium arc is especially rich in the ultraviolet region; relatively less of its spectral output falls in the visible range. The xenon arc is rich in the ultraviolet, and it has a very significant component in the visible. Because of these characteristics, deuterium arcs are ordinarily employed only in the ultraviolet region, whereas xenon arcs may be used as either ultraviolet or visible sources. Bear in mind that neither the deuterium nor the xenon arcs give anything like a uniform output over their useful spectral ranges.

Excitation of deuterium or xenon requires fairly high voltages (in the neighborhood of several hundred volts) at considerable current. This requires an accessory apparatus (1) to convert line power to the required value, (2) to convert the alternating voltage to a direct voltage, and (3) to regulate the direct voltage to as nearly as possible a constant value; a more modest source of alternating current power also is needed initially to heat the vapor to the excited state. These requirements necessitate the use of bulky components which must dissipate a large amount of heat. Consequently, the power supplies for these vapor sources are ordinarily packaged in cases external to the rest of the spectrophotometric equipment.

Incandescent Sources

It is common experience that any solid, when heated sufficiently, emits radiation. Two fundamental laws govern the radiation process. The Stefan–Boltzmann law states that a black body radiates energy in proportion to the fourth power of its temperature (neglecting the surroundings, which we assume here to be the ambient). Obviously then, emission rises sharply as the temperature of the body is increased. Wien's law states that the wavelength corresponding to maximum energy is inversely related to the temperature of the radiator. Thus, the light of a bonfire and the light of the sun differ qualitatively as well as quantitatively.

Figure 2–6 is a graphic presentation of the significance of the laws of Stefan–Boltzmann and Wien. The curve at 500°K might represent the output of an iron rod heated until "red-hot." It is the kind of source that might be employed in an infrared spectrometer (Globar), or even in a cooking stove. Ordinary daylight falls between curves 3 and 4. Daylight has a color temperature of about 5500°K. The area under each curve increases sharply with temperature, and the wavelength of the maximum output falls with increasing temperature. Other things being equal, it is clearly advantageous to operate spectrophotometric sources at the highest practical temperature. The usual incandescent source is a tungsten filament device. The operating temperature is limited by the melting point of the glass envelope, and glass has a poor ability to transmit the shorter ultraviolet wavelengths (even though the filament may generate them). The very high operating temperature of more powerful tungsten lamps causes vaporization of metal from the filament and deposition of the tungsten on the relatively cooler glass. This leads to lamp blackening and defeats the very purpose of higher-temperature operation. Recent development of the tungsten–halogen lamp has gone some distance in solving these problems. In these lamps, some iodine vapor is added to the lamp envelope, which is made of fused quartz instead of glass. The iodine prevents tungsten deposition on the envelope and prevents lamp blackening. Use of quartz in place of glass allows a higher operating temperature and

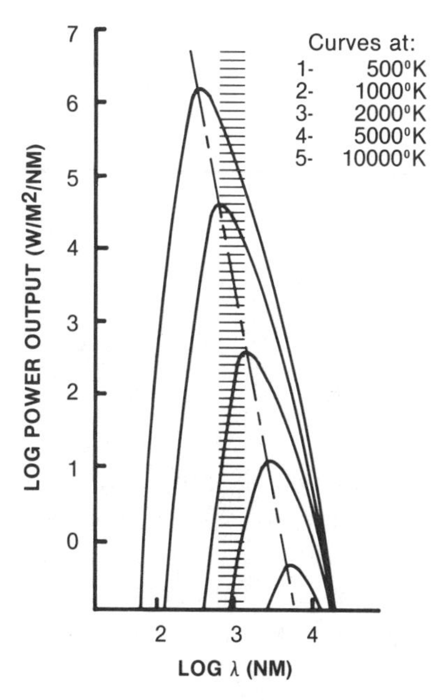

Figure 2–6. Theoretical output curve for black-body radiators operated at various temperatures. An expression of the Stefan–Boltzmann and Wien laws dealing with the energy distribution of thermally excited radiators. The shaded bar marks the approximate limits of the visual range.

also allows good transmission of the shorter ultraviolet wavelengths. The filaments are commonly made of heavier-gauge tungsten wire, which allows a higher current flow through the lamp. Since the heating power is given by the quantity I^2R, a given power rating can easily be obtained at lower voltages by utilizing the capability of incandescent lamps of modern design to employ a single tungsten–halogen lamp to cover the visible and a useful portion of the ultraviolet range.

This development simplifies construction and operation of the instrument.

SENSING RADIANT ENERGY

Photodetectors

The earliest photodetector was the human eye, which is still used to match the color of indicator papers with standard color charts, and so forth; however, it does not discriminate equally well between colors, fatigues rapidly, and responds only to a limited range of wavelengths.

The development of practical photodetectors dates from the pioneering work of Albert Einstein. He demonstrated that photons striking the surface of many materials could cause ejection of electrons from the photosensitive substances, including many metals. Under proper circumstances, the ejected electrons could be collected and measured in the form of a

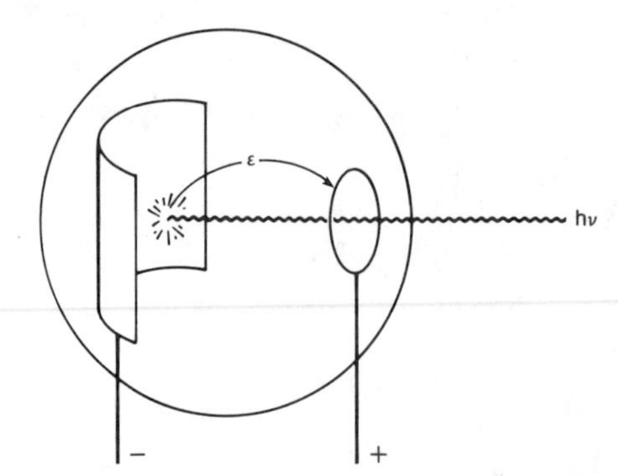

Figure 2–7. A single-stage, Einsteinian photoemissive detector. Although Einstein's original work was based on photocathodes made of a sheet of pure metal, modern detectors frequently have some base metal coated with a photoactive material or mixture of materials. The composition can be varied to provide peak sensitivity in important areas of the visible and/or ultraviolet spectral regions.

current flow. Einstein's original photodetector consisted of an evacuated transparent envelope in which two electrodes were mounted. One was in the form of a wire ring, the other in the form of a curved plate, as shown in Fig. 2–7. A source of potential was applied to the electrodes as shown. When a photon striking the surface of the plate ejected an electron, the potential accelerated the electron until it was attracted to the positively charged ring. Modern production detectors are surprisingly similar to those hand-made in Einstein's laboratory, but modern production models employ fairly sophisticated coatings of the plate electrode in order to modify spectral sensitivity quantitatively. Einstein summed up his studies of the photoelectric phenomenon in the equation

$$E_{max} = h\nu - W \tag{2–14}$$

where E_{max} = the maximum energy of the emitted electron; $h\nu$ = the energy of the photon; and W = the so-called work function, a quantity characteristic of the electrode material. Note that E_{max} is entirely a function of the frequency of the photon; it has nothing to do with the intensity of the radiation. The intensity affects only the number of electrons emitted.

Photomultipliers

The basic Einstein photodetector can be logically extended as a cascade device known as a photomultiplier. A diagrammatic sketch of a photomultiplier is shown in Fig. 2–8. Light strikes a photosensitive surface of a photocathode, PC, and causes emission of electrons. The electrons are attracted to the first of a series of positively charged dynodes, D_1. The positive charge on D_1 not only collects the electrons, but also accelerates them. Since the accelerated electrons may be regarded as photons of a particular energy range, when they pass to the next dynode they cause the emission of a still larger number of electrons. In this way the original signal is cascaded, or multiplied, by a factor that may be as large as 5×10^5. Note that the guard ring, GR, here serves to direct emitted electrons away from the entry window and the photocathode toward the first dynode.

In theory, every photon that strikes the photocathode should lead to emission of one electron; that is, the quantum efficiency should be unity. In practice, this kind of efficiency is never observed, either with single-stage detectors or with photomultipliers. The reasons are quite complex, and we shall not go into them here. Photoemissive detectors are also sensitive to slight changes in the applied voltage. In practical instruments, fairly elaborate means are required to stabilize the high voltage power supply. These detectors are also easily shocked by exposure to intense light. In the case of photomultipliers, for example, the high currents caused by exposure to bright light may cause destruction of the photoemissive surfaces. Care must be taken never to expose the window of a multiplier to bright light with the high voltage applied. Thermal noise, caused by random events with the detector, is responsible for the so-called *dark current,* which flows even when the detector is not exposed to light. Thermal noise can be minimized by operating at temperatures below the ambient; the most elaborate spectrophotometers are equipped to do just this. Photomultipliers employed in liquid scintillation spectrometers frequently are housed in cabinets that are cooled to 20–30°C below the ambient. Newer developments in formulating photoemissive surfaces make this precaution less stringent than it previously was.

Because photoemissive detectors typically operate at voltages in the range of 100–1000 volts, the current flow must be measured through a load resistor of quite high value. In older instruments, leakage currents in humid weather were an annoyance; in severe conditions they may still pose a problem. In newer equipment the problem has been much abated, but it must still be kept in mind as a possible cause of erratic operation.

Solid-State Photodetectors

As electronics moved from the era of vacuum tube technology into the era of solid-state technology, photoemitting and photodetecting devices were part of the overall development. Solid-state technology ultimately

Figure 2–8. A diagrammatic sketch of a photomultiplier. The guard ring, GR, serves to deflect emitted electrons away from the entry window and toward the first dynode, D_1. Each dynode in the series is given an increasingly positive charge of 50–100 volts, with the photoanode, PA, being the most positively charged element in the system. Each successive stage of the cascade multiplies the number of electrons emitted. A typical modern photomultiplier has from 9 to 10 dynodes. The internal connections in the multiplier are omitted from the diagram.

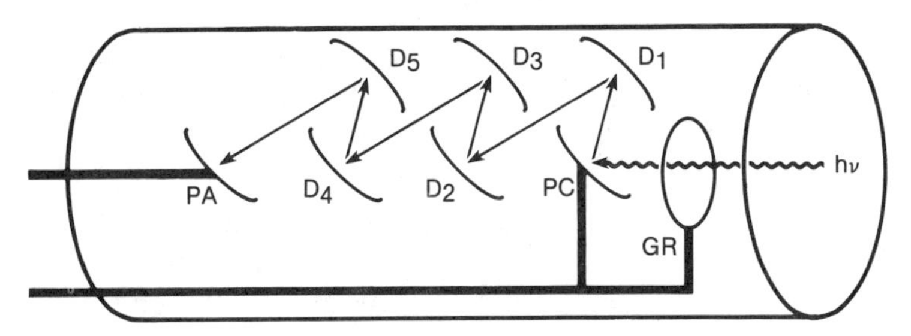

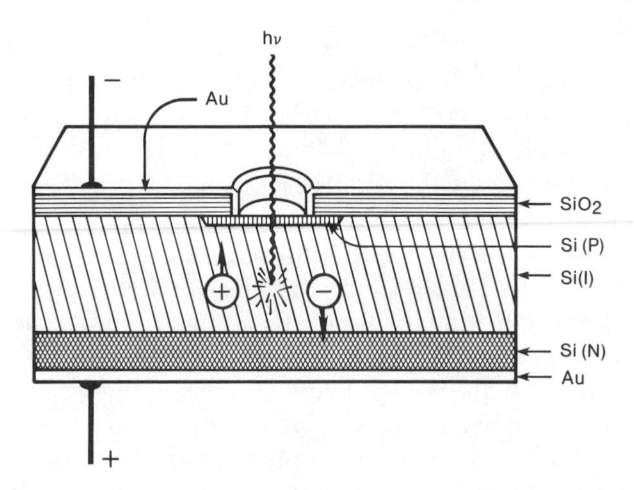

Figure 2–9. A solid-state (PIN) photodetector. See text for detailed discussion.

depends on the properties of semiconductors in the single crystal form. The single crystals are "doped" or deliberately contaminated with impurities that give the semiconductor a defective crystal form; there is either an excess or a deficiency of electrons in the defective crystal. In engineering jargon, a deficiency of electrons is regarded as an excess of "holes." Without going further into the rationale of this approach, we can proceed directly to Fig. 2–9, which diagrams a cross section of a solid-state photodetector. This diagram is not drawn to scale; the thickness of the several layers is grossly distorted for clarity of presentation.

The device shown is known as a PIN detector because it contains three distinct forms of silicon. On a thin gold base there is coated a layer of Si (N), which has an affinity for free electrons because its structure is deficient in electrons. Over this is a layer of Si(I), so called because it is uncontaminated and has the intrinsic properties of pure silicon. On top of this is a very thin, semitransparent layer of Si(P), doped to have an affinity for "holes." The sandwich is covered by a thin, nonconducting layer of SiO_2 except for a small window over the center of the Si(P) layer (which is contacted by a second thin gold electrode), which also forms the energy-admitting window of the photodetector. Radiant energy striking the Si(I) layer after passing through the Si(P) layer, causes release of a free electron and generates a "hole" in the crystal lattice of the Si(I). These move as shown, and constitute the flow of current which can be amplified and quantitated.

Power supplies for PIN photodetectors, as for solid-state devices in general, operate at voltages in the range of ± 20 volts. The currents are also fairly low, up to the last stage of amplification. Regulation of these low-voltage suppliers is easily and quite economically accomplished. Solid-state devices can have rapid response times; thus they are able to follow transient changes faithfully. They are, in general, orders of magnitude smaller and frequently are at least as sensitive as the components they replace. PIN detectors have a broad spectral response which does not extend too far into the ultraviolet, and which has a peak output at about 800 nm. They hold great promise for future developments in chemical instrumentation.

SAMPLE CUVETS: PRECAUTIONS IN THEIR USE

Samples are introduced into the spectrophotometric light path in containers called cuvets. These may be fabricated of Pyrex or Corex glass (for visible light transmission) or of fused and polished quartz (for the best ultraviolet transmission). Cuvets normally are of rectangular cross section, and the standard units have an optical path length of 1.00 cm. One pair of opposite sides is very carefully polished; this pair of sides should not be touched with the fingers, since they serve as the surfaces through which the light beam enters and leaves the cuvet. The opposite pair of sides is not so finely polished, or may even be ground to a rough finish; these sides are the surfaces by which the cuvet may be handled. The usual cuvet is approximately 4.5 cm in height and can contain about 3.0 ml, although the required volume for proper use is ordinarily somewhat less. Semimicro- or microcuvets are made by increasing the thickness of the unpolished sides, reducing the required volume while maintaining constant outer dimensions. The microcuvets that are commonly available hold about 1.0 ml, but can be used with as little as 0.60 ml. Precision quartz cuvets are expensive and should be handled with care. Glass units are somewhat less expensive but should also be handled with care. For routine use in the visible range, inexpensive disposable plastic cuvets may be used; their cost is about 1% that of quartz cuvets.

Some spectrophotometers hold only a single cuvet, some hold only two, whereas others hold four or more, any one of which may be placed in the optic path by moving a carrier at right angles to the light beam. The carrier may be manually shifted by an external lever, or it may be motor driven semiautomatically. The carrier is properly positioned by a series of detent stops. In instruments of the sorts described above, the cuvets are filled, emptied, and washed by hand after they have been removed from the instrument. Manufacturers provide an alternative configuration, in which the cuvet (usually a single cuvet) is fixed in the optic path. Filling, emptying, and washing operations are performed by means of a pump-driven aspiration system, described as a "sipper" system. This may be an advantage and a time saver in situations where a large number of measurements per hour is demanded.

Machines that carry two cuvets, fixed in their relative positions, are known as double-beam instruments (e.g., Hitachi-Perkin Elmer, Perkin-Elmer Lambda 3). One of the two cuvets always must contain a "blank" or "reference" solution, against which some "unknown" solution is compared for 60–500 times/sec. The output signal of double-beam instruments reports the difference in properties of the two solutions. Machines that carry one, or more than two, cuvets are usually single-beam instruments (e.g., Gilford 120, Beckman DU-5 and DU-7). In single-beam instruments no provision is made for continuous and simultaneous comparison, and reference solutions must be separately observed. One can marshal a variety of arguments in favor of one operating mode or the other, but there is no overriding theoretical reason to prefer either.

Cuvets and their carriers are mechanically simple devices, but they can easily be misused. A light beam passing through a cuvet face is *always* subject to some refraction, since the value of n for the cuvet material differs from the value of n for air. The beam is also *always* subject to some reflection, but according to Snell's law, reflective losses will introduce no stray light if the instrument is properly aligned. However, if the surface of the cuvet is scratched, or greasy with fingerprints caused by improper handling, then some scattering and/or reflection will occur. Scratches or greasy films on the inner cuvet surfaces will have exactly the same effect. If the contained solution is not perfectly limpid, suspended or dispersed matter will also cause some reflection and/or scattering. Indeed, very asymmetric biopolymers in solution may be estimated by their effect on light scattering in specialized spectrophotometers. Users should also keep in mind that air bubbles are very effective light-scattering entities. Avoid trapping air in solutions placed in spectrophotometer cuvets. Where mixing by inversion is required, this should be done gently.

If the level of the contained solution is too low, then the meniscus falls in the light path of the cuvet and causes some scattering and reflection at angles not parallel to the beam. This problem can be eliminated simply by increasing the contained volume. Manufacturers are not always careful to describe the exact size or position of the light beam in passage through the cuvet, but the path can usually be located by a few simple experiments.

If a movable microcuvet is incorrectly located or the detent stops are not properly set, only part of the light beam passes through the solution. The remainder passes through the thickened side wall of the cuvet. Obviously, reliable absorbance measurements depend on the uniformity of the medium through which the light passes. In using machines with movable cuvet carriers, one is advised to check the positioning mechanism periodically.

Even when a microcuvet is properly positioned, the size of the monochromator exit aperture may be too large for the cuvet in use. Again, the light beam does not pass through a uniform medium. One must reduce the aperture, or exchange it, depending on the manufacturer's instructions. In some instruments, this is a matter of moving a lever; in others more extensive efforts may be required. One can easily visualize the light path through a cuvet by filling the cuvet with a solution of a dichroic dye and looking down through the open cuvet compartment at some appropriate wavelength. (**Note:** *If this is necessary, protect the open compartment from bright-light shock by putting an opaque cloth over your head and the open compartment!*)

QUANTITATIVE ABSORPTION SPECTROPHOTOMETRY

Basic Principles

The analysis of absorption spectra depends on some simple rules summarized as follows:

1. *The interaction between radiant energy and other forms of matter is quantized.* Such interactions occur between a photon and one or more electrons of a molecule. As a result of the interaction, the energy state(s) of the electron(s) increase(s). Any real sample contains an enormous number of molecules, each of which contains many electrons. Consequently the discrete, quantized energy increments are "smeared out" so that a real spectrum is usually a smooth curve. In accord with the Boltzmann distribution, a plot of absorbance vs. wavelength for a single-electron system would give a Gaussian curve.
2. *The magnitude of energy absorbance is a colligative property; it depends solely on the concentration of absorbing particles in the light paths.* This rule is the basis for quantitative analysis by absorption spectrophotometry. It tells us also that the energy absorbance is proportional to the area under the spectral curve. Note that concentration has no effect on the wavelength of maximum absorbance of a given molecular population.
3. *In a mixture, each molecular species absorbs independently of any other.* To be separated by absorption spectrophotometry, substances A and B must have different wavelengths of maximum absorbance. If the wavelength maxima are sufficiently separated, two distinct peaks will appear in the spectrum of a mixture. If the wavelengths of maximum absorbance are not sufficiently separated, the result will be one distorted (asymmetric) peak, as shown in Fig. 2–10. Note that the wavelength of maximum absorbance *may* be characteristic of a particular substance. At any wavelength of the spectral curve of a mixture, the observed absorbance is the sum of absorbances due to A, B, C, . . . etc., at that specified wavelength.

Figure 2–10 shows, at the left, a symmetric peak obtained from a solution of pure substance A. In the middle, a similar peak is shown for pure substance B. Note that even if both solutions are at identical concentrations, the principles cited above do *not* require that the absorbance of A and B be the same. Since A and B are distinct, with different electronic configurations, it would be most unusual if both absorbed to an identical degree. To the right, in Fig. 2–10, is shown the spectrum obtained from a mixture of A and B at the same concentrations. Note that, because the wavelengths of maximum absorbance are not sufficiently separated on the arbitrary scale, the solid curve of the observed spectrum is notched. Given slightly different conditions of peak separation, the observed curve could have showed a single, asymmetric peak, quite broadened, or a peak with a

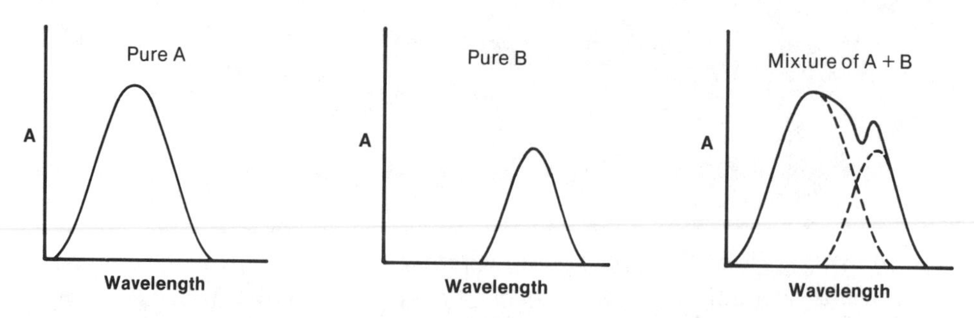

Figure 2–10. Individual and combined absorbance spectra of two solutes exhibiting overlapping spectral peaks.

"shoulder." The dashed curves indicate the separate absorbances due to A and B, in relation to the total absorbance at any wavelength.

The case of overlapping spectral peaks can be analyzed further. Fig. 2–11 shows three spectra. Curve A represents data obtained with a solution of pure A; curve B, data obtained with a solution of pure B; and curve A + B, data obtained with a mixture of A and B each at the same concentration as in the pure solutions. Several features of these spectra are worth comment:

1. The wavelength at which the mixture shows maximum absorbance is shifted with respect to the values for pure A or pure B, and the curve for A + B shows a distinct shoulder. Resolution of the complex curve for A + B can be done by hand, but the effort is very tedious. Instruments exist which can greatly expedite such analysis. Because the resolved curves will always cross each other at one point (represented by the arrow on the abscissa) where each component has the identical value of absorbance, this point has been given a special designation. It is known as the *isosbestic* point.
2. It is possible to determine B in the presence of A by quantitative absorption spectrophotometry because at λmax_B there is virtually no absorbance due to A. The reverse analysis, determination of A in the

Figure 2–11. Absorbance of a binary fixture indicating summation of the two spectra.

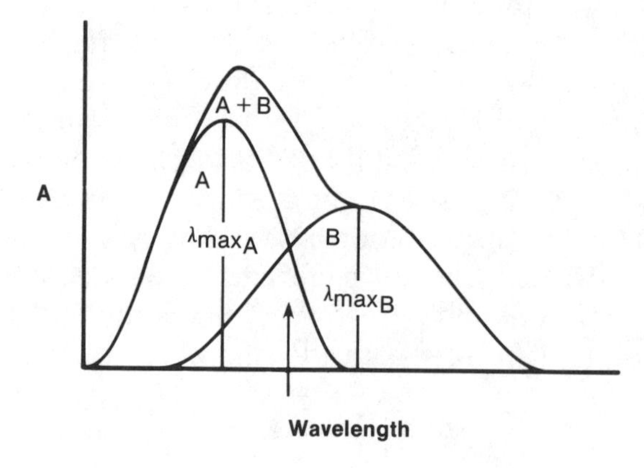

presence of B, would be erroneous because B absorbs significantly at λmax_A.

3. By solution of two simultaneous equations in two unknowns, it is possible to obtain correct analytical values for A in the presence of B in a binary mixture of the type shown above. In fact, the solution of the simultaneous equations is perfectly general and can be employed for analysis of any pair of intersecting curves.

The absorbance (A) of a given solution depends on the absorption per unit concentration, on the actual concentration, and on the length of the path through which absorption occurs. These facts can be summarized in the expression:

$$A = \epsilon c L \tag{2-15}$$

where c is the concentration, L is the light path length, and the constant, ϵ, is the absorption or *extinction coefficient* (less commonly, the absorptivity). We shall discuss extinction coefficients in greater detail later.

To solve for the individual concentrations of A and B, we must know absorbances (A) of the mixture at the wavelength maxima for A and B, $A_{\lambda maxA}$ and $A_{\lambda maxB}$, and the extinction coefficients (see below) for A and B at the two wavelengths, that is, $\epsilon^{A}_{\lambda maxA}$, $\epsilon^{A}_{\lambda maxB}$, $\epsilon^{B}_{\lambda maxA}$, $\epsilon^{B}_{\lambda maxB}$. We may set up a pair of simultaneous equations:

$$\begin{aligned} A_{\lambda maxA} &= (\epsilon^{A}_{\lambda maxA} \cdot [A]) + (\epsilon^{B}_{\lambda maxA} \cdot [B]) \\ A_{\lambda maxB} &= (\epsilon^{A}_{\lambda maxB} \cdot [A]) + (\epsilon^{B}_{\lambda maxB} \cdot [B]) \end{aligned} \tag{2-16}$$

These simultaneous equations represent a perfectly general solution; they can be used to analyze any curve that is the sum of two intersecting or overlapping curves. This analysis is the basis of the method of Warburg and Christian for determination of proteins in the presence of nucleic acids, a method that depends on absorbance measurements at 260 and 280 nm.

The Beer–Lambert Law and Its Consequences

A beam of radiant energy, after passing through a uniform absorbing medium, will be diminished in intensity by virtue of absorption. As noted in Eq. 2–15, the actual absorbance will be determined by the length of the light path, by the concentration of the absorbing substance(s), and by the extinction coefficient. This is essentially what the Beer–Lambert law states.

The initial intensity of the incident beam can be represented by I_0 and the intensity of the transmitted beam by I_t. The total path length is represented by L. If one imagines a series of thin slices cut at right angles to the beam, each of thickness, ΔL, then the average decrease in beam intensity is

$$\Delta I = I_0 - I_t = \kappa I_0 c \Delta L \tag{2-17}$$

or, expressed as differentials: $dI = -\kappa I_0 c \, dL$, where κ is a constant. Rearranging and integrating gives the equation:

$$\int_{I_0}^{I_t} dI/I = \int_0^L -\kappa c \, dL \qquad (2\text{–}18)$$

The solution of this integral equation is:

$$\ln I_t/I_0 = \kappa c L \qquad (2\text{–}19)$$

which is very similar in form to Eq. 2–15. The ratio I_t/I_0 is defined as the *transmittance* of the sample, and it is generally expressed at $\%T$, where T represents I_0 for the system. It is customary to convert the natural logarithmic expression to Briggsian logarithms, as

$$\log I_t/I_0 = -\frac{\kappa c L}{2.303} \qquad (2\text{–}20)$$

Because graphic presentation of logarithmic functions is at best awkard, and in order to remove the negative sign, it is common to convert the transmittance to absorbance, where A is defined as $-\log T$. This gives rise to the by now familiar equation:

$$A = \epsilon c L = \log T \qquad (2\text{–}21)$$

where ϵ is the extinction coefficient, and the other symbols have their previously assigned meanings.

Manipulation of Extinction Coefficients

Knowledge of extinction coefficients is a great help in quantitative analysis by absorption spectrophotometry, but the values are not always expressed on the same basis. It is important to be able to convert one form of the expression to another. In the published literature, values of the coefficient may appear in one of several ways. In some cases they are written as

$$E^{1\%}_{280\ \text{nm}} \text{ or } \epsilon^{1\%}_{280\ \text{nm}} \text{ or } \epsilon(280\ \text{nm}, 1\%)$$

The notational style is unimportant; what matters is that the wavelength of measurement and a measure of the concentration must *both* be specified. In the examples above, the wavelength of observation is clear. The concentration statement indicates that the sample contained 1 part in 100 (w/v is assumed unless otherwise specified), and *it is assumed that the length of the light path was 1.00 cm, unless otherwise specified.* Conversion to a more rational concentration basis requires knowledge of the molecular weight of the material. Thus, the millimolar extinction coefficient relates to the solution containing 1 mmol/L in a 1.00-cm cuvet at some designated wavelength.

A few examples follow:

1. A solution containing 0.515 g of compound X in 2.30 ml of buffer absorbed too strongly for measurement in a 1.00-cm cuvet at a wave-

length of 411 nm. The solution was placed in a cuvet with a 2.0-mm light path; it then gave a value for A, of 0.442. What is the value of $\epsilon^{1\%}_{411nm}$?

$$\text{Concentration correction:} \quad 100 \times \frac{0.515\ \text{g}}{2.30\ \text{ml}} = 22.39\%\ (\text{w/v})$$

$$\text{Light path correction:} \quad \frac{10.00}{2.00} \times 0.442 = 2.21$$

Therefore,

$$\epsilon^{1\%}_{411nm} = 2.21/22.39\% = 9.87 \times 10^{-2}$$

2. The molecular weight of X is known to be 13,500. If $\epsilon^{1\%}_{411nm}$ is known to be 9.87×10^{-2}, what is the value of the millimolar extinction coefficient?

   In a 1% solution,

$$\text{mmol/L} = \frac{10{,}000\ \text{mg/L}}{13{,}500\ \text{mg/mmol}} = 0.741\ \text{mmol/L}$$

Then,

$$\epsilon/\text{mmol} = \epsilon^{mM}_{411nm} = \frac{9.87 \times 10^{-2}}{0.741} = 1.33 \times 10^{-1}$$

Note also that one can immediately write down the *molar* extinction coefficient as 133.0.

3. At pH 7.5, a solution of a certain compound had a value of $\epsilon^{mM}_{259nm} = 17.8$. A standard solution, made for a particular assay, gave $A = 1.25$ in a 1.00-cm cuvet. What was the concentration of the standard solution, expressed in μmoles per milliliter?

   Since $A/\epsilon = c$ (expressed in mmol/L),

$$\frac{1.25}{17.8} = 7.02 \times 10^{-2}\ \text{mmol/L}\ \textit{or}\ \mu\text{mol/ml}$$

These typical illustrations show the importance of ϵ values in quantitative applications of absorption spectrophotometry. Note, however, that the values of ϵ depend critically on such exernal factors as the pH, the oxidation–reduction state, or on anything else that may affect the structure of the energy-absorbing centers in the molecular species; always use the value of ϵ that is appropriate for the circumstances. Furthermore, with strongly absorbing species, one must be certain that the absorbance value on which calculations are based is a reasonable one in terms of the dynamic range of the spectrophotometer, and that no interfering substances are present. This is the subject to which we next turn.

Limitations of the Beer–Lambert Law

Figure 2–12 shows three separate plots of absorbance vs. concentration. Curve *a* depicts a system that obeys the Beer–Lambert law perfectly. For

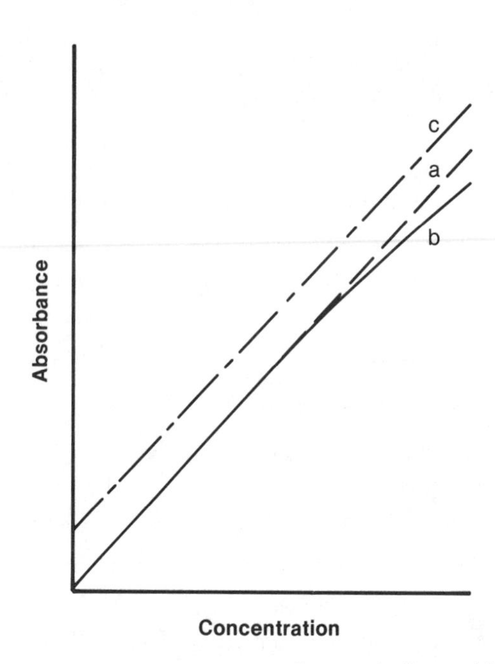

Figure 2–12. Absorbance-concentration plots. Curve *a* indicates ideal Beer–Lambert law behavior; curve *b* shows negative deviation at high concentrations; curve *c* indicates possible positive error introduced by an absorbing contaminant.

the system represented by curve *b,* the measured absorbance at higher concentrations is less than predicted by the law. For the system depicted by curve *c,* absorbance is linear over the entire concentration range, but all of the absorbance values appear to be higher than predicted by the law.

An explanation of curve *c* is that there is present in the system some unknown component that also absorbs energy at the same wavelength. This is comparable to the case described in Fig. 2–11, where if one were attempting an analysis of A in the presence of a *constant* contamination by B, data like those of curve *c* in Fig. 2–12 would result. Such a situation might arise, say, from a dirty water supply, or from a contaminated buffer.

The explanation of curve *b* is more complicated and demonstrates that the Beer–Lambert law is not obeyed over an infinite range of concentrations. Failure of this sort may derive from several sources, discussed in the remaining sections.

Chemical Concentration Effects

1. If the chromophore-containing molecules participate in a dissociation equilibrium, the dissociated and undissociated species may have different spectral characteristics. The absorption maximum may shift so that the wavelength used for low concentration would be incorrect for higher concentrations. As an example, concentrated cupric chloride solution is green, due to $CuCl_2$, whereas dilute solutions are blue, due to Cu^{2+}. If protons are involved in the dissociation, then the equilibrium would be pH dependent and precautions should be taken for adequate buffering of the system.

2. Molecular associations or aggregations may result in stacking of the absorbing molecules with possible alteration in the electronic transitions. This could alter the position of the absorption maximum, the extinction coefficient, or both.
3. The refractive index changes with concentration and this would affect the observed absorbance since, scrupulously, it is $\epsilon \cdot [n/(n^2 + 2)^2]$ that is a constant and not ϵ. For most measurements, however, especially below a concentration of 0.01 M, this effect is negligible.

Relative Photometric Error. This type of error is a function of the instrumentation employed and of the error intrinsic in all measurements. Older instruments were designed to perform in a more or less linear manner over the absorbance range of 0–2; newer machines are designed to operate well over a range of −0.2 to +3. Regardless of the practical limits imposed by the hardware, it can be shown that some relative photometric error always exists. It is therefore a good idea to make important measurements somewhere near the center of the instrumental absorbance range. Put in other terms, one may ask the question: Is there some point on the absorbance scale where the relative photometric error is least? Existence of such a point is simply demonstrated. Minimal photometric error occurs when

$$A = \log e \; (A = \log 2.718)$$

This proof is easier to comprehend if we begin in terms of transmittance rather than absorbance. Recall that the concept of absorbance was introduced from the beginning only as a convenience; the fundamental argument is better expressed in terms of transmittance as a physical phenomenon. Let

$$c = \frac{-1}{\epsilon L} \log T, \text{ so that } \log T = -\epsilon L c \tag{2–22}$$

Then

$$dc = \frac{-1}{\epsilon L} d(\log T) \tag{2–23}$$

But by definition,

$$d(\log T) = \log e \cdot \frac{dT}{T} \tag{2–24}$$

Therefore

$$dc = \frac{-1}{\epsilon L} \cdot \frac{\log e}{T} \cdot dT \tag{2–25}$$

To get the *relative error* in c, divide both sides of Eq. 2–25 by c, giving

$$\frac{dc}{c} = -\frac{1}{\epsilon L c} \cdot \frac{\log e}{T} \cdot dT \tag{2–26}$$

where $e = 2.718$. Recall from Eq. 2–22 that $1/\epsilon Lc = -1/(\log T)$. Therefore

$$\frac{dc}{c} = Q = \frac{\log e}{T \log T} \cdot dT \tag{2–27}$$

When the expression equivalent to Q, on the right-hand side of Eq. 2–27, is minimized by differentiation and set equal to zero, the relative photometric error will also be minimized. Thus, when $dQ/dT \rightarrow 0$, then

$$e + \log T = 0 \tag{2–28}$$

(**Hint:** Rewrite $Q = (\log e)T^{-1}(\log T)^{-1}dT$. Let $T^{-1} = u$, and $(\log T)^{-1} = v$, then apply the formula $d(kuv) = k[u\,dv + v\,du]$). Since $-\log T = A$, it follows that minimal relative photometric error in concentration is observed when

$$A = \log e = 0.434 \tag{2–29}$$

REFERENCES

Bauman, R. P. *Absorption Spectroscopy.* John Wiley and Sons, New York (1962).

Brown, S. B. *An Introduction to Spectroscopy for Biochemists.* Academic Press, New York (1980).

Clayton, R. K. *Light and Living Matter,* Vol. 1. McGraw-Hill, New York (1970).

Jackman, L. M. *Applications of Nuclear Magnetic Resonance Spectroscopy in Organic Chemistry.* Pergamon Press, New York (1959).

Jaffe, H. H., and Orchin, M. *Theory and Applications of Ultraviolet Spectroscopy.* John Wiley and Sons, New York (1967).

Marshall, A. *Biophysical Chemistry.* John Wiley and Sons, New York (1978).

Meites, L., and Thomas, H. C. *Advanced Analytical Chemistry.* McGraw-Hill, New York (1958).

Mellon, M. G., ed. *Analytical Absorption Spectroscopy.* John Wiley and Sons, New York (1950).

Moore, A. D. Henry A. Rowland. *Sci. Am. 246:*150–161 (1982). [A rewarding commentary on Rowland the man, and a nontechnical look at the "engine" he designed and built.]

Nakanishi, K. *Infrared Absorption Spectroscopy.* Holden-Day, San Francisco (1964).

Pecsok, R. L., and Shields, L. D. *Modern Methods of Chemical Analysis.* John Wiley and Sons, New York (1968).

Peske, A. J., et al. *Fluorescence Spectroscopy: An Introduction for Biology and Medicine.* Marcel Dekker, New York (1971).

Robinson, J. W. *Undergraduate Instrumental Analysis,* 3rd ed. Marcel Dekker, New York (1982).

CHAPTER THREE

Nuclear Instability and Radioactive Transformations

Atomic nuclei are very small (diameter $\cong 10^{-12}$ cm); for our purposes we may regard them as composed of positively charged *protons* and uncharged *neutrons.* The forces that hold these particles together are incompletely understood, although it is clear that they somehow overcome the repulsive effects of the charge on the protons; current physical theories are clarifying this problem. Protons and neutrons have very nearly the same mass when at rest, although, according to Einstein's equation, their masses increase when they are accelerated. The mass of a proton is approximately 1.007 atomic mass units, and the mass of a neutron is 1.008. The atomic mass unit is defined as $\frac{1}{12}$ the mass of that carbon species with mass number = 12.0000; therefore, the atomic mass unit (amu) is approximately 1.660×10^{-24} g.

Every atomic species can be characterized by three quantities, A, Z, and N, where:

A = *atomic mass,* taken as the least integer closest to the ordinary chemical atomic weight.

Z = *atomic number,* the number of protons in the nucleus. Values of Z run from 1, for hydrogen, to 92 for uranium, although several man-made elements have larger values, up to more than 100.

$N = A - Z$ = the number of neutrons in the nucleus. As we shall see, the instability of a nuclear species is related to the ratio of protons to neutrons and thus accounts for the phenomenon of radioactive decay.

ISOTOPES: BASIC PRINCIPLES

One theory of the genesis of chemical elements suggests that at one time the universe contained nothing but gaseous hydrogen. For reasons we cannot go into here, nuclei of hydrogen atoms condensed to form the heavier elements but in the process a small amount of nuclear mass was released as energy. This accounts for the so-called *mass defect,* based on the observation that atomic weights of the heavier elements are not exact multiples of the hydrogen mass. Indeed, the mass defect is the basis of attempts to produce useful free energy by atomic fusion. It is assumed that in the primordial fusion process the ratio of protons to neutrons was also altered, giving rise to the elements as we know them today. Consequently, many naturally occurring elements can exist in the form of two or more *isotopes;* these are species that have the same value of *Z,* because they contain the same number of protons, but they have different values of *A* because they contain different numbers of neutrons. Thus, the three hydrogen isotopes all have the same atomic numbers, but each has a different mass and different stability. They are commonly written as

$$^{1}H_{1},\ ^{2}H_{1},\ \text{and}\ ^{3}H_{1}$$

The three mass numbers are written to the left of the symbol as superscripts, whereas the *Z* value for each is written to the right of the symbol as a subscript. Ordinary hydrogen, or *protium,* is stable, as is *deuterium,* the isotope with mass number = 2. *Tritium,* the heaviest isotope, is unstable and radioactive. Although these isotopes are named as shown, it is not customary to name the isotopes of any heavier elements.

The notation just described, in which the superscript before the chemical symbol represents the mass number and the subscript following it represents the atomic number, is designed to clarify discussion of transformation reactions involving nuclei, but it is unduly cumbersome for biological applications of isotopes, where it is usually sufficient to identify the mass number of the isotope employed. To the degree that the symbol defines the atomic number, the subscript is redundant. Current convention allows one to describe a labeled molecule as, for example:

$$[1\text{-}^{13}C]CH_3COOH$$

by what is meant acetic acid labeled in the carboxyl group with $^{13}C_6$, the stable isotope of carbon with $A = 13$. One might also prepare

$$[2\text{-}^{14}C]CH_3COOH$$

which would identify a species of acetic acid bearing the radioactive isotope of carbon in its methyl group. A doubly labeled form would be

$$[2\text{-}^{3}H,2\text{-}^{14}C]CH_3COOH$$

in which both ^{3}H and ^{14}C were introduced into the methyl group of acetic acid. Such species require specific means of synthesis to insert the isotope exactly where it is wanted. If this is not feasible, then a species might be prepared by some exchange process that delocalizes the isotope. Such materials are frequently identified as

$$[^{3}H(U)]C_6H_5NH_2$$

describing a species of aniline in which the tritium is *u*niformly distributed among the hydrogens of the molecule. Depending on the supplier, the U is sometimes replaced by G, for *g*eneral distribution.

Distribution of Isotopes

Not all of the known isotopes occur in nature. Those that are known in natural materials are far more uniformly distributed. Thus, for carbon, seven isotopes have been prepared, identified as

$$^{10}C,\ ^{11}C,\ ^{12}C,\ ^{13}C,\ ^{14}C,\ ^{15}C, \text{ and } ^{16}C$$

but only those with mass numbers of 12, 13, and 14 are found in nature. The remainder are too unstable to persist for more than a few seconds or minutes. For this reason they are not practical for use in biology. In general, the number of possible isotopes increases with increasing values of Z, but this tells us nothing of the natural abundance of isotopes. Thus, both ^{12}C and ^{13}C are stable, but the natural abundance of ^{13}C is only about 1.1%, whereas the natural abundance of ^{14}C is less than 0.1%.

Unstable Nuclei

The term *nuclide* refers to any nuclear species with a known number of protons and neutrons. When the collection of all known nuclides is considered, they can be divided into three classes. One class, represented by ^{12}C, is indefinitely stable, because of the ratio of protons to neutrons in that nucleus is, in terms of nuclear binding forces, a favorable state. A second class, represented by ^{14}C, is unstable, or radioactive, as a result of an excess of neutrons over protons, or a lower than favorable proton to neutron ratio. However, the degree of instability is such that the isotope has its characteristic long half-life ($t_{1/2}$), that is, the time required for exactly half of any given quantity of ^{14}C to undergo radioactive decay. The third class, represented by ^{10}C, is also unstable. It is too unstable to exist in nature and is, in fact, an artificially produced radioisotope with a half-life of seconds.

As the value of Z increases, the value of $A - Z$ also increases, but a plot of $A - Z$ against Z is not a straight line. This is shown in Fig. 3–1, in which the stable isotopes are shown as solid squares and the naturally occurring radioisotopes are shown as squares with black dots in their centers. Artificially radioactive isotopes are shown as open squares. The solid, straight line in the figure indicates what might be expected if each addition of a proton required addition of a neutron. The difference between the theoretical values of $A - Z$ and the actual values is a measure of the mass defect resulting from nuclear condensation reactions; it represents energy released during the condensation process. It shows, perhaps, why nuclear fusion is being intensively studied as a potential supply of energy to replace fossil fuels. The masses of the neutron and the proton are known quite precisely. When one examines nuclear transformations, it becomes clear that nuclear masses are not exactly equal to the sums of the masses of the neutrons plus protons. Only if mass and energy are considered together does the conservation principle apply. From Einstein's equation $E = mc^2$, one can show that the disappearance of 1 g of matter releases

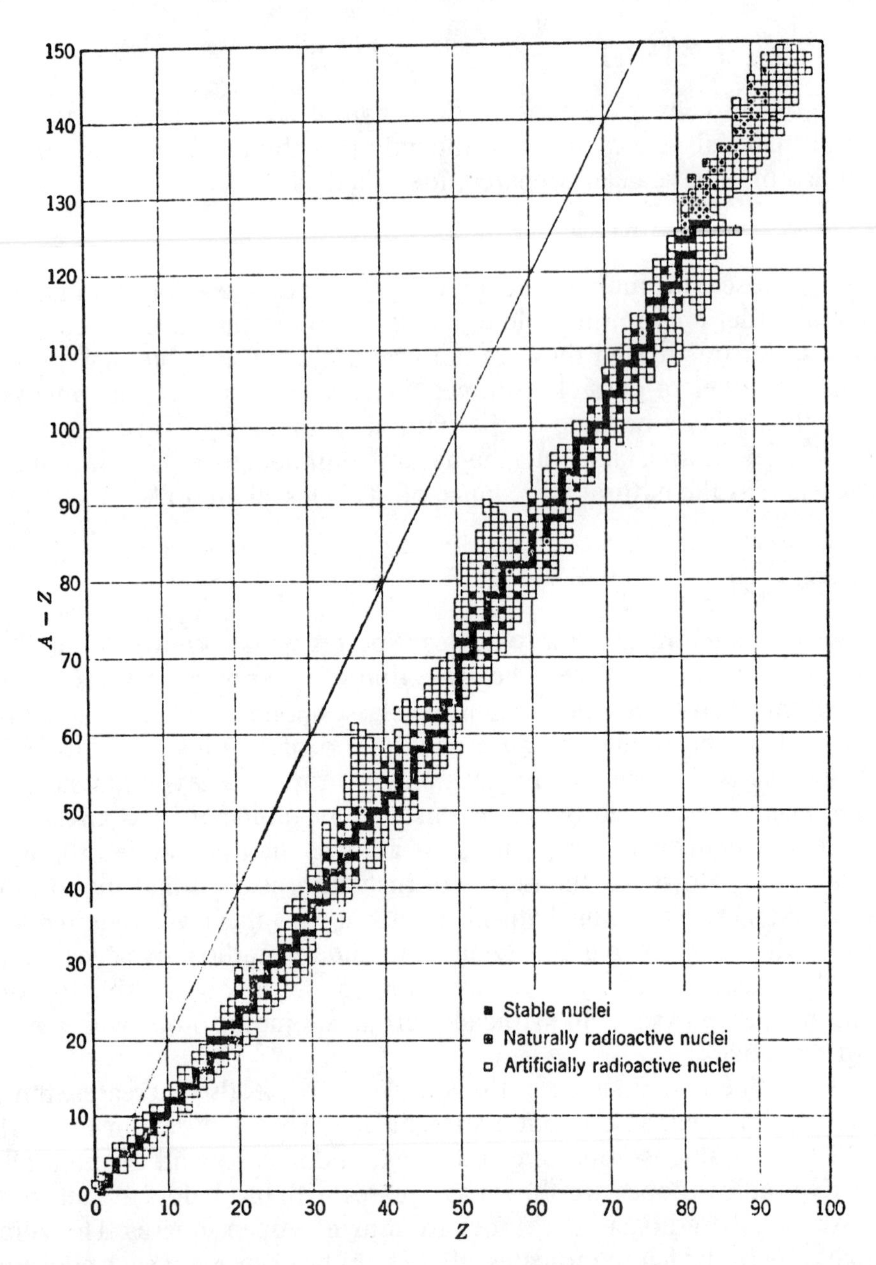

Figure 3–1. Plot of number of neutrons, $A - Z$, against atomic number, Z. The filled squares, indicating the stable nuclei, do not lie on a straight line. The light line to the left of the arranged nuclides shows the theoretical line predicted by the theory that nuclei were condensed from equal numbers of protons plus neutrons.

8.968×10^{20} erg, or about 2×10^{10} kcal/g, if the heat equivalent of the energy is considered.

Even when the stable isotopes are set aside, and only the unstable situations are considered, Fig. 3–1 shows clearly that unstable nuclei at any value of Z occur with proton/neutron ratios greater than *and* less than the stable values. Further, it shows that when $Z > 84$, all nuclides appear to be unstable and to undergo one or another kind of transformation to a more favorable configuration.

Modes of Radioactive Decay

Nuclear transformations may occur in nature, or they may be the result of deliberate experiment, that is, the result of artificial isotope production. For most biologically interesting purposes, the isotope must then be introduced into some metabolite of interest. Because the fate of the metabolite can be followed, or traced, by measurement of the radioactivity, the kinds of experiments that can be performed have become known as "tracer" methodology. The consequences of nuclear transformation are at least two fold. First, energy is frequently emitted either as x-rays or as γ rays; β particles (accelerated electrons, β^-; or positrons, β^+) or α particles (He nuclei) account for the remainder of the energy. Second, the unstable nuclei are converted to more stable nuclei that are not, in general, chemically identical with their precursors. In other words, product nuclei do not have the same Z as the precursor nuclei. It is customary to speak of nuclear transformations collectively as the processes of *radioactive decay,* even though a variety of processes may be involved, and more than one process may be involved in a particular nuclear transformation.

Because energy is so frequently released in the form of charged particles, it is useful to characterize it in terms of *electron volts.* An electron volt (eV) is defined as the energy acquired by an electron being accelerated by a potential difference of one volt. In view of the scale of nuclear events, practice dictates that the better units are thousands of electron volts (keV) or millions of electron volts (MeV). The charge on an electron is 1.602×10^{-12} erg, so 1 MeV $= 1.602 \times 10^{-6}$ erg. The atomic mass unit, noted earlier, is $[6.02 \times 10^{23}]^{-1}$ g, so the energy released by 1 amU, expressed as MeV, can be calculated as 931 MeV. This is a large amount of energy, and gives a considerable penetrating power to particles thus accelerated. Because x-rays and γ rays also have considerable penetrating power, it is usual to speak of them also in terms of electron volt equivalents. Some of the characteristics of radiated forms of energy are presented in Table 3–1.

Table 3–1. Characteristics of Particles Released in Nuclear Transformations

Identity of Particles	Symbol	Resting Mass (amu)	Charge	Typical Range of Energies (MeV)
Alpha[a]	α	4	+2	2 –10
Beta				
Electron	β^-	5.5×10^{4}	−1	0.02– 5
Positron	β^+	5.5×10^{-4}	+1	0.02– 5
Gamma	γ	0	0	0.1 –10

[a]Normally encountered only in isotopes of high Z. Alpha emission is not typically a feature of isotopes employed in biological studies.

Table 3–2. Modes of Decomposition and Decay Characteristics of Some Commonly Employed Isotopes

Isotope	$t_{1/2}$	Modes of Decay	Maximum Decay Energy (MeV)	Particle Energy (MeV)
^{3}H	12.43 yr[a]	β^-	0.0186	0.0186
^{14}C	5730 yr	β^-	0.156	0.156
^{125}I	60 d	EC[b] (γ)	0.149	
^{131}I	8.07 d	β^-	0.970	0.257 (1.6%) 0.333 (6.9%) 0.487 (0.5%) 0.606 (90.4%) 0.806 (0.6%)
^{35}S	88.2 d	β^-	0.167	0.167
^{32}P	14.3 d	β^-	1.71	1.71
^{45}Ca	165 d	β^-	0.252	0.252
^{40}K	1.28×10^9 yr	β^-, β^+, EC[b]	1.35(β^-) 1.51(β^+)	

[a]Many sources still cite $t_{1/2}$ for ^{3}H as 12.26 years, but a recent convention (Vienna, September, 1979) has increased the value as shown here.
[b]EC represents the process of orbital electron capture; see text for details.

Table 3–2 describes some decay characteristics of a few isotopes commonly employed in biology. Note that most of these are β^- emitters. The half-lives are quite variable, ranging from a few days to many thousands of years. Long half-lives are an asset when one is preparing substrates or metabolites, but can become a nuisance when waste disposal problems must be considered. The maximum decay energy values noted reflect the penetrating power of the radiation. Low values, such as noted for ^{3}H, mean that detection of the isotope in such media as electrophoretic gels or in thick tissue slices may be difficult. At the same time, handling of ^{3}H is safe since even ordinary glass vessels prevent exposure to the radiation. More energetic particles, such as the β^- of ^{32}P, are much more penetrating.

Recent data indicate that workers can be needlessly exposed through the walls of glass containers or even through open air at the top of a bottle of ^{32}P. This isotope should be handled, insofar as is possible, behind a plastic shield.

The last column in Table 3–2 indicates that β^- radiation may be emitted over a range of energy values up to the noted maxima. The data for ^{131}I are given in some detail to show that in some instances there are peaks in the spectrum of radiation intensities; in other instances, the radiation forms a continuum.

α Particle Emission. As noted earlier, emission of α particles occurs naturally only in the heavier nuclides. The classic case, first studied by Marie and Pierre Curie, is that of radium. A nuclear equation can be written as

$$^{226}Ra_{88} \rightarrow {}^{222}Rn_{86} + {}^{4}He_2 \text{ (an } \alpha \text{ particle)}$$

This conversion of radium to radon releases 4 mass units and decreases the atomic number from 88 to 86.

β Particle Emission. Unless specified to the contrary, β emission implies release of a negative electron, or negatron. It is the pattern commonly observed in nuclei that have a low ratio of protons to neutrons and that are unstable for that reason. Stability is gained by conversion of a neutron to a proton plus an electron (β^-). The electron is released from the nucleus with some characteristic range of velocities, along with a *neutrino,* which carries the remainder of the liberated energy. The nuclear reaction (neglecting the neutrino) can be written as

$$^{14}C_6 \rightarrow {}^{14}N_7 + \beta^-$$

The mass remains unchanged, but the atomic number increases by 1. Because the mass of the emitted β^- particle is very small ($5.485+ \times 10^{-4}$ amU, or about $1/1847$ the mass of the hydrogen atom), the product nucleus has virtually the same mass as its precursor, but conversion of the neutron to a proton has increased the value of Z by 1. In other words, carbon has been transformed into nitrogen.

An inverse case occurs when the proton/neutron ratio is too high for stability. Here a proton is converted into a neutron plus a positron, or β^+ particle. Positron emission is much less common than β^- emission in elements located in the lighter half of the periodic table. A typical nuclear reaction can be written

$$^{13}N_7 \rightarrow {}^{13}C_6 + \beta^+$$

This reaction may well be the source of the stable ^{13}C isotope found in nature. Isotopes that emit β^+ particles tend to be quite short-lived and as a result are not practical tools in biology.

Positrons are themselves short-lived; when they have lost their energy by collision with other atoms, they interact with electrons. The mass of the paired particles, β^- and β^+, is annihilated by conversion into two quanta of γ radiation, each having energy of 0.51 MeV, and moving in opposite directions.

$$\beta^- + \beta^+ \rightarrow 2\gamma \text{ (0.51 MeV)}$$

This is the so-called *annihilation radiation* typical of β^+ emission; the most readily discernible consequence of positron decay is the emission of γ rays, equivalent in energy to the mass that was annihilated.

Orbital Electron Capture. The scheme just described for positron decay requires that the energy difference between precursor and product nuclei be at least equal to $2 \times 0.51 = 1.02$ MeV, in order to account for the pair of emitted γ photons. Complications arise when this much energy is not available. A competing process is known as orbital electron capture, or simply *electron capture* (EC). In this mechanism a nucleus gains stability by capturing an electron from one of the innermost orbital shells, usually the K shell but sometimes the L shell. As in previous examples, capture of the electron does not significantly affect the nuclear mass, but the value of Z is reduced by 1. The vacancy left by the captured electron is filled by rearrangement of the remaining orbital electrons, so the detectable consequence of electron capture is the emission of x-rays. A typical nuclear reaction may be written as

$$^{22}Na_{11} + \beta^- \rightarrow {}^{22}Ne_{10} + \text{x-rays}$$

Positron emission and electron capture frequently are competing processes in the transformation of unstable nuclei. As shown in Table 3–2, both are known to occur in certain commonly used isotopes.

Artificial Production of Isotopes

When Lise Meitner and Otto Hahn demonstrated fission of uranium, the atomic age began in earnest. Subsequent to their work, atoms were bombarded with accelerated neutrons, protons, γ rays, and electrons in a program of isotope production. For the latter, the aim is to bombard target atoms with particles in such a way that the ratio of protons to neutrons is shifted from a stable to an unstable or radioactive ratio. Neutron bombardment of stable isotopes has been particularly useful.

In a shorthand notation to describe isotope production reactions, the bombarded element is shown first, with its mass number. The notation in parentheses specifies first the bombarding particle, then the emitted particle or radiation. Finally, the isotope product is shown by its mass number and symbol as in these examples:

$^{32}S(n,p)^{32}P$ $^{23}Na(n,\gamma)^{24}Na$
$^{14}N(n,p)^{14}C$ $^{35}Cl(n,\alpha)^{32}P$
$^{6}Li(n,\alpha)^{3}H$ $^{16}O(\gamma,n)^{15}O$

In the first example, ^{32}S is bombarded with neutrons, protons are emitted, and the product is ^{32}P. A very elaborate technology has been developed, involving production of particle accelerators, neutron sources, and processes for product isotope recovery. Both radioactive and stable mass isotopes are now available at relatively low costs.

Radioactivity: Detection, Quantitation, and Kinetics

DETECTION OF RADIOACTIVITY

Radioactivity is directly detectable only by some effect due to emission of energy either as charged or uncharged particles (protons, electrons, or neutrons) or uncharged photons (x-ray and/or γ ray photons). Two particular effects are frequently observed. The first of these is fluorescence, caused when target atoms are excited by highly energized particles or photons. The second effect is ionization of target atoms when they are bombarded

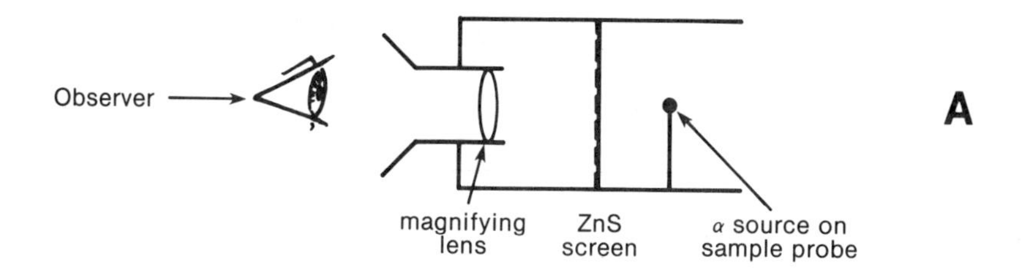

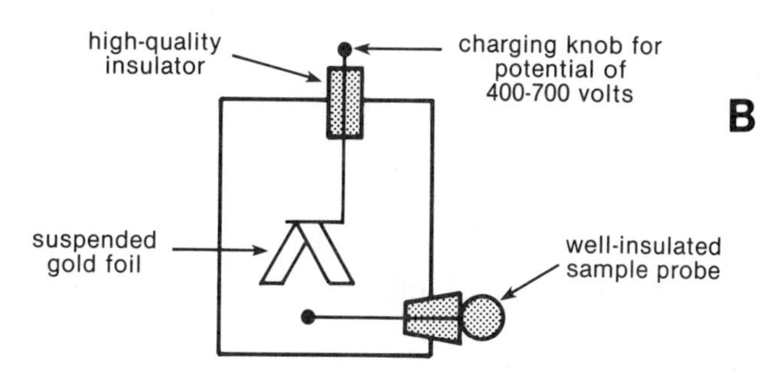

Figure 3–2. Early radiation detectors. (A) A sample spinthariscope. The isotope, $^{209}P_o$, has $t_{1/2}$ = 100 years and emits α and γ radiation. If 1 femtomol (fmol) of this isotope was placed on the sample probe, it would give a decomposition rate of ~6 dpm. This number of decompositions can readily be detected by eye as separate flashes. Because some of the α particles would not impinge on the ZnS screen, the observed rate of flashes would be somewhat less than 6 dpm. (B) A gold leaf electrometer. Charging the foil to a moderate voltage with respect to ground causes its free ends to repel each other. An ionizing sample allows the charge to leak to ground, and the ends of the foil to approach each other again. The rate of leakage is a measure of the radioactivity.

by emitted radiation. The ionized molecules are usually those of the circumambient atmosphere, which may be air or a different mixture of gases. Some simple detectors, based on these principles, are diagramed in Fig. 3–2.

Spinthariscopes (Fig. 3–2A) are small tubes, fitted with a magnifying lens at the focus of which is a screen coated with a thin layer of ZnS. On the opposite side of the screen is a support for an α-emitting source. An observer sees a brief flash of fluorescence as each α particle strikes the ZnS screen. At one time, before the hazards of radiation were completely understood, spinthariscopes were sold as toys for the amusement of children. The principle persists in modern machines that contain crystals of NaI deliberately contaminated ("doped") with thorium iodide (ThI). When bombarded, these crystals also fluoresce; in modern γ counters the flashes of light are detected by photomultiplier tubes and counted by electronic circuitry.

Electrometers were the precursors of modern ionization chambers. The principle of a gold-leaf electrometer is shown in Fig. 3–2B, which depicts a very thin gold leaf allowed to hang of its own weight over a metallic support rod contained in a glass envelope. By means of the charging knob, protruding through a high-quality insulator, a charge of several hundred volts can be applied to the gold leaf. This causes the two ends of the flexible gold foil to repel each other, so that they hang in an inverted "V" over the support rod. If one now introduces a source of radioactivity through the side port of the electrometer, the radiation will cause ionization of some molecules in the contained air. These ionized molecules provide a path by which the charge on the gold foil can be dissipated to ground potential (the potential of the envelope). As a result, the repulsion of the two ends of the suspended gold foil is dissipated, and they hang, limply, in the uncharged state. The rate of discharge can be observed through a calibrated viewing telescope and is a measure of the radioactivity of the sample.

Ion Chambers

The electrometer just described is the prototype of more modern ion chambers. In the simplest form, an ion chamber may consist of a metal tube filled with a gas. Through the center of the tube runs a thin rod or wire, carefully insulated from the chamber walls. A potential difference is applied so that the central thin electrode is positive with respect to the walls. Depending on the purpose intended, the potential difference may range from 100 to 300 volts, and the circuit includes a current-measuring device.

When energetic radiation penetrates the wall of the chamber, it causes ionization of some of the gaseous molecules, just as in the electrometer. Positive ions and free electrons are produced, many of which promptly recombine after passage of the radiation. A few of the ions are drawn to the negatively charged wall while the corresponding electrons are drawn to the central electrode. This situation constitutes a current flow which can be detected and quantitated and is proportional to the intensity of the radioactive source. Sensitivity of ion chambers depends on the geometry of the device and on the applied potential. If the potential is too high, the device passes to an undesirable, continuously conducting state. Sensitivity of ion chambers also depends on the nature of the gas with which they are filled, since some gases are more easily ionized than others.

Geiger–Müller Detectors

One advanced form of the ion chamber detector was designed originally by Geiger and Müller, both of whom worked in Rutherford's laboratory. Modern versions of the GM tube are made with very thin end windows of mica, so as not to impede the relatively low energy radiation from isotopes such as ^{3}H. The thinness of the end window makes the devices quite fragile. The tubes are filled with a mixture of vapors such as ethane or benzene, which are known to ionize readily. A drawback to GM tubes is their tendency to go into a continuously conducting state. This is usually inhibited by addition of a small quantity of Br_2 or H_2O vapor, either of which "quenches" or suppresses conduction after the initial passage of the radiation. With appropriate circuitry, GM tubes may be made to record the actual radioactive intensity, or a signal proportional to it. These instruments are still widely used to monitor laboratory uses of isotopes or in prospecting for radioactive mineral deposits. Because they are slow detectors, they are no longer widely used in quantitative work.

GM counters suffer from an additional drawback, which relates to their geometry. As ordinarily constructed, GM tubes are in the form of cylinders; radiation may enter either across the solid wall or, primarily, through the thin end window. However, radiation is emitted from a source in random directions through the 4π radians of solid space, so that much of the emission will never enter the GM tube.

Gas Flow Counters

This device takes the form of a hollow metal sphere, divided through its equator. Suspended through a high-grade insulator is a ring electrode, and there is also provision for placing a sample on a very thin film of plastic, or similar material that will not absorb weak radiation. The sphere is also pierced by two small tubes, one for inflow of a gas mixture, the other for its escape. Commercially available counting gas mixtures may contain 10% methane in argon, or 5% CO_2 in argon. These mixtures are carefully dried and are also air-free. They were devised to promote ionization and to minimize absorption of radiation.

The gas flow counters hark back to the gold leaf electrometer in that the sample is placed inside the counting chamber. Their symmetry is such that 4π efficiency is easily achieved. By adjusting the applied voltage to suitable values, one can obtain stable systems of very high precision. In spite of the cost of the counting gas, and the nuisance of thoroughly flushing the counting chamber to remove air and water vapor, gas flow counters still enjoy an important status for specialized uses. In particular, slightly modified chamber designs are used to detect and to quantify radioactivity on paper chromatograms, strips of which may be fed through the chamber in a continuous mode. The paper drive is commonly geared to a strip chart recorder, which then provides a tracing of radioactivity as a function of position on the chromatographic strip. Narrow glass or plastic thin-layer chromatography (TLC) plates may also be so counted. Tracings of this kind are commonly provided by suppliers of labeled metabolites as proof of purity of their preparation. However, for more routine counting purposes, gas flow systems have been largely replaced by scintillation spectrometers, designed to count either β- or γ-emitting isotopes, or even x-rays.

Solid Scintillation Detectors

Earlier reference to the spinthariscope made mention of the fact that certain substances emit brief flashes of fluorescence when bombarded by certain sorts of radiation. More up-to-date versions of this principle are embodied in solid scintillation detectors. A number of crystals, including (1) NaI doped with ThI; (2) CsI; (3) LiI doped with EuI; or even (4) anthracene, all share the property of emitting fluorescence. Appropriately prepared materials can be commercially grown to single crystals of sizes up to 2 × 3 × 2 cm, then bored to receive sample holders of glass or plastic. Opposite ends of such crystals are ground flat and polished, then placed in good optical contact with flat surfaces of photomultiplier tubes. As radioactive decomposition occurs within the sample, radiation causes the crystal to emit flashes of light which are sensed by the photomultiplier tubes. The signals are processed and amplified by the appropriate circuitry and the results read out as a function of time or number of counts recorded. Crystal scintillators are widely used in machines designed primarily for γ counting, but versions are available for x-ray or even for α counting. They are also used for β counting, but function in this case by virtue of the γ radiation which so often accompanies β emission.

Liquid Scintillation Detectors

The energy levels of most β emissions are suitable for excitation of fluorescence in some organic molecules. Two of these organic substances are 2,5-diphenyloxazole (known as PPO) and 1,4-bis(5-phenyloxazol -2-yl)benzene (known as POPOP). These materials are soluble in various mixtures of toluene, dioxane, and water; in such mixtures they form "scintillation cocktails" which may be added directly to small glass or plastic vials containing the sample. The treated samples are then placed in counting chambers fitted with paired photomultiplier tubes that sense the flashes of emitted light in a manner quite like the system already described for γ-counting instruments. The primary distinction is that β counters of the liquid scintillation variety need not involve a solid crystal; instead the fluorescent species is in solution mixed with the samples to be counted and is therefore in direct and immediate contact with the radiation source.

A major drawback is that the scintillation cocktail must be added to each and every sample to be counted, thereby rendering it useless for any other purpose. Until quite recently, liquid scintillation counters could operate only in a discrete mode, with a finite sample volume to which the scintillant had to be added. The latest developments have solved this problem by affixing the scintillant to inert beads, and packing these coated beads into a tube of small bore. Now it is possible to flow a continuous stream of fluid containing radioactivity, perhaps from a high-pressure liquid chromatographic (HPLC) column effluent, through the packed tubes. As the β-emitting species contacts the immobilized fluors, the latter become excited and it is possible to record radioactivity as a function of effluent volume. Thus the discrete sample requirement has been eliminated, although the working sensitivity of these detectors is not yet as high as might be desired. Because the need for toluene and dioxane has been eliminated, operating costs are reduced and some health hazards are eliminated. Lastly, the samples remain uncontaminated.

Solid-State Detectors

Semiconductors based on deliberately contaminated chips of silicon, germanium, and a few other elements have revolutionized electronics, and now they may bring about similar profound changes in detection of radioactivity. It is possible to prepare what are known as lithium-drifted germanium detectors of high sensitivity. The term *drifting* refers again to deliberate introduction of contaminant ions into a different crystalline solid. Ge(Li) detectors have some peculiar physical properties. They operate well only at temperatures close to that of liquid nitrogen, and they are very sensitive to destruction by other contaminant ions. For this reason they are generally encased in thin metallic envelopes and kept in liquid nitrogen throughout their working life. They are especially good for detection of γ rays. Other promising materials are being studied such as gallium arsenide, already in commercial use as a sensing element in photographic light meters and the like. A useful advantage of solid-state detectors is that they can be made in a very small physical configuration, possibly small enough to permit implantation in living organisms allowing the course of radiolabeled metabolites to be followed in vivo.

Noninstrumental Detectors: Autoradiography

Photographic film, a mixture of silver halide crystals suspended in a gelatin emulsion, is by nature sensitive to light and some other parts of the electromagnetic spectrum. If radioactive atoms are contained in thin-layer or paper chromatograms, in dried electrophoresis gels, or even in thin tissue sections, the radioactive decompositions can be detected by laying a sheet of x-ray film over the surface of the radioactive source. This process is known as autoradiography, meaning that the specimen itself provides the radiation. Probably the first successful reported example of autoradiography is that by Leblond, who studied radioiodine uptake by thyroid tissue. We will limit discussion, however, to autoradiography of chromatograms and gels. After a suitable exposure period, the film is developed in the usual way, producing deposits of metallic silver corresponding to deposits of radioactivity in the original. Weak β radiation, such as that from ^{3}H, is difficult to detect by simple autoradiography because of its limited pentrating power. Unless quite thin electrophoresis gels are used, much of the radiation may not be able to reach even the surface of the gel, and even with lengthy exposure periods and/or increased emulsion thickness the results may not be satisfactory without certain modifications. Two variations of the simple procedure have been developed to improve detectability.

Chromatograms or electrophoresis gels may first be soaked in solutions

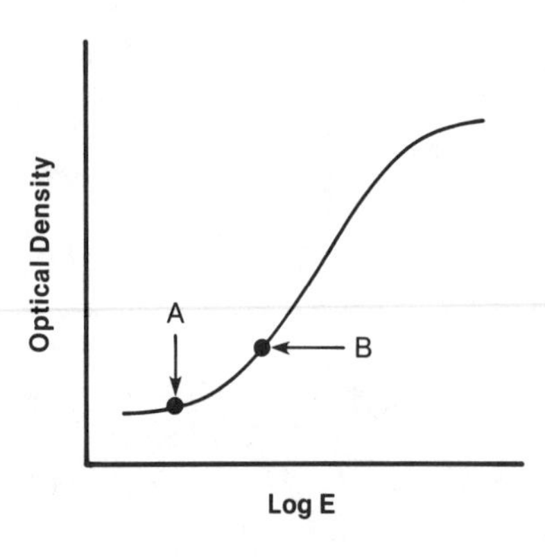

Figure 3–3. Hurter–Driffield Curve for a photographic film. Note the "heel" (lower left) and the "shoulder" (upper right) joined by a nearly linear segment. The letter A marks a point without preflashing whereas the letter B marks the same exposure after preflashing.

of scintillants similar to those employed in liquid scintillation counting, then carefully dried in a vacuum. Care must be taken in the drying process to avoid cracking or buckling of thick gels or distortion of their dimensions. More recently, the process has been somewhat simplified by substituting a solution of solium salicylate for the more complex scintillation cocktails. Sodium salicylate is a sufficiently effective scintillator for the longer exposures of autoradiography, compared to liquid scintillation applications. Regardless of the solution used, the effect of the treatment is to put scintillant into immediate contact with the source of radioactivity *within the structure* of the chromatogram or the gel. It is the light flashes from the excited scintillant species which then affect the photographic film. Films used for autoradiography are ordinarily x-ray films, which are designed for use in cassettes that commonly contain fluorescent screens to enhance imaging of bones and of soft tissues.

A second variation relates to the properties of photographic films in general. It is known that the optical density (O.D.) that can be generated in a photographic film is a logarithmic function of the exposure (E). The plot of O.D. against log E is known as the Hurter–Driffield curve; a typical example is shown in Fig. 3–3.

At very low exposures, the density is not at all proportional to the exposure; this portion of the curve is known as the "heel" (or sometimes as the "toe") of the curve. At very high exposures, when all of the available silver grains have been exposed, the density again bears no functional relation to the exposure; this portion of the curve is known as the "shoulder." Between these extremes, there is a nearly linear portion of the curve, the useful range for photographic purposes, in which the density generated is proportional to the log of the exposure, E. It is also true that even without exposure, development of many films will produce some slight blackening, since a few of the silver halide grains will be reducible even in the dark. Photographers speak of this phenomenon as "fog."

One can preexpose a sheet of film to very faint sources of light. By itself, such a flash might produce a generalized fog equivalent to an optical density of about 0.05. Since the effect of successive exposures is roughly additive, the preexposure by flashing *plus* the autoradiographic exposure has the effect of moving the total exposure from some point on the toe of the curve to a point on the linear portion of the curve. Although there is a faint fog over the entire surface of the film, this usually does not prevent detection of as few as 3000 radioactive decompositions/cm^2 of gel or chromatogram surface when ^3H is employed. Once a suitable preflashing light source is provided, this technique is rapid, simple, and far less expensive than the use of any scintillant. Furthermore, it does not involve addition of any foreign substances to the sample, which can then be recovered for other uses if required. The method is a general one and may be used with any isotope.

QUANTITATION OF RADIOACTIVITY

Quantitation requires some discussion of the units of radioactivity measurements. For many years, the standard unit has been the *curie* (Ci), defined as the number of disintegrations per second occurring in 1 g of pure radium. The curie = 3.7×10^{10} disintegrations per second (dps); more usefully, 1 Ci = 2.2×10^{12} disintegrations per minute (dpm). The curie is an entirely arbitrary unit, and depends on the quality of radium, whether or not it is in equilibrium with radon and on other factors. It is also a large unit, and it is far more customary to deal in milli- or microcuries (mCi or μCi).

Under a recently adopted convention, a new system of scientific units has been established, known as Système International (SI) units. In this scheme the new unit of radioactivity is the *becquerel* (Bq), defined as a disintegration rate = 1 sec^{-1}. Accordingly, 1 Bq = 27 pCi. However, it does not seem likely that the curie will easily be replaced although the major scientific journals are committed to adoption of the SI units. Note also that both the curie and the becquerel address only the rate of radioactive decomposition. Neither addresses the quality of the resulting radiation.

Specific Radioactivity

This term refers to the amount of radioactivity incorporated into some mass of another compound. For example, if ^{14}C has been incorporated into acetic acid so that 1 mmol of the acid contains 0.8 mCi, we would state that the specific radioactivity of the material is 0.8 mCi/mmol. It would be equally correct to state that the specific radioactivity is 800 μCi/mmol, 8×10^5 μCi/mol or, 13.3 μCi/mg.

KINETICS OF RADIOACTIVE DECAY

Given any quantity of radioactive material, there is a steady and an inexorable loss of activity with the passage of time. The rate at which activity declines varies widely and depends on the nuclear instability of the isotope in question. The rate of decline can be analyzed simply by reaction rate theory.

Radioactive decay is a monomolecular, first-order process. If N_0 represents the number of radioactive molecules at the initial time, t_0, then the number of molecules remaining at any later time, t, is proportional only to t and to N_0, so that:

$$N = \lambda N_0 t \tag{3-1}$$

The proportionality constant, λ, is known as the *decay constant* for the isotope. The rate equation can be written as:

$$-dN = \lambda N_0 dt, \quad \text{or} \quad dN/N_0 = \lambda dt \tag{3-2}$$

of which the solution is:

$$\ln[N/N_0] = -\lambda t \tag{3-3}$$

The special case where $N/N_0 = ½$ is important because it defines $t_{1/2}$, the *half-life* for the isotope, as the time required for the initial activity to be reduced by a factor of 2. Equation 3–3 then becomes

$$0.6931 = \lambda t_{1/2}, \quad \text{or} \quad \lambda = 0.6931/t_{1/2} \tag{3-4}$$

Either λ or $t_{1/2}$ may be used to characterize the decay rates, but use of $t_{1/2}$ is far more common because it is the directly measurable parameter. When N/N_0 is plotted as a semilog function of time a straight line is obtained (Fig. 3–4).

Significance of the Half-Life

Experimental Design Considerations. Radioisotope decay begins at the moment the unstable atoms are produced and continues with the passage of time. If the half-life is fairly long, measured in months or years, the decay encountered in the relatively shorter time scale of most experiments is not a problem. The situation is far different when isotopes with half-lives measured in days or weeks are used.

For technical reasons, radioisotopes are usually produced as inorganic substances. The uses made of them most frequently require incorporation into metabolic intermediates, such as acetate, alanine, or glucose, or as labels for more complex structures such as hormones or nucleic acids. Production of the labeled materials can require sophisticated syntheses and elaborate purification schemes, all of which take time. The iodination of insulin, in preparation for a radioimmunoassay of the hormone, is a good example. Iodination and cleanup of the hormone ordinarily take not less than 36 hours, equivalent to about 0.18 half-lives. If this preparation is used only during the first half-life, nearly 20% of the potential working time has already disappeared or been consumed while the product was in preparation. If one purchases iodinated insulin from a supplier, it will be delivered with a statement that the specific radioactivity had a certain

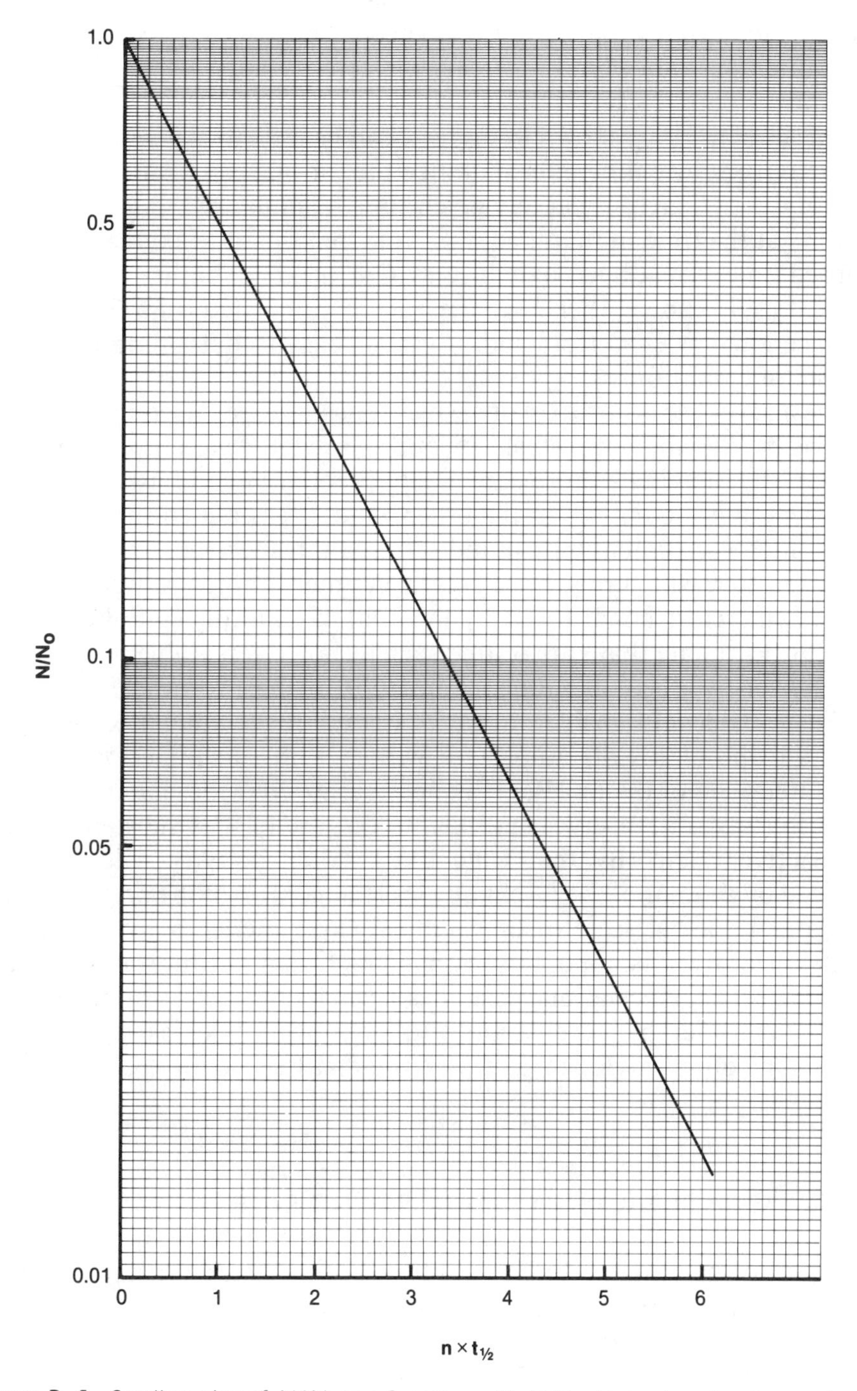

Figure 3–4. Semilog plot of N/N_0 as a function of half-life. Note that the time scale can be read in elapsed time as well as in half-lives.

value at a specified time on a specified date. If one does not receive the sample until three days later, significant decay will have occurred. This requires that, if a certain minimum specific radioactivity is needed for an experiment, a somewhat higher specific activity must be purchased. How can this be determined?

Curves of the type shown in Fig. 3–4 can be used to estimate the remaining activity at any time after manufacture, or one can, from a determination of the radioactivity at any given time, read backward to estimate the activity at the time of manufacture. Note that the time axis can be read in elapsed time as readily as in half-lives. One need not bother with the curve itself; if $t_{1/2}$ is known, then λ can be quickly calculated from Eq. 3–3 or 3–4. Once λ has been determined, it is then easy to determine N/N_0 under any set of conditions.

The above discussion explains why ^{131}I has been replaced in many applications by ^{125}I, which has a half-life approximately seven times longer than that of ^{131}I, allowing a more convenient preparation and use of key radiochemical intermediates. On the other hand, ^{131}I is a β emitter, which can be counted conveniently by liquid scintillation spectrometry, whereas ^{125}I is a γ emitter, which requires a special counter. Choice between these isotopes is therefore conditioned by the kind of counting equipment available. It is fortunate that with iodine such a choice is possible; with many of the shorter-lived isotopes, one is limited to a single form.

Waste Disposal Considerations. A second important use of the half-life relates to problems of radioactive waste. With short-lived isotopes ($t_{1/2}$ measured in months or less) it is possible to hold the material in a safe place for, say, 10 half-lives and then to dispose of it as substantially nonradioactive waste. By that time the residual activity will have been reduced by a factor of $(\frac{1}{2})^{10}$—slightly less than 0.01% of the original activity and, on the usual scale of biochemical experiments, a trivial quantity. With longer-lived isotopes this is not a practical procedure, and the wastes must be sent to special disposal sites permanently set aside for this purpose, at considerable expense and with unknown ecological consequences, since none of us would survive even the first half-life of ^{14}C.

NATURAL RADIOACTIVITY: BACKGROUND RATES

There is a very small but measurable radioactivity due to natural sources which finds its way into glass and/or plastic containers, reagents, and the ambient air. In addition, we are constantly exposed to cosmic radiation. The sum of these sources constitutes a source of counts, from 20 to 50 per minute, detectable by sensitive, modern instruments, whatever their type. In experiments involving low levels of radioactivity it is important to measure the background rate to correct counting observations. Obviously, when total counting rates reach 30–50 $\times$ 10^3 per minute, the background counting rate becomes negligible and is then frequently ignored. Background rates tend to be quite constant, but the ever-present possibility of spillage or other overt contamination requires that instruments be frequently checked for this potential source of error. This is especially true for instruments other than liquid scintillation counters.

Liquid Scintillation Spectrometry

INSTRUMENTATION

The working components of a modern, three-channel liquid scintillation spectrometer are shown in Fig. 3–5. Two photomultipliers are arranged on either side of a sample well so that their end windows can detect light flashes generated by interaction of β-particles, released by decomposition of unstable nuclei, with some dissolved scintillator(s). The photomultipliers are connected to a common high-voltage supply, which ensures that most irregularities in the supply voltage are canceled out. Each decay event produces a small current pulse in the multipliers. There may also be some randomly generated pulses due to thermal noise or to cosmic rays which strike the multipliers, but it is unlikely that such random events will occur simultaneously in both photomultiplier tubes. The purpose of the coincidence detector is to reject any such nonsimultaneous events, and to pass on to the summing amplifier only those current pulses seen simultaneously by both multipliers.

The summed current pulses vary in their lifetimes and in their intensities, corresponding to the energy contents of the individual light flashes as generated by radioactive decompositions. The summing operation serves two purposes: First, it amplifies the signals to a moderate degree; second, it restores a one-to-one correspondence between the number of current pulses and the number of decompositions.

The summed pulses are then fed to the three channel amplifiers. In most modern instruments these are logarithmic amplifiers, meaning that the outputs of the amplifiers are proportional to the logarithms of the input signals. In any event, each of the channel amplifiers has an adjustable gain, or ratio of output to input voltages. The gain can be varied by the user so that the actual voltages attained by each channel amplifier fall within some predetermined span. This produces the signals which must be further manipulated independently of the energies of the β^- particles themselves. The significance of this will shortly become clearer.

Recall that the light pulses generated by radioactive decay have variable intensities and lifetimes. A second effect of the channel amplifiers is to integrate these individual pulses and to reshape the signals so that the output pulses have a constant time base and a magnitude (pulse height) that is proportional to the energy content of the light flashes. Each of the channel amplifiers produces exactly the same array of pulses, but these may be

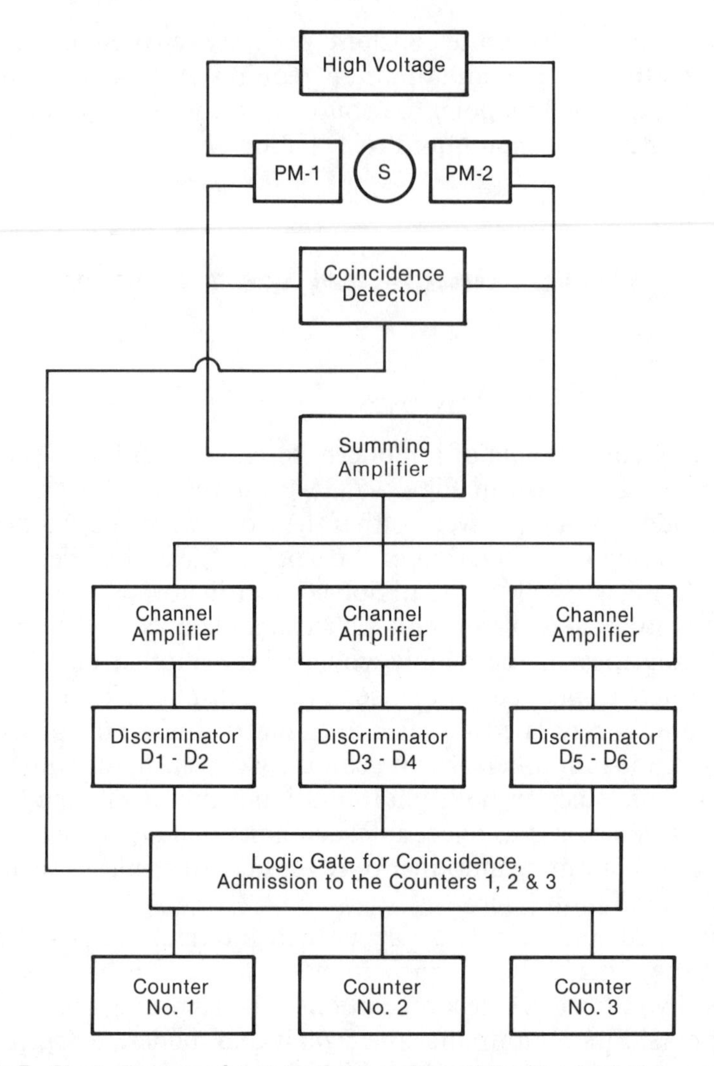

Figure 3–5. Block diagram of a typical three-channel liquid scintillation counter. Two photomultipliers are used to minimize thermal noise and extraneous impulses. Coincident energy pulses are noted as such and summed. These then are amplified by the three channel amplifiers, and the signals are properly shaped to a common time base. Discriminators then sort out the array of shaped pulses into three predetermined categories. These are passed to the channel counters only if they pass through the logic gate designed to count only pulses simultaneously noted by both detectors. See the text for further details.

differently magnified (amplified) by means of the gain controls. The train of signals then passes into separate discriminators that set the width of each channel.

The discriminators are sorting devices, controlled by operator-set variable resistors. The function of the discriminators is to reject pulses of less than a certain height or greater than another height. The rejection limits are determined by settings of the variable resistors just mentioned. In effect, the discriminators function in a manner analogous to light filters in a visible-light spectrometer. They "transmit" signals of some predetermined energy content selected from the total energy that reaches them.

A second check for coincidence is provided by the logic gate. This eliminates spurious signals that might be generated in the earlier electronics. The gate is normally closed; it opens only on receipt of a check signal from the coincidence detector. When the logic gate is opened, it allows pulses to enter the scaling counters from the corresponding discriminators. The counters drive the printout or readout devices.

Modern instruments obviously contain many other refinements, but the fundamental elements are those discussed above. There are three reasons why liquid scintillation counting is so widely used in biological and biochemical measurements of radioactivity: First, the primary means of detection is interaction between the scintillant species, PPO or the like, and the common β-emitting radionuclides. Second, available instruments are abridged spectrometers, most of which contain three channels. This means that more than one isotope can be counted at once and that experiments can be performed with doubly labeled metabolites, or with two metabolites each bearing a different isotope. Third, modern instruments are equipped with automatic sample changers so that 100–300 samples can be sequentially counted without need for operator intervention. Samples may be counted for a predetermined time or until some predetermined number of counts has been accumulated in each channel register or until a selected percent counting error has been attained. Operational details vary with manufacturer and with model design, but the options indicated here are available in virtually all machines.

CAUSES OF REDUCED EFFICIENCY

The transfer of energy in liquid scintillation counting runs, in a fairly linear manner, from β^- particle → solvent molecules → scintillator molecules. The generated light flash is converted to an electrical impulse and processed in the manner already described. In theory, each radioactive decomposition should result in a single light flash and a corresponding electrical impulse. In other words, the counts registered per minute should be identical with the decompositions per minute. Unfortunately, this is rarely the case; for a number of reasons the *efficiency* of scintillation counting is almost never 100% and is frequently as low as 20%. This is acceptable as long as one knows or can determine the efficiency in any given situation. Causes of reduced efficiency can be grouped into three classes, next discussed in order of increasing significance.

Instrumental Basis

Even though the electrical impulses can be manipulated with a time base of approximately 1 μsec, there is an upper limit to the counting rate; that is, there is a "dead" time during which the counter is inoperative. Within usual limits of experimental design this is not a major problem, but it is possible to exceed machine response times. The simple solution is to use samples of lower radioactivity.

Effects of Radioactive Source(s)

The energy spectrum of β^- emitters ranges from low values up to a maximum characteristic of the nuclide in question (see Table 3–2). Low-energy particles excite fewer molecules of scintillant, generating light flashes that are more feeble and of shorter duration than those generated by more energetic β^- particles. When these light flashes are transformed by the signal-shaping channel amplifiers, the low-energy emissions will generate pulses of lower height. If one is counting only a single nuclide, the discriminators can be set at "infinite" limits so that all of the pulses will be counted regardless of their heights. However, when more than one nuclide is being counted, it is necessary to set the discriminators to defined (noninfinite) limits in order to sort out counts due to one nuclide from those due to another.

The energy spectra of ^{3}H, ^{14}C, and ^{32}P are shown below. It is clear that there is a considerable overlap on the pulse height axis. In counting any two of these isotopes in the same sample, it is virtually impossible to avoid "spillover," or assignment of counts to an inappropriate channel. One could, by proper discriminator settings, completely eliminate ^{3}H or ^{14}C counts from the ^{32}P channel, but in no other case are counts due to different nuclides cleanly separable.

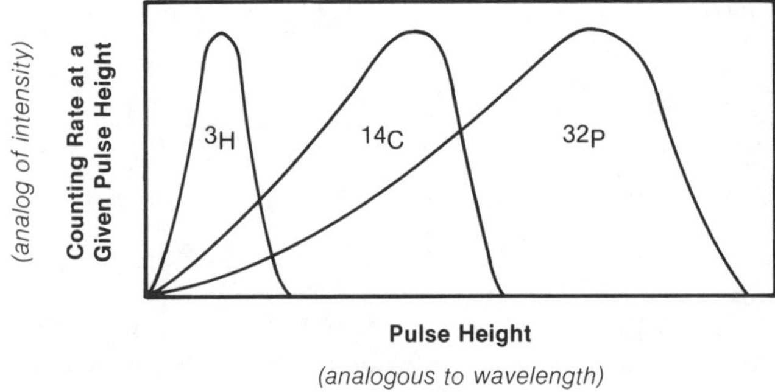

The spectral distribution here is entirely analogous to overlapping spectra in visible light spectrophotometry, and the analogy is emphasized by the italicized designations shown on the axes of the graph. In visible light spectroscopy the problem was solved by determination of the specific absorbances of the pure compounds (here, the counting rates per curie of the isotopes taken individually) at some wavelength (here, the pulse height under similar conditions).

In visible light spectroscopy, one employs narrow-band passes in order to minimize extraneous absorbances. In scintillation counting, the discriminators (in conjunction with the gain controls) can be regarded as determining an electrical "window" or "filter" in much the same way. Using a window narrower than the total energy distribution increases selectivity *at the expense of efficiency.* Hence, the counts per minute (cpm) registered will not be identical with the disintegrations per minute (dpm) of the sample. In a mixture of isotopes, one can set up two simultaneous equations in two unknowns. This is the analog of similar equations used in resolving mixtures of two absorbing substances in visible light spectroscopy.

Quenching

Effects of Sample Composition. The best solvents for liquid scintillation counting are alkylated aromatic hydrocarbons such as xylene or toluene. These provide efficient energy transfer between the β^- particles and the scintillator species in common use, and they are capable of dissolving reasonable concentrations of the scintillators such as 2,5-diphenyloxazole (PPO) and 1,4-bis(5-phenyloxazol-2-yl)benzene (POPOP). Oxygenated molecules (alcohols, ketones, and water) are poor solvents for exactly the opposite reasons. Unfortunately, solvents regarded as "good" from the viewpoint of scintillation efficiency are "poor" in their ability to dissolve tissue extracts and similar biological samples, and in practice, mixed solvents are used as a compromise. They provide a moderate counting efficiency in conjunction with a reasonable ability to dissolve or "carry" sample materials.

Many solvent systems have been proposed, based largely on toluene, with smaller amounts of dioxane, ethanol, or methanol and water. In addition, most mixtures contain a detergent, frequently Triton X-100. This is an alkylated benzene derivative which has the following structure:

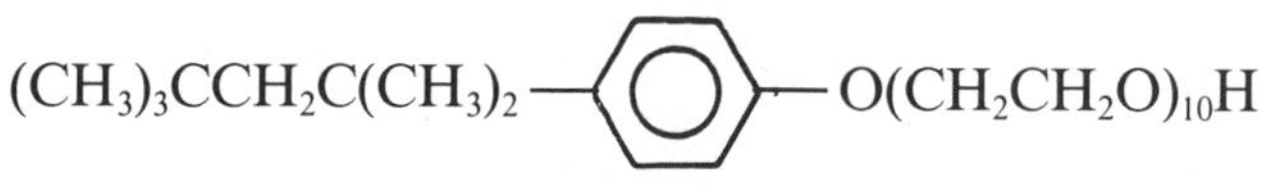

$$(CH_3)_3CCH_2C(CH_3)_2-C_6H_4-O(CH_2CH_2O)_{10}H$$

The alkylated benzene ring makes one end of the molecule hydrophobic whereas the extended polyethoxy chain with its terminal OH group makes the other end hydrophilic. Similar detergents are marketed under other trade names. These mixed, detergent-containing solvents are capable of dissolving sufficient concentrations of scintillators and at the same time can retain homogeneity up to a total water concentration (solvent water + sample water) of about 10% with only modest loss of counting efficiency. When the content of water or other oxygenated molecules rises much above this figure, the counting efficiency drops off sharply; such systems are said to be highly "quenched."

Quenching results when energy is transferred from β^- particles to some oxygen-rich molecular species in the solvent. In general, these species are not aromatic and cannot be excited to emit the captured energy as visible light; instead they emit it as heat. Because the transferred energy is released in a form not detectable by the photomultipliers, the counting efficiency is reduced. Understanding of this mechanism explains why quenching is so dependent on the mole fraction of water in the solvent mixture, and why considerable ingenuity has been expended on design of solvent systems. It also demonstrates why it is virtually impossible to prevent some degree of quenching in biological applications of scintillation counting. Since it is so difficult to prevent quenching, one must correct for its presence. Means of quench correction will be discussed later on.

Effects of Scintillant Structure. Molecules that contain several conjugated rings connected in a linear fashion make good scintillators. Terphenyl and biphenyl are examples of such species. The same number of rings fused together (naphthalene or anthracene) are less efficient scintillators. The oxazole ring is also a good basis for construction of scintillators. The compounds already mentioned, diphenyloxazole (PPO) and bis(phenyloxazolyl)benzene (POPOP), are commercially available at moderate cost. Others are available but are usually too expensive for large-scale use.

Scintillants can be excited not only by β radiation, but also by visible light. Whereas the excited state produced on collision with accelerated electrons is quite short lived, that produced by visible light may take a considerable period of time to decay. This property of scintillants is known as *chemiluminescence.* To minimize counting errors due to chemiluminescence, scintillation cocktails are generally stored in low-actinic (brown) glassware. Filled scintillation counting vials should be allowed to stand in the dark for 30–60 min before counting begins so that chemiluminescence may subside. A few of the most elaborate counting instruments have built-in features to measure chemiluminescence, and to delay the start of counting operations until chemiluminescence has been minimized, but this feature is not found on most of the equipment in general use. Chemiluminescence may also be promoted by use of tissue solubilizers, which are usually strong organic alkalis added to digest (solubilize) tissue fragments or homogenates when it is necessary to count such samples. A typical tissue solubilizer is known as Hyamine hydroxide. This is similar in structure to Triton, except that it has a terminal quaternary nitrogen in place of the OH found in Triton.

Photon Loss. Hyamine and similar compounds have a second effect on scintillation cocktails. The strong alkalis can react with scintillators to form colored compounds, which act as optical filters in the system, reducing the intensity of the light flashes sensed by the photomultipliers, and inducing a degree of quenching. Similar effects are produced by any substance that can absorb energy from the sample. Thus, counting labeled hemoglobin is complicated by the light absorbance due to the sample material itself, and appropriate corrections must be made.

Photon loss is not totally dependent on the color of a dissolved substance. For example, some workers add Cab-O-Sil, a very finely divided form of silica, to absorb water or immiscible components of a sample. The theory is that by absorption on the surface of the silica particles, a more uniform distribution of the radioactivity can be obtained. Although this may be true, the particles of silica acts as opacities and may prevent some of the light flashes from being simultaneously sensed by both photomultipliers. As a result, the coincidence gates will block the corresponding electrical impulses from the counters, a form of quenching.

Summary. Among the problems that cause a disparity between the true dpm and the recorded cpm, quenching is the most troublesome. It can be caused in greater or lesser degree by a variety of mechanisms briefly discussed above. It can never be entirely eliminated so corrections must be made for truly quantitative results. However, if only qualitative results are needed (i.e., the presence or absence of radioactivity), then the effects of quenching are less troublesome and may even be ignored.

DETERMINATION OF COUNTING EFFICIENCY: QUENCH CORRECTION

A number of methods have come into general use for correcting measured cpm to obtain the true dpm, some of which require only a single channel of the scintillation counter and some of which require two channels. The more precise methods also require, in addition to the experimental samples, a separate source of the radionuclide that has been accurately characterized in terms of its specific activity and date of preparation, whereas others of slightly lower precision do not demand the standard preparation but do require that one devote two counting channels to the question of quench correction. These methods will be described in order of their increasing conceptual complexity, which is not necessarily the order of their utility.

Sealed Radioactivity Standards

One needs first of all some chemically and physically stable substance, the radioactivity of which is accurately known, along with the date of preparation. Either [^{14}C]- or [^{3}H]toluene is frequently used as a standard since it is compatible with virtually all scintillation cocktails in the absence of quenching additives such as water. However, toluene is somewhat volatile, and some workers prefer to use labeled hexadecane instead. For a considerable fee, manufacturers provide sealed standards, containing toluene in a nonquenched cocktail and packaged in vials that have the same dimensions as standard scintillation vials. Sealed standards can be used only as reference standards; that is, one cannot alter them in any way. They provide information about the performance of the counting instrument but not about quenching in experimental samples. Their major function is to assist in establishing proper settings for gain and discriminator controls and to ensure that the counter is functioning properly. Clearly, sealed standards employ only the single channel devoted to the isotope in question.

Internal Standards

Internal standards are those that may be added in known amounts to experimental samples. Ideally, an internal standard should be as well characterized as the materials incorporated into sealed standards, but in actual practice many workers are content to employ materials less well defined. For example, if the experiments involve [1-^{14}C]oleic acid for which the exact date of manufacture is not usually known, some of this material may be dissolved in toluene to make an approximate internal standard. The implicit assumption is that the nominal specific activity is the same as the actual activity. The error introduced by this assumption is probably very small. The greater problem concerns the use of substances that are insoluble in toluene cocktails, and which therefore require introduction of some solvents that may also act as quenchers. An example of this problem is the use of [1-^{14}C]glucose, or [2-^{14}C]alanine. Obviously, a proper internal standard should be soluble in toluene, but the merits of the case are often overridden by costs. A quench correction curve may be prepared with internal standards by either of two procedures, as described in the following two sections.

Indirect (Reference) Method

To each of a series of scintillation vials one adds carefully measured aliquots of whatever standard is to be used. These aliquots should more or less match the dpm expected in the experimental samples; generally they contain slightly more radioactivity than the experimental vials. To each of these vials are then added graded and increasing aliquots of the quenching components known to be present in the experimental samples—for example, water, buffer salts, and the like. The volumes added should, in the most quenched standard vial, somewhat exceed the expected values for the experimental vials. The same volume of scintillation cocktail is added to all vials which are then counted.

Example. A sample of estradiol (MW = 288.4) with a specific activity of 0.7 μCi/μmol was made into a solution containing 1.46 mg/ml in 95% ethanol. No data were available regarding the actual age of the estradiol preparation. Aliquots of this solution were employed in a metabolic experiment. The product of the experiment was recovered as estradiol 17β-maltoside, a water-soluble form of the steroid.

In order to determine a suitable correction for quenching due to the presence of water in the extracted estradiol maltoside, the experimenters put carefully measured aliquots (10 μl) of the original steroid solution into each of a series of empty scintillation vials, then removed the solvent by evaporation in a gentle stream of N_2. To all but one of these dried residues, they next added small, known, serially increasing volumes of water. Finally, they delivered into all of the vials 10 ml of scintillation cocktail containing 4 g/L of PPO, and 10 g/L of Triton in a mixture of toluene and dioxane (1:1, v/v). After thorough mixing, all of the vials were counted.

The 10-μl aliquots of radiolabeled estradiol each contained 0.05 μmol; if the nominal specific activity was accepted as the actual value, this should have produced a counting rate of 77,000 dpm. The observed cpm's for the series of standards are tabulated below, and the same data are plotted in Fig. 3–6.

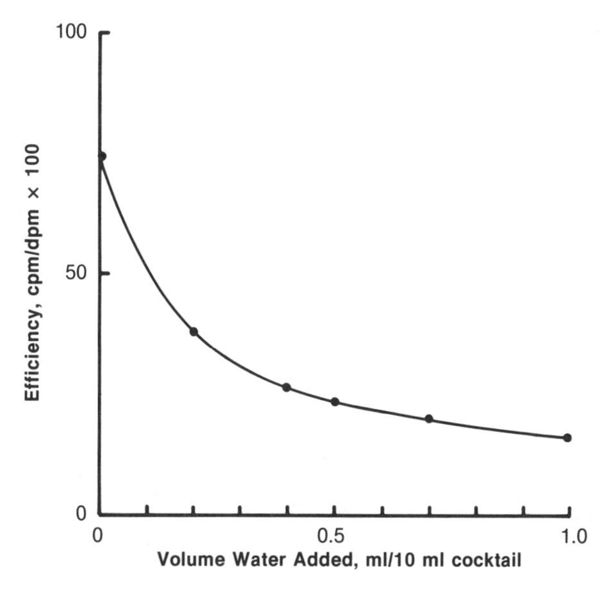

Figure 3–6. Quench correction curve based on reference standards. Quenching is due only to added water.

Water added (ml)	Observed cpm	Efficiency (%)
0	57,750	75.0
0.2	30,030	39.0
0.4	20,020	26.0
0.5	18,865	24.5
0.7	15,785	20.5
1.0	11,940	15.5

From the table, and even more clearly from the graph, the effect of added water on the counting efficiency is made evident; comparable effects can be seen in any similar experiment. The data emphasize that efficiency of scintillation counting is almost never 100%. Notice that even in the absence of added water, the efficiency is only 75%. This is due to a number of factors. First, there is some dioxane in the cocktail, and dioxane by itself induces some quenching. Second, the maltose moiety of the estradiol maltoside may induce self-absorption of radiation, an additional source of quenching. The scintillator, PPO, may itself absorb some of the light energy from adjacent scintillating molecules of PPO, further inducing a form of quenching. The sum of these effects quench, even in the sample from which water was omitted.

Knowing the volumes of sample taken and scintillation cocktail added, one can read from curves similar to Fig. 3–6 an estimate of counting efficiency, and the dpm values can be calculated from the relation

$$\text{dpm} = \text{cpm/efficiency}$$

This calculation is subject to the assumption that only the water contained in the sample makes a significant contribution to quenching, an assumption that is not always justified. Reference standards are sometimes poor replicas of experimental samples, especially if the latter contain pigments or good quenching agents such as proteins.

Direct (Sample Addition, or "Spiking") Method

As the name implies, an aliquot of standard is added to the unknown, or experimental, samples. One first measures each of the experimental samples to obtain a value identified as cpm_1. To each vial is then added a small, accurately measured aliquot of the standard that has a known value of dpm. All vials are counted a second time, giving cpm_2. The difference, $\Delta\text{cpm} = \text{cpm}_2 - \text{cpm}_1$, is a measure of the counting rate due to the added quantity of standard. The efficiency is calculated as

$$\text{Efficiency} = (\text{cpm/dpm}) \times 100$$

A major advantage of this method is that each experimental sample serves as its own control. Correction is simultaneously made for color quenching as well as for chemical quenching. The method is simple in concept and in performance. Its disadvantages include the need for doubled counting time, and the errors that may result if the aliquots of standard that must be added are not all identical. If the counting rates of the experimental samples are sufficiently high, it is possible to divide the samples into two equal parts, adding standard to only one part and making only a single counting run through the doubled series.

The Channels Ratio Method

This method is conceptually more complicated than the methods previously discussed, and it requires that two channels of the spectrometer be devoted to determination of counting efficiency for a single radionuclide.

The channels ratio method depends on the fact that quenching shifts, or displaces, the energy spectrum of β^- particles characteristic, say of ^{14}C. In an unquenched sample, the spectrum for ^{14}C ranges from 0 to 156 keV. In quenched samples the maximum energy output will be reduced (shifted downward), and some pulses of very low energy will be so attenuated that they may not be registered by the counters. The peak value in the spectrum (maximum counting rate as a function of β^- particle energy) will also be shifted downward. These facts are displayed in Fig. 3–7, where curve A represents an unquenched sample, and curve B, a quenched sample of

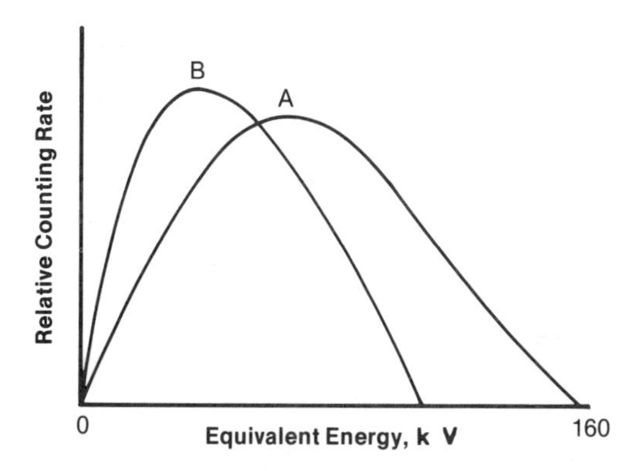

Figure 3–7. Unquenched (A) and quenched (B) spectra for ^{14}C. Distribution of counting rate as a function of B^- particle energy values.

the same dpm value. The exact magnitude of spectral shift will depend on the degree of quenching.

By proper manipulation of the gain and discriminator controls it is possible to count all of the pulses included in the energy range 0–156 keV, in channel A of the spectrometer. It is equally feasible to count only one-third, or some other arbitrary fraction of the total, which fall in the energy range 0–X keV, in channel B. This is accomplished by setting the upper discriminator of channel B so that the registered count in that channel is just one-third the rate registered in channel A. The value of X is not important; what does matter is that some preselected portion of the lower end of the total energy spectrum is simultaneously recorded in channel B while the total spectrum is recorded in channel A. It is intrinsic to the argument that *for any single sample* the rate recorded in channel B will be less than the rate recorded in channel A. However, the *channels ratio,* defined as the counting rate in channel B divided by the counting rate in channel A, will tend to approach unity as the degree of quenching increases. In effect, one uses a measure of the spectral shift as a measure of quenching. This is the underlying theory behind the channels ratio method. The following example will clarify these principles.

Example. A solution of a certain ^{14}C-labeled substance had a presumed decomposition rate of 10,000 dpm per 0.10 ml. Aliquots (0.10 ml) of this solution were added to each of a series of scintillation vials. Small, graded aliquots of acetone were added to the series to increase quenching, and all of the vials were made to constant volume with a scintillation cocktail. The vials were then counted in channels A and B of a spectrometer adjusted as described above; that is, channel A was adjusted to include the entire energy range, and channel B was adjusted to count only the lower third of the range. The resulting data are displayed in Table 3–3.

The counting rate in channel A decreased with increasing volume of added acetone, as would be expected because all of the pulse heights would be diminished by progressive quenching. Electrons with the very lowest

Table 3–3. Data for the Channels Ratio Quench Correction

Acetone Added (ml)	Counting Rate, Ch. A (cpm)	Counting Rate, Ch. B (cpm)	Efficiency, Ch. A	Channels Ratio, B/A (cpm)
0	8440	2810	84.4	0.3329
0.1	8380	3125	83.8	0.3729
0.2	8120	3590	81.2	0.4421
0.3	7970	4375	79.7	0.5489
0.4	7810	4750	78.1	0.6082
0.5	7500	5000	75.0	0.6666
0.6	7345	5310	73.5	0.7229
0.7	7185	5430	71.9	0.7557
0.8	7095	5790	71.0	0.8146
0.9	7000	5800	70.0	0.8357
1.0	6875	5940	68.8	0.8640

energies might not elicit pulses under quenching conditions, and these would be lost to the counter. The counting rate in channel B actually rises with the volume of added acetone because quenching shifts the more energetic pulses downward into the lower third of the unquenched spectral region. Again, electrons with the very lowest energies might not elicit detectable pulses, but that loss is proportionately less than is the quenching effect on more energetic electrons.

The counting rate in channel A is used to determine the efficiency, because it included the entire energy spectrum. The last column of Table 3–3 shows the calculated values of the channels ratio, B/A. It is clear that the limiting value for an unquenched sample is 0.33, whereas the limiting value at the other end is 1.00, because as quenching increases, a larger number of counts per minute will move from channel A to channel B. In theory, at some point along the quench curve, counts in A will equal counts in B.

One can plot the counting efficiency in channel A as a function of the channels ratio, as shown in Fig. 3–8. As long as the instrumental readings are not disturbed, and as long as one uses the same scintillation cocktail, the curve shown in Fig. 3–8 may be employed to determine counting efficiency for ^{14}C-containing unknowns. One determines the channels ratio from simultaneous counts in the two channels, then simply reads the counting efficiency from the graph.

In the example just presented, channel A was set to include the entire β^- particle energy spectrum whereas channel B was set to include only its lower third. There is no reason why channel A could not have been set to include only the upper two-thirds of the spectrum, or some other fraction. In other words, the rationale behind the channels ratio method is a perfectly general one. Whatever the ratio selected, it is essential that the same value be maintained throughout the experiments for which the curve is to be used. It is also essential that the scintillation cocktail be the same throughout the experiments. How the spectrum is best divided depends only on the β^- particle energy spectrum of the given isotope. Once pre-

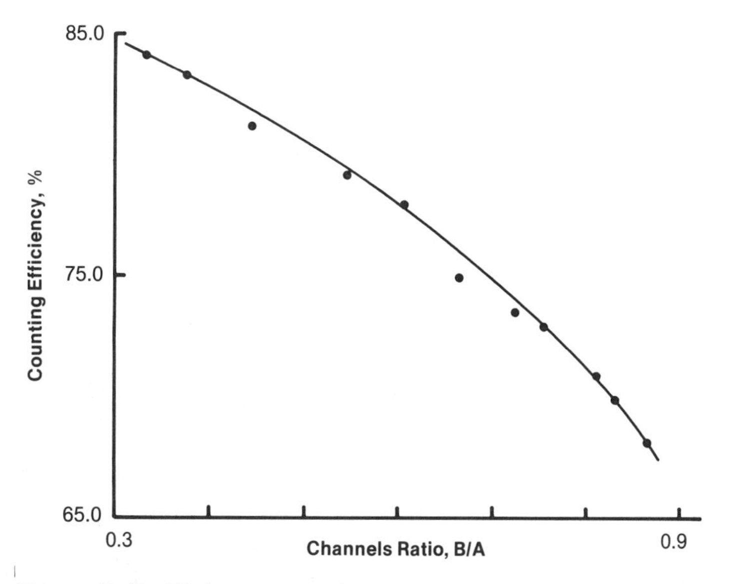

Figure 3–8. Efficiency curve derived by the channels ratio method.

pared, an efficiency–channels ratio curve may be used indefinitely as long as instrumental parameters remain unchanged and the scintillation cocktail has a constant composition.

A major advantage of the channels ratio method is that the quenched standards need to be prepared only once. This can represent a considerable saving in time and material when compared to the use of internal standards by the direct addition method. A major disadvantage is that the method is based deliberately on only a small portion of the total cpm in channel B. With samples of low activity, it may be necessary to count for somewhat extended times in order to get measurements of adequate reliability. With a lengthy series of unknown samples, the extended counting time may be irksome. Also, the channels ratio method does not work well in samples that contain solid matter. It is important that all samples be dissolved by whatever means are required.

Use of Automatic External Standards

All of the methods described thus far have one thing in common; they depend on the same β-emitting radionuclide as is found in the unknown samples, and the methods described can be used with any two- or three-channel spectrometer.

An entirely different kind of external standard is a pellet of some fairly long-lived γ emitter. Among the radionuclides that have been used for external standards, few are pure γ emitters, but all have been selected by virtue of rather energetic γ radiation. Thus, Beckman instruments employ $^{137}Cs_{55}$, with a half-life of thirty years. It emits some β^- radiation and 0.661 MeV γ radiation. Packard instruments frequently use $^{226}Ra_{88}$, with a half-life of 1622 years, some α emission, and 0.64 MeV γ radiation. When used as an external standard, a small pellet of the isotope is sealed in a jacket that blocks transmission of all except the γ radiation. The sealed pellet is inserted into a pneumatically operated transport system which, on instrument command, can move the pellet from a heavily shielded area to a point close to the counting well, and, after a finite counting interval, back into the shielded area.

When 400–600 keV γ photons interact with matter, either of two competing processes may occur. The distribution of radiant energy between these depends on the precise energy of the radiation and on the electron density (or, in other words, the atomic number) of the target atoms.

1. *The photoelectric effect:* an interaction by which the photon energy is totally transferred to an orbital electron of a target atom. The energized electron is driven out of the atom and thereafter behaves like any other β^- particle of comparable energy, that is, it produces energetic scintillations easily recorded by the spectrometer. At the upper end of the usual working range, this may be the predominant process of the interaction.
2. *Compton scattering:* a process essentially similar to photoelectric stimulation of β emission, except that the energy of a photon is not totally transferred to one orbital electron. Instead, a photon of reduced energy results from the first collision. Since the energy required to eject an orbital electron is in the order of only 40 eV, many photons undergo multiple collisions, ejecting and accelerating numerous orbital electrons from numerous atoms before its total energy is dissipated. Compton electrons therefore tend to have somewhat lower and more disperse velocities than photoelectric electrons, and the range of pulse heights they generate in scintillation counters is somewhat broader.

The two processes just described are the major mechanisms by which typical external standards generate intense pulses when the standard pellets are in close proximity to a scintillation vial. A third process, ion pair production, is theoretically possible but ordinarily plays little part because ^{137}Cs and ^{226}Ra produce only a few photons with energy sufficient to generate ions pairs.

Instruments equipped for automatic external standardization (AES) contain an additional amplifier channel devoted to this purpose. Unlike the normal counting channels, the AES channel gain and discriminator controls may not be adjustable by the user. They are often preset at the factory to pass only the high-intensity pulses generated by the external standard and to reject the less energetic pulses due to the sample. Because the counting rate of the AES pellet is very high, only 20- or 30-sec counting periods are required to accumulate a very large number of counts in the AES printing register; thus most instruments also have factory preset counting periods for the AES channel, regardless of how long the normal channel counting times may be. Ordinarily, the experimental sample is counted first in the absence of the AES pellet and the rate is recorded. The pellet is then automatically drawn into the sample well, and a second rate is recorded for the preset number of seconds; then the pellet is returned to its storage position. Both counting rates are subject to identical quenching, since both count rates are measured in the same medium.

The AES mode is thus a special case of the channels ratio method. Instead of dividing an experimental energy spectrum into two fractions, which requires two channels, the entire energy spectrum is measured in one channel and a separate, fixed channel is devoted to the higher-energy spectrum produced by the AES pellet. The sample counting rate provides one number of the channels ratio, and the AES channel rate provides the other. One must first determine the AES rate with a series of quenched standards, from which the efficiency can be calculated. Thereafter, the AES rate is related to this efficiency just as in the case of ordinary channels ratio procedures. Once this is done, the AES counting rate of experimental samples can be employed to determine the efficiency of counting in the unknown samples, as previously described.

Incorporation of AES hardware into scintillation counters has made quench correction simpler and faster. All three channels remain available for experimental use. One can be sure of the constancy of the external standard source. Instrumental settings for a particular isotope need not be disturbed. The method does suffer from certain drawbacks. It obviously depends on precise and reproducible placement of the AES pellet with respect to the sample in the counting well. It also depends on the assumption that the scintillation vials are as nearly as possible of constant dimensions and that they contain the same fluid volumes. This is to ensure that the penetrating γ rays sweep through identical volumes and concentrations of scintillant. To check on reproducibility of pellet placement, it is useful to prepare and preserve a series of quenched standards and to determine, from time to time, that these have constant counting rates. The pneumatic systems for translocation of the AES pellet may become clogged with foreign matter, necessitating service of the equipment by trained personnel. Uniformity of vial dimensions is ordinarily not a problem if the vials are purchased from a single source; however, there may be significant differences in the dimensions of glass and plastic vials, and therefore, these should not be mixed in any single run where quenching is determined by AES. Constancy of volume is largely a matter of reliable dispensing equipment, properly used.

CALIBRATION OF LIQUID SCINTILLATION SPECTROMETERS

Manufacturers provide detailed manuals for operation and calibration of their spectrometers, but the language used in these manuals frequently does not give a clear understanding of the principles involved. As used here, the term "calibration" refers to adjustments of the gain and discriminator controls in order to optimize the usual counting operations. Since the basic ideas are independent of the source of the hardware, it is worth reviewing the fundamentals here.

At this point, our concerns pass to the realm of electronics. All of the events that occur in the scintillation vials and the photomultipliers can be lumped together and considered only as a source of current pulses, the intensities of which reflect the energies released by decomposing atoms. If these signals are simultaneously generated by the multipliers, they pass through the coincidence gate as a corresponding train of varying voltages. It is therefore reasonable to deal from here on only with pulse voltages, since these are what potentiometers (gain and discriminator controls) adjust.

When the discriminator potentiometers are set at their widest limits, the output voltage of a channel amplifier covers some finite range, usually 0–10 volts. Increasing the lower discriminator setting and/or decreasing the upper discriminator setting clips off, or rejects, pulses of low and/or high voltage. This reduces the effective band width of the pulses passed on to the counting registers. The first step in spectrometer calibration is therefore to make sure that the maximum number of output pulses falls within the 0- to 10-V range. Excessively amplified pulses that exceed the 10-volt limit would be rejected, and no output pulse can have a value <0. One proceeds as follows: With the discriminators set at their widest limits, an unquenched sample is placed in the counting well. The gain control(s) is then varied by constant increments from its lowest to its highest settings, and the counting rate is recorded at each setting. The optimum setting will be that at which the counting rate is highest (highest efficiency). The optimum value will vary with the energy range of the β^- emissions of the isotope being counted; lower energies will require a greater gain (amplification), and higher energies a lower gain. As the gain is increased, the calculated efficiency will exhibit a more or less flat region, or plateau, and then it will begin to decrease. The decrease is caused by a relative shift upward of all pulse voltages, as shown in Fig. 3–9. The solid curve in the figure shows that, at the optimum gain setting, the entire output window from 0 to 10 volts of the channel amplifier is used to register pulses. When the gain is increased much beyond the optimum, the dashed curve results. The total counts recorded in either case is proportional to the areas under these curves, and the area under the dashed curve is less than the area under the solid curve. A similar argument can be made for gain settings much below the optimum. When the gain control(s) are adjusted for maximum efficiency, the amplifier is said to be balanced, and the particular gain setting is known as the *balance point*. The balance point is unique for each isotope, and it is the setting that provides greatest counting sensitivity (because it is also the point of highest efficiency). Provided that the output windows are not changed, the balance point remains constant, but it does shift slightly when the window is readjusted.

Multichannel spectrometers have become popular because they permit the simultaneous counting of several isotopes in the same sample. It is frequently desired to designate one channel exclusively for counting of ^{3}H and another for ^{14}C, these being the most commonly used isotopes. In other circumstances, channels might be designated for ^{14}C and ^{32}P, etc.

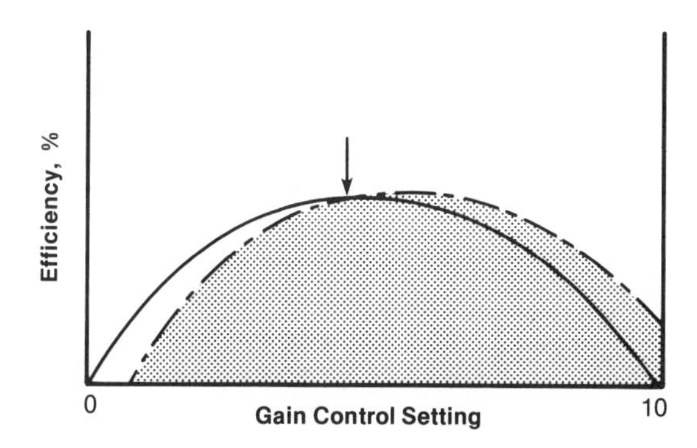

Figure 3–9. Relation between counting efficiency and balance point. When gain control is at the balance point, indicated by the arrow, counting efficiency is highest and the total window is used. When the gain is set too high (or low), some counts are lost, as shown by the shaded area.

Whatever the case, the remainder of the calibration process involves separating counts due to one isotope from counts due to another. This sorting process is done by adjustment of the output windows by manipulation of the discriminator controls. In some spectrometers, this can be conveniently done by insertion of solid-state modules, preprogramed at the factory for maximum efficiency, into receptacles provided on the front of the console panel. Other instruments require that the user performs the entire calibration by means of permanently installed but variable potentiometers.

The energy spectra of most β emitters show some degree of overlap (see also previous section, Causes of Reduced Efficiency). This principle is illustrated by the case of ^{3}H and ^{14}C, shown in Fig. 3–10. At some cost in counting efficiency, it is possible to eliminate counts due to an isotope with lower energies from the channel committed to an isotope with higher energies, but the reverse is not usually possible. Total elimination of overlap, or spillover, may involve an unaffordable efficiency cost, and some compromise is often necessary. Provided the degree of compromise can be calculated, it can be taken into account.

Although the problem is discussed here in terms of ^{3}H and ^{14}C, the issue is a general one, and might well be discussed in terms of X and Y, with the understanding that Y produces more energetic radiation than X. Thus, X might be ^{14}C and Y might be ^{32}P, etc. To optimize the counting of any pair, two standards are required. Except for the source of radioactivity, composition of the standards must be identical so that the degree of quenching is the same in both. Throughout the calibration procedure all counting times must be the same. The designated channels should be set at their balance points, and discriminators should be set to provide the widest windows. Subsequent steps in the procedures are aimed at setting the discriminators for most efficient sorting of counts from either isotope into the proper designated channel.

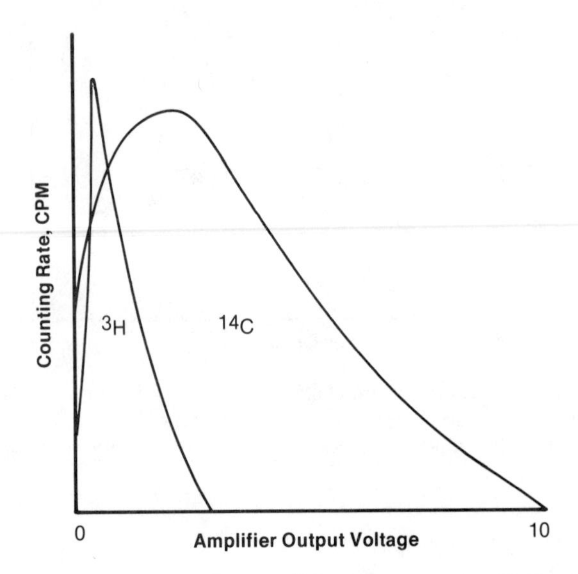

Figure 3–10. Spectral overlap of ^3H and ^{14}C energy spectra. If the lower discriminator were set at 3 or more volts, counts due to ^3H would be excluded from the counting rate. There would be a considerable reduction of efficiency in ^{14}C counting. Note the impossibility of excluding counts due to ^{14}C from estimations of ^3H. A similar situation would exist with any comparable pair of radionuclides suitable for scintillation counting.

The ^{14}C Channel

With the *^3H standard* in the *^{14}C counting channel,* increase the setting of the lower ^{14}C discriminator until the counting rate due to ^3H is very nearly zero. Without changing any of the controls, replace the ^3H standard with the ^{14}C standard and again measure the counting rate. From this measured rate, calculate the efficiency of ^{14}C counting. This will be a measure of the radioactivity due to ^{14}C, substantially free of ^3H spillover. Reference to Fig. 3–10 shows, for the example on which that figure was based, that the lower ^{14}C discriminator must be set at a point equivalent to 2.8 volts. Only if a ^{14}C-derived β^- particle has an energy greater than any ^3H-derived particle, will it be registered in the ^{14}C channel as it is now set.

The ^3H Channel

The procedure here is slightly different, and it has two parts. Place the *^3H standard* in the *^3H counting channel,* the discriminators of which are still set to give the widest window, and record the initial counting rate. Lower the upper ^3H discriminator setting by some convenient decrement and again record the counting rate. Continue the stepwise decrease in upper discriminator setting, recording the counting rate at each step, until the counting rate approaches zero.

Return the upper discriminator to its maximum setting, replace the ^3H standard with the ^{14}C standard, and record its initial counting rate. Repeat the sequence of lowering the upper discriminator in the same stepwise fashion described above, recording the counting rate at each step, until the counting rate again approaches zero. This second set of data is a measure of the ^{14}C-derived counts which are necessarily included in the ^{3}H channel. It is spillover due to spectral overlap.

Determination of the Crossover Point

Recall that the initial counting rate for either isotope was obtained at a maximal discriminator setting. The initial counting rates can be represented by the symbol C_i. Subsequent values in each series, obtained by stepwise reduction of the discriminator settings, can be represented by C_x. Since all counting times were the same, one can use C_x and C_i to determine *fractions of the total counts* sorted into the separate channels under the experimental conditions described. Thus, one arrives at the expression

$(C_i - C_x)/C_i$ = fraction of total ^{3}H counts lost from the ^{3}H channel as the discriminator was lowered

C_x/C_i = fraction of the total ^{14}C count included in the ^{3}H channel due to spectral overlap

When the two sets of fractional count values are plotted as a function of the discriminator settings the two curves will intersect. The point of intersection is known as the *crossover point;* it corresponds to the discriminator setting at which the fraction of ^{3}H counts lost is just equal to the fraction of ^{14}C counts gained by spillover. It is the discriminator setting which, in theory, should be used for samples doubly labeled with ^{3}H and ^{14}C because it is the point of equal efficiency in terms of spectral overlap. This conclusion can be proven, but the proof will not be shown here (see Bush, 1964). As a matter of common practice, the discriminator for ^{14}C is usually set just a little above the crossover point to be sure that ^{3}H is totally excluded.

This concludes discussion of the mechanics of the calibration process.

STATISTICS OF SCINTILLATION COUNTING

The decomposition of unstable nuclei is a random event. In any sample of radioactive atoms a given atom has a certain probability, p, of decomposing at some instant. The probability that it will not decompose at that instant is, therefore, $1 - p$. Since only these two states are possible, all decay processes must be analyzed by two-valued, or binomial, statistics, the algorithms of which may be derived from coin-flipping or similar experiments. Because the total number of atoms potentially available for decomposition is very large compared to the number that actually decompose per unit time, radioactive decay is a limiting case of the binomial distribution. The probability curve for such situations was described by the French mathematician, Poisson. He showed that, in general, its properties were not those of the Gaussian, or normal error curve.

If a radioactive sample is counted for 1 min, a certain counting rate will be obtained. If the measurement is repeated a large number of times, the counting rates, here represented by x, will exhibit some range of values, just like any other experimental measurement. A mean of the large number of values, here represented by γ, would result. Poisson's law states that the independent probability of finding any counting rate, or $P(x)$, is given by:

$$P(x) = \frac{e^{-\gamma} \cdot \gamma^{x}}{x!}$$

where $x! = (x)(x - 1) \cdots 1$, and γ is the population mean. A property of the Poisson distribution is that the standard deviation is equal to the square root of the mean or

$$\sigma = \sqrt{\gamma}$$

which is quite different from the definition of σ for Gaussian distributions. A second property of the Poisson distribution is that as x gets larger ($x >$ 1000), the Poisson curve becomes increasingly close to the normal error curve. As a rule, γ is not known. Furthermore, one wishes to avoid the necessity of repetitive counting procedures. By taking advantage of the fact that at higher counting rates the Poisson distribution approaches the normal, it is possible to apply Gaussian solutions to the problem, and to operate on the assumption that 95.5% of the area under the curve includes the range of values represented by $\gamma \pm 2\sigma$, and 99.7% of the area contains the range $\gamma \pm 3\sigma$, and so forth. Confidence limits may then be calculated in the usual way, or one can calculate relative or percentage errors based on the total counts accumulated. Tables of relative errors, expressed as a function of total counts, can be constructed, or nomograms designed, to facilitate analysis of counting data. An example is Table 3–4.

To summarize this argument, the data of Table 3–4 say that when the total counting rate falls between 10,000 and 40,000 cpm, the error of the single count lies between 1% and 2%, which is generally negligible. When the total counts falls below 1000 cpm, the error becomes substantially greater and may no longer be negligible. Experiments should therefore be designed to ensure that each sample contains sufficient radioactivity to avoid large errors of counting. When low counting rates are inescapable, counting times must be increased to offset the statistical errors of measurement. If one wishes to reduce relative counting error only by longer counting periods, it can be shown that the long time necessary is inversely proportional to the square of the desired accuracy. Thus, if a 1% error is wanted in place of a 5% error, then the counting times are related by the factor $(5/1)^2$. This is a powerful argument to find some corrective other than counting time, especially if a lengthy series of samples is to be counted. Another reason for proper experimental design relates to background counting rates. In modern scintillation spectrometers, properly maintained, the background rate is quite constant at 20–35 cpm. Ordinar-

Table 3-4. Relative Error (2σ) Corresponding to Total Counting Rate

Relative Error	Total Counts	Relative Error	Total Counts
0.004	250,000	0.02	10,000
.0045	200,000	.0283	5,000
.005	160,000	.03	4,400
.006	111,000	.04	2,500
.0063	100,000	.05	1,600
.007	81,600	.06	1,100
.008	62,500	.0631	1,000
.0089	50,000	.07	820
.009	49,400	.08	630
.01	40,000	.0893	500

Note: Total counting rate equals the sum of sample counting rate plus background counting rate. Multiplying the relative error by 100 gives the percentage error.

ily this is a negligible quantity, but in instances where the total counting rate is low (<500 cpm), it becomes an appreciable part of the total. Special precautions are then necessary when one is estimating counting error (these problems are taken up in the references cited and thus will not be discussed here).

Note: In addition to the references cited below, see the series, International Symposia on Scintillation Counting; these were originally published by Heyden and Sons, Inc., but since have been taken over by so many other publishers there is no regular procedure by which the individual volumes may be conveniently recognized. In particular, see the volume edited by E. D. Bransome, Jr., which is entitled *Current Status of Liquid Scintillation Counting* (Grune and Stratton, New York, 1970). Papers in this volume cover some important topics and are very well done.

REFERENCES

Birks, J. B. *Theory and Practice of Scintillation Counting.* Pergamon Press, New York (1964).

Bush, E. T. Liquid scintillation counting of doubly-labeled samples. *Anal. Chem. 36:*1082–1089 (1964).

Fox, B. W. Techniques of sample preparation for liquid scintillation counting. In *Laboratory Techniques in Biochemistry and Molecular Biology* (T. S. Work and E. Work, eds.), *5:*1–334 (1976).

Herberg, R. J. Statistical aspects of liquid scintillation counting by the internal standard technique. *Anal. Chem. 35:*786–791 (1964).

Herberg, R. J. Statistical aspects of double-isotope liquid scintillation counting by internal standard technique. *Anal. Chem. 36:*1079–1082 (1964).

Kobayashi, Y., and Maudsley, D. V. Liquid scintillation counting. In *Methods of Biochemical Analysis* (D. Glick, ed.), *17:*55–134 (1969).

Laskey, R. A., and Mills, A. D. Quantitative film detection of ^{3}H and ^{14}C in polyacrylamide gels by fluorography. *Eur. J. Biochem. 56*:335–341 (1975).

Leblond, C. P. Localization of newly administered iodine in the thyroid gland as indicated by radio-iodine. *J. Anat. 77*:149–152 (1943).

Neame, K. D., and Homewood, C. A. *Liquid Scintillation Counting.* John Wiley and Sons, New York (1974).

Snyder, F. *Advances in Tracer Methodology.* Plenum, New York (1968).

CHAPTER FOUR

Buffers and pH Regulation

IONIZATION OF WATER

Water is an *amphiprotic* substance, serving as a proton donor or acceptor. In pure water and in all aqueous solutions there exists an equilibrium written as:

$$2H_2O \rightleftharpoons H_3O^+ + OH^- \tag{4-1}$$

The *ion product* of water, at ordinary temperatures, is given by the expression:

$$K_w = (\alpha_{H_3O^+})(\alpha_{OH^-}) = [H_3O^+][OH^-](f_{H_3O^+})(f_{OH^-}) \tag{4-2}$$

In Eq. 4–2, α represents the true *activity* of the ions in question, and f represents the *activity coefficients,* values determined by experiments to bring the formal concentration values into numerical accord with the observed activities. Values of f are dependent on the temperature and on the concentrations of the ions.

Note that no expression for the concentration of water appears in the ion product equation. The reason is that the activity of water, especially in dilute solutions, is essentially the same as its molar concentration. The latter is, effectively, a constant, equal to 55.56 mol/L, and is incorporated in the dissociation constant. The notations for this constant, when the equation takes into account the activity coefficients, varies; for example, it may be κ or K_w. Since, in the vast majority of biochemical calculations, concentrations (usually molar) rather than activities are used, we will use K_w to represent the ion product or dissociation constant for pure water.

Equations 4–1 and 4–2 take proper cognizance of the fact that the unhydrated proton is a highly improbable entity. It is clear that the hydronium ion itself, H_3O^+, is hydrated and should be written as $H_3O^+(H_2O)_n$, yet the fiction of discussing ionic equilibria in aqueous solutions as involving H^+ is too deeply embedded in the biochemical literature to ignore. As long as one understands the convention, the simpler style does no great harm. Thus, we shall speak of (and write of) "hydrogen ion" with the clear understanding that in the physical sense we refer to something quite different.

If one bubbles HCl gas into a sample of pure water, a great increase in $[H^+]$ occurs. Since Eqs. 4–1 and 4–2 must still be satisfied, it follows that a corresponding decrease in $[OH^-]$ must take place at the same time. Experience indicates that at moderate concentrations (0–150 mM), addition of most uni-univalent electrolytes give values of $f \cong 1$, so that $\alpha Y \cong [Y]$, where Y is any solute. However, there are some notorious exceptions to this generalization. Thus, solutions of XH_2PO_4 or X_2HPO_4 (X = Na or K) exhibit activity coefficients that are very concentration dependent. Solutions of tris(hydroxymethyl)aminomethane (Tris) or its salts exhibit activity coefficients that are extremely temperature dependent. Because both of these compound types are important biochemical buffers, it is essential to understand the effects of such properties. We shall return to this subject later on.

Pure water is arbitrarily defined as the standard of neutrality. By careful experiment, it has been determined that at 24°C, $K_w = 10^{-14}$, so that in pure water, $[H^+] = [OH^-] = 10^{-7}$. Therefore, any solution in which $[H^+] > 10^{-7}$ M is said to have an acid reaction. If $[H^+] < 10^{-7}$ M, the solution is said to have a basic reaction. In this sense, "reaction" refers to response toward some indicator, usually an ionizable dyestuff such as litmus.

THE pH NOTATION

In biological experimentation, $[H^+]$ may vary from about 10^{-1} M (in fairly acid solutions) to about 10^{-10} M (in fairly basic solutions). It is clumsy to work with nonintegral values of exponents, but it is frequently necessary. An invention credited to the Danish chemist, S. P. L. Sörensen, avoids the difficulty. Sörensen defined the "p" value of any quantity as the negative logarithm of that quantity. For the $[H^+]$ in particular,

$$pH = -\log [H^+], \quad \textit{or} \quad [H^+] = 10^{-pH} \tag{4–3}$$

Using this definition plus Eq. (4–2) (assuming $f \simeq 1$), we can write that:

$$pK_w = pH + pOH = 14 \tag{4–4}$$

Equation 4–4 is also satisfied in every aqueous solution; from this it follows that any acid solution has pH < 7 and every basic solution has pH > 7. The pH notation has become the universal convention for describing acidic or basic reactions of solutions, but it is important to keep in mind that the "p" notation may also be employed in dealing with ions other than H^+ and OH^-, or even with quantities other than ion concentrations.

STRONG AND WEAK ELECTROLYTES

Substances that, when dissolved in water, yield a mixture of ions are known as electrolytes. However, all electrolytes are not equal in this respect. The distinction between strong and weak electrolytes may be operationally defined by the following two hypothetical experiments:

If one adds known amounts of HCl to water, making a series of solutions ranging in concentration from 1 mM to about 1 M, the HCl dissociates to form H^+ and Cl^-. One can write that:

$$HCl + H_2O \rightarrow 2H^+ + Cl + OH^- \quad (4\text{–}5)$$

The OH^- derives from the dissociation of water; the H^+ from this dissociation joins the H^+ pool from the HCl.

Analytical measurements—for example, conductivity measurements or titrations with alkali—demonstrate the following:

1. $[H^+] = [Cl^-] = [HCl]$
2. $[OH^- <<< [H^+]$, as [HCl] increases, $[OH^-]$ rapidly decreases
3. Over the range of [HCl] examined, one finds no constancy for the value of K, where

$$K = \frac{[H^+][Cl^-]}{[HCl][H_2O]}$$

True chemical equilibrium between the HCl and its product ions lies so far to the right that we represent the situation of Eq. 4–5 by an unidirectional arrow. This is precisely what is meant by the term *strong electrolyte.*

If one repeated these experiments by addition of acetic acid to water (we shall represent CH_3COOH by the symbol, HOAc) so that:

$$HOAc + H_2O \rightleftharpoons H^+ + OAc^- + OH^- \quad (4\text{–}6)$$

somewhat different results would be observed.

Analytical measurements in this instance demonstrate the following:

1. $[H^+] = [OAc^-] \ll [HOAc]$
2. $[OH^-] <<< [H^+]$ (As [HOAc] increases, $[OH^-]$ rapidly decreases.)
3. Over the range of [HOAc] examined, one finds reasonable constancy for the value of K, where

$$K = \frac{[H^+][OAc^-]}{[HOAc][H_2O]}$$

True chemical equilibrium does exist between the undissociated HOAc and its product ions; it lies well to the left in Eq. 4–6, and for this reason one represents the situation by a directional arrow.

By such titrations (or related measurements) the nature of an acid (or a base) may be determined, at least in part. Since such measurements do not ordinarily damage or destroy the sample, they are of particular value to chemists. For example, a titration curve of an unknown acid with a standardized base can be used to estimate the equivalent weight of the sample, since the equivalents of base consumed ($N_b V_b$), where N is the normality and V is the volume, must be exactly the same as $N_a V_a$; that is, $N_b V_b = N_a V_a$. Furthermore, the dissociation constants of many carboxylic acids fall fairly close to one another. Sulfonic acid groups similarly lie close to one another, but fall at a different point on the pH scale. Sulfhydryl groups are still different. A complete titration curve can detect the presence of any or of all these groups. When the curve is extended in the opposite sense by titration with a standardized acid, similar evidence for the presence of basic groups can be adduced.

In polyprotic molecules, such as proteins, the number of ionizable groups can be very large. In an aqueous solution at any given pH, all of the ionizable groups must adopt a state consistent with that pH. Thus, at a pH of 8.6, all ionizable groups with p*K* values of 7 or less are essentially deprotonated and carry a negative charge. All groups with a p*K* of 9 or more are essentially protonated and carry a positive charge. The net charge on the molecules is the algebraic sum of the positive and negative charges. The net charge determines the electrophoretic mobility, or how the molecules move in a charge field. The net charge may also determine whether or not an enzyme can exert its catalytic ability. These sensitivities to charge (ionization) explain why it is so essential to regulate the acidity of most biochemical systems. Such regulation is generally accomplished by means of *buffer solutions.*

BUFFER SOLUTIONS: COMPOSITION AND PROPERTIES

Buffer solutions are composed of a weak acid (or a weak base) and one of its salts, usually taken in aqueous solution.

The properties of a buffer can be defined in terms of two properties, an *intensive* property and an *extensive* property. The intensive property is a function of the p*K* value of the buffer acid (or base). For reasons which we shall explore shortly, most simple buffers function well only over that part of the pH scale represented by the span, p*K* ± 1. This explains why biochemists need and use so many buffer systems. The extensive property is sometimes known as the *buffer capacity;* it is a measure of how well a given buffer protects against change in pH. Buffer capacity depends largely on the concentration of the buffer solution. Given two buffer solutions at the same pH, one twice as concentrated as the other, the more concentrated of the two is said to have a buffer capacity twice as great as the other. **Note:** *This statement does not imply that the buffer capacity is the same everywhere in the* pK *± 1 range.*

It is instructive to prepare a series of mixtures of HX and NaX which meet the following requirements. First, the moles of HX and NaX must sum to a constant in all the mixtures. Second, the ratio $[X^-]/[HX]$ is systematically varied from 0.05 to 95. When the pH of the solutions is plotted as a function of the ratio, the results appear as shown in Fig. 4–1. Note that the curve is symmetric about the p*K* value. The effect on pH of a given change in the ratio is clearly greater at either extreme than at more central values. These observations provide an explanation of why, for any buffer, the maximum buffer capacity is always found at pH = p*K*. Conversely, they also explain why the effective range of a buffer is generally limited to a range represented by p*K* ± 1. Beyond this range, defense

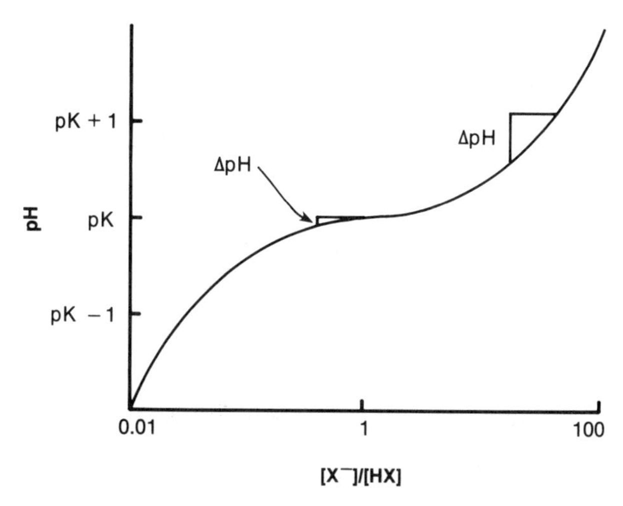

Figure 4–1. Change in pH as a function of $[X^-]/[HX]$. Note that when the ratio = 1, pK is numerically equal to pH. This curve is described by the Henderson-Hasselbach equation.

against pH change becomes doubtful since either $[X^-]$ or $[HX]$ is so reduced.

SOLUTIONS OF WEAK ACIDS OR OF THEIR SALTS

Taken alone, solutions of weak acids or of their salts do not appear to comply with the earlier definition of a buffer system. However, solutions of both the acid and the salt contain potentially ionizable HX, so both must satisfy the equilibrium condition defined by the value of K_a. This is just as true of the solutions alone as in any mixtures. It is important to understand how to estimate the pH of such solutions and to appreciate the effects of K_a and concentration (if any) on the pH of a solution.

pH of Weak Acid Solutions

Given a solution of a weak acid with known K_a, at a concentration of c mol/L, what is the pH of the solution? A logical attack on the problem begins with the expression:

$$K_a = \frac{[H^+][X^-]}{c - [H^+]} \tag{4–7}$$

Because $[H^+] \ll c$, we can effectively neglect $[H^+]$ in the denominator. Typical values of K_a are about 10^{-4} or less, and the concentrations of even dilute buffers are one or two orders of magnitude larger. Thus, the error introduced by this approximation is tolerable. In the numerator, $[H^+] = [X^-]$, so that Eq. 4–7 can be expressed as:

$$K_a \cong \frac{[H^+]^2}{c} \quad \text{or} \quad [H^+]^2 \cong K_a c \tag{4–8}$$

Taking the logarithms of both sides, then solving for $-\log [H^+]$, it follows from the Sörensen definition that:

$$pH \cong \frac{1}{2}(pK - \log c) \tag{4–9}$$

From this useful relation we conclude that, for any weak acid, the pH of a solution increases directly as a function of the pK and the concentration.

Classification of Weak Acid Salts

Salts of weak acids may be divided into two major types:

Type 1: derived from strong bases and weak acids

Type 2: derived from weak bases and weak acids

Salts of both types confront us with the problems of *hydrolysis.* In the sense used here, hydrolysis is the interaction of water with one or more kinds of ions produced in salt solutions. In type 1 salts, water interacts with the weak acid ion. In type 2 salts, water interacts with both the weak acid and weak base ions. Hydrolysis effectively reverses ionization; like the latter, it is an equilibrium process.

Type 1 Salts. As before, let X represent some weak acid. Then, when NaX is dissolved in water at some concentration, c, we can write that:

$$Na^+,X^- + H_2O \rightleftharpoons Na^+ + OH^- + HX \tag{4–10}$$

The salt is written as Na^+,X^- to emphasize its ionic nature, even in the solid state. In solution, the Na^+ finds its counter ion in OH^- supplied by the water. H^+ and X^- remain, but these must satisfy the weak acid properties of HX, most of which remains in the undissociated state. It is written as such in Eq. 4–10. The hydrolysis constant, K_h, is given by:

$$K_h = \frac{[OH^-][HX]}{[X^-]} \tag{4–11}$$

No change is introduced if the right-hand side of Eq. 4–11 is multiplied by $[H^+]/[H^+]$, giving:

$$K_h = [OH^-][H^+] \cdot \frac{[HX]}{[H^+][X]} = K_w \cdot \frac{1}{K_a} \tag{4–12}$$

The degree of hydrolysis, x, can be defined as the fraction of the dissolved salt that is hydrolyzed at equilibrium; the concentration of hydrolyzed salt is given by cx, and that of the unhydrolyzed salt by $c(1 - x)$. If we neglect the small change in [HX] due to its ionization, we can write, using Eq. 4–11, that $[OH^-] = [HX] = cx$. Then:

$$K_h = \frac{c^2x^2}{c(1 - x)} = \frac{cx^2}{1 - x} \tag{4–13}$$

By neglecting very small quantities and applying Eq. 4–12, one obtains a useful *approximate* solution of the quadratic equation solved for x:

$$x \cong \sqrt{\frac{K_h}{c}} \cong \sqrt{\frac{K_w}{K_a c}} \tag{4–14}$$

From our earlier assumption that $[OH^-] = cx$, and from the definition of K_w, we can write that:

$$[H^+] = K_w/cx \tag{4–15}$$

The approximate value of x, obtained in Eq. 4–14, can now be inserted into Eq. 4–15 to give:

$$[H^+] \cong \frac{K_w}{c\sqrt{\frac{K_w}{K_a c}}} \tag{4–16}$$

Finally, Eq. 4–16 can be expressed in the Sörensen notation to give:

$$pH \cong \frac{1}{2}(pK_w + pK_a + \log c) \tag{4–17}$$

Approximately, then, the pH of a type 1 salt solution increases with increasing pK of the weak acid and with the concentration of the solution.

Type 2 Salts. Predicting the pH of type 2 salt solutions is a little more complicated than predictions for type 1 salts, but good approximate values can be generated. If the weak base is represented by B and the weak acid by HA, the salt can be shown as BH^+, A^- for reasons already mentioned.

At equilibrium, the overall hydrolysis reactions can be written as:

$$BH^+ + A^- + H_2O \rightleftharpoons H^+ + OH^- + B + HA$$

but since the equilibrium between water and its ions would have to be met in any case, we can omit it from consideration here, or subtract it to give:

$$BH^+ + A^- \rightleftharpoons HA + B \tag{4–18}$$

and the hydrolysis constant K_h can be written as:

$$K_h = \frac{[HA][B]}{[BH^+][A]} \tag{4–19}$$

Equation 4–19 can be simplified by introducing K_a and K_b, the equilibrium constants, for the acid and the base, respectively, so that:

$$K_h = \frac{K_w}{K_a K_b} \tag{4–20}$$

If the concentration of the solution is taken as c, and the degree of hydrolysis as x, then $[HA] = [B] = xc$ and $[BH^+] = [A^-] = c(1 - x)$. Inserting these equivalent expressions into Eq. 4–19 gives:

$$K_h = \frac{x^2}{(1 - x)^2} \quad or \quad \sqrt{K_h} = \frac{x}{1 - x} \tag{4–21}$$

From the definition of K_a one can write that:

$$[H^+] = \frac{[HA]}{[A^-]} K_a = \frac{cx}{c(1 - x)} K_a = \frac{x}{1 - x} K_a \tag{4–22}$$

Substituting the result of Eq. 4–21 into Eq. 4–22 gives:

$$[H^+] = K_a\sqrt{K_h} \tag{4–23}$$

Equation 4–20 provides an expression for K_h in terms of the known dissociation constants. When substituted into Eq. 4–23, this gives:

$$[H^+] = \left(\frac{K_w}{K_a K_b}\right)^{1/2} K_a \tag{4–24}$$

When terms are combined and the results expressed in the Sörensen notation, Eq. 4–24 becomes:

$$pH = \frac{1}{2}(pK_w + pK_a - pK_b) \tag{4–25}$$

This result leads to some interesting conclusions. As a first approximation, the pH of a type 2 salt is independent of the solution concentration. If the acidic and basic dissociation constants are equal, Eq. 4–25 predicts that the solution will be exactly neutral. If $K_a > K_b$, the solution will have an acid reaction, and vice versa. Because type 2 salts can defend against pH change in either direction, they serve as better buffering agents than do other types of salts, taken alone.

Certain procedures call for buffer systems such as pyridine–acetic acid or ammonium hydroxide–formic acid mixtures. You should understand that these buffers involve type 2 salts.

pH OF A MIXTURE OF A WEAK ACID AND ITS SALT: THE HENDERSON-HASSELBACH EQUATION

The pH of a buffer, made by mixing varying quantities of a weak acid and a Type 1 salt of this acid, may be predicted approximately as follows.

At all times, the equilibrium equation for the acid dissociation must be satisfied:

$$K_a = \frac{[X^-][H^+]}{[HX]} \tag{4–26}$$

where X^- is the anion derived from the acid dissociation plus that contributed from the added salt. Taking logarithms of both sides, we have:

$$\log K_a = \log \frac{[X^-]}{[HX]} + \log [H^+] \tag{4-27}$$

Multiplying through by -1, using the Sörensen notation, and solving for pH, we have:

$$\text{pH} = \text{p}K_a + \log \frac{[X^-]}{[HA]} \tag{4-28}$$

This is the Henderson–Hasselbach equation, which also describes the curve in Figure 4–1.

At least for pure, comparatively dilute solutions, *it is quite possible to predict the pH without any measurements whatsoever;* one can establish precise secondary standards for pH control by means of simple chemicals and rather modest laboratory skills. Difficulties may arise when the solutions to be controlled are not limited to the buffer components but contain extraneous materials such as proteins, nucleic acids, or even simple salts. All of these may introduce errors in measurement, depending on the measurement technique, or they may affect the underlying assumptions on which the foregoing discussion was based. This is the subject to which we next turn.

pH EVALUATION: SOURCES OF ERROR

Whether the pH of a solution is estimated on theoretical grounds or determined by experiment (glass electrode, indicators), certain sources of error need to be understood. Two kinds of errors are frequently encountered. External errors originate outside of the buffer system and internal errors relate to the buffer system itself.

External Errors

Most biochemically useful pH measurements are made on complex systems. Typically, they contain moderately high concentrations of total solutes of diverse nature, many of which will be ionic. These ions, simply by their presence, affect the activity coefficients of the buffer components as well as of each other. If there is a common ion in the sample and the buffer, such as the sodium ion, this will affect the buffer equilibrium. The *common ion effect* contributes to a slight (but detectable) shift from the nominal pH of the buffer. This is one of several reasons for preparing cellular homogenates or subcellular fractions on a 1:10 dilution basis.

The most glaring case of ionic strength effects is probably the fractional precipitation of proteins from solution by ethanol at low temperature or by sodium or ammonium sulfate at controlled pH. In none of these methods is it likely that the pH of the buffer in the separation mixture is the same as in the stock buffer solution. The behavior of the buffer in 30% ethanol or in 3 M ammonium sulfate is simply not the same as in water. The errors are largely due to altered activity of the water and of the buffer ions.

Proteins are present in most biochemical systems at low concentrations; even so, proteins may introduce error in measurement of pH. Glass electrodes depends on an equilibrium between the $[H^+]$ in the solution to be measured and the surface of the ion-sensitive glass membrane. If that membrane is coated with a heavy layer of adsorbed protein, true equilibrium is not attained. The response time of the electrode will also be increased, a frequently overlooked fact. An unresponsive electrode may sometimes be rejuvenated by overnight immersion in 1 N HCl or in a *mildly* alkaline detergent. An extreme measure would be brief (30–60 sec) immersion in dilute ammonium fluoride, followed by extensive rinsing in water.

Indicators—weak acids (or bases) that have one color in the undissociated state and another color when ionized—are also subject to error in the presence of protein. Some indicators are strongly bound to certain proteins, with color shifts that foil correct pH estimates whether made by eye or by spectrophotometer. Obviously, indicators are also sensitive to the same external effects of extraneous ions as are glass electrodes, or the buffer ions themselves.

Some simple precautions will minimize external errors. Glass electrodes should not be left immersed in experimental solutions any longer than is necessary. They should not be left exposed to air any longer than is necessary, lest solutions evaporate and deposit solutes on the sensitive membrane. *Glass electrodes should always be standardized as nearly as possible under the conditions of use.*

Internal Errors

In the earlier analysis of buffer properties, two key assumptions were made. One was that the activity coefficients of the buffer ions were approximately equal to unity over the useful range of buffer concentrations. The second was that the value of K_a (or, of course, K_b) was essentially constant over the usual range of temperatures. Although these two assumptions give results that are frequently adequate, one must be aware of the important exceptions.

There are good theoretical reasons, which we shall not go into here, for believing that K depends on temperature. Pitzer showed that

$$\log K = A + B/T - 20 \log T \tag{4–29}$$

where A and B are constants characteristic of water. Likewise, the Debye–Hückel theory clearly indicates that f, the activity coefficient, is a function

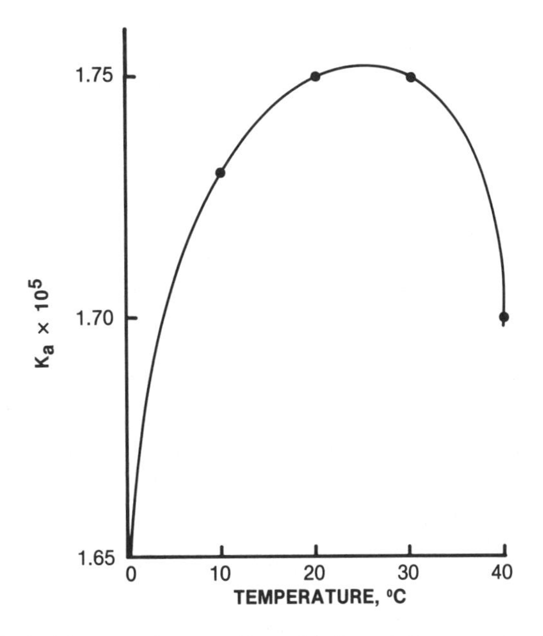

Figure 4–2. Effect of temperature on the acid dissociation constant, K_a, of acetic acid.

of the temperature. Unfortunately, there is no good way of predicting the properties of individual substances; these must be learned by trial. The form of Eq. 4–29 is such that one expects a maximum value of K at some specific temperature. Figure 4–2 illustrates the effect of temperature on the K_a of acetic acid over the range of usual biochemical interest. In going from 25°C to 0°C, the temperature effect alone introduces an error of about 5% for acetic acid, which is far from being the worst known case. Figure 4–3 shows the temperature effect as calculated for a 25 mM phosphate buffer.

One might assume, from the Henderson–Hasselbach equation, that a concentrated buffer, for example, 1 M, of a given pH could be diluted to 0.05 M without change in pH. Thus, in setting up some experiment, one ought to be able to add small volumes of a stock concentrated buffer to a mixture without fear of error. Indeed, many "recipes" for laboratory procedures call for just that. Electrophoresis buffers, for example, are commonly made up as "5×" or "10×" stocks, to be diluted when filling the electrode chambers. Unfortunately, the nonideal behavior of weak and strong electrolytes alike introduces errors. These errors may frequently become significant because of the change in α as a function of change in concentration. From the Debye–Hückel theory we can write an equation relating the true, or thermodynamic, value K, as opposed to the approximate value, k, and the concentration. The equation states that

$$\log K = \log k - 1.02(\alpha c)^{1/2} \qquad (4\text{–}30)$$

The numerical coefficient is a constant characteristic of water. Equation 4–30 shows that only at quite dilute solution is there good agreement

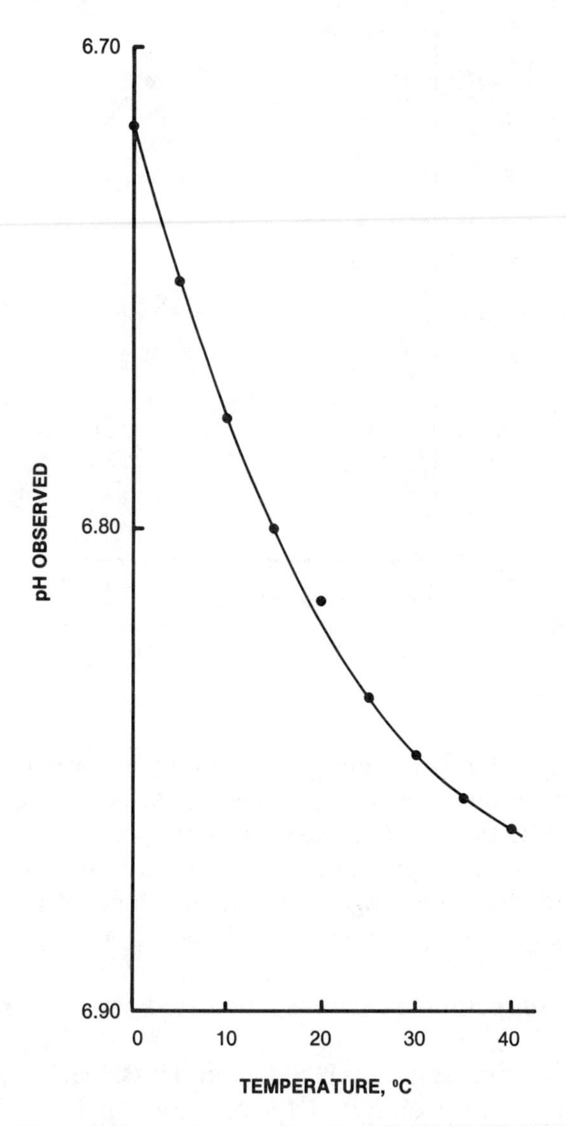

Figure 4–3. Effect of temperature on the pH of 25 mM sodium phosphate buffer.

between *K* and *k*. As the concentration increases, there will be a slowly increasing error due to change in activity of the component ions.

The intent of this discussion is to instill a certain sense of caution regarding the use of buffers to regulate pH. Much trouble could be avoided by observing the rule which requires standardization under conditions of temperature and concentration that as nearly as possible mimic the conditions of contemplated use. Thus, there is nothing wrong with preparing buffers as 5× stocks *provided that the five-fold dilution, NOT the stock, is at the desired or required pH value.* Good compilations of the necessary data for proper buffer preparation are readily available (see Dawson et al., 1986; Good et al., 1966).

SPECIFIC EFFECTS OF BUFFERS ON FACTORS OTHER THAN pH

Some weak acids (or bases) are unsuitable for use as buffers in certain cases. For example, citrate or phosphate buffers are ill-advised in systems that are highly calcium dependent. Citric acid and its salts are powerful chelators of Ca^{2+}, and the calcium phosphates are usually insoluble. Use of either of these buffers could so lower the $[Ca^{2+}]$ as to thwart the system being studied. These examples are obvious, but more insidious cases have been reported. In a brief but important paper, Mahler cited evidence that tris(hydroxymethyl)aminomethane (Tris), can chelate Cu^{2+} and perhaps other essential metals. Tris may also act as a competitive inhibitor of some enzymes, and may serve as an artificial substrate for others. The irony of the situation is that Tris was initially employed as a substitute for phosphate and citrate buffers so as to avoid interference with calcium-related or -dependent phenomena. To allow for possible specific buffer effects, more than one buffer system should be used with overlapping ranges whenever possible, especially if the pH dependency of a reaction system is being evaluated.

CHOOSING A BUFFER

A weak acid (or base) suitable for buffer use should have certain desirable properties. First and foremost, its pK should lie as close as possible to the middle of the working range required. The buffer should have a reasonably high molecular weight (100–200+) and be reasonably soluble, to allow preparation of solutions with precision and at useful strengths. The compound should also be stable and commercially available in high purity at moderate cost. Finally, it is desirable that the compound show no significant light absorption in the interval from 240 to 700 nm to allow its use in spectrophotometric measurements. Of the many weak acids and bases known, only 50–75 are in common use. A fair number of these are synthetics that were deliberately designed for buffer usage. Some of the synthetics are acids, some are bases, and some are zwitterionic materials. Good and his colleagues have developed and characterized a series of these covering the pH range from approximately 5.5 to 10.9, which embraces much of biochemical interest; the properties of these compounds are well documented.

WATER AS A REAGENT

Water is the most widely used reagent in biochemistry, and it is critical, therefore, that adequate supplies of acceptable purity must be maintained. Like any other reagent, the water that we use in the laboratory need not always be of the highest possible purity. In general, there are four types of water that are in use or are available: (1) ordinary tap water, which essentially is untreated since it left its central municipal or other source; (2) demineralized water, which is tap water passed through an ion exchange column so that ionic components have, with varying degrees of efficiency, been removed; (3) distilled water, to prepare which tap water has been boiled, and the vapor therefrom condensed in some fashion and collected; and finally (4) water that has been subjected to a combination of the latter two processes.

Tap Water

Tap water is either hot or cold (an obvious statement), but the point to stress is that besides not being of equal temperature, they are not of equal purity. Cold water is the least pure of all the waters available. It contains minerals, such as magnesium and calcium ions as well as many other cations together with their counter ions, including sulfates and possibly chlorides, nitrates, etc. Organic materials are present as well, which vary widely, particularly with the time of the year. For instance, in some parts of the country, the springtime melt-off from snow, together with the low oxygen tension in the water, results in the generation of a motley collection of odorous and nonodorous substances, which may consist of aldehydes, acids, and other unidentified materials. Tap water is obtained mainly from city water purification systems. Municipal treatment of water to remove impurities generally includes addition of alum (aluminum potassium sulfate) for aggregation, lime (calcium oxide) for pH adjustment, a cationic polymer, followed by chlorine or chlorine dioxide treatment and finally filtration. Products of chlorine treatment, which could include chlorates, chlorine itself, and perhaps some chlorinated organic compounds, may persist in the final product. Not surprisingly, of all the different types of water available, that with the lowest bacterial count is tap water. Hot water is often somewhat purer than cold tap water since it has been partly demineralized before entering the boiler. During the heating process, some carbonates may have been destroyed, and CO_2 and some of the chlorine used in its treatment are driven off. Tap water obviously cannot be used in the preparation of reagent solutions and should not be used in water baths since the minerals would precipitate out during the slow evaporation that occurs in these vessels.

Demineralized Water

Fundamentally, cation- or anion-exchange treatment of water involves the exchange on an ion-exchange column of an undesirable ion for one that is innocuous. For example, water can be "softened" by exchanging Mg^{2+} and Ca^{2+} for Na^+. Complete removal of all ions (preferable in the laboratory) can be effected by exchanging the metallic cations for H^+ and the anions for OH^-, resulting in the formation of water and the complete removal of the ions. Such water is prepared by passing tap water through an ionic exchanger such as Permutit® or, in more limited but more efficient systems, mixed-bed ion exchangers such as sulfonated polystyrene cation exchangers and quaternary ammonium anion exchangers. Demineralized water may have a high content of organic compounds not usually removed by the ion-exchange column; bacterial growth in such water systems is often quite extensive. To remove some of these organic materials, the water can be passed over an adsorbent such as activated charcoal or a non-ionic resin which quite effectively removes the remaining contaminants. This process is discussed in more detail in the section entitled High-Purity Water, below).

Distilled Water

One of the more common, and in past years most widely used, procedures for purification of water is distillation. In the simplest form of this process, water is boiled in an enclosed vessel; the water vapors are led up through the neck of the apparatus, through a cooled condensing system; and the condensed water is then collected and used. Ideally, such a system would leave all solids behind in the boiling chamber. Usually the initial and final fractions condensed are discarded and the middle fractions are used. It is possible, however, that volatile impurities may still carry over in the distillate. Higher-boiling materials (those with lower vapor pressures) might not distill as rapidly, but they do carry over; what may also occur is the formation of an azeotropic mixture of the water and another liquid so that the impurity appears in the distillate. Such azeotropic mixtures (which represent deviations from Raoult's law) cannot be separated by simple distillation. Depending upon the efficiency of the condenser system used—that is, the number of theoretical plates—even materials with high boiling points or low vapor pressures might be carried over as aerosols with the vaporized water and condensed with the water being collected. Modifications of the simple distillation of water are:

1. Distillation from acid. For example, sulfuric acid is added to the water being boiled, effecting the oxidation of some impurities, perhaps converting them into volatile materials that are distilled off in the first discarded fraction of the condensate.
2. Distillation from acid permanganate and sulfuric acid, both added to the water. This mixture vigorously oxidizes impurities, rendering some of them volatile and removable in the first collected fraction. This is clearly a procedure that must be carried out batchwise, and the boiling vessel must frequently be cleaned out.

Here again, it is critically important that none of the sulfuric acid from the boiling mixture is carried over in the distillate by aerosol formation; columns used in acid distillation of any type must be of high efficiency.

Quite frequently, doubly, triply, or even quadruply distilled water using quartz boiling vessels and condensers is preferred, especially for use in critical physical chemical studies. The efficiency of distillation must be carefully controlled, and only a center "cut" of the distillate used; a disadvantage is that only small quantities of water can reasonably be prepared by this process.

High-Purity Water (Distilled and Further Purified)

In order to obtain large quantities of water of high purity, a combination of the above techniques is used. Water is first distilled and then subjected to five steps in the purification process.

1. The first step may be a membrane filtration which removes the bulk of suspended material, that is, any solid materials or slime that would interfere with the efficiency of subsequent purification units.
2. The second step involves a column of granular activated carbon and cellulose fibers which present about 1.1 million square-meters of surface to incoming water. In this step, dissolved organic contaminants are removed by adsorption to the carbon. Alternatively, a nonionic polymer with high adsorption capability may substitute for the charcoal. In some systems, both the charcoal and resin columns are used in tandem.

3 and 4. In the third and fourth steps, two mixed-bed (strong acid/strong base) ion-exchange resin columns are presented to the water stream.

5. The fifth step is passage of the water through a membrane filter which efficiently removes any microscopic particles or microorganisms that might escape retention upstream of this filter, or particles of the column materials themselves. The membranes consist of cellulose esters and will remove all particles larger than 0.22 μm.

The water obtained by such a system is of high purity and may be used in demanding operations such as tissue culture and fluorescence measurements.

Any water purification system must be properly maintained for it to deliver a product of the expected purity. For example, in batch distillations, the residue in the boiling flask must be frequently removed, and, in the "high-purity" four-cartridge system described above, not only must the ion-exchange, charcoal, and particle filters be regularly replaced, but also the compartments containing the cartridges must be rigorously cleaned to remove possible slime mold accumulation.

Evaluation of Water Purity

The conductivity of the water may be measured as the water is being drawn, and the electrical resistance of the water is highest if the concentration of ionic solutes is lowest. Water in such systems is never drawn until the resistance meter reads at least 10 and usually 18 megohm-cm in some systems. Any organic impurities would have to be analyzed for by other techniques, for example, by HPLC.

Storage of Water

Water stored for long periods of time (i.e., >2 weeks) in glass or plastic containers may be suspect with respect to its continued purity, because of the possibility of contamination from three sources:

1. Although unlikely in a properly protected water vessel, particulate contamination from the air may occur.
2. The water is not sterile, and thus, although the nutrient concentration *should* be very low (essentially zero), some slime mold or bacterial growth is possible.
3. Depending on the nature of the container used, some organic (from plastic) or ionic (from glass) materials may be extracted.

To obviate these possibilities, therefore, stored water should be replaced frequently.

Gases Dissolved in Water

It is difficult to prepare and store water without including dissolved gases. These would be primarily the usual components of air as well as CO_2 and ammonia; their presence can lead to problems of varying severity.

Nitrogen, oxygen, and other gases in air will have a solubility in water (and all liquids) described by Henry's law: $C_l = K \cdot p$, where "C_l" is the concentration of the gas in liquid, "K" is the solubility coefficient, and "p" is the partial pressure of the gas over the liquid. This relationship may also be expressed in terms of the Bunsen coefficient, α, which is the volume of gas (at S.T.P.) dissolved in one volume of liquid. Generally, the Bunsen coefficient, α, varies inversely with the temperature. For example, for CO_2 in water, $\alpha = 1.73$ at 0°C, whereas $\alpha = 0.555$ at 38°C. The contamination of reagent solutions with gases may be minimized by boiling the water beforehand, and thereafter agitating it as little as possible. Alternatively, one might boil the reagents after their preparation if this would not destroy the solutes. A third alternative is to place the solution to be degassed in a filtration flask and apply a vacuum, usually adequately supplied by a water aspirator. This is continued until no more *dissolved* gas bubbles appear. If this procedure is continued too long, however, the solution may boil away at the reduced pressure in the flask (Henry's law).

If gases are *not* removed from certain solutions, some problems may arise:

1. If the solution phase of a chromatographic column has not been degassed, the dissolved gases may come out of solution, especially if the temperature rises even slightly, forming gas pockets in the bed or

matrix materials and thereby seriously interrupting the homogeneous flow of elution fluid through the column. Such columns are ruined and have to be repoured.

2. Dissolved oxygen interferes with the polymerization of acrylamide to polyacrylamide gels since oxygen is a free radical inhibitor. The acrylamide and buffer solutions used for gel formation should therefore be degassed.
3. In some optical measurements made at wavelengths below 215 nm (e.g., in spectrophotometry or circular dichroism), oxygen interferes since it absorbs light from that spectral transition region on down.
4. In fluorescence measurements, oxygen and nitrogen may have quenching effects.
5. A different type of problem is presented by the presence of CO_2 in H_2O. This gas is ubiquitous and, in solution:
   a. It decreases the pH of water by formation of carbonic acid. Although the pH of "pure" water varies with temperature, the usual acidic pH of the water we use (i.e., $\sim$ 6–6.5) is primarily due to CO_2.
   b. It causes the precipitation of Ba^{2+} or Ca^{2+} as $BaCO_3$ or $CaCO_3$ if these cations are added to the water in a reagent.
   c. If present in sufficient concentrations, it may interfere with accurate measurements in CO_2-evolving systems.
6. Ammonia may also be encountered as a contaminant of water. Its presence would obviously interfere with nitrogen determinations and with determination involving Ca^{2+} complexation (e.g., biuret tests), and may also affect pH (increasing it) if in excessive concentrations.

REFERENCES

Bates, R.. G. *Determination of pH: Theory and Practice,* 2nd ed. John Wiley and Sons, New York (1964).

Dawson, R.M.C., Elliott, D. C., Elliott, W. C., and Jones, K. C. *Data for Biochemical Research,* 3rd ed. Oxford University Press, New York (1986).

Good, N. E., Winget, G. D., Winter, W., Connolly, T. N., Izawa, S., and Singh, R.M.M. Hydrogen ion buffers for biological research. *Biochemistry 5:*467 (1966).

Mahler, H. Use of amine buffers in studies of enzymes. *Ann. N.Y. Acad. Sci. 92*(2):426 (1961).

Pitzer, K. The heats of ionization of water, ammonium hydroxide, carbonic, phosphoric and sulfuric acids. The variation of ionization constants with temperature and the entropy change with ionization. *J. Am. Chem. Soc. 59:*2365 (1937).

General References

Bull, H. B. *Introduction to Physical Biochemistry,* 2nd ed. F.A. Davis, Philadelphia (1971).

Marshall, A. G. *Biophysical Chemistry.* John Wiley and Sons, New York (1978).

Montgomery, R., and Swenson, C. A. *Quantitative Problems in the Biochemical Sciences,* 2nd ed. W.H. Freeman, San Francisco (1969).

CHAPTER FIVE

General Principles of Ligand Binding Equilibria

An important biological phenomenon is the noncovalent interaction between one molecule known as a *ligand* and a second molecule known as an *acceptor* or *receptor*. Many ligand–acceptor interactions are related to problems of cellular transport, including the binding of free fatty acids or bilirubin to serum albumin, the binding of free hemoglobin to haptoglobin, or the binding of certain steroid hormones to transcortin. Some metals, including iron and copper, are also carried by specific proteins, from which they can be released to cells as needed. Other examples illustrate forms of cellular regulation. The interaction of the hormone, insulin, with specific receptors on cell surfaces is such a case. Yet another class, the interaction between antigens and antibodies, addresses the problem of recognizing self versus nonself.

Most of the above illustrations have in common a surprisingly high degree of specificity between the ligand and the binding site on its acceptor, but this need not always be true. Thus, the binding of thyroxine to serum albumin is reduced by ingestion of aspirin, because salicylate competes with thyroxine for a common binding site on albumin. One may therefore conclude that albumin is clearly of a lower specificity than is, say, the insulin receptor.

From even these few examples it is obvious that ligands may vary greatly in molecular size and complexity, ranging from small inorganic ions to small organic molecules to macromolecules. Acceptors also vary in complexity and may exist as free molecules (serum albumin) or as a constituent protein (more probably, as an assemblage of proteins and phospholipids) in cell membranes or organelles. Preliminary work indicates that the insulin receptor can be solubilized and reconstituted into artificial membranes without great loss of specificity (but with considerable loss of activity). Some steroid receptors have been localized and isolated from both cytoplasmic and nuclear preparations of hormone-sensitive cells; whether or not similar internal receptors exist for insulin is still not completely resolved.

Ligand binding is also known in the nonbiological realm, but there are some distinct differences. What is termed a "receptor" in biological systems may be termed an "adsorbent" in the more general chemical systems. Adsorbents bind ligands also by noncovalent forces, but are, as a rule, very much less specific. Powdered charcoal, chalk, silica, and even glass are all effective adsorbents. As such, they form the basis of adsorption chromatography, which will be considered in Chapter 7. For the moment, it is necessary only to understand that the basic principles of the two kinds of ligand binding are essentially the same.

QUANTITATIVE ANALYSIS OF LIGAND BINDING PHENOMENA

The analysis that follows owes much to the fundamental studies of George Scatchard, who was interested in the binding of dyes (many of them pH indicators) to serum albumin as illustrative of proteins in general. In recognition of Scatchard's work, we shall use the notation he employed, although other sources may employ a different notation. We start with the following definitions:

1. C_B^0 = total concentration of acceptor/receptor binding sites
2. C_{BL} = concentration of binding sites occupied by the ligand
3. C_B = concentration of unoccupied binding sites
4. C_L^0 = total ligand concentration
5. C_L = concentration of free (unbound) ligand
6. $C_{BL}/C_B^0 = \nu$ = fraction of available binding sites that are occupied
7. $C_B/C_B^0 = \alpha$ = fraction of available binding sites that are unoccupied.
8. n = number of binding sites per molecule of acceptor/receptor
9. $B + L \rightleftharpoons BL$, the reversible reaction between the ligand and the acceptor; the rate constants need not be the same for both directions
10. $K = [BL]/[B][L]$, the equilibrium constant for the system, based on the law of mass action

In addition to the above definitions, three conservation equations must apply:

1. $C_B^0 = C_{BL} + C_B$ (binding sites are either occupied or unoccupied)
2. $C_L^0 = C_{BL} + C_L$ (ligand can be only bound or free)
3. $\alpha + \nu = 1$ (sum of fractions of unoccupied and occupied sites must be unity)

From the law of mass action, and assuming that the acceptor/receptor is univalent ($n = 1$), we can write an expression for the *association constant:*

$$\frac{C_{BL}}{(C_B)(C_L)} = \frac{C_{BL}}{(C_B^0 - C_{BL})(C_L)} = K_a \tag{5-1}$$

which can be rearranged to give:

$$C_{BL} = K_a(C_L)(C_B^0 - C_{BL}) = K_a(C_B)(C_L) \tag{5-2}$$

Equation 5-2 is the simplest form of what is commonly identified as the Scatchard equation. It relates the ratio, bound/free ligand, to the association constant and the concentration of unoccupied binding sites. Frequently we do not know very much regarding C_{BL} or C_B; in fact, what is most commonly sought is information concerning just these parameters. It would be helpful to have expressions for α and ν in terms other than those given above, and which involve only C_{BL}, C_L, and K_a, since these quantities generally are known or can readily be measured. From definition 7 and conservation equation 1 we can redefine α as:

$$\alpha = \frac{C_B}{C_B^0} = \frac{C_B}{C_B + C_{BL}}$$

From Eq. 5–2, we note that $C_{BL} = K(C_B)(C_L)$; therefore

$$\alpha = \frac{C_B}{C_B + K(C_B)(C_L)} = \frac{1}{1 + K(C_L)} \tag{5-3}$$

In a similar way we can redefine ν from definition 6 and conservation equation 5–1, as follows:

$$\nu = \frac{C_{BL}}{C_B^0} = \frac{C_{BL}}{C_B + C_{BL}}$$

From Eq. 5–2, $C_{BL} = K_a(C_B)(C_L)$, so that:

$$\nu = \frac{K_a(C_B)(C_L)}{C_B + K_a(C_B)(C_L)} \tag{5-4}$$

which rearranges to:

$$\nu = \frac{K_a(C_L)}{1 + K_a(C_L)} \tag{5-5}$$

Note that the sum of Eqs. 5–3 and 5–5 reduces to conservation equation 3. Note also that α and ν can be interpreted as *probabilities* that a given fraction of the binding sites are unoccupied or occupied, respectively. Finally, and this is most useful, note that α and ν can now be evaluated in terms of the association constant and the concentration of free ligand.

The ability to analyze quantitatively any ligand binding system depends on methods that can distinguish between bound ligand and free. The procedure used depends upon the molecular size of the ligand, the affinity of binding, and the method of determining the concentration of the ligand. For low molecular weight ligands (MW < 1000) equilibrium dialysis is probably fundamentally the most sound. This technique involves two compartments, 1 and 2, separated by a semipermeable membrane. In the absence of a binding macromolecule, the ligand will diffuse reversibly through the membrane until C_L in both compartments is the same, that is, $C_L^1 = C_L^2$.

If a nondiffusible binding molecule, for example, a protein, is introduced in compartment 2, it will bind ligand molecules, thereby reducing C_L^2. This creates a temporary inequality between the two compartments, that is, $C_L^1 > C_L^2$, which must be redressed by diffusion of ligand from the compartment 1 to 2. This process continues until once again at equilibrium, $C_L^1 = C_L^2$. In this new equilibrium, however, the free ligand concentration in both compartments has been reduced *by the amount bound to the macromolecule* in compartment 2. Quantitating this reduction in C_L^1 therefore permits calculation of C_{BL}. In addition, C_L^0 in compartment 2 has increased also by the amount bound, and analysis of this increase yields an additional evaluation of C_{BL}. A fine discussion of this procedure is found in Cantor and Schimmel (1980).

Other techniques for distinguishing free from bound ligand include gel filtration (discussed by Ackers), biphasic aqueous polymer systems (Gray), and various DEAE-disc binding procedures (see Expt. 16, DEAE-Disc Binding Assay for CBG–Cortisol Binding).

There are many methods for determining ligand concentrations; the choice depends upon the properties of the molecule. The most popular of these methods employs radiotagged ligands, although for practical reasons these must be available at quite high specific activities. Where commercial material is available, high specific activity is easy to accomplish. Where commercial material is not available, one must prepare derivatized ligands by labeling with ^{125}I or some other strongly emitting isotope. Using radiolabeled insulin, for example, and vesicles containing its receptors, one can count the radioactivity remaining after equilibration of the vesicles with insulin, followed by filtration or centrifugation to remove the vesicles. Alternatively, one can count the filtered vesicles.

If the ligand absorbs light at different λ_{max} values in the free and bound states, simple or difference spectrophotometry may suffice for estimation of the two states. This is frequently the case in dye-binding systems. In other instances, fluorescence properties of the free and bound forms differ, and useful measurements can be made by spectrophotofluorometry. Specific chemical analysis has also been employed, although for ligands other than inorganic ions methods of sufficient sensitivity may not be at hand. Whatever the approach taken, information must be collected over a considerable range of ligand concentrations in order to generate a useful graphic analysis from which the binding parameters may be extracted. Care must also be exercised to ensure that a true equilibrium has been reached before the free and the bound forms are separated by the procedures used. Since the entire analysis depends on satisfying the association constant, failure to observe this precaution can give only invalid results. The reversibility of the binding equilibrium must be demonstrated. Some systems are quite sensitive to extraneous factors, such as the ionic strength, or to the presence of some particular ion. Scrupulous care must be taken to control constancy of the environment in which the binding takes place, and to regulate the temperature of the system closely.

THE SIMPLEST GRAPHIC DATA ANALYSIS: ν AS A FUNCTION OF C_L

The data points may be plotted by hand or by a computer program. The generalized form of Eq. 5–5 can be written as $y = ax/(1 - ax)$, from

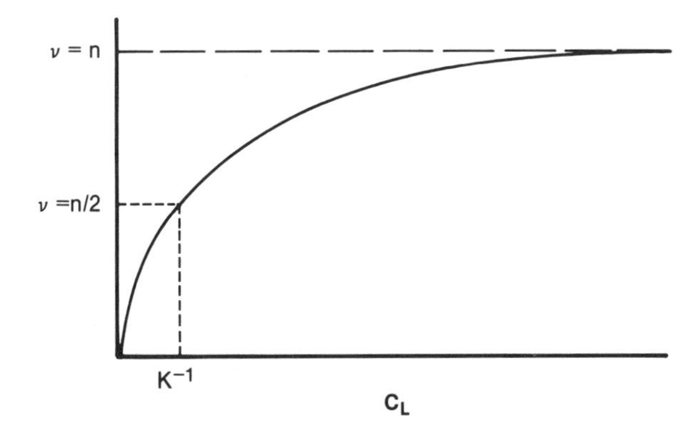

Figure 5–1. Direct plot of ν as a function of C_L. As C_L increases, the slope of the curve approahces a limiting value. At the point where $\nu = n/2$, the value of $C_L = K^{-1}$, where K = the *association constant*. Note the general similarity of this curve to the Michaelis–Menten curve.

which it follows that the graph of this function is a hyperbola. The abscissa represents C_L and the ordinate ν, as shown in Fig. 5–1.

This method has all the disadvantages of working with nonlinear graphs and therefore is not the most widely used. However, the advent of hyperbolic fitting programs for small computers has made it more practical than it once was, and it is now possible to extract the association constant or the dissociation constant (recall that $K_d = K_a^{-1}$) without even plotting the data.

THE DOUBLE RECIPROCAL PLOT: $1/\nu$ AS A FUNCTION OF $1/C_L$

The reciprocal of Eq. 5–5 can be written as:

$$\frac{1}{\nu} = \frac{1 + K(C_L)}{K(C_L)} = \frac{1}{K(C_L)} + 1 \tag{5-6}$$

Figure 5–2. Double reciprocal plot of $1/\nu$ as a function of $1/C_L$. This plot is a means of linearizing the hyperbolic plot of Fig. 5–1, but see the text for required precautions in interpretation of transformed plots.

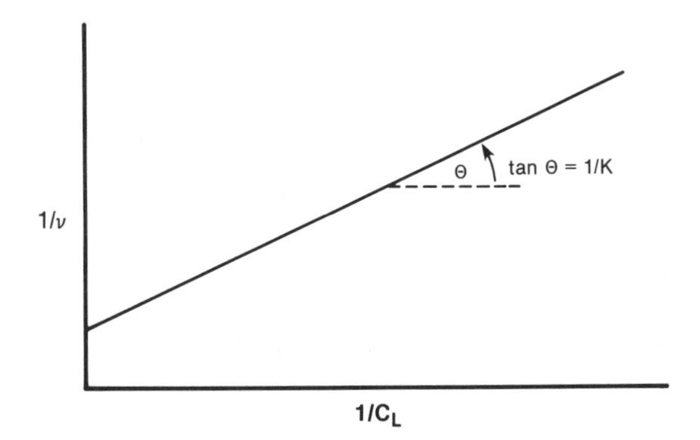

The generalized form of Eq. 5–6 is $y = ax + b$, so the reciprocal transformation has the effect of linearizing the hyperbolic plot, as shown in Fig. 5–2. This transformation is convenient, but it is not ideal. Elsewhere (see Chapter 12, on data analysis by statistical methods) the hazards of reciprocal transformations in altering the statistical weight of data points have been noted. At high values of C_L, double reciprocal plots will suffer from some nonlinearity. It may be difficult to determine, from experimental values, if the nonlinearity is due to experimental error or to the effects of the transformation itself. Unless it has been subjected to rigorous statistical evaluation, this method can at best be regarded as approximate.

THE CASE OF MULTIPLE, EQUIVALENT BINDING SITES

Thus far, the analysis has been based on the assumption that each macromolecule contains only a single binding site. Suppose now that the acceptor/receptor species contains more than one binding site per molecule, and that each binding site is equivalent and noninteracting (noncooperative). At any given site, i, the extent of binding averaged over all of the macromolecules is given by the expression:

$$\nu_i = 1 - \alpha_i = \frac{K_{ai}C_L}{1 + K_{ai}C_L} \tag{5–7}$$

Since all of the sites are equivalent (have the same value of K_a), we can write that

$$\nu = \sum_{1}^{n} \nu_i = n\nu_i \tag{5–8}$$

where ν is the average number of occupied sites per macromolecule and n = number of binding sites per macromolecule. The probability that any randomly chosen site is occupied becomes:

$$\frac{\nu}{n_i} = 1 - \alpha = \frac{K_a(C_L)}{1 + K_a(C_L)}, \quad \text{and} \quad \nu = \frac{nK_a(C_L)}{1 + K_a(C_L)} \tag{5–9}$$

Using the expression for ν in Eq. 5–9, we can also write an expression for $n - \nu$, which states that:

$$n - \nu = n - \frac{nK(C_L)}{1 + K_a(C_L)} = \frac{n}{1 + K_a(C_L)} \tag{5–10}$$

We can now divide Eq. 5–9 by Eq. 5–10 to get the expression:

$$\frac{\nu}{n - \nu} = \frac{nK_a(C_L)}{1 + K_a(C_L)} \cdot \frac{1 + K_a(C_L)}{n} = K_a(C_L) \tag{5–11}$$

Equation 5–11 is a generalized form of the Scatchard equation; it no longer depends on the limiting assumption that each macromolecule contains only a single binding site. It still contains the assumption that all of the binding constants are identical. This restriction can be eliminated by using an expression for ν quite similar to that derived in Eq. 5–9, but which involves the sum of a series of terms, as:

$$\nu = \frac{n_1 K_{a1} C_L}{1 + K_{a1} C_L} + \frac{n_2 K_{a2} C_L}{1 + K_{a2} C_L} + \frac{n_3 K_{a3} C_L}{1 + K_{a3} C_L} \cdots \qquad (5\text{–}12)$$

If Eq. 5–12 (instead of Eq. 5–9) is divided by Eq. 5–10, the completely generalized Scatchard equation is obtained; it takes into account multiplicity of binding sites and possible differences in the affinity constant for each class of sites. That result is not given here, but the reader should have no difficulty in writing it out.

EXPRESSION OF LIGAND BINDING DATA IN TERMS OF LIGAND CONCENTRATIONS ONLY

One problem remains. The left-hand term of Eq. 5–11 expresses the ratio of occupied to unoccupied binding sites. This terminology, of course, defines the situation with respect to the acceptor/receptor molecules. Although that ratio is what we ultimately wish to know, experimental constraints require that it be determined by measurements of bound or free *ligand,* since it is the ligand that is analyzed. It would be useful to find some expression equivalent to the left-hand side of Eq. 5–11, that did not contain the term, $n - \nu$. Each side of Eq. 5–11 can be multiplied by the quantity $(n - \nu)/C_L$, with the result that Eq. 5–11 becomes:

$$\nu/C_L = K_a(n - \nu) \qquad (5\text{–}13)$$

A comparable trick can be worked on the equation for cases with classes of binding sites having different values of K_a, based on the development with Eq. 5–12.

The effect of converting Eq. 5–11 to Eq. 5–13 is significant. From definition 6, ν is a measure of the occupied sites. Since site occupation reflects ligand binding with a one-to-one correspondence (i.e., each site can be occupied only by a single ligand molecule) it follows that ν is also a measure of bound ligand. C_L is, by definition 5, a measure of the free ligand. Therefore, the left-hand side of Eq. 5–13 can be interpreted as:

$$\frac{\nu}{C_L} = \frac{\text{bound ligand}}{\text{free ligand}}$$

The right-hand side of Eq. 5–13, $K_a(n - \nu)$, is obviously a function of the bound ligand. All of the algebraic manipulation has brought us to the point where the behavior of a macromolecular binding process can be precisely described in terms of the bound and free ligand concentrations. The number of binding sites per macromolecule, or the presence of multiple classes of sites with independent affinity constants, can be experimentally

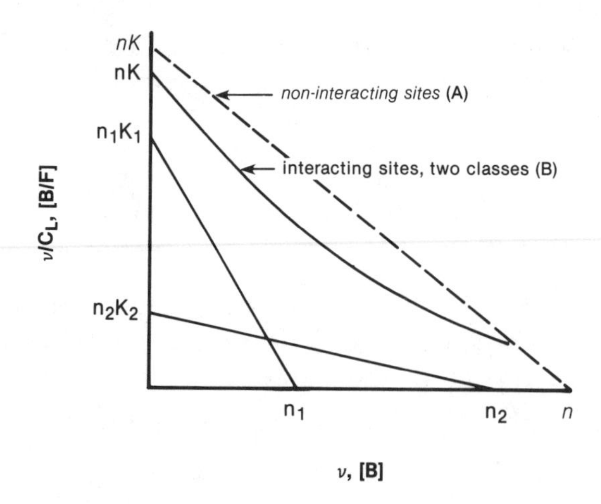

Figure 5–3. Typical Scatchard plots for macromolecular-ligand interactions: (A) Noninteracting sites; (B) two classes of sites, or interacting sites. Each axis is marked first in terms of Eq. 5–13; to emphasize the meaning of each axis in terms of the ligand, each axis is also marked as [B/F] or [F]. Intercepts on the axes have the indicated values; if curve B is interpreted as the resultant of two classes of sites, then the intercepts would be n_1K_1, and n_2K_2, and $n_1K_1 + n_2K_2 = nK$.

observed from the shape of the generalized Scatchard plot. For these reasons, Eq. 5–13 is the most widely used version of the Scatchard equation. It yields plots like those shown in Fig. 5–3.

Curve A in Fig. 5–3 is a ligand binding curve for macromolecules that contain a single class of binding sites; the experimental data are represented by the solid portion of the curve. Extrapolation of the curve, until it cuts both axes, allows estimation of *n* (for clarity, the value that pertains to curve A is italicized). On the ordinate, the point of intersection represents *nK*. Curve B describes macromolecules that contain two classes of binding sites with different affinity constants. By a double extrapolation of the linear (or nearly linear) portions of the curve, it is possible to evaluate both sets of constants. Because the affinity constant of the first class is larger than the constant for the second class, the curve bends as shown. The sharpness of the bend is related to the difference in magnitude of $K_1 > K_2$. When $K_1 \gg K_2$, the bend in the curve will be more gradual, as is the case in the illustration.

OTHER GRAPHIC METHODS

It would be incorrect to leave the impression that the Scatchard plot is the only means of analyzing information on the ligand binding properties of macromolecules. Other kinds of plots exist and have some specialized uses, but they are much less practical for the majority of cases. The interested reader will find descriptions of these in the references cited at the end of this chapter.

NONSPECIFIC LIGAND BINDING

Implicit in all of the foregoing discussion was the assumption that a ligand binds only to some specific site, but Nature is seldom that simple. Specific binding has the useful property of a high affinity with a low to moderate capacity. Nonspecific binding usually has a low affinity with a moderate to high capacity. Another way to express these differences is to describe specific binding as *saturable* and nonspecific binding as *nonsaturable.* The saturability of specific binding has already been illustrated in Fig. 5–1, which is an idealized illustration, because nonspecific binding does not appear. A more realistic set of data would look like that shown in Fig. 5–4. The lowest, linear curve, marked NS, is a measure of nonspecific binding. It is linear because it is, for practical experimental purposes, nonsaturable and increases steadily as C_L increases. The specific binding curve, marked S, is already showing signs of saturation; it has come close to its limiting value of ν. The uppermost curve, marked T, represents the total or observed binding. It has the property that, for any value of C_L, the corresponding value of ν is the sum of the ν values for the first two curves. An example of this is shown in the figure. Looking only at the total binding curve, which is what one first sees in the laboratory, the approach to saturation is harder to appreciate. This is always a problem in experimental systems that involve very much nonspecific binding. There is no absolute way to correct for nonspecific binding but a quite good approximation can be made.

The correction method depends on the affinity difference between a specific and a nonspecific binding site. As an example, consider a membrane vesicle preparation known to contain receptors for a hormone identified only as X. The vesicles can be exposed to a preparation of [^{125}I]X, with the result that the radiolabeled hormone will be bound to both the specific and

Figure 5–4. Nonspecific ligand binding as a function of C_L. NS, nonspecific binding; S, specific binding; and T, total (observed) binding. The small arrows, all at the same value of C_L, indicate the corresponding value of ν for each curve.

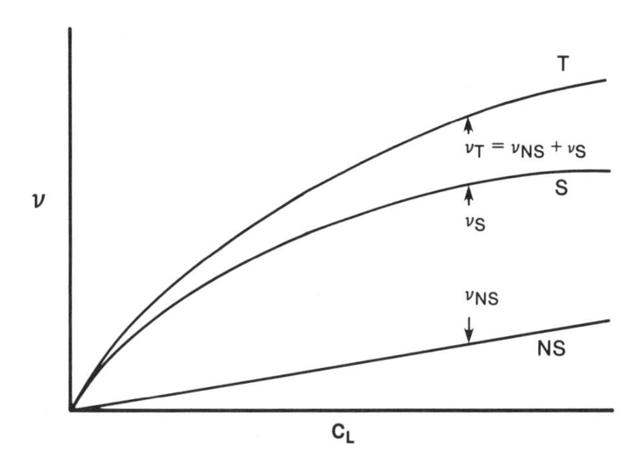

the nonspecific sites. If the mixture is next treated with unlabeled X, the "cold" hormone will exchange with [^{125}I]X, in proportion to the concentrations of [^{125}I]X and X, *at the specific sites.* Because the affinity for ligand of the specific sites is so very large, little or none of the nonspecifically bound [^{125}I]X will be exchanged, even though the labeled hormone be diluted with unlabeled X by as much as 1:500. One is never sure, of course, that *none* of the nonspecifically bound [^{125}I]X was exchanged; to the extent of that uncertainty, one must regard the correction as an approximation, but the error is certainly very small. In sum, one first determines the total binding, then one estimates the nonspecific binding by large dilution with the unlabeled ligand. The difference is taken as a measure of the specific binding.

For some ligands the rates of the "on" (binding) and "off" (exchange) reactions are quite different. This explains why certain procedures call for different times of equilibration, and sometimes quite different temperatures, in making proper assessments of labeled ligand binding and exchange with cold. Radioimmunoassays are elegant examples of ligand binding systems, and these frequently contain instructions that are puzzling unless the rate differences are understood. The on and off reactions are examples of measurable chemical equilibria that are reversible by reasonable shifts in relative concentrations and conditions. Only the nonspecific binding behaves irreversibly within any reasonable changes in concentrations and conditions. Ultimately, this is the basis on which the nonspecific binding correction depends. Competition between ligands has virtually no effect on nonspecific binding.

REFERENCES

Ackers, G. K., Studies of protein ligand binding by gel formation techniques. In *Methods Enzymol. 27:*441–455. Academic Press, New York (1973).

Bergeron, J.J.M., Posner, B. I., Josefsberg, Z., and Sikstrom, R. Intracellular polypeptide hormone receptors. *J. Biol. Chem. 253:*4058–4066 (1978).

Bonne, D., Belhadj, O., and Cohen, P. Calcium as modulator of the hormonal receptors–biological response coupling system. *Eur. J. Biochem. 86:*261–266 (1978).

Cantor, C. R., and Schimmel, P. R. *Biophysical Chemistry,* Part 3: *The Behavior of Biological Macromolecules.* W.H. Freeman, San Francisco (1980).

Dahlquist, F. W. The meaning of Scatchard and Hill plots. In *Methods Enzymol. 48:*270–299. Academic Press, New York (1978). [A discussion of Scatchard plots.]

Edsall, J. T., and Gutfreund, H. *Biothermodynamics.* John Wiley & Sons, New York (1983).

Binding of Insulin to Its Receptors

Cuatrecasas, P. Unmasking of insulin receptors in fat cells and fat cell membranes: perturbations of membrane lipids. *J. Biol. Chem. 246:*6532–6542 (1971).

Cuatrecasas, P. Insulin receptor of liver and fat cell membranes. *Fed. Proc. 32:*1838–1846 (1973).

De Meyts, P., Biano, A. R., and Roth, J. Site–site interactions among insulin receptors: characterization of the negative cooperativity. *J. Biol. Chem. 241:*1877–1888 (1976).

Gutfreund, H. Equilibria and kinetics of protein–ligand interaction. In *MTP Int. Rev. Sci.,*

Biochem. Series, No. 1, Vol. 1 (H. Gutfreund, ed.). University Park Press, Baltimore (1974).

Huse, P.D.R., Poulis, P., and Weidemann, M. J. Isolation of a plasma membrane subfraction from rat liver containing an insulin-sensitive cyclic AMP phosphodiesterase. *Eur. J. Biochem. 24:*429–437 (1972).

Lineweaver, H., and Burk, D. Determination of enzyme dissociation constants. *J. Am. Chem. Soc. 56:*685–686 (1934).

Marshall, A. G. *Physical Biochemistry,* Chap. 12. John Wiley and Sons, New York (1978).

Munson, P. J., Rodbard, D., and Klotz, I. M. Number of receptor sites from Scatchard and Klotz graphs: a constructive critique. *Science 220:*979–981 (1983).

Scatchard, G., Attractions of proteins for small molecules and ions. *Ann. N.Y. Acad. Sci. 51:*660–672 (1949).

Schrader, W. T., and O'Malley, B. W. *Laboratory Methods Manual for Hormone Action and Molecular Endocrinology,* Chaps. 2 and 3. Houston Biological Assoc., Houston, TX (1981).

Schultz, G. E., and Schirmer, R. H. *Principles of Protein Structure,* Chap. 10. Springer Verlag, New York (1979).

Tanford, C. *Physical Chemistry of Macromolecules,* Chap. 8. John Wiley and Sons, New York (1961).

CHAPTER SIX

Electrokinetic Phenomena

CONCEPTS OF ELECTROPHORESIS AND ELECTROOSMOSIS

Unless otherwise constrained, charged particles exposed to an electrostatic field move toward the electrode of opposite charge. When charged particles are suspended in a liquid, across which a field may be applied, the resultant movement of the particles is known as *electrophoresis.* There may also be movement of the bulk liquid, giving rise to the associated process of *electroosmosis.* In biochemical practice, electrophoresis is a very useful tool whereas electroosmosis is a process that one generally attempts to suppress as much as possible. The reasons for this will become clearer in the following discussion.

ELECTROPHORESIS AS A SURFACE PHENOMENON

Although recent preoccupation with electrophoresis concerns analytical- or preparative-scale separations of peptides, proteins, and nucleic acids dissolved in various buffers, it is important to understand that the electrophoretic phenomenon is far more general. For example, bits of iron oxide, cellulose, or linen fibers; fine droplets of oil; erythrocytes; and even *Escherichia coli* cells all have a surface charge. When dispersed in a buffer across which a field is applied, all of the above materials will move toward one electrode or the other. Proof that electrophoretic movement is a surface-related phenomenon comes from the following experiment. All of the above particles can be coated with a film of neutral gelatin. When this is done, all of the particles move toward the same electrode as would tiny bits of gelatin itself, even though, before coating, they might have moved in the opposite direction. Furthermore, when the pH of the buffer is varied, the rates of movement of the coated particles vary in a manner that parallels the rate of movement of gelatin itself, as the ionization of the exposed -NH_2 and -COOH groups on the gelatin is affected. When the *net* charge on gelatin is zero, movement ceases; the same is true of the coated particles.

If, instead of dispersing a solid as fine particles, one organizes it into an extended, semirigid gel, application of an electrostatic field will cause movement of fluid through the gel. This movement is known as *electroosmosis,* and its direction is almost always opposite to that of electrophoresis. The magnitude of electroosmotic movement depends on the surface charge of the gel relative to the fluid, as well as on the applied potential. Liquid can be pumped against the force of gravity by electroosmosis, until the hydrostatic pressure is just equal to the opposing force of gravity. Certain starches have excellent gelation properties, which make them potential candidates for particular electrophoretic applications, but they suffer from a considerable degree of electroosmosis by virtue of the -COOH groups they contain. Reduction of these groups with $NaBH_4$ converts the -COOH groups to CH_2OH moieties, reducing electroosmosis without damage to gelation properties. Similar treatment is sometimes applied to columns packed with finely divided cellulose.

CLASSIFICATION OF ELECTROPHORETIC METHODS

Many technologies have been developed for electrophoretic separations either as analytical or as preparative tools. These can be divided into two major classes, *frontal* and *zonal* procedures, or they can be classified as *one-dimensional* or *two-dimensional* procedures. For practical reasons, frontal methods are one-dimensional, but zonal methods may be one- or two-dimensional. As in other forms of separation science, the method is divided into three stages: (1) separation of a mixture, (2) detection of the separated materials and, where appropriate, (3) recovery of the separated products. Whereas the separation principle is the same for all forms of electrophoresis, principles governing detection and recovery differ with the technology employed.

BASIS OF ELECTROPHORETIC SEPARATION

In solution, and at low ionic strength, the mobility, u, of a charged particle is given by:

$$u = u_0 - u' = \frac{QD}{kT} - u' \tag{6–1}$$

where u_0 is the mobility in the absence of any other electrolyte, u' is the electroosmotic mobility, Q is the net charge on the particle and D is its diffusion coefficient, k, is Boltzmann's constant, and T is the absolute temperature. Since D is a function of other molecular properties, the electrophoretic mobility can be employed to estimate molecular weights, although this method has more recently been supplanted by other, simpler procedures. Equation 6–1 applies in all forms of electrophoresis.

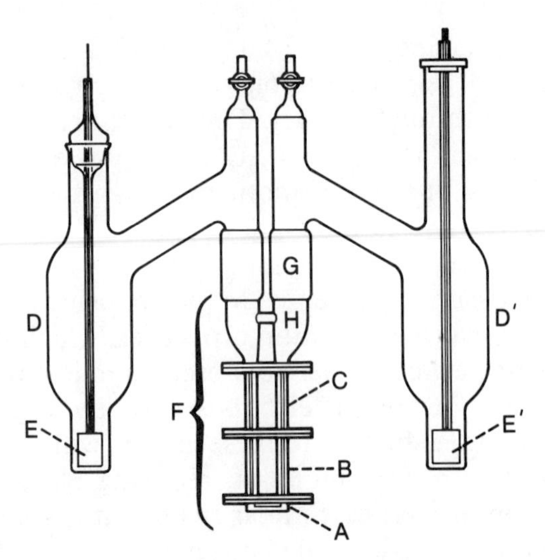

Figure 6–1. Complete Tiselius apparatus for free solution electrophoresis (frontal electrophoresis). The electrode vessels (D,D′) and electrodes (E,E′) are connected to the electrophoresis cell (F) by watertight rubber cuffs (G).

The cell segments (A, B, C, and H) end in plates ground flat and lubricated with a stopcock grease. The entire apparatus is mounted in a rack fitted with levers that can slide one cell section with respect to the others. With the cell segments aligned as shown in the figure, the U-shaped channel comprising A, B, and C are filled with a protein solution. A, B, and C are then moved laterally, as a unit, with respect to H, forming a sharp boundary between the plates joining C and H. Excess protein solution is removed from H, which is then filled with buffer and joined to the rest of the apparatus. The segments A, B, and C are then realigned.

By careful pumping of a little buffer into the top of vessel D, the sharp boundary is forced downward into the left leg of C, and current is then applied. The migration of individual molecular species from the sharp frontal boundary is followed by application of special optical methods known as *schlieren* techniques. Note that each species migrates into the protein solution.

FRONTAL ELECTROPHORESIS

The earliest versions of electrophoresis were performed in free solution and were applied almost exclusively to solutions of proteins. Tiselius, and later, Longsworth, developed an elaborate apparatus by means of which sharp fronts, or concentration boundaries, could be formed between a buffer solution free of protein and one containing protein (see Fig. 6–1). Protein species moved in a sharp band down through the original mixture. Detection of the separated fractions was accomplished by means of an

Figure 6–2. Three modes of electrophoretic separation of proteins. (A) Frontal electrophoresis; a typical Tiselius plasma protein pattern. (B) One-dimensional acrylamide gel autoradiograph of an extract of *E. coli* which were incubated with [^{35}S]methionine. (C) Two-dimensional acrylamide gel autoradiograph; isoelectric focusing separation in the horizontal direction, and acrylamide gel electrophoresis in the vertical direction. The sample in C is from a whole-cell lysate of cultured mouse smooth muscle cells incubated with [^{35}S]methionine. (Displays B and C are reproduced by courtesy of Dr. Peter Rubenstein, Dept. of Biochemistry, Univ. of Iowa.)

A

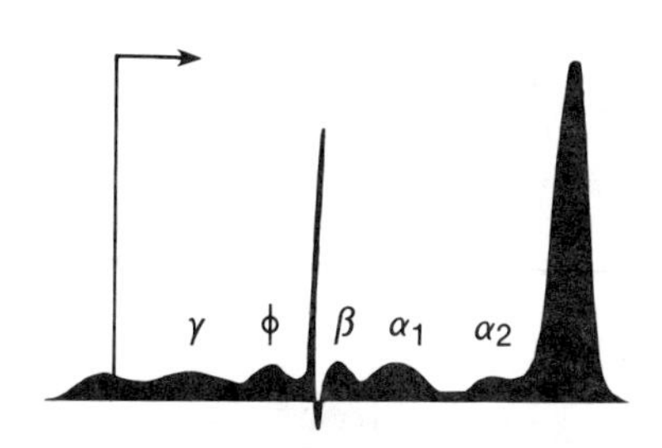

B

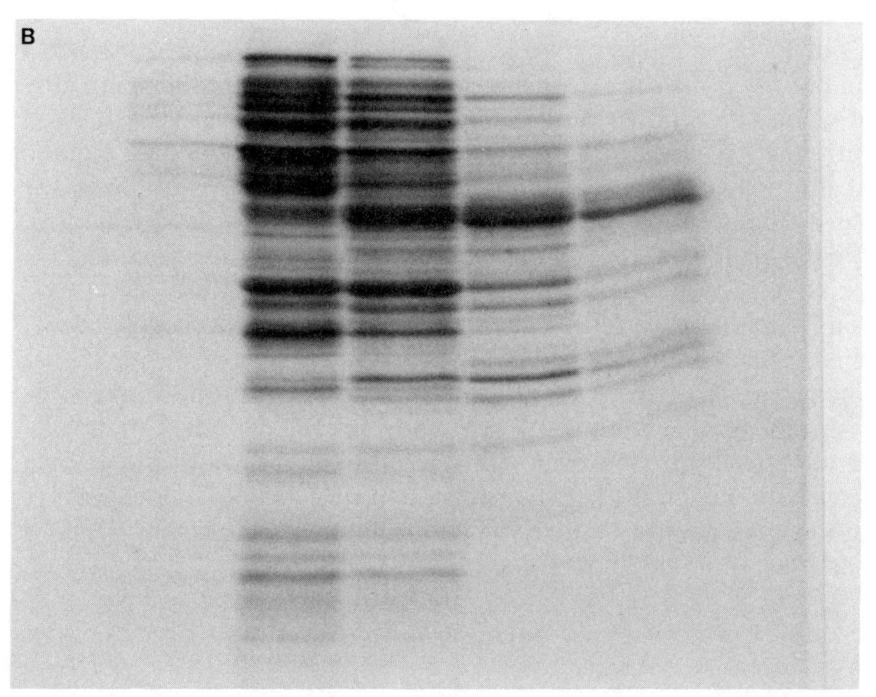

C

optical technique that depended on the refractive index gradient associated with the concentration gradient. When a finely corrected lens was placed immediately in front of the electrophoresis cell, the refractive index gradient in the solution acted as if it were a defect in the lens, causing dark bands to appear on the image of a light source projected through the cell and the lens. The term *frontal electrophoresis* takes cognizance of the dependence of this method on the front, or gradient, as opposed to the bulk of the protein solution.

Prior to the pioneering work of Tiselius, very little progress had been made in fractionating plasma or serum proteins into other than albumin and globulin fractions. With his apparatus, Tiselius was able to show distinctly the existence of albumin, α_1 and α_2 globulins, β and γ globulins, and fibrinogen (ϕ) in plasma—names by which these fractions are still known. Figure 6–2A shows the resolution of a typical Tiselius pattern. Note the "spike" in the β globulin, due to lipoproteins, which have a high specific refractive index as compared to nonlipoproteins.

The Tiselius approach, based on frontal analysis, has largely been abandoned. The apparatus was quite expensive, the cells were fragile, and the protein solution had to be of quite high concentration to give sufficient sensitivity to the method of observing refractive index gradients. The overall resolution was quite low because of diffusion and convection effects within even the best thermostatted cells. Where it does survive, the method is used almost entirely in connection with molecular weight determinations, as noted above.

Electrophoresis in Stabilized Media (Zonal Electrophoresis)

There is no theoretical reason why electrophoresis must be performed in free solution, and experiments quickly demonstrated that the adverse effects of diffusion and convection could be minimized by "stabilizing" the medium, that is, filling the system with a porous support through which the buffer ions and the solutes to be separated could penetrate. At first, strips of filter paper or cellulose acetate were employed as the mechanical support in which electrophoresis was performed. These media still enjoy a modest usage, but they have been superseded for most purposes by gels of various sorts. Gels suffer less than paper from the problems imposed by electroosmosis, and they frequently have more desirable physical properties.

Methods that rely on stabilized media are known as *zonal* methods, as distinguished from *frontal* methods, because detection is based on some physical characteristic of the mass of material separated, or on some specific chemical property. Some very ingenious detection schemes have been developed, a few of which are briefly sketched below.

1. One can measure the absorbance in either the visible or the ultraviolet regions of the spectrum, provided the background absorbance of the stabilizing medium is not too high. Special gel scanners are made for this purpose. The method works reasonably well with cellulose acetate strips or acrylamide gels, but not so well with other supports.

2. One can apply nonspecific stains that differentially bind to a protein or nucleic acid analyte, but not to the gel substance itself. Two dyes widely used with proteins are Amido Schwarz (also known as Buffalo Black NBR or Naphthol Blue Black, Color Index No. 20470) and the Coomassie Brilliant Blues (Brilliant Blue G, Color Index No. 42655 or Brilliant Blue R, Color Index No. 42660, of which the G form gives much the best results). For nucleic acids, good results can be obtained with Brilliant Green G (Malachite Green 1, Color Index No. 51010), or one can rely on fluorescence in the presence of ethidium bromide (see Chap. 11). Note that none of these is specific for any given protein or nucleic acid; they merely react with general structural features of either.

3. If the analytes can be labeled by feeding a radioactive amino acid to cultured cells, the isolated proteins can be detected by autoradiography; this process may require extensive pretreatment and drying of the gel.

4. If suitable antibodies to one or more of the separated proteins are at hand, immunoprecipitation methods can be employed for detection. Antibodies are sometimes tagged with fluorescent adducts to enhance detection procedures.

5. If one of the separated proteins has a known enzyme activity, it is sometimes possible to detect that activity by means of a specific reaction. The formazan procedure, based on use of tetrazolium salts, is typical of the specific reactions that are applied to certain dehydrogenases.

6. Of special interest to molecular biologists are the newer "blot" techniques, first developed by E. M. Southern. Single-stranded DNA is separated on a gel, which is then laid over a thick piece of filter paper soaked in a buffer. Atop this assembly is placed a sheet of nitrocellulose paper, also soaked in the buffer. The entire package is then covered with dry paper towels, or similar absorbent material. Buffer is drawn by capillarity from the bottom to the top of the stacked sheets, and the paper towels are frequently replaced.

The DNA flows with the buffer from the gel to the nitrocellulose to which, for reasons not clearly understood, it binds tightly. The cellulose acetate is then dipped into ^{32}P-labeled complementary DNA, which interacts with the DNA as expected. The sheet is washed, dried, and subjected to autoradiography. If there is not great homology between the DNA and the radiolabeled probe, there is no binding, and no spot on the autoradiogram.

When used with RNA fragments and a ^{32}P-DNA probe, the procedure is called a "Northern" blot. If proteins are separated on the gels, then transferred to nitrocellulose paper, and detected by the use of antibodies labeled in some fashion (see Chapter 10), the procedure is called a "Western blot."

APPARATUS FOR ZONAL ELECTROPHORESIS

Compared to frontal methods, zonal electrophoresis requires only the simplest apparatus, although many commercial models are fitted with a wide range of elegant accessories. Where paper or cellulose acetate strips are used as stabilizers, the strips can be suspended between two beakers that serve as electrode vessels, and are connected to a power supply. To dissipate the heat generated by passage of the current, the entire apparatus may be placed in a refrigerator or immersed in an oil bath provided with external cooling.

When gels are used as stabilizers, they may be prepared in glass tubes of 3- to 4-mm internal diameter and any convenient length, or they may be cast as thin (0.5–3.0 mm) slabs of any desired dimensions. The cast gel, still supported between two flat glass plates, may then be clamped to a stand fitted with electrode vessels that apply current across one or the other of the larger dimensions. The basis of a typical tube gel assembly is shown in Fig. 6–3. Ordinarily, a tube gel rack might hold 10–15 tubes, arranged about a central core through which the connections to the electrodes are made. (See also Fig. E7-1, p. 369.)

Several more sophisticated versions are also possible. For example, tube gels performed under isoelectric focusing conditions (i.e., with a pH gradient along the length of the gel) can provide one dimension of a two-dimensional separation. The gel is removed from its glass envelope, then held in place along one edge of a slab gel by a little molten agar. When current is applied, proteins move from their loci in the gel cylinder to new loci in the slab. Because the directions of movement are at right angles to each other, such systems are known as two-dimensional. A number of variations on this theme are now widely used, and a typical application is shown in Fig. 6–2C. Note the enormous increase in resolving power of the two-dimensional technique, compared to the methods discussed earlier.

Second only to the two-dimensional technique, and requiring only the very simplest of apparatus, are the "disc" methods of Ornstein (1964) and Davis (1964), who cast running gels of a moderately high concentration, then overlaid them with thinner segments of a gel at lower concentration. The term "disc" refers to the fact that the gels are *dis*continuous in concentration. Disc methods tend to give very finely discriminated zones of separated materials, and the bands appear to be compact and discoid in shape. The explanation of why disc methods work so well is somewhat complicated, and will be deferred until the generalities of gel behavior are discussed.

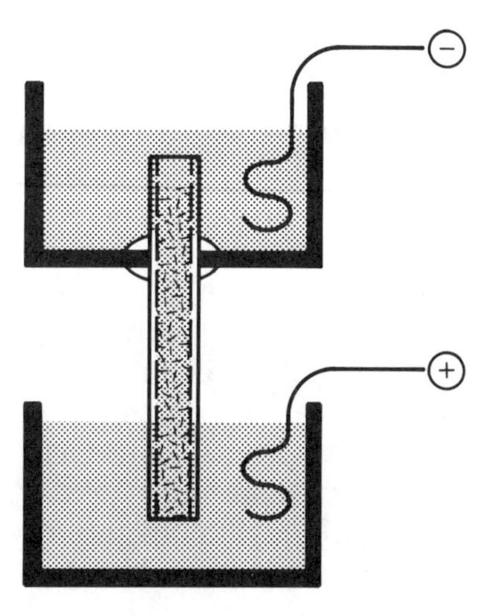

Figure 6–3. A simplified gel electrophoresis (zonal electrophoresis apparatus). A glass tube is filled with a gel, or stabilized medium, then passed through a watertight rubber grommet in the bottom of the upper electrode vessel. The lower end of the tube dips into the lower electrode vessel. When both vessels are filled with a buffer and the electrodes are inserted, a complete circuit is formed when power is applied. Note that the gel does not reach all the way to the top of the tube. The empty space in the tube, above the top of the gel column, is used to apply the sample after it has been made more dense than the buffer by addition of glycerol or sucrose. Commercial versions of this apparatus frequently have ten or more tube-holding grommets arranged in a circle about a core that keeps the two electrode vessels properly aligned.

GELLING AGENTS

A good many potential gelling agents were investigated as replacements for filter paper or cellulose acetate. Of these, only three remain prominent. Amylose and agarose are, respectively, the uncharged polysaccharides found in starch and agar. Polyacrylamide is a purely synthetic material, based on the controlled polymerization and cross-linking of a mixture of acrylamide and *N,N*-methylenebis(acrylamide) or *N,N*-diallyltartardiamide.

Amylose

Amylose and amylopectin are both components of starches. Amylopectin, like amylose, is composed of α-D-glucose units, joined by $\alpha(1\rightarrow4)$ glycosidic bonds, but amylopectin also contains $\alpha(1\rightarrow6)$ branch points, whereas amylose is a linear polymer. In addition, amylopectin carries a number of charged groups, and gels made from amylopectin show considerable electroosmosis. The proportion of amylose and amylopectin in a typical starch varies with the source. For electrophoresis one of the best sources is the potato, and suitable products are now commercially available. In its native state, the starch is packaged in granules, the size of which is actually less than the length of the amylose chain which, since it contains approximately 1000 glucose units, is approximately 500 nm long. Part of the processing is required to break open the granules, part is to remove the membrane debris, and part is to separate the amylose and amylopectin.

Amylose is not cross-linked, but it forms a left-handed helix containing six glucose residues per turn and with an inner diameter of about 14 nm. The blue color formed when amylose is exposed to iodine probably results from ordering of iodine molecules within the cavity of the amylose helix. Because the ability of amylose to form gels does not depend on cross-linking, it is argued that the length of the chains is the determining factor. It is surely true that partially hydrolyzed starch forms gums, but not gels. At concentrations of 6–10%, starch forms friable gels, and on cooling these become quite translucent rather than transparent. A second disadvantage of starch is that it requires fairly high temperatures (water virtually at the boiling point) to form the solution from which gels may be cast.

Agarose

This material is one of several components that can be separated from agar, the major source of which is certain species of seaweed. Agarose, like amylose, is a linear polymer that is nearly free of charged groups. Unlike amylose, it is not a glucose polymer, but contains alternating units of galactose and 3,6-anhydrogalactose, the repeating unit of which can be written as

$$[\beta\text{-D-Gal-}(1\rightarrow4)\text{-3,6-anhydro-}\alpha\text{-L-Gal-}(1\rightarrow3)]_n$$

Agarose has an average molecular weight of about 120,000, and thus its chains are not quite as long as amylose chains; nevertheless, agarose chains also form left-handed helices that appear to intertwine with one another. Perhaps this feature is what gives to agarose its truly remarkable gelling property. Solutions as dilute as 1% agarose give a fairly stiff gel, and concentrations of only 3% are used quite commonly. Agarose gels are less friable than amylose gels, and the softening point is also considerably lower. However, agarose gels of even 3% are completely transparent when cooled to room temperature.

Purification of agarose from raw agar is not quite as cumbersome as purification of amylose from potato starch. The ionic components and other impurities can be eliminated by fractionation with polyethylene glycol, but it is far easier to purchase the refined agarose as such.

Both agarose and amylose are natural products, and their properties may vary from lot to lot. Both are unbranched, linear polymers that share a helical structure with a rather large central diameter. Under normal conditions, neither agarose nor amylose gels show significant sieving activity, and virtually any molecular species added to an agarose gel will penetrate its very large pores. This is an advantage when one is working with nucleic acids, but it may be a disadvantage when one is working with many proteins.

Polyacrylamide

Acrylamide, $CH_2{=}CHCONH_2$, is the monomer on which polyacrylamide gels are based. Polymerization can be initiated by any of several free-radical mechanisms, two of which will be described below. In the absence of cross-linking reagents, long head-to-tail chains would result. Ordinarily, gels are cross-linked by addition of *N,N*-methylenebis(acrylamide) or *N,N*-diallyltartardiamide. The ultimate degree of polymerization is controlled by the concentration of acrylamide, and the degree of cross-linking by the proportion of acrylamide to bis-acrylamide or tartardiamide.

Acrylamide itself is reportedly a neurotoxin, and should be handled carefully, but the polymer is nontoxic. The gels in the range of 5–20% are quite firm and are completely clear, which is a distinct advantage. The ultraviolet absorbance is also quite low, provided the gels are cast from fresh solutions (aged solutions contain unknown materials that have a high absorbance in the ultraviolet, but are still clear in the visible range). Acrylamide gels adhere tightly to wettable surfaces such as glass. This is a useful property in preventing electrical short-circuits between a gel and its support, but can be something of a nuisance when one is attempting to remove gel cylinders from narrow tubes. Even more useful is the fact that polyacrylamide gels retain their transparency when they are dried, thus simplifying gel scanning and gel photography, especially of slabs. High-purity reagents for gel formation are commercially available.

CATALYSTS AND THE CHEMISTRY OF ACRYLAMIDE POLYMERIZATION

The polymerization of acrylamide is initiated by addition of, or generation in situ, of free radicals; the two most common catalysts, acting as the source of free radicals, are ammonium persulfate and riboflavin or riboflavin monophosphate.

Ammonium Persulfate

This reagent would be better identified as diammonium peroxydisulfate. It has the structure:

$$NH_4SO_3\text{-O-O-}SO_3NH_4 \quad \text{or} \quad (NH_3)_2S_2O_8$$

It is a powerful oxidizing reagent, and should be kept cool, dry, and away from organic matter. It is moderately unstable, and is preferably purchased in small quantities. When it is reduced in a two-electron reaction, persulfate is converted to sulfate, as:

$$S_2O_8^{2-} + 3\epsilon \rightarrow 2SO_4^{2-} \qquad \mathcal{E}^o = 2.01 \text{ volts}$$

Persulfate can also undergo a one-electron reduction, as:

$$S_2O_8^{2-} + 1\epsilon \rightarrow SO_4^{2-} + SO_4^{\dot{-}}$$

where $SO_4^{\dot{-}}$ represents a free radical. This is the effective species in persulfate catalysis of polymerization. The decomposition of persulfate is base-catalyzed, and is usually promoted by addition of the base *N,N*-tetramethylenediamine (TEMED), which has the structure:

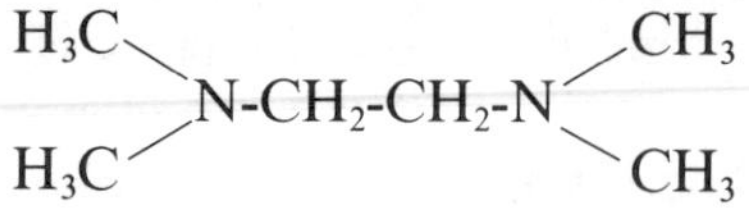

$$(H_3C)_2N\text{-}CH_2\text{-}CH_2\text{-}N(CH_3)_2$$

Riboflavin

Free riboflavin (or the more soluble riboflavin phosphate) has long been known to be light sensitive; riboflavin is also unique in that it acts as its own photo-reducing agent. In this process the ribityl side chain is cleaved, with production of lumiflavin and lumichrome. In the course of a complicated reaction sequence, a semiquinoid free radical is formed, as:

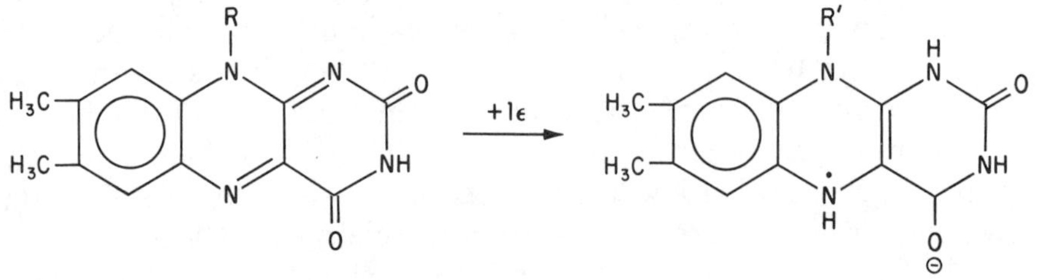

$$\xrightarrow{+1\epsilon}$$

When a riboflavin solution is irradiated, most of the intermediates can be detected by thin-layer chromatography, including the free radical (indicating that it is relatively stable), which makes riboflavin a good initiator of acrylamide polymerization at very low concentrations (as low as 1 mg/100 ml). Consequently, the breakdown products of the initiator have little or no effect on electrophoresis which is to follow. This is in marked contrast to the use of ammonium persulfate, where the accumulated products should be removed by preelectrophoresis prior to use of the gels. Preelectrophoresis avoids ionic effects of excess sulfate ion on the electrophoresis buffers and potential damage to the proteins by any undecomposed persulfate.

MECHANISM OF POLYMERIZATION AND CROSS-LINKING

On the basis of the discussion just presented, neither persulfate ion nor riboflavin is properly termed a catalyst; that role is reserved for the unpaired electron associated with a free radical. Furthermore, both persulfate ion and riboflavin are consumed in the polymerization reaction, which is inconsistent with the definition of a catalyst. It is certainly more correct to speak of persulfate ion or of riboflavin as an initiator; one can then generalize the polymerization mechanism in terms of the monomer(s), M, and the initiator, I.

The simplest case to consider is a mixture of acrylamide and some initiator. A set of reactions can be written, as:

$$\mathrm{M} + \mathrm{I} \rightarrow \dot{\mathrm{R}}$$
$$\dot{\mathrm{R}} + \mathrm{M} \rightarrow \mathrm{R}\dot{\mathrm{M}}$$
$$\mathrm{R}\dot{\mathrm{M}} + n\mathrm{M} \rightarrow \mathrm{R}\dot{\mathrm{M}}_{n+1}$$
$$\mathrm{R}\dot{\mathrm{M}}_{n+1} + \mathrm{R}\dot{\mathrm{M}} \rightarrow \mathrm{M}_{n+2}$$

where the dot over a symbol represents an unpaired electron, resulting in formation of a free radical. In specific terms, where the monomer is acrylamide, the chain will grow in a head-to-tail fashion, represented as:

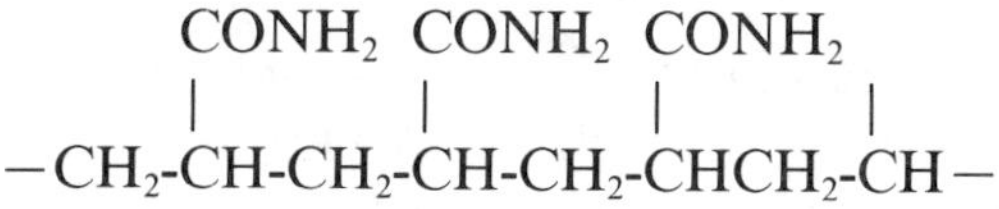

```
      CONH2   CONH2   CONH2
      |       |       |       |
-CH2-CH-CH2-CH-CH2-CHCH2-CH-
```

until all of the monomer has been incorporated into lengthy strands. Since acrylamide has only a single reactive site, there is no opportunity for cross-linking, and the result will be a very loose mass rather than a firm gel. Its properties are not suitable for electrophoresis. This explains the need for a bifunctional cross-linking reagent, such as bisacrylamide.

When acrylamide, bisacrylamide, and an initiator are mixed, both head-to-tail chain extension and cross-linking will occur on a probabilistic basic, conditioned by the relative concentrations of acrylamide and bisacrylamide, and by the total quantity of acrylic groups of either form in the mixture. A microscopic bit of the gel might be represented as follows:

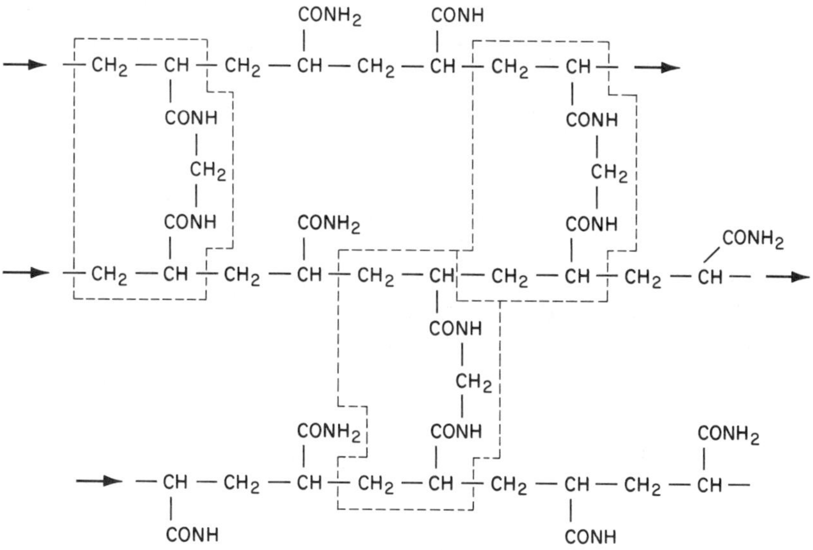

```
                             CONH2        CONH
                               |            |
→  CH2 — CH — CH2 — CH — CH2 — CH — CH2 — CH —  →
          |                                  |
        CONH                               CONH
          |                                  |
         CH2                                CH2
          |                                  |
        CONH          CONH2                CONH                CONH2
          |             |                    |                /
→  CH2 — CH — CH2 — CH — CH2 — CH — CH2 — CH — CH2 — CH —  →
                                 |
                               CONH
                                 |
                                CH2
                                 |
                  CONH2        CONH                      CONH2
                    |            |                         |
→ — CH — CH2 — CH — CH2 — CH — CH2 — CH — CH2 — CH —
     |                                  |
   CONH                               CONH
```

The arrows show the direction of acrylamide chain lengthening, and the boxed areas show the cross-links caused by incorporation of the bisacrylamide in the polymerization process.

CONDITIONS AFFECTING ELECTROPHORETIC SEPARATIONS

Effect of Acrylamide/Bisacrylamide Ratio

When mixtures of the two monomers are polymerized, the resulting gels can be examined for pore size by the exclusion principle. That is, molecules of known size can be percolated through the gel, and, from the retention data, the pore range can be estimated. Careful studies of acrylamide gels have shown that the pore size range is an exponential function of the ratio of bisacrylamide to total acrylamide when expressed on a weight basis. Furthermore, the exponential relation has a distinct minimum when the bisacrylamide constitutes about 5% of the total. For this reason, most gel recipes call for a 1:20 ratio of bisacrylamide to acrylamide, no matter what the total concentration might be. A plot of the weight ratio effect is shown in Fig. 6–4. Since cross-linking occurs on a random basis, the pore sizes noted are averages, and many of the individual pores might be distinctly smaller or larger. The data are therefore expressed as an average pore size for which half of the gel volume is penetrable by a given molecular species.

Effect of Total Acrylamide Concentration

Morris and Morris, in a comprehensive study, examined eight typical proteins with molecular weights that ranged from 17,000 to 160,000, in twenty different formulations of polyacrylamide gel. Both T, the concentration of total acrylamide, and C, the proportion of cross-linking reagent (bisacrylamide), were systematically varied. At constant values of C, the measured mobility for any of the proteins is related to T by the equation:

$$\ln u' = \ln u_0 - K_r T \qquad (6\text{–}2)$$

or

$$u'/u_0 = U_r = e^{-K_r T} \qquad (6\text{–}2a)$$

where u' is the measured mobility under the experimental conditions, u_0 is the mobility calculated by extrapolating T to zero (in other words, a value essentially identical with the mobility in free solution), K_r is a constant for gels of constant cross-linking, U_r is known as the reduced mobility, and T is the total acrylamide concentration in the gel, as already defined. Equation 6–2a, or something very much like it, has been independently confirmed by others, and it applied to starch gels as well as to acrylamide gels.

The sense of Eq. 6–2a can be put very simply: For any protein, examined in a series of gels with constant cross-linking, the mobility decreases as an exponential function of T. Increased total acrylamide concentration results in diminished pore size. The path of a given molecule through the gel is therefore more tortuous, and that molecule will take longer to travel

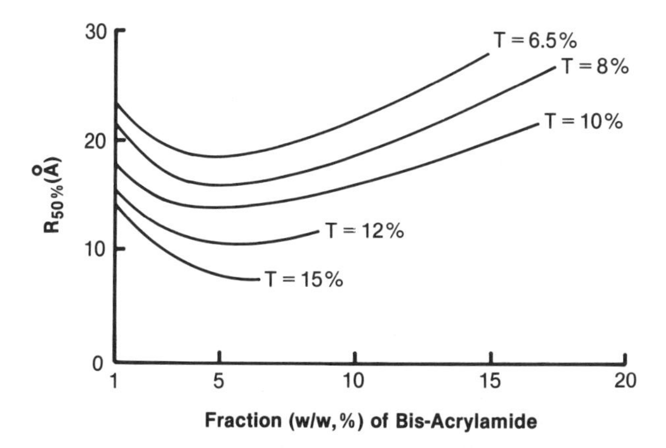

Figure 6–4. Effect of bisacrylamide/acrylamide ratio on gel pore size range. $R_{50\%}$ is the size of molecules to which half of the gel volume is available, and the T value associated with each curve is the total concentration of acrylamide groups in the monomer mixture. The minimal values of $R_{50\%}$ at about a 5% bisacrylamide concentration are clearly apparent. (Modified from data presented by Morris and Morris, 1976.)

through the length of the gel. Excessive cross-linking may reduce the pore size to the extent that very large molecular species may not be able to penetrate at all.

There is a fundamental homology between electrophoresis and exclusion chromatography. Morris, and others, have attempted to develop a universal theory that would include both separation modes. The argument is that electrophoresis could be regarded as a special case of exclusion chromatography where the force of gravity is replaced by an electrical force. The constant, K_r of Eq. 6–2a has been likened to the partition coefficient, K, of exclusion chromatography. For values of C from 5% to 20%, the agreement between K_r and K is quite good, but for $C < 5\%$ the agreement breaks down. In any event, it is clear that the structure of the gel in electrophoresis has a profound effect on separations, as is true in exclusion chromatography.

Effect of Additives

To solubilize and/or to denature certain proteins, it is common to add reagents such as urea or guanidine hydrochloride in high concentrations (4–8 mol/L), or detergents such as sodium dodecylsulfate. These reagents have only very slight effects on the polymerization time of acrylamide. Published reports attest to the fact that detergents may generate anomalous bands by reacting with lipids present in some membrane preparations. The buffer ions selected for electrophoretic systems may also produce artefactual results by combining with some component of the unknown mixture. For example, borate buffers can interact with the carbohydrate moieties of glycoproteins. One must constantly be aware of these possibilities.

Acrylamide Gradient Gels

In the discussion thus far, the total acrylamide concentration of the gels has been kept constant throughout the gel. It is quite useful to prepare gels with a continuously varying acrylamide concentration, thereby imposing a continuously changing sieving effect in addition to the charge effect. This is especially useful when one is studying preparations that may contain proteins ranging in size from very small to very large.

The preparation of gradient gels is reminiscent of density gradient preparation for centrifugation. A mixture of acrylamide, bisacrylamide, and an appropriate initiator is mixed at five times the normal concentration. This mixture is placed in the stirred chamber of a gradient maker. The unstirred chamber is filled with buffer, which dilutes the acrylamide mixture gradually as material is withdrawn to the gel tubes or slab. It is not at all difficult to prepare acrylamide gels where the concentrations, over a height of perhaps 15 cm, range from 5% at the top to 20% at the bottom. It is necessary, of course, to calibrate each gel individually by running a known series of marker proteins along with the unknown samples. Once the best gradient for a particular task has been estimated by preliminary experiments, one can tailor the gradient to cover most of the tube or slab length for the finest resolution.

Gel tubes or slab holders filled with gradient mixtures should be allowed to stand undisturbed until the gel has completely set, and they should be filled only in draft-free environments to avoid convection mixing.

Effect of Heat

Heat is the enemy of electrophoresis. In frontal electrophoresis, heat causes convection and destroys the sharp refractive index boundary which is the usual basis of measurement. In zonal electrophoresis, heat causes increased diffusion, with distortion of the separated zones. In agarose gels, it may even cause melting of the gel, since agarose softens at quite low temperatures. Polyacrylamide gels are not likely to melt under most circumstances, but sensitive proteins may be denatured in the event of severe overheating.

According to Ohm's law, $E = IR$, where E is the applied voltage, I is the current flowing through the system, and R is the resistance of the medium. R is determined by the conductivity of the medium and, for practical purposes, may be regarded as constant for any given experimental condition. The energy passed through the gel is $EI = I^2R$, virtually all of which must be dissipated as heat. Therefore, one must control the current flowing through the gel by adjustment of the voltage, and this poses some stringent limits on the dimensions of gels in the absence of external cooling. Other things being equal, the current through a gel is proportional to the cross section of the gel. It is possible to use gels of large diameter for preparative purposes, but only if they are provided with elaborate cooling systems. Preparative columns are sometimes designed with annular

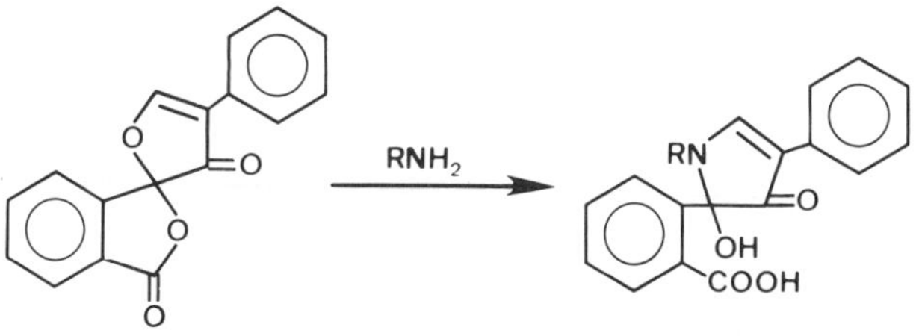

Fluorescamine Fluorescent Product

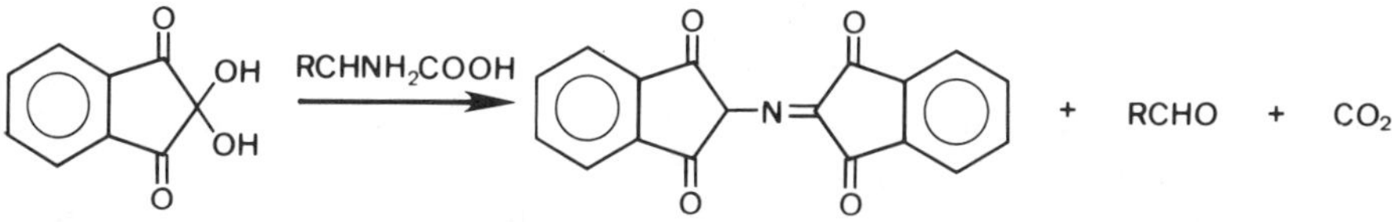

Ninhydrin Purple Pigment

Figure 6–5. Fluorescamine and ninhydrin reactions. Fluorescamine reacts with primary amines or with primary amino groups of peptides and proteins to give an intensely fluorescent product. It can detect a few nanograms of material and reacts especially well with the ε-amino group of lysine.

Ninhydrin is a powerful oxidant and causes decarboxylation of amino acids and peptides. An intermediate, hydrindantin, reacts with NH_3 and a second mole of ninhydrin to give the purple or bluish pigment. The remainder of the amino acid is converted to an aldehyde and CO_2, as shown. Ninhydrin will react with amino acids and peptides on the skin, staining it as well as any unknown sample; thus one should avoid contact with solutions or sprays.

cross sections, and cooling water can be pumped through the hollow center as well as around the outer surface. Such systems are expensive and are too complicated for ordinary analytical use.

The more sophisticated power supplies now sold for electrophoresis frequently include current-limiting (constant current) circuitry; this is especially convenient for isoelectric focusing (see below) in which the pH gradient is established by electrical separation of the ampholytes. To be done most reproducibly, two-dimensional electrophoresis requires precise knowledge of the watt-seconds of energy passed through the isoelectric focusing gel.

High-Voltage Electrophoresis

In spite of the limitations imposed by ohmic heating in starch, agarose, or polyacrylamide gels, high-voltage electrophoresis is still a valuable tool when employed with large paper sheets. Small peptides, frequently generated by limited proteolysis of proteins, can be regarded as "fingerprints" of the original protein. Study of the peptide products is frequently helpful in establishing the primary sequence of the protein, or in determining its amino acid concentration by subsequent amino acid analysis of the recovered fragments. After electrophoretic separation, the peptides can be detected by staining with fluorescamine or with ninhydrin. The reactions with these reagents are shown in Fig. 6–5.

Small molecules such as peptides diffuse more rapidly than larger ones, so it is essential to perform the electrophoretic separation as rapidly as possible. To do so, it is not unusual to apply voltages of up to 5000 volts across sheets of filter paper up to 50 × 50 cm in size. With dimensions of this magnitude the resistance is usually high enough to limit the current within safe bounds even at the high voltages employed. In theory, gel layers as thin as filter paper would be equally effective, but such thin gels would be difficult to prepare and would lack the physical strength of paper sheets. Fingerprint studies may be performed by electrophoresis along one dimension of the paper, followed by chromatographic separation along the other dimension, or they may be done by electrophoresis along both dimensions. In the latter case, they amount to two-dimensional electrophoresis on paper.

Heating is still a problem, even with paper sheets. Consequently, high-voltage electrophoresis is performed in tanks filled with a hydrocarbon oil to help dissipate the heat. In spite of this precaution, heating may cause water to distill off the paper sheet, raising the buffer concentration and increasing the current flow. Accidents are not uncommon; it is unfortunate and potentially quite dangerous to have a critical separation char for lack of proper cooling, or have the oil catch fire. With the advent of high performance liquid chromatography (HPLC) methods, high-voltage electrophoresis is not as widely employed as it once was. HPLC methods are faster, more sensitive, and are free of the hazards noted above. Nevertheless, high-voltage electrophoresis remains useful in skilled and experienced hands.

GEL ELECTROPHORESIS UNDER DENATURING CONDITIONS (SDS-PAGE)

Typical proteins may have thirty or more centers of positive and negative charge, contributed by ionizable groups dependent on the primary structure. The net charge on the protein is a function of the pK_a values of these groups and on the pH of the solution. Most globular proteins are folded so that the majority of the ionized groups, being hydrophilic, face the exterior, whereas the uncharged, hydrophobic residues tend to be buried in the interior. Additionally, hydrogen bonds and disulfide bonds contribute to specific folding patterns and to forces that hold subunits in the aggregated state.

The detergent sodium dodecylsulfate (SDS) forms complexes with most proteins, coating the protein surface with many SDS molecules. The precise explanation for this is not clear; it is presumed to involve attractive forces between the alkyl moiety of the detergent and apolar portions of the protein surface.

Weber and Osborn were the first to note what has since been confirmed by many others: that the charges introduced by the dodecylsulfate ions vastly exceed the limited number of charges intrinsic to the protein, so that the SDS–protein complexes, when subjected to electrophoresis, act as if they all had a large, negative charge. The only factor remaining to separate such proteins is their *size* relative to the size of the gel pores. The size of proteins frequently correlates well with their molecular weight, and the method of Weber and Osborn has been quickly adopted as a means of estimating (with an accuracy of about $\pm 5\%$) the molecular weights of proteins and large peptides. This method, known as SDS–polyacrylamide gel electrophoresis, is frequently identified by the acronym SDS–PAGE. The detergent breaks hydrogen bonds and causes a considerable degradation of tertiary structure, but it does not affect disulfide bonds. Addition of either β-mercaptoethanol or dithiothreitol reduces disulfide bonds, completing the denaturation process for most globular proteins. Molecular weight estimations therefore include heating of an unknown sample in a mixture of SDS and one of the thiol reagents, followed by electrophoretic separation. This method is suitable for peptides or proteins with molecular weights in the range 12,000–70,000. The native weights of the parent proteins may, of course, be considerably larger.

A word of caution is necessary. Although molecular weight estimation by SDS–PAGE is quite useful, it is not absolute. Some proteins, especially glycoproteins, show anomalous behavior, for reasons not totally clear. Especially at the high end of the molecular weight range, some curvature of the otherwise linear relation between the mobility in the gel and the log of the molecular weight may be noted. It is always necessary to calibrate a given gel with samples of known molecular mass as markers, and kits of especially purified proteins are sold for this purpose. Only SDS of the highest-purity is suitable for SDS–PAGE, and if there is any doubt about the quality of the reagent it should be recrystallized from hot ethanol.

At reduced temperatures (below about 10°C) and at high concentrations, SDS may crystallize out; lithium dodecylsulfate (LDS) may be substituted, although at increased expense.

Estimation of Mobility: Use of Tracking Dye

The rule proposed by Weber and Osborn can be formalized as follows:

$$U = \frac{u_x}{u_0} = \log \mathrm{MW} \tag{6–3}$$

where u_x is the distance traveled by the unknown protein; u_0 is the distanced traveled by some standard substance, the nature of which will shortly be discussed; and MW is the molecular weight of the unknown.

The standard used for mobility measurements according to Eq. 6–3 is commonly bromthymol blue (MW = 624). Its molecular weight is very much lower than the molecular weights of proteins; for practical purposes one may assume that the indicator travels through the gel with no impediment whatever, and that the distance it travels reflects the travel of any small ion front. Proof of this argument comes from the fact that many other substances of about the same small molecular mass can be substituted for bromthymol blue.

When used for this purpose, bromthymol blue is described as a *tracking dye,* since it allows the observer to observe the rate of travel of one of the fastest ions in the gel, or to "track" the movement of the colorless proteins which are bound to move more slowly. As long as the tracking dye remains visible in the gel, one can be certain that none of the proteins has migrated out of the gel and into the electrode vessel. Without some fast-moving dye, it would be impossible to determine when a suitable end of electrophoretic movement had been reached, since most proteins and peptides are colorless. In SDS–PAGE, the use of a tracking dye is of fundamental importance, as shown by Eq. 6–3, but visualization of migration is of almost the same importance in all forms of electrophoresis, and the addition of a tracking dye to sample mixtures is therefore a commonplace procedure. If all of the proteins are chromoproteins, such as hemoglobins, then the use of a tracking dye is less important for visualization, but is still essential for molecular weight estimations.

ISOELECTRIC FOCUSING AND RELATED VARIANTS OF ELECTROPHORESIS

Isoelectric Focusing (IEF) is a procedure for separation of proteins that takes advantage of the differences in net charge on protein molecules due to differences in their isolectric pH. Our previous discussion of electrophoresis showed that negatively charged particles migrate toward the anode in an electric field; positively charged particles move oppositely. In a solution of fixed pH, proteins generally find themselves on the alkaline or the acid side of their isoelectric points: If on the alkaline side, the protein will be negatively charged; if on the acid side, the protein will be positively charged. If, however, the pH of the solution coincides with the isoelectric point of a protein, that protein will not migrate at all, since it lacks the net charge necessary to interact with the applied field. This is the basis of IEF.

Imagine a series of glass compartments, connected by membranes permeable to proteins. Visualize the chambers filled, sequentially, with a series of buffer solutions, each at a different pH, uniformly covering the pH range 2–10. A positive electrode is placed in the most acid solution and a negative electrode in the most basic. If a protein is now placed in any of the chambers, it will assume a certain charge, depending on the exact relation between its isoelectric point and the pH within the compartment. If a field is applied, the protein will migrate through the series of compartments until it reaches its isoelectric point, and there it will stop. Such an imaginary system would work to separate protein mixtures: The larger the number of compartments and the smaller the pH difference between adjacement chambers, the more efficiently the system could be used to fractionate protein mixtures as a result of even slight differences in their isoelectric points. A system somewhat like the one described above has been constructed and applied to separation problems, with limited success.

If, instead of using buffers with discrete pH differences and instead of separating them by membranes, we could remove the barriers and make a stable and continuous pH gradient, the resolution achievable in separating proteins would be enormously increased. Imagine now a different system. A mixture of ampholytes, such as the natural amino acids, is dissolved in water and poured into a long, horizontal glass tube or trough. One end of the container is connected to a dilute solution of a strong acid, the other to a solution of a strong base. The acid contains an anode and the base a cathode. When an electric field is applied, the ampholytes migrate, depending on their net charge, until they come to a region of the trough where they are at their isoelectric points. With a sufficient number of ampholytes at adequate concentrations, the entire chamber will contain a more or less continuous distribution of these species along its length. The gradient will be reasonably stable as long as the field is applied. A protein introduced into this gradient also migrates until it comes to its own isoelectric point; then it remains stationary. A mixture of proteins is similarly separated. The quantity of protein added must not be so great as to override the buffering effects of the smaller ampholyte species, or the pH gradient will be significantly distorted. Systems of this sort have also been constructed and have been moderately successful.

In either of the theoretical cases described above, as well as in the form of IEF now widely used, three major problems must be carefully addressed:

1. *Diffusion or convective disturbance:* What happens when a molecule of a separated protein diffuses some slight but finite distance from a small volume element where its isoelectric point is identical with that of the solution? If it moved into a more acid region, it would promptly assume a net positive charge, causing it to move back toward the cathode. A similar but opposite result would occur if it moved toward a more basic pH. The protein, in effect, is restricted to a very narrow band within the column or gel. This is how the process first came to be known as "focusing" and it explains why the resolution of IEF can be made so great as compared to other separation methods. Note, however, that once the field has been removed, diffusion and/or convection can operate to degrade the resolution of the separated proteins. For these reasons, IEF gels designed for qualitative analysis should be fixed promptly, to preserve the integrity of separation. Preparative-scale IEF separations should also be cut promptly into segments for elution and recovery of isolated materials.
2. *Stability of the pH gradient, once formed:* Stability can be improved by forming the gradient in solutions made viscous with sucrose. More practically, diffusion and convection effects may be minimized by

forming the gradient in a polyacrylamide gel; this is the alternative now in widest use.

3. *Selection of suitable ampholytes:* Use of the natural amino acids was discussed previously. Unfortunately, the isoelectric points of the natural amino acids do not fall in a way that makes it practical to cover the pH range 2–10. Furthermore, some of the natural amino acids interact with certain proteins, which is undesirable. For these reasons, and others, a series of synthetic polyelectrolytes, or *ampholytes,* has been developed. These are commercially available in broad-range (6–8 pH units) mixtures or in narrow-range (2–4 pH units) mixtures, as needed. The ampholytes in most general use are polyaminopolycarboxylic acids (LKB Ampholines®) or polyaminopolysulfonic acids (BioRad Bio-Lytes®). Details of these preparations are proprietary secrets.

Desirable Characteristics of Ampholytes

Suitable ampholytes should have the following properties:

1. They must be quite soluble in order to provide adequate buffering capacity.
2. They must not interact with proteins in a way that could affect protein migration.
3. They must have titration characteristics such that, at zero net charge, they have little buffering action. On either side of the zero charge state, however, they must provide sufficient buffering to affect protein charge and must have sufficient conductivity to "carry" the protein.
4. They must be available as a series of molecules that provide adequate and fairly uniform coverage of a pH interval.
5. They should be compatible with materials in use to stabilize the pH gradient. This means that, in the main, they must be compatible with the components of acrylamide gels, with agarose gels, and with sucrose.
6. They should not have appreciable light absorbance at wavelengths higher than 260 nm, to avoid interference with the detection and determination of most proteins.
7. They must have suitable molecular weights. The molecular weights should be low enough for the ampholytes to be removed from proteins by dialysis or gel filtration, and be high enough to make diffusion of the ampholytes themselves a minimal problem when the gels containing separated proteins are handled. (The typical molecular weight of the commonly available commercial preparations is in the range of $3–5 \times 10^3$.)

Resolvability of Proteins

The theory of IEF electrophoresis is reasonably well understood, and the major factors that determine ability to resolve one or more proteins with similar isoelectric points have been identified. This is illustrated by the following equation (for derivation of this equation, see H. Haglund, 1971):

$$\Delta \text{pH} = K \sqrt{\frac{D\, d\text{pH}/dx}{-E\, du/d\text{pH}}} \tag{6-4}$$

where the symbols are defined as follows:

ΔpH = the difference between the isoelectric points of two proteins that is necessary for them to be sufficiently distinguished from each other

K = a system constant, usually 3.07

D = the diffusion constant of the protein (in cm^2/sec)

x = coordinate along the direction of travel (in cm)

E = field strength at the isoelectric pH of the focus region (in volts/cm)

u = electrical mobility at a pH close to the isoelectric pH (in cm^2/volt · sec)

The term, dpH/dx, depends on the selection of ampholyte. A narrow-range preparation yields a higher resolving power than a broad-range preparation. The smaller the value of ΔpH, the greater the resolvability of one protein from another.

Properties of the Stabilizing Medium

Mention has already been made of the use of sucrose or polyacrylamide gels. The gel technique has virtually replaced the sucrose method except for the most unusual situations (see specific references at the end of this chapter). The polyacrylamide gel concentration may be varied, as in the standard PAGE technique, and either ammonium persulfate or riboflavin may be used to induce polymerization. Gels may be cast in tubes or in flat slabs of varying size. Additionally, a finely granulated form of polyacrylamide gel (e.g., Bio-Gel® beads) may be loaded into flat-bed slab trays.

Cooling of IEF Systems

Because of the high voltages and not insignificant currents (compared to standard PAGE), cooling of IEF gels is quite important. Cooling must be efficient, particularly in the preparative-scale systems. Ingenious apparatuses have been developed, including annular columns, through the center of which cooling liquid may be circulated. It is important that the concentration and range of ampholytes present be sufficient to prevent formation of any regions of low conductance (and corresponding high resistance); such a condition would greatly increase the heat load that would have to be dissipated. Good cooling also assists in stabilizing the proteins during the focusing period.

Handling the Proteins

Some proteins tend to precipitate in focusing systems, in part because proteins are usually least soluble at their isoelectric pH. To prevent precipitation, solvents such as glycerol, formamide, dimethylformamide, or moderate concentrations (2–4 M) of urea may be added to a gel component. These additions diminish the likelihood of precipitation, which may be accompanied by denaturation.

Proteins may be recovered from IEF gels after focusing is complete by slicing the gel and macerating the slices in a suitable buffer to extract the protein. If a granular gel bed is employed, the beads are simply scooped out of the gel tray and similarly extracted. Unfortunately, some proteins are incompletely extracted from the gel beads. More recently, electrodialysis methods have proved effective. In either case, the proteins recovered must be freed of ampholytes, (See also below, "Recovery of Material from Electrophoresis Gels.")

Measuring the pH Gradient

It cannot be assumed that the gradient actually achieved in a given gel is precisely determined by the nominal pH range of the chosen ampholytes. Direct measurement is necessary and may be performed as outlined below. A sample of the gel, usually in an extra tube or an empty slab lane reserved for this purpose, is sliced into strips 1–2 mm wide. These slices, numbered from the end where material would normally be applied, are extracted with minimal volumes of freshly boiled and cooled water. The pH of these extracts is then measured in the usual way. Preboiling of the water is necessary to avoid effects due to dissolved carbon dioxide.

Special pH electrodes are available with small, flat ends. By means of these, applied directly to the surface on intact gels, the pH gradient may be estimated quite conveniently especially with slab gels.

Quantity of Protein that Can Be Separated

For analytical purposes, small tube or slab gels may be prepared. Typical loads range from 10 to 50 μg of sample, as in standard PAGE. For preparative purposes, large slabs or columns may be prepared, to which 10–20 mg may be applied at a time.

Two-Dimensional Electrophoresis (IEF + SDS–PAGE)

IEF electrophoresis exploits the differences in protein isoelectric points, whereas SDS–PAGE electrophoresis exploits the differences in protein molecular weights. These properties of proteins are essentially independent of each other, so a combination of the two forms of electrophoresis constitutes a powerful tool for the very high resolution of proteins in various biological systems (see previous section, Apparatus for Zonal Electrophoresis).

In the technique first described by O'Farrell, a protein mixture is initially separated by IEF performed in a long, small-diameter glass tube. This provides the first dimension of the separation desired. The IEF gel is removed from its glass tube and placed along one edge of an SDS–polyacrylamide gel slab. Good physical and electrical contact is ensured by pouring molten agarose gel between and over the two. When an electrical field is applied to this system, separation occurs as the proteins, initially classified in the IEF gel by differences in isoelectric point, now migrate as a result of differences in their molecular weights. This provides the second dimension of the separation. The final separation may then be visualized by any of the usual methods—autoradiography, fixation, and staining—or even by UV absorbance measurements. Many hundreds of discrete spots can often be seen in two-dimensional displays. A modification of the O'Farrell system suitable for separation of membrane proteins has been described by Ames and Nikaido. Machinery has been developed for automatic scanning of two-dimensional gels; this will greatly enhance its applicability. Unfortunately, identification of the dispersed spots seriously lags behind ability to detect them.

DISCONTINUOUS GEL ELECTROPHORESIS SYSTEMS

In order to understand fully the behavior of discontinuous gel systems, it is necessary to examine the distribution of small ions (buffer ions) in an electrical field in the absence of any gelling agent whatever. Even before the turn of this century, Kohlrausch and others demonstrated that when one electrolyte solution was layered over another, forming a sharp boundary, the passage of current through the system could cause that boundary to move, more or less unchanged, in one direction or the other. This phenomenon can be usefully applied as a zone-sharpening device in discontinuous electrophoresis. For simplicity, the analysis that follows is couched in terms of uni-univalent electrolytes, but the reader must understand that the basic theory applies equally to Tris as well as to K or Li ions, and to acetate, borate, glycinate, and so forth as well as to Cl ions; that is, to all of the ions likely to be encountered in electrophoresis buffers.

Transferance Numbers and the Migration of Ions

The quantity of electricity, q_i, carried through a unit volume of an electrolyte solution by ions of the ith kind is proportional to the concentration, c_i; to the charge as the ion, z_i; and to mobility of the ion, which is its velocity per unit potential gradient, u_i. Accordingly, for any single ion:

$$q_i = kc_iz_iu_i \tag{6–5}$$

while for all of the ions taken together:

$$Q = k\Sigma c_i z_i u_i \tag{6–6}$$

The fraction of the total current carried by the *i*th ion is therefore:

$$q_i/Q = t_i \tag{6–7}$$

where t_i is known as the *transferance number* of that ion. In the case of a simple electrolyte, such as KCl, $t_+ + t_- = 1$, whereas in the case of electrolytes $\Sigma t_+ + \Sigma t_- = 1$, but in neither case does this mean that t_+ is necessarily the same as t_-. This follows from the fact, set forth in Eqs. 6–5 and 6–6, that the velocities of ions are not all the same. At the same time, it is clear that one cannot have great charge separations in a solution, and that on average the number of positive and negative charges must be equal. This principle of charge equality can be reconciled with Eq. 6–5 by what is known as the Kohlrausch regulatory function, which states that:

$$\frac{t_+}{c} = \frac{t_-}{c'} \tag{6–8}$$

where c is the concentration of the positive ion and c' is that of the negative ion. In other words, the concentration of an ion at a boundary is self-regulatory in terms of the transferance number, which tends to maintain the boundary during the passage of current. Since, in electrophoresis, proteins act as ions, it is not at all surprising that they are constrained to obey the Kohlrausch regulatory function as well.

Figure 6–6 shows an apparatus used for determination of transferance numbers. The similarity of this device to apparatuses for tube gel electrophoresis (Fig. 6–3) is readily apparent. Suppose one wished to study the transferance numbers of K^+ and Cl^-; the lower bulb of the apparatus would be filled with a solution of potassium acetate (KOAc), and this would be carefully overlaid with a solution of KCl to form a sharp boundary between them. The upper bulb would then be filled with a solution of LiCl until the electrode was immersed. For simplicity, the solutions could all be of the same concentration, although this is not necessary. For convenience in formation of the boundaries, the densities of the solutions could be adjusted by addition of glycerol or by adjustment of the electrolyte concentrations themselves. The initial boundaries are commonly observed by noting differences in refractive index or by some other suitable scheme. If certain conditions are met, passage of current through the system will result in shifting boundary *a* to *a′* and boundary *b* to *b′*:

First condition: The velocity of Li^+ must be less than that of K^+; similarly, the velocity of the OAc^- must be less than that of Cl^-. If this condition is met, the indicator ions do not catch up with K^+ or Cl^-. For this reason, the indicator ions are sometimes known as "following" or "trailing" ions, whereas the ions of interest are known as "leading" ions.

Second condition: In the immediate neighborhood of the boundary, *a′*, the cations Li^+ and K^+ must have the same velocity, and a similar situation

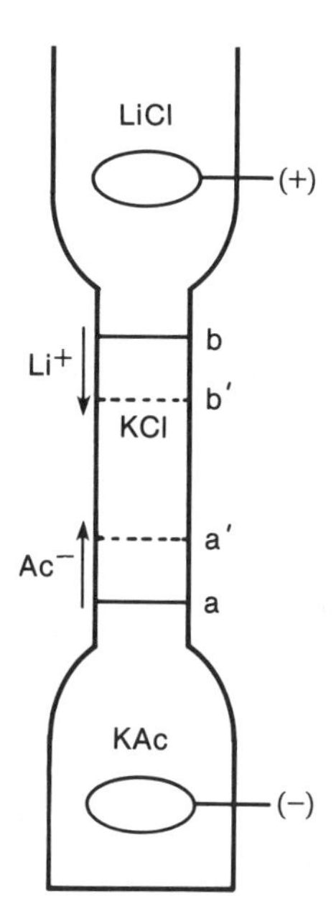

Figure 6–6. Transferance number apparatus. To study the transferance numbers of KCl, the apparatus is filled first with KOAc (acetate) to the point marked *a*, then with KCl to the point marked *b*, and finally with LiCl until the upper electrode is immersed. The LiCl and KOAc solutions are known as "indicator" solutions, used to follow movements of the K^+ and Cl^- ions. As current flows through the system, the Li^+ and AOc^- ions move in the directions shown by the arrows to the left of the central tube. The boundaries move to the points marked *a′* and *b′*.

must apply to the anions OAc^- and Cl^- in the neighborhood of the boundary, *b′*.

From Eq. 6–6, the transferance of an ion, divided by its equivalent concentration (c_iz_i), is proportional to the mobility of that ion. Hence, when t_i/c_i is the same for both ions at a boundary, their velocities will be equal. This means that in the immediate neighborhood of the boundary, concentrations of the ions adjust themselves according to Eq. 6–8, so that the initial boundary moves, in one direction or the other, as a concentration difference that is self-propagating, and is marked by a potential difference that also is self-propagating. Note that elsewhere in the apparatus both concentrations and potential drop are uniform; only at the boundaries can one find discontinuities of concentration or potential drop, brought about by the transferance properties of the ions in an electric field.

Effects on Macromolecular Ions

Any macromolecular ion introduced into the central segment of an apparatus such as shown in Fig. 6–6 is subject to the same forces as the smaller ions. Indeed, frontal electrophoresis of the Tiselius type always includes a small ion front as well as fronts due to separation of proteins. A kind of zonal electrophoresis, in free solution, can be performed using the scheme

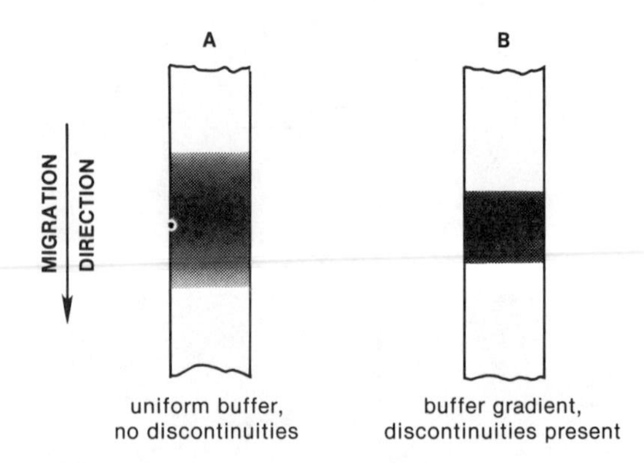

Figure 6–7. Zone appearances with (A) and (B) without buffer discontinuities. Note how the absence of discontinuities allows diffusion to blur the zone edges, causing band broadening, and how the potential drop that results from discontinuities narrows and sharpens the band edges. Band sharpening results in greater resolving power, which is the major advantage of disc gel electrophoresis.

shown in Fig. 6–6, provided that the sample is introduced appropriately and as a separate segment of solution. Although the disadvantages of free solution separations still apply, it is useful to see just how this method maintains the sharp boundaries in the face of tendencies to diffuse. In the absence of the concentration and potential drop discontinuities, a macromolecular particle is subject to random, Brownian motion in any conceivable direction, causing the edges of the zone to blur. Figure 6–7A shows a view across the direction of migration in the absence of a field and of the resultant discontinuities. A similar situation would exist if the system contained only a single, uniform buffer. The blurring of the zone edges is apparent. Figure 6–7B shows a similar view with the concentration and potential discontinuities in effect. The zone is much thinner and much sharper. It is entirely characteristic of disc electrophoresis, especially when performed in gels.

In disc electrophoresis,* the major electrolytes are buffers designed to control the pH of the system. The minor electrolyte is a mixture of proteins or peptides. Not only are the proteins very much larger, they are usually polyvalent. If a few protein molecules diffuse ahead of the main sample mass, they encounter an additional charge and an ion deficit, and both of these factors tend to pull the stragglers back into the main zonal mass. Exactly the reverse situation applies to sample molecules that might diffuse against the direction of electrophoresis. As a result, the band is sharpened, as shown in Fig. 6–7B. The proteins are, in effect, trapped between a volume of higher ionic concentration and one of lower concentration. Additionally, forces due to possible pH gradients and potential drop across the interface add to the concentration effects, as already described.

*As briefly mentioned previously under Apparatus for Zonal Electrophoresis, "disc" here does not refer to the bands produced by use of the technique, but rather to the voltage *dis*continuity resulting from the pH discontinuity.

The Ornstein and Davis Disc Gel Technique

Although Ornstein and Davis were not the first to capitalize on the disc gel approach, they certainly gave it renewed vigor; their method has been widely quoted and adapted to many problems, because it combines the high resolving power of disc systems in general with the sieving action of acrylamide gels of differing porosities. There is a discontinuity of buffer ions, of the sort just discussed, and a discontinuity of gel pore sizes. The gel column is divided into three segments: (1) the *running gel,* or small-pore gel; (2) the *stacking gel,* or large-pore gel; and (3) the *sample application gel,* which may have the same porosity as the stacking gel. Each of these gel segments is independently polymerized within the gel tube, as shown in Fig. 6–8.

In their initial work on blood plasma and serum protein separations, Ornstein and Davis first prepared the small-pore running gels (zone I in Fig. 6–8), then polymerized over them the large-pore stacking gels (zone II in Fig. 6–8). The sample was diluted with the monomer mix that was then polymerized in gel tubes to become the sample application gel (zone III in Fig. 6–8). Although it was argued that the small amount of ammonium persulfate required to polymerize the sample application gels did not appear to harm the proteins, there may well be more sensitive proteins in

Figure 6–8. Three views of a typical disc-gel electrophoresis system. View (a) shows initial conditions, with the proteins in the sample application gel (III); view (b) shows the stacking effect, with the proteins tightly banded in the large-pore stacking gel (II); and view (c) shows protein separation and migration through the small-pore running gel (I). The leading and trailing buffer ions are indicated by L^- and T^-, respectively, and B^+ represents the buffer base. In view (c), the ion front between L^- and T^- is shown by the dotted line, *xy*. In the original Ornstein–Davis procedure, B^+ was Tris, T^- was glycine, and L^- was chloride, but many other buffer combinations can be selected.

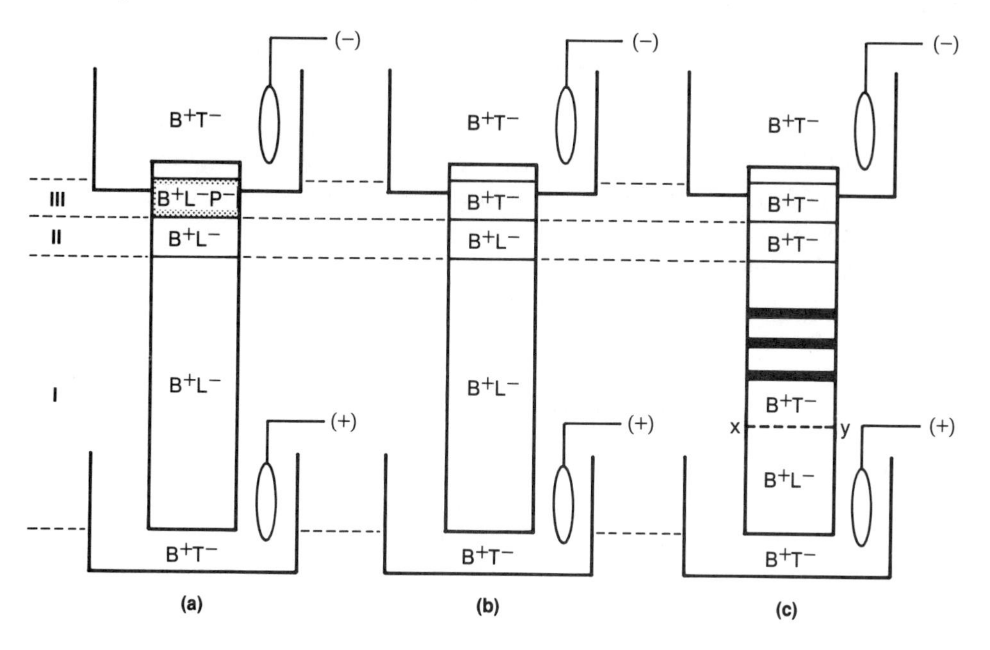

other kinds of samples. For this reason, most workers prefer to employ a two-zone system, composed of running gel and stacking gel, adding the free liquid sample to the top of the column in the usual way, increasing the sample density with sucrose or glycerol where necessary. All of the original Ornstein and Davis experiments were performed in tubes, but there is no practical reason why the system must be so arranged, and disc-slab gels are becoming increasingly popular. A single 8 × 8-in. slab can easily be loaded with up to ten samples. For simplicity and convenience, disc-slab gels are almost always run with stacking and running zones only.

By careful selection of buffer ions and their concentrations, and by altering the total acrylamide concentrations of the stacking and running gels, a variety of controls can be manipulated, but certain principles must be observed. First, the mobility of the trailing ions, u_T, in gel layers II and III should be less than that of the slowest protein in layer III. The leading ion should be selected to have a mobility, u_L, that is large and independent of the pH. Thus, $u_T < u_p < u_L$, where u_p is the mobility of protein. To function properly, all of the proteins must have a net charge of the same sign, which is why Ornstein and Davis worked with plasma or serum proteins at pH 8.3–8.6. However, one can work with other protein mixtures and with nucleic acids at more acidic ranges, in the neighborhood of pH 3–4. One can also manipulate the pH differences between the stacking and running gels, just as one can adjust the porosity of these gels by alteration of the acrylamide concentrations.

In commenting on the flexibility of discontinuous systems, Hjerten et al. (1965) pointed out a very much simplified scheme for zone sharpening, based on a conductivity discontinuity only. They used a uniform gel column of porosity comparable to the Ornstein–Davis running gel and a Tris-glycine buffer, pH 9.5. They decreased the conductivity of the sample by diluting 1:5 with water, then applied the liquid sample to the top of the column. When a protein zone is passed by a conductivity discontinuity, it will become narrowed, or sharpened, because the rear of the zone will migrate faster than the front of the zone. After the discontinuity has passed the protein zone, the protein will be in a region of constant conductivity again, and will migrate at a uniform velocity. Limited evidence suggests that this technique gives results equal in resolving power to the more complicated Ornstein–Davis procedure; however, it has not yet achieved the widespread popularity of the latter.

RECOVERY OF MATERIALS FROM ELECTROPHORETIC GELS

Situations frequently arise in which it is desired to recover one or more zones from electrophoretic gels. If, for example, it was necessary to measure the radioactivity of a zone, it might not matter if the protein was denatured. If, on the other hand, one wished to determine the presence of some enzymic activity in a recovered zone, care would have to be taken to avoid denaturation. The means of recovery must therefore be shaped by the purpose in mind.

Polyacrylamide gels can be dissolved by incubation in 30% (v/v) hydrogen peroxide. Solution may be hastened by gentle heating, but the process is still not a rapid one. Further, few proteins will escape considerable damage from the powerful oxidizing effects of the peroxide. This method is usually limited to preparation of stained zones for scintillation counting or for similar purposes in which the biological quality of the recovered protein is unimportant.

A second and much milder process is available, but it requires use of *N,N*-diallyltartardiamide as a cross-linking reagent in place of bis-acrylamide. The tartaric acid moiety of the diamide contains vicinal OH groups, and it is well known (Malaprade reaction) that vicinal diols undergo cleavage of the C—C bond when exposed to periodate. A 2% (w/v) solution of KIO_4 will dissolve gels cross-linked by tartardiamide in just a few hours even at room temperature, and periodate is much less damaging to most proteins than is peroxide. Although first proposed as recently as 1970, the use of tartardiamide has become widespread because it so facilitates gel dissolution.

For some purposes it is not even necessary to remove the gel. If one wishes to generate antibodies against some protein fractionated by gel electrophoresis, one may simply cut out the separated gel zone, grind it up with Freund's adjuvant, and inject the whole mixture into a host animal. The gel fragments do not seem to inhibit immune system response. They may, in fact, serve as an additional stimulus to the reticuloendothelial system, much as other adjuvants do. Clearly, this method is of specialized interest and cannot be generally employed.

Electrical recovery of proteins has been accomplished by first gently breaking up the gel, transferring it with some buffer to dialysis tubes, and passing a current through the system. The isolated protein is moved from the gel pores to the free buffer but cannot cross the dialysis membrane. The gel particles can then be filtered or centrifuged off the free solution, and the recovered protein concentrated as needed.

A few "brute force" methods have also been described. These depend on a variety of syringes or micropresses by means of which the gel is, quite literally, squeezed until its porous structure has collapsed. This forces the protein into a free solution that can be recovered.

Blot transfer techniques have also been developed for special recovery problems. The Southern, Northern, and Western techniques have already been briefly mentioned, and others exist.

All of the above remarks apply with particular force to analytical-scale electrophoresis. For truly preparative-scale isolation, a variety of specialized (and quite expensive) apparatuses are available. Information concerning these may be found in the references cited at the end of this chapter, or in brochures supplied by the equipment manufacturers.

Formats of Gel Electrophoresis

Electrophoretic gels may be prepared by casting polyacrylamide gels in the form of cylinders, each in its own glass tube and each devoted to separation of a single sample. Electrical contact to the gels is made by inserting the tubes through tightly fitted grommets arranged in a circle around the bottom of the upper buffer reservoir, as shown in Figs. 6–3 and E7–1 (see Expt. 7). At the end of the experiment, the glass tubes are removed from the rack and each gel cylinder is removed from its glass container prior to staining or other examination. Removal of the gels from the rack can be troublesome, and recovery of the gels from the glass tubes is tedious. Most gel tubes have an inner diameter of 4–5 mm; experience indicates that with smaller tubes the likelihood of gel breakage during the recovery process becomes very high.

Staining and destaining of the gels is diffusion limited, and the rather large diameter of the gel cylinders makes these operations time consuming. For similar reasons, it is difficult to perform autoradiography on the usual cylinder gels, since longitudinal slicing is impractical, and it is not easy to obtain good physical contact between an x-ray film and a cylindrical object such as a gel cylinder. Slab gels provide the solution to these problems.

SLAB (FLAT SHEET) ACRYLAMIDE GEL ELECTROPHORESIS

Acrylamide solutions can be polymerized as flat slabs of virtually any desired dimensions by forming a "box" from two flat, rectangular glass plates. On three sides, the plates are separated by strips of some inert plastic, which are generally fastened to one of the plates by a suitable adhesive. The entire assembly is made leakproof by use of a 1.5-mm-diameter silicone rubber tube running around and between the edges of three sides of the glass plates. The entire setup is then secured by spring clamps applied around the edges. The acrylamide mixture is added through the open edge; the chamber is then allowed to stand undisturbed until polymerization is complete. It is convenient to form a desired number of sample wells in the free edge of the gel by inserting a plastic "comb" into the upper edge before polymerization commences. These wells need be only a few millimeters deep, and it is easy to cast as many as twelve sample wells across the top of a gel slab measuring 150 mm or so in length.

In one of the more popular versions, both the front and back plates of the gel "box" have a notch cut into their upper edge, above the top of the volume actually occupied by the gel. This is the back plate of the final assembly. A diagram of the assembly is shown in Fig. E10–1 (Expt. 10).

Plastic spacers of various thickness may be used or they may be fastened to the back plates with a double-sided adhesive tape.*

Alternatively a light application of silicone grease may be used or, by exercising great care, the plastic spacers may, simply by the use of water, be made to adhere to the glass plate surface long enough for the plastic tubing to be clamped in place.

When polymerization is complete, the holding clamps and the plastic tubing are removed and the plate assembly is mounted on a vertical rack (see Reid and Bieleski, 1968), as shown in Fig. E10–3. The rack is made of a vertical "table" to which is attached an upper and a lower buffer chamber, each fitted with an electrode. The upper chamber also bears a notch corresponding to the notch in the back plate of the gel assembly. A gasket of closed-cell sponge rubber is glued to the front surface of the vertical table. By means of tight spring clamps, the gel assembly can be held tightly against the rubber on the vertical table (shown by vertical stippling between the gel assembly and the table itself) to provide a leakproof seal. The upper and lower buffer chambers with their associated electrodes allow good contact with the top and the bottom of the gel that has been cast between the glass plates. The top of the gel ordinarily comes just below the bottom of the notch of the back plate. Once the apparatus has been assembled as shown, and tested for freedom from leaks, samples may be applied to the wells cast into the gel. Gels can be modified by addition of SDS or isoelectric focusing ampholytes just as with cylinder gels; one can even cast gels with acrylamide concentration gradients running from top to bottom of the gel slab.

Advantages

There are three distinct advantages to the use of slab gels, as compared to cylindrical gels:

1. Reduction of gel thickness to approximately 1 mm (or less, depending on the spacers used) greatly speeds the processes of staining and destaining whether by Coomassie Blue or by other stains. At the same time, the reduction in gel thickness makes it easier to dissipate the coulombic heat. (But see p. 147.)
2. Because more than one sample can be run on a single gel slab, it is much easier to compare several samples. Obviously, all samples run on a given gel have been run under *almost* identical conditions of electrophoresis. This is particularly useful in following the process of multistep protein purification.
3. The thin gels can be treated quickly with scintillation enhancing reagents, then dried quickly under vacuum. Autoradiography is thus

*The special tape, made by 3-M, is 10 mil × 34 in. It is available from R. S. Hughes, 1076 40th Street, Oakland, CA 94710, Catalog No. Y-9373.

simple and direct, and contact between the dried slab and an x-ray film makes for efficient, short-term exposures.

For these reasons slab gels are generally preferred to cylinder gels; precast gels are also being sold by a number of vendors.

Intrinsic Problems

The extended areas of slab gels and their thinness pose some mechanical problems in handling. With gels of 5% or lower concentration, the gels are particularly fragile. At 7% concentration or above, the gels are less likely to tear when they are being removed from the glass support. When it is necessary to transfer these gels from one container to another, they should be supported by a piece of heavy filter paper, or better, by nonwoven cloth of the sort sold for household cleaning. Bending at acute angles or kinking will almost certainly cause the gels to fracture.

When polyacrylamide slab gels are dried, they will certainly suffer dimensional change and cracking unless they are placed on a mechanical support of heavy filter paper or modified plastic films. These films have one surface treated by a proprietary process that permits covalent bonding of polyacrylamide to the treated side. These films are expensive, but they have the advantage of preserving transparency during the drying process, as ordinary filter paper does not. Use of these gel-bonding films makes it easy to scan stained gels by densitometry.

A simpler procedure is to soak the gels in 5% glycerol (to maintain flexibility after drying), sandwich the gel between the two halves of a folded cellulose acetate sheet (also soaked in 5% glycerol), and then mount the sheets and gel between two Lucite® frames held tightly together with spring clamps. The setup is air-dried overnight, for example, in the door of a fume hood. After demounting, the excess cellulose acetate is trimmed from the edges of the gel; the mounted gel is placed between the pages of a large book for 1 or 2 hours to prevent curling, after which it can be photographed, scanned, or stored in a laboratory notebook.

Alternative Staining Methods for Electrophoretic Gels

Regardless of how an electrophoretic separation is performed, the problem of visualizing the results remains. Traditional methods involve (1) prelabeling of the original mixture with one or another radioactive label, and, after separation by autoradiography, (2) staining the separated materials with a suitable dye such as Buffalo Blue Black (also known as Amido Schwarz) or with Coomassie Blue. It is also possible to avoid stains altogether, and to scan the length of a cylinder gel for UV absorbance, but this is not very practical, owing to high background absorbance and low sensitivity.

Labeling with radioactive atoms is not always feasible and it is often difficult to attain sufficient activity of the separated components. Unless the label is directly incorporated into a precursor of the proteins or peptides, there is a likelihood of altering the net charge on the components by some of the chemical modifications used in labeling procedures. The emission characteristics of available isotopes may require autoradiographic exposures of as long as a week, even when the gels are pretreated with a mixture of fluors similar to liquid scintillation cocktails. Since the x-ray film acts roughly as an integrator of radioactivity, labeled bands of highly concentrated proteins will usually be grossly overexposed while bands of low concentration are barely detectable.

Staining methods based on dye binding yield a deeply colored gel matrix from which unbound dye must be removed, either by free diffusion or by electrophoresis of the charged, free dye molecules. In theory, this technique leaves bands of stained protein, fixed in the gel matrix by acid denaturation. In practice, it is not easy to obtain perfectly clear backgrounds. The destaining process is time-consuming and usually requires frequent changes of relatively large volumes of destaining solutions. The extended manipulations sometimes lead to loss of minor protein bands. If the gel matrix is not completely cleared of unbound dye, there will be a significant loss of contrast between the matrix dye background and faintly stained protein bands.

SILVER STAINING PROCESSES

It has long been known that Ag^+ binds to certain organic tissue components, including the double bonds of fatty acids and other molecules, to -SH, -COOH, and perhaps -CHOH groups of proteins. On a more or less empiric basis, silver has been employed in a number of special histochemical stains, especially for the structures of the nervous system. In these formulations, the silver is generally presented to the tissue to be stained in the form of a silver-ammine complex* in order to keep the concentration of the free Ag^+ at acceptably low values. This ensures that the Ag^+ will stain by combination with "reaction centers" of the sorts described above.

Ordinary photography also depends on the interaction between Ag^+, produced from microcrystalline granules of silver halides dispersed in a layer of gelatin which composes the so-called film. In this instance, the concentration of free Ag^+ is kept low by the low solubility of silver halides. When exposed to light in a camera, the activated silver halide grains are more readily reduced, forming the desired image. Suspension of the silver halide in nonprotein dispersing agents, such as collodion, gives films of

*Coordination complex between silver and ammonia or amine.

much lower sensitivity to light. Here too, the exact chemistry of the "reaction centers" is not clearly understood.

Switzer et al. (1971) proposed the use of analogous methods for detecting proteins dispersed in acrylamide gels after electrophoresis. The proteins were fixed in the gels by means of methanol and acetic acid and then subjected to mild oxidizing conditions by immersion in a dilute solution of potassium dichromate. This treatment appears to generate additional "reactive centers." The washed gels are then treated with a dilute solution of a silver-ammine complex, from which Ag^+ binds to the oxidized proteins. The bound silver is then converted to Ag^0 by a reducing agent (analogous to the photographic developer) to give a brownish-black image of the protein bands. The process requires a total of approximately 3 hours from start to finish and gives results reportedly up to 100-fold more sensitive than the Coomassie Blue methods.

The original silver staining methods had some inherent drawbacks. First, because of the cost of silver, the method is not inexpensive for routine use. Second, silver-ammine solutions, when unduly exposed to the air, oxidize to form silver nitride, Ag_3N. When dry, silver nitride (and other compounds also possibly formed by oxidation) is explosive. The risk of explosions can be avoided, however, by promptly disposing of the silver-ammine solution after use or by treatment with an excess of sodium hydrosulfite. The third drawback is that extraneous silver deposits may be formed on the surface of the gels, obscuring the detail intrinsic in the bands of faintly stained proteins. Finally, the silver methods must be optimized, in terms of reagent concentrations and treatment times, for the thickness of the gels, and these must be empirically determined.

"SECOND GENERATION" SILVER STAINING METHODS

Refinements of the silver staining methods have been directed toward eliminating or minimizing the drawbacks mentioned above.

Switzer proposed that surface silver deposits could be removed by solution with a well-known "image reducer" consisting of cupric chloride plus sodium thiosulfate. (**Note:** The terminology used here deserves the quotation marks; the solution is, from the chemical viewpoint, an oxidizing reagent. It converts Ag^0 to AgCl, which then forms soluble complexes with the thiosulfate.) Although this solution worked, it had an unfortunate tendency to attack the desired image as well as the surface contaminants. Furthermore, it was difficult to remove excess reagent from the gel when image reduction had proceeded to the required stage. It has since been claimed that the action of the image reducer can be more effectively arrested by treatment of the gels with a persulfate solution (the photographer's "hypo eliminator"). Oxidation of the thiosulfate effectively prevents removal of silver from the gel by stopping soluble complex formation.

Surface silver contamination can be largely eliminated by passing all of the reagents through a 0.2-μm filter, since it was determined that in part the problem lay in contamination by microscopic dust particles that adhered to the gel surface by static charges, then served as nuclei for the formaton of "reaction centers." A similar source of contamination was found in the powder used to dust plastic gloves commonly worn when the gels and staining reagents were handled. Such dust particles were found to be the source of fingerprint-shaped deposits sometimes noted on the gel surfaces.

It is preferable to use reagents of the highest quality and to prepare acrylamide/bisacrylamide solutions weekly. While stored, they should be shielded from light. Such precautions eliminate or minimize the yellowish background often seen in silver-stained gels.

The diffusion-limited movement of reagents into (or out of) gels depends on the gel thickness, the concentration of acrylamide in the gel, the viscosity and concentration of solutes in the reagents, and the temperature. It was originally held that cylinder gels, commonly cast at diameters of about 3 mm, could not be satisfactorily stained by the silver methods because of their thickness. These conclusions were reached in earlier studies where the silver-ammine reagent was based on solution of silver nitrate in ammonia. These methods were employed at ambient temperature, presumably close to 25°C. By substituting methylamine for ammonia, and by warming the reagents to 60°C, it is possible to obtain high-quality results with gels up to 3 mm in thickness, and with an acrylamide concentration ranging from 4% to 20%.

Another variant employs a reducing reagent that contains $NaBH_4$ in addition to the usual NaOH and HCHO. After an appropriate reduction time, which generates a typical silver deposit over the protein bands or spots, the gels are "enhanced" by immersion in a solution of Na_2CO_3. The enhancement process produces an astonishing play of colors ranging from blue through green, yellow, and red. The background is frequently slightly amber-colored. The exact nature of the pigments is not known, but the play of colors is reportedly reproducible from gel to gel and from protein to protein. In complex two-dimensional gels of samples such as whole tissue extracts, which may contain many hundreds of spots, the addition of color aids in resolution of overlapping or very closely adjacent spots. Reagents for this procedure are now available from commercial sources.

NONSILVER METALLIC STAINING METHODS

The most recent entry into this field was proposed by J. S. Yudelson, of the Eastman Kodak Company, and is now marketed under the trade name of "Kodavue Kit®. The kit contains five reagents, all of which have an indefinite shelf life at room temperature, a great advantage over some of the silver-based staining methods in which the reagents frequently become unusable within several hours or less. The manufacturer regards certain aspects of the chemistry of the process as proprietary information, so it cannot be explained fully. This staining method is fairly fast, requiring slightly less than 1 hour from the beginning to the finally washed gel. As a rule, destaining is not required, since little or no metal is deposited in the absence of protein.

As designed, this method is useful only with acrylamide gels; use of agarose, cellulose nitrate strips, or other support media is not recommended. The Yudelson method will *not* work in the presence of ampholytes, or dithiothreitol- or trichloroacetic acid-containing fixatives. Traces of certain heavy metals anywhere in the system (or in wash water) may cause failure of staining. If the gel pH is <6, the method does not work well, but the gel pH may be shifted upward by a sodium carbonate presoak. These limitations rule out the use of this method in isoelectric focusing gels and in gels run under denaturing conditions.

The Yudelson method depends on a principle long known to photographers—the principle of *physical development.* It involves three discrete steps:

1. *Prenucleation:* Catalytic precursors are introduced into the surface of the protein embedded in the polyacrylamide matrix by soaking the gel in a suitable reagent solution. The exact metal salt employed for this prupose is a proprietary secret, but it is probably a salt of rhodium, platinum, iridium, palladium, or gold.
2. *Nucleation:* The precursor is reduced to form a catalytic development center on the surface of the prenucleation centers.
3. *Development:* The oxidant metal and reducing agent react at the catalytic center and produce reduced metal. This freshly deposited metal is a catalyst. The reaction proceeds into a autocatalytic phase, and its velocity accelerates. The deposited metal is, in this case, nickel; the finely divided nickel has a grayish-black appearance that contrasts sharply with the clear background of the empty gel matrix. Since the noble metal is deposited only on the protein bands, and not on the empty gel, destaining is not ordinarily required.

Physical development is a surface-dependent phenomenon, which accounts for some of the properties of this staining method. The method is reported to have a sensitivity conservatively rated as 50–100 times greater than that of other nonsilver staining methods. The manufacturers claim that the Yudelson method can detect 1–2 ng of protein. Although these values can be reached under ideal conditions, it is not likely that they can be routinely reached by use of nonselected protein mixtures. Furthermore, the cited values apply only to fairly thin gels, of 1.0 mm or less in thickness. It is only in such thin gels that the surface concentration of protein binds sufficient catalyst to provide adequate staining. It must be remembered that the sensitization of the deposited protein does not extend through its mass, but is limited to a thin film at its surface. For these reasons, the Yudelson method cannot be recommended for tube gels that are ordinarily several millimeters in diameter.

Note: The literature on electrophoresis and related topics is enormous. Citations in the reference list were selected for their particular connections with matters presented in the textual discussion. Many other important papers could have been cited and many other topics could have been entertained. As examples, starch gel and agarose gel electrophoresis were barely mentioned, and no references to these topics were cited. This does not imply that these media and their uses are unimportant, but only that the basic principles of electrophoresis could be adequately presented without them. The reader will find ample discussion of much that was omitted here in the general references cited, and in the original literature.

REFERENCES

General Sources

Abramson, H. A. Moyer, L. S. and Gorin, M. H. *Electrophoresis of Proteins and the Chemistry of Cell Surfaces.* Reinhold Publishing, New York (1942).

Bier, M. *Electrophoresis: Theory, Methods and Applications.* Academic Press, New York (1959).

Freifelder, D. *Physical Biochemistry: Applications to Biochemistry and Molecular Biology,* Chap. 9. W. H. Freeman, San Francisco (1976).

Goldman, D., and Merril, C. R. Silver staining of DNA in polyacrylamide gels: linearity and effect of fragment size. *Electrophoresis 3:*24 (1982).

Kodak Lab. Chem. Bull. 54: No. 1 (1983).

Marshall, T., and Latner, A. L. Incorporation of methylamine in an ultrasensitive silver stain for detecting protein in thick polyacrylamide gels. *Electrophoresis 2:* 228 (1981).

Merril, C. R., Goldman, D., Sedman, S. A., and Ebert, M. N. Silver staining of electrophoretic gels. *Science 211:*1437 (1981).

Merril, C. R., Goldman, D., and Van Keuren, M. L. Simplified silver protein detection and image enhancement methods in polyacrylamide gels. *Electrophoresis 3:*17 (1982).

Morris, C.J.O., and Morris, P. *Separation Methods in Biochemistry,* 2nd ed. John Wiley and Sons, New York (1976).

Poehling, H. M., and Neuhoff, V., Visualization of protein with a silver stain: a critical analysis. *Electrophoresis 2:*141 (1981).

Sammons, D. W., Adams, L. D., and Nishizawa, E. E. Ultrasensitive silver-based color staining of polypeptides in polyacrylamide gels. *Electrophoresis 2:*135 (1981).

Switzer, R. C., Merril, C. R., and Shifrin, S. A highly sensitive silver stain for detecting proteins and peptides in polyacrylamide gels. *Anal. Biochem. 98:*231 (1971).

Wieme, R. J. Theory of electrophoresis. In *Chromatography, A Laboratory Handbook of Chromatographic and Electrophoretic Methods* (E. Heftmann, ed.). Chap. 10. Van Nostrand Reinhold, New York (1975).

Willoughby, E. W., and Lambert A. A sensitive silver stain for proteins in agarose gels. *Anal. Biochem. 130:*353–358 (1983).

Young, D. S., and Anderson, N. G., eds. *Clinical Applications and Developments in Two-Dimensional Electrophoresis. Clin. Chem. 28:*737 (1982). [Detailed symposium papers occupying an entire special issue of the journal. Devoted to generation and analysis of two-dimensional electrophorograms.]

Gel Electrophoresis

Anker, H. S. A solubilizable acrylamide gel for electrophoresis. *FEBS Lett. 7:*293 (1970). [First mention of *N, N*-diallyltartardiamide as a cross-linking reagent to facilitate gel dissolution.]

Carraway, K. L., Lam, A., Kobylka, D., and Huggins. J. Anomalous staining of membrane lipids on acrylamide gels. *Anal. Biochem. 45:*325–331 (1972). [Warns of formation of artefactual complexes between phospholipids and SDS.]

Davis, R. H., Copenhaver, J. H., and Carver, M. J. Quantitation of stained proteins in polyacrylamide gels. *Anal. Biochem. 58:*615–623 (1974).

Gordon, A. H. Electrophoresis of proteins in polyacrylamide and starch gels. In *Laboratory Techniques in Biochemistry and Molecular Biology,* Vol. 1 (T.S. Work and E. Work, eds.). American Elsevier, New York (1970).

Oster, G. K., Oster, O., and Prati, G. Dye-sensitized photopolymerization of acrylamide. *J. Am. Chem. Soc. 79:*595–598 (1957).

Oster, G., Bellini, J. S., and Holmström, B. Photochemistry of riboflavin. *Experientia 18:*249–253 (1962).

Pusztai, A., and Watt, W. B. Polyacrylamide gel electrophoresis of proteins in phenol–ethanediol–water (3:2:3, w/v/v) buffers at various pH values. *Anal. Biochem. 54:*58–65 (1973).

Raymond, S., and Weintraub, L. Acrylamide gel as a supporting medium for zone electrophoresis. *Science 130:*711 (1959). [First reference to the use of acrylamide in electrophoresis.]

Weber, K., and Osborn, M. Reliability of molecular weight determinations by dodecylsulfate–polyacrylamide gel electrophoresis, *J. Biol. Chem. 244:*4406–4412 (1969). [The single most widely quoted paper on molecular weight determinations by electrophoretic methods; a classic report.]

Weidekamm, E., Wallach, D.F.H., and Flückiger, R. A new, sensitive, rapid fluorescence technique for the determination of proteins in gel electrophoresis and in solution. *Anal. Biochem. 54:*102–114 (1973). [Suggests prestaining with *o*-phthalaldehyde, forming fluorescent complexes with proteins prior to electrophoresis.]

Preparation and Uses of Ampholytes for IEF

Haglund, H. *Methods Biochem. Anal. 19:*1–104 (1971).

Vesterberg, O. Synthesis and isoelectric fractionation of carrier ampholytes. *Acta Chem. Scand. 23:*2653–2666 (1969).

Vesterberg, O. Isoelectric focusing of proteins. *Methods Enzymol. 22:*389–412 (1971).

Two-Dimensional Electrophoresis

Ames, G. E., and Nikaida, K. Two-dimensional gel electrophoresis of membrane proteins. *Biochemistry 15:*616–623 (1976).

Anderson, N. L., and Anderson, N. G., High resolution two-dimensional electrophoresis of human plasma proteins. *Proc. Natl. Acad. Sci. U.S.A. 74:*5421–5425 (1977).

Brown, W. T., and Ezer, A. A computer program using Gaussian fitting for evaluation of two-dimensional gels. *Clin. Chem. 28:*1041–1044 (1982). [This issue is devoted entirely to two-dimensional gel electrophoresis of various body fluids in health and disease, and to the technology of gel data reduction.]

O'Farrell, P.H. High-resolution two-dimensional electrophoresis of proteins. *J. Biol. Chem. 250:*4007–4021 (1975). [First reported resolution of 1100 proteins from *E. coli* on a single gel; a fundamentally important paper.]

Reid, M. S., and Bieleski, R. L. A simple apparatus for vertical flat-sheet polyacrylamide gel electrophoresis. *Anal. Biochem. 22:*374–381 (1968). [The hardware used by O'Farrel and many others ever since.]

Disc Electrophoresis

Davis, B. J. Disc electrophoresis—II: Method and application to human serum proteins. *Ann. N.Y. Acad. Sci. 121:*404–427 (1964).

Hjerten, S., Jerstedt, S., and Tiselius, A. Some aspects of the use of "continuous" and "discontinuous" buffer systems in polyacrylamide gel electrophoresis. *Anal. Biochem. 11:*219–223 (1965).

Ornstein, L. Disc electrophoresis—I. Background and theory. *Ann. N.Y. Acad. Sci. 121:*321–349 (1964).

Relationship Between Gel Chromatography and Gel Electrophoresis

Morris, C.J.O., and Morris, P. Molecular sieve chromatography and electrophoresis in polyacrylamide gel. *Biochem. J. 124:*517–528 (1971).

Rodbard, D., and Chrambach, A. Unified theory for gel electrophoresis and gel filtration chromatography. *Proc. Natl. Acad. Sci. U.S.A. 65:*970–977 (1970).

Blot Transfer Techniques in Electrophoretic Analysis

Davis, L. G., Dibner, M. D., and Battey, J. F. *Basic Methods in Molecular Biology.* Elsevier, New York (1986).

Maniatis, T., Fritsche, E. F., and Sambrook, J. Southern transfer. In *Molecular Cloning: A Laboratory Manual,* pp. 382–389. Cold Spring Harbor Laboratory, Cold Spring Harbor, N.Y. (1982).

Southern, E. M. Detection of specific sequences among DNA fragments separated by gel electrophoresis. *J. Mol. Biol. 98:*503–517 (1975). [See also *Methods Enzymol. 69:*152 (1980).]

CHAPTER SEVEN

The Many Facets of Chromatography

The Nobel Prize in chemistry for 1952 was shared by A. J. P. Martin and R. L. Synge for their pioneering work in developing the theory and practice of chromatography. The origins of chromatography can be traced back to the turn of this century. The Russian botanist, Tswett, and the American geologist, Day, were among its earliest practitioners and proponents. It was their work on plant and petroleum pigments, respectively, that led to the coining of the term "chromatography," since the only feasible means of realizing that some separation had occurred was direct visualization of pigmented bands. Chromatography is, quite literally, "writing by color."

Martin defined the process as "the uniform percolation of a fluid through a column of more or less finely divided substance, which selectively retards, by whatever means, certain components of the fluid." Even today, more than thirty years after Martin's definition was written, there is no reason to alter it in any way, despite the fact that the field of chromatography has greatly expanded.

SELECTION OF A SUITABLE PROCESS

Several considerations apply:

1. *The efficiency of the process.* Major factors affecting efficiency include the required capacity of the method, the required resolution of the separation to be made, and the speed of the method in question.
2. *The basis of a proposed separation.* The choice of adsorption, ion-exchange, or another form of chromatography must be based upon the nature of the solute(s) of major interest and the matrix in which they are contained. The requirements of a system suitable for a total analysis differ from the requirements of a system suitable for purification of a single protein.
3. *The detection of solute(s).* Many detection methods are available, including thermal decomposition (charring of TLC plates), spray reagents, change in ultraviolet absorption, and change in refractive index.

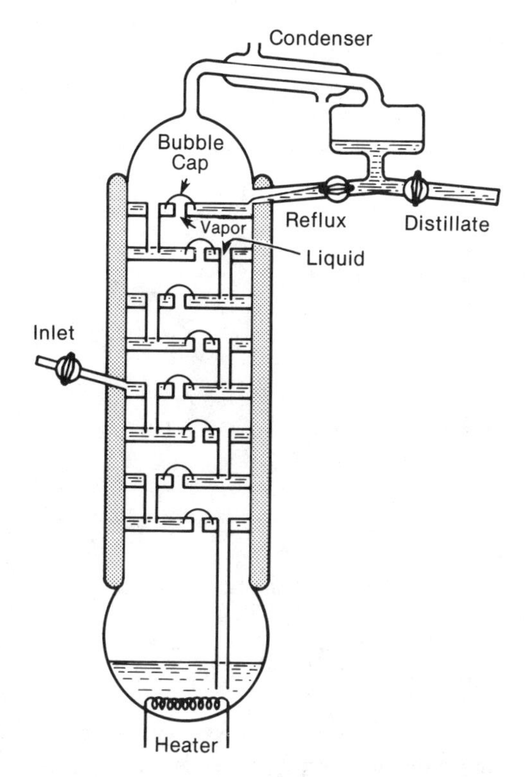

Figure 7–1. Details of a bubble-cap still. (Reproduced by permission from Daniels and R. A. Alberty, *Physical Chemistry*, 2nd ed., John Wiley and Sons, New York, 1961.)

4. *The recovery of solute(s).* Solute recovery is closely linked to solute detection method and to the scale of the separation.
5. *Equipment and operating costs.*

EFFICIENCY OF CHROMATOGRAPHIC PROCEDURES: ANALOGY WITH DISTILLATION THEORY

A proper concern in chromatography is the efficiency of separation. With the possible exception of affinity procedures, most chromatographic methods are fractionation methods by which some number of solute components are subjected to differential retardation. In this respect, chromatography is similar in function to fractional distillation of a mixture, where the retarding influence can be expressed in terms of vapor pressure or boiling point, and the separating force is the thermal input to the still pot. It is therefore not surprising that a review of distillation theory can provide rewarding insight into aspects of chromatographic theory. In order to clarify the connectedness of distillation and chromatographic processes, a cross section of a bubble-cap still is shown in Fig. 7–1. This fairly high efficiency device is frequently employed in industrial operations, ranging from solvent recovery to petroleum fractionation. Bubble-cap stills operate at nearly equilibrium conditions.

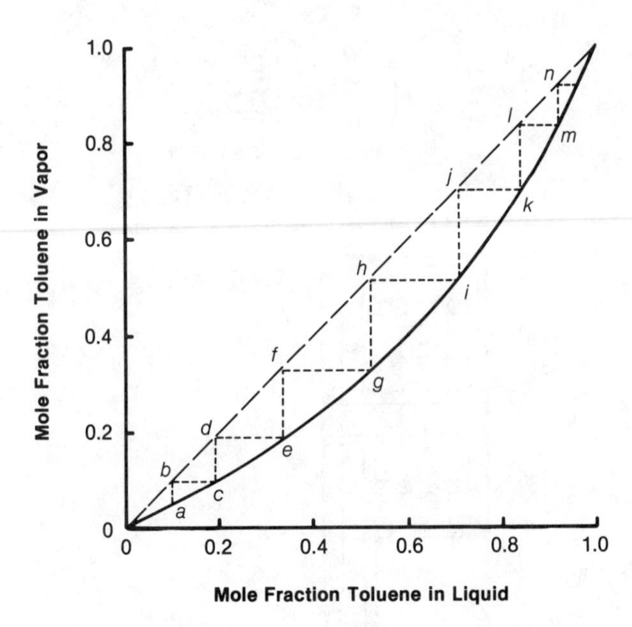

Figure 7–2. Plot of distribution of toluene between vapor and liquid phases in a toluene:benzene mixture.

Vapor leaving the still pot has a mixed composition; each of the components has a vapor phase concentration directly proportional to its vapor pressures (the sum of the vapor pressures = atmospheric pressure). Vapor then condenses, filling in turn each of the condensation plates with which the column is filled. Excess condensate from any one plate is returned to the plate below, through the liquid return lines shown. Uncondensed vapor reaches the higher plates only by forming bubbles which escape through the bubble caps. This means that vapors come virtually to equilibrium with liquid phase in each plate. When the system is operated at nearly total reflux (i.e., with removal of little distillate), a steady state is quickly reached. The operating efficiency of a column is best expressed in terms of the number of *theoretical plates,* that is, the number of successive, infinitesimal vaporizations at equilibrium required to give the separation actually observed. No real device operates at 100% efficiency, but any real device can be characterized in terms of the *height equivalent to a theoretical plate (HETP),* provided that one knows the composition of the liquid and vapor phases at a given temperature and pressure.

One can in fact calculate HETP without reference to any physical apparatus, as shown in Fig. 7–2 for a mixture of benzene and toluene, if the phase compositions are known.

The diagonal broken line bisecting the graph is what one would observe had no fractionation occurred. The solid curve is a plot of the mole fraction of the less volatile component in the vapor phase versus its mole fraction in the liquid phase. In other words, if a solution with toluene mole fraction equal to *a* was boiled and the vapors condensed, the mole fraction of toluene in the condensate would be equal to *b,* and so forth. In the diagram, each of the rectangular steps represents a theoretical plate. From such a plot, the number of theoretical plates can be determined for any real device, and, together with the dimensions of the device, can be used to calculate the HETP.

Returning now to chromatogrpahy, we can define the HETP as an imaginary segment of a chromatographic bed such that the fluid exiting from it has a composition in equilibrium with the average composition of the retarded solute held by the retarding medium. The HETP is larger in systems removed from equilibrium, which are inefficient systems, and is correspondingly smaller in systems operating at or near equilibrium. Good HPLC or GLC columns frequently contain 2500 or more theoretical plates, but this figure is rarely reached by less sophisticated forms of chromatography on the bench. The large plate number is a function of the fineness and uniformity of the packing and freedom from voids in the columns.

The relation between chromatography and distillation was carefully examined first by Mayer and Tompkins, who also provided a method to approximate the number of theoretical plates for a chromatographic process. On the assumption that the profile of an eluted chromatographic peak would have a nearly Gaussian shape (Fig. 7–3), the peak width at 0.606 (maximum peak height) should be a good estimate of its standard deviation. If the peak is approximately Gaussian, then:

$$N = (V_e/\sigma)^2 \tag{7–1}$$

where N is the number of theoretical plates for the particular solute in the system under examination, V_e is the elution volume, and σ is ½ the width of the curve. From the known length of the chromatographic bed, the HETP can be calculated as L/N, where L is the length of the bed through which separation occurs. More elaborate equations have been developed by others, especially by Van Deemter, but we shall not pursue the matter further here. Instead, we shall address the uses made of N and HETP in chromatography.

Chromatography can take many forms, including adsorption, exclusion, ion-exchange (IEC), affinity, gas liquid (GLC), high-performance liquid

Figure 7–3. Profile of an approximately gaussian chromatographic peak. The peak is traced by a recorder pen from left to right. The broken lines show that the standard deviation is given approximately by a line drawn at $0.606C_{max}$. This is the value used in estimation of plate number by the method of Mayer and Tompkins.

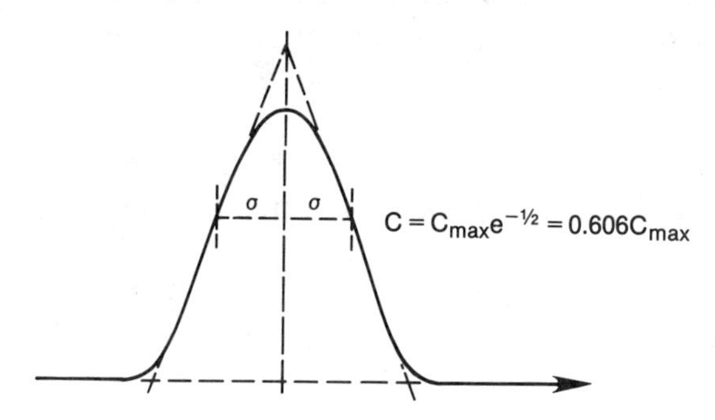

chromatography (HPLC), reverse phase, and, more recently, supercritical fluid chromatography. We discuss adsorption chromatography first since, to a large or small degree, it is involved in many other chromatographic modalities.

ADSORPTION CHROMATOGRAPHY

Nature of Adsorptive Forces

The forces that bind sorbed solutes to the surface of an adsorbent may be of various kinds. The forces are frequently electrostatic in nature or involve hydrogen bonding between the solute and some oxygen-containing group on the adsorbent. Water is an excellent partner in hydrogen bond formation; therefore, the ability of a surface to adsorb solute is inversely related to its degree of hydration. For this reason, most inorganic adsorbents are "activated" by heating to drive off bound water; however, the heating must be neither so intense nor so prolonged as to degrade the surface chemically or to cause melting. The usual upper limit for activation of silica gel is about 120–150°C, whereas for alumina the limit is usually about 350–400°C. For inorganic carbonates and for organic adsorbents the limits are usually lower. In actual practice, the activation temperatures employed may in fact be much lower since many solvent systems inherently contain some quantity of water. The point to remember is that the adsorptive capacity of many potential adsorbents is significantly affected by bound surface water; we shall return to this point later on.

Principles of Adsorptive Separations

In the simplest case, adsorptive separations employ single-component solvents and a solid surface to which solute(s) can adsorb. Common adsorbents include SiO_2, Al_2O_3, $CaCO_3$, powdered sucrose, charcoal, hard rubber, and many others. Some surfaces may be further modified by incorporation of metal ions (Ag^+, Cu^{2+}, or Hg^{2+}) or acidic materials (H_3BO_3) to alter the adsorptive properties of the surface. In more complex cases, the solvent systems may include more than one component, but the fundamental basis of separation remains the same. For our purposes, *adsorption* may be regarded as the *reversible adherence of solute(s) to some solid phase.*

When a solute mixture is exposed to a solid phase, solute molecules can be bound; that is, solute molecules leave the free solution (here defined as the mobile phase, M) and pass to the solid phase (here defined as S). As the concentration in the mobile phase (C_m) is depleted, the concentration in the solid phase increases until an equilibrium is reached.

In a system at constant temperature, and with an initial solute concentration represented by C_m, the distribution of solute between the phases S and M can be analyzed in terms of the following symbols:

Let R = total number of adsorption sites on solid phase
R_u = total number of unoccupied sites on solid phase
R_o = total number of occupied sites on solid phase
C_m = concentration of solute in mobile phase
k_1 = rate constant for desorption
k_2 = rate constant for adsorption
ν = proportion of total sites occupied

$$R = R_u + R_o \text{ (conservation equation)}$$

$$\text{desorption rate} = k_1 \cdot R_o$$
$$\text{adsorption rate} = k_2 \cdot C_m \cdot R_u$$

At equilibrium:

$$k_1 \cdot R_o = k_2 \cdot C_m R_u$$

or,

$$\text{or, } k_1 \cdot R_o = k_2 \cdot C_m \cdot (R - R_o) \tag{7-2}$$

Rearranging:

$$\frac{R_o}{R} = \frac{k_2 \cdot C_m}{k_1 + k_2 \cdot k_2 \cdot C_m} \tag{7-3}$$

If:

$$\nu = \frac{R_o}{R},$$

and letting

$$\frac{k_2}{k_1} = \kappa$$

then:

$$\nu = \frac{\kappa \cdot C_m}{1 + k \cdot C_m} \tag{7-4}$$

Note that the form of Eq. 7–4 here is identical with that of Eq. 5–5 for ligand binding in systems with noninteracting binding sites. In fact, adsorption can be regarded as a special case of "ligand" binding, where nonspecificity or low specificity is the hallmark. Further, κ can now be interpreted as the overall equilibrium constant for the sorption–desorption process.

Mechanism of Zone Migration: Differential Solute Movement

In a developing chromatographic system at some instant after separation begins, the solvent will have moved a distance Δm from the origin to some other point. At the same instant, a solute will have moved some lesser distance, Δs, since its movement has been retarded by the separation process. The *relative frontal mobility,* R_f, where

$$R_f = \Delta s/\Delta m \tag{7-5}$$

is the usual way of quantitatively describing the movements of solutes with respect to one another and the origin. The R_f values of a chromatogram are a fair measure of the system's resolving power.

At the molecular level, migration occurs as a random process, each molecule behaving independently of another. Solute migration is essentially a statistical process. Each molecule is retarded some part of the time by sorptive forces, and when it is adsorbed its motion ceases. One mobile molecule may pass an adsorbed molecule at time t_x, only to be overtaken at $t_x + \Delta t$, when it may be transiently adsorbed and the first molecule desorbed. The statistical nature of the process accounts for some solute band spreading (diffusion), the extent of which is a direct function of the duration of the separation process. One can therefore express the fundamental migration parameter, R, in terms of the time spent in the mobile phase, t_m, and the time spent in the adsorbed state, t_a, as:

$$R = \frac{t_m}{t_m + t_a} \tag{7-6}$$

This equation emphasizes the kinetic nature of the separation process in terms of single sorption-desorption steps.

PARTITION CHROMATOGRAPHY

Basic Principles

To understand the theoretical aspects of partition chromatography, we need only slightly modify the theory just developed for adsorption chromatography. In the latter case it was assumed that adsorbed molecules were taken up by a solid, rigid surface, and that while they were adsorbed, solute molecule movement ceased. In partition chromatography this way may or may not be the case. Thus, in GLC, a high-boiling liquid is coated in a thin film on an inert support. Ideally, this liquid film does not move. A solute molecule dissolved in the liquid film may be regarded as motionless with respect to gas flow. In paper chromatography with butanol–water–acetic acid solvent systems, the situation is quite different. Here the water is more tightly adsorbed to the paper than is the butanol, so there is some movement of butanol with respect to the water, but both components actually flow through the paper. (The acetic acid is regarded as a means of pH control, being compatible with both water and butanol, just as is the NH_4OH employed in alkaline solvent systems. Neither affects the sense of this discussion.) Close examination of the solvent front of such systems shows a slight "fuzziness" of the front due to real physical difference in flow rates of the water with respect to the butanol. In sum, compound solvent systems are always selected for partition chromatography so that one component moves more rapidly than another, which accounts for the solute separation.

One analytical approach emphasizes the equilibrium point of view and considers the total collection of solute molecules all at once. In some arbitrarily small region of a column (or a thin layer), the amount of solute in the mobile phase is given by the expression, $C_m V_m$, and the amount of solute in the stationary (or less mobile) phase is given by $C_s V_s$. Accordingly, the fraction of solute in the mobile phase is given by:

$$R = \frac{C_m V_m}{C_m V_m + C_s V_s} = \frac{V_m}{V_m + (C_s/C_m) V_s} \tag{7-7}$$

But by definition, $C_s/C_m = K_d$, where K_d is the distribution or partition coefficient. Consequently, it is proper to write a final form for R as:

$$R = \frac{V_m}{V_m + K_d V_s} \tag{7-8}$$

Experiments indicate that K_d is a function of the temperature and pressure at which the system operates, the nature of the solvent system, and the solute nature.

MEASUREMENT OF OPERATING CHARACTERISTICS

For a given column or TLC system that depends either on adsorption or on partition, the practical questions are: How large a sample can be loaded, how long will it take to elute a given solute, and how good will be the resolution between the particular solute and one that runs closely adjacent to it? These questions can be answered, provided one has some knowledge of the values for K_d, the distribution or adsorption coefficients.

Let us begin by defining the following:

V_R = retention volume of a given band or peak (the volume required to just elute the substance in question from a column or lane of a TLC plate)

V_0 = retention volume required to elute a nonretarded substance (equivalent to use of a tracking dye in electrophoresis)

W = weight of packing or of a thin layer lane

t_R = retention time for a particular solute

t_0 = retention time for a nonretarded substance

Q_S = capacity or loading factor for the solute, S

N = number of "theoretical plates" (from distillation or GLC theory)

It can then be shown that $Q_S = (W/V_0)K_d$. It can also be shown that the retention time for a particular solute is given by

$$t_R = t_0(1 + Q_S) = t_0(1 + W/V_0) \tag{7-9}$$

Given corresponding information for more than one solute, the resolution, or separation between centers of bands or peaks, expressed in terms of retention volumes, is as follows:

$$\Delta V_R = \left(\sqrt{\frac{N}{R}}\right)\left(\frac{Q_2}{Q_1} - 1\right)\left(\frac{Q_1 + Q_2}{2}\right)\left(1 + \frac{Q_1 Q_2}{2}\right) \tag{7-10}$$

Implicit in the above analysis is the assumption that the adsorption isotherms for the solutes is linear; then the emerging peaks will be symmetric. Nonlinear adsorption isotherms give rise to skewed peaks, which will suffer distortion of the leading or trailing peak edge. Under such conditions the expression for ΔV_R cannot apply.

ADSORBENTS

Properties of a Selected List

Powdered Sucrose. Powdered sucrose is one of the earliest organic adsorbents, used by Twsett and others to separate certain carotenoid and chlorophylloid plant pigments. Its use is no longer in vogue, except to demonstrate the historical separations mentioned above. However, the adsorptive properties of powdered sucrose should not be overlooked.

Cellulose. Cellulose is still widely employed in chromatography, either as paper sheets or as thin layers. A considerable body of data on paper sheet chromatography can be carried over directly to TLC. Paper sheets possess a troublesome grain, imposed by the milling process in manufacturing. Migrations in the direction of the grain almost always differ from rates running against the grain, and the grain direction is not always easy to determine with dry sheets of paper. The problem can be minimized by always cutting paper sheets in the same direction. Cellulose processing sometimes causes oxidation of end groups to -COOH groups, which interfere with absorption. To avoid this problem some workers pretreat the cellulose fibers with $NaBH_4$, reducing any -COOH groups to $-CH_2OH$ groups. Unmodified cellulose is not ordinarily employed in column chromatography because it has a tendency to pack so tightly as to give very unsatisfactory flow rates.

Ground Hard Rubber. Because of the variable nature of the starting material, ground hard rubber is no longer used much, but it was previously widely employed in separation of certain hydrophobic materials. It has been largely supplanted by synthetic resins.

Polystyrene Beads. Polystyrene polymers or polystyrene-divinyl+benzene copolymers can be cast in the form of tiny beads, without any charged groups being attached to the surfaces. By controlled cross-linking, the beads can be made with known porosity and swelling properties. Because they lack ionized groups, these beads function by adsorption only. They are becoming increasingly useful to adsorption chromatography of a wide range of substances, from alkaloids to steroids, etc. These beads are now commercially available, designated as XAD resins.

Fuller's Earth. Sold under one trade name as Florisil®, Fuller's earth consists of a mixture of minerals obtained from certain clay deposits. The minerals include kaolinite, montmorillonite, and Halloysite. Fuller's earth is still widely used to decolorize food oils. It frequently contains heavy metals that must be removed by acid treatment. Fuller's earth is also used for removal of pigments from urine samples prior to analysis for other substances. It probably has no tremendous advantage over other adsorbents, other than its very low cost.

Powdered Charcoal. Powdered charcoal is a very powerful adsorbent, but its properties are markedly dependent on its origin and on the process by which it was prepared. Bone, blood, and coconut charcoals are especially useful. Depending on the origin, charcoal may be more or less contaminated with heavy metals, and these can be removed only by drastic treatment with acid. Reproducibility from lot to lot of charcoal may also be a problem. Finely divided charcoal is also quite hard to wet, and preparing a column packed with charcoal can be a most frustrating experience. Charcoal is one of the few adsorbents that may be regenerated by heating to a red glow. Obviously, some of the substance will be consumed in this process, but ordinarily the loss is not a major factor.

Hydroxyapatite (Durapatite). Hydroxyapatite is a basic calcium phosphate, with the probable formula of $3Ca_3(PO_4)_2 \cdot Ca(OH)_2$, which must be very carefully prepared and stored. Its preparation is tedious, and the product frequently varies from batch to batch. It is now commercially available. The gel must not be allowed to dry out (which makes it of very little use for TLC), and it must not be allowed to freeze. Drying or freezing grossly alters the degree of hydration and so changes the gel structure. It has proven quite useful for protein separations and is still widely used for that purpose.

Alumina, or Aluminum Oxide Gel. Alumina is a powerful adsorbent that is widely used. It is prepared by precipitation of aluminum hydroxide by addition of base to some soluble aluminum salt. The gelatinous precipitate is then washed, dried, and sized. Depending on the conditions of precipitation, or sometimes by later treatment, it may be made "acidic, basic, or neutral." These terms, of course, relate to the nature of certain groups left bound to the surface. Alumina is mechanically quite strong, and has the advantage of not crumbling to still finer particles when manipulated for packing into columns or for casting thin layers. It must be strongly heated to bring it to the anhydrous state, represented by Al_2O_3.

Alumina is also available as a true gel, especially in the form known as alumina C-γ. This gel is employed in protein separations, although since the advent of newer protein separation methods it is not as popular as it once was. Its preparation is tedious, and its properties vary with age and conditions of storage. Like hydroxyapatite, it must not be allowed to freeze or to dry out. Commercial preparations are not always of good quality.

Silica, or Silica Gel. Silica is a popular adsorbent because it binds a diverse range of substances with moderate strength. The form known as silica gel is prepared by precipitating hydrated silica from sodium silicate solution, under carefully controlled conditions, on addition of acid. The technology of its manufacture for chromatographic purposes is complex. Silica gel must be exhaustively washed, dried, and sized precisely. It is more friable than alumina and so should be handled carefully to avoid change in size by mechanical damage. Silica is also convenient because it can be dried at relatively modest temperatures as compared to many other possible adsorbents.

Silica has also been used in the form of crushed and powdered glass, which has very strong adsorptive properties for some proteins. Indeed, the adsorption may be so strong that sometimes the proteins cannot be desorbed without denaturation. This is true, for example, for some of the blood serum α globulins. Porous glass beads are commercially available, made with known pore sizes. These enjoy some use in chromatographic separations based on adsorption and gel filtration principles.

Standardization of Adsorbents

Earlier mention was made of the effect of water on the ability of an adsorbent to bind solutes; it should be clear that the maximum sorptive power is demonstrated only when the adsorbent is heated sufficiently to remove the bound water. When this is done, it may happen that a given solute or group of solutes is so strongly adsorbed that acceptable chromatographic separations cannot be performed. One must then select a less powerful adsorbent, or one may elect to inactivate the adsorbent partially by deliberately adding a known amount of water. No single standard has been adopted for this purpose, but the ideas first put forth by Brockmann have been most widely quoted. Initially, Brockmann worked with alumina but later went on to examine the properties of partially hydrated silica gel and other adsorbents.

Brockmann's procedures may be explained as follows. A sample of the alumina was dried to constant weight at 350–400°C. The heated material was allowed to cool in an atmosphere of dried inert gas, such as nitrogen, or by storage in a vacuum dessicator. Weighted aliquots were transferred to each of a series of five closed jars. Water was added to each of the samples in an amount to provide partially hydrated materials, as shown in Table 7–1.

The samples were allowed to stand until equilibrium was reached (more commonly, the samples are gently shaken or tumbled in a closed jar con-

Table 7–1. Effect of Added Water on Relative Mobility of Azobenzene

Brockmann Grade	Water Added, % (w/w)	Relative Mobility of Azobenzene
I	3	—
II	6	0.59
III	9	0.74
IV	10	0.85
V	15	0.95

taining small pieces of glass rod). The samples were then prepared for chromatography, azobenzene (or one of a series of azobenzene derivatives) was used as a test solute, with the results shown above. The original solvent used was dry carbon tetrachloride.

Notice that Brockmann's procedures were entirely arbitrary. There was no theoretical justification for addition of water in the amounts quoted, nor for use of azobenzene and its derivatives as standardizing substances, nor for use of carbon tetrachloride as the developing solvent. Nevertheless, the ideas put forth by Brockmann and his collaborators were of great utility, which explains why even today one may purchase lots of alumina identified as Brockmann I, Brockmann II, etc. These are samples pretreated by the vendor according to Brockmann's initial scheme. The reader should take note of the fact that Brockmann's standardization applies, strictly speaking, only to the defined solute and the defined solvent. There is no guarantee that other solutes or other solvent systems will produce a corresponding order of relative migration. The designations reflect only the degree to which the adsorptive power of the alumina (or silica, etc.) has been diminished by addition of water. This is exemplified by Brockmann's later work, in which the effects of substituent groups in a series of azobenzene derivatives was examined. In this list, R = $\phi N{=}NC_6H_4^-$, the compounds studied can be represented as shown:

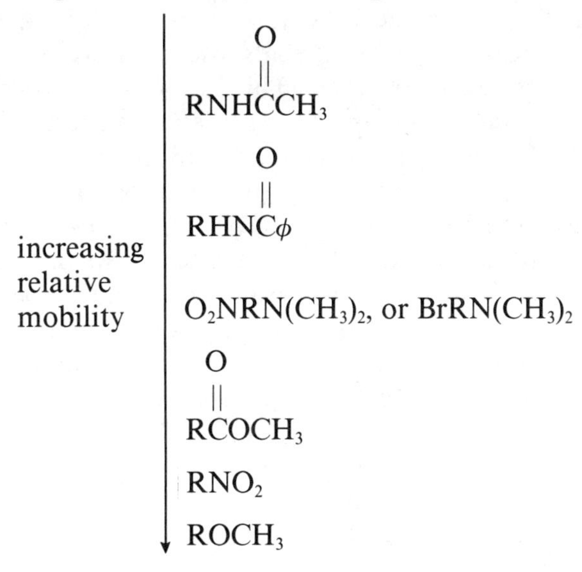

If these compounds had been studied on a differently hydrated alumina, or on a different adsorbent altogether, the results might not have been the same. In modern biochemical studies, one might elect to use as standards a series of phospholipids with different polar head groups, or a series of nucleotides with different bases, or whatever meets the need. The fundamental idea remains unchanged.

SOLVENT SYSTEMS AND THEIR MANIPULATION

In the experiments described above, the development solvent was dry carbon tetrachloride. More commonly, especially in biochemical applications, solvent systems of greater complexity are required. These may contain as many as four reagents, such as chloroform, methanol, acetic acid, and water, or butanol, isopropanol, acetic acid, and water. Furthermore, modest changes in the proportions of the solvent system may have significant effects on the quality of the separations achieved. As before, we seek some underlying body of experience or theory to guide solvent system selection. In entering this discussion, it is essential to keep in mind the fact that all mixed solvent systems are described strictly on a volume basis; thus a system described as dioxane/water (9:1) refers to 9 *volumes* of dioxane *plus* 1 *volume* of water. After mixing, however, the result may *not* equal 10 volumes.

To be useful in chromatography, a solvent must have two properties. *First,* it must be an effective solvent for a given solute in bulk solution. Experience indicates, for example, that benzene would not be a good solvent for some substance like glucose; we operate on the intuitive notion that "like dissolves like." Pertinent terms that are used more or less loosely include "hydrophobic" and "hydrophilic," "polar" and "nonpolar," and so forth. All of these refer primarily to the interaction between a certain solvent and a certain solute. *Second,* one must consider the interaction between the solvent and the adsorbent itself: To desorb a solute, the solvent must displace it from the adsorbent surface, implying that the solvent must be strongly adsorbed. This is commonly described as the *eluting power* or *solvent strength* of the solvent with respect to the adsorbent in question. Solvents and solvent systems are frequently arranged in *eluotropic series;* in general, these will apply to most common adsorbents, but the ordering can depend very much on the solutes tested. Lloyd Snyder has shown that the strength of a solvent can be theoretically predicted from the individual Langmuir adsorption isotherm, but the necessary calculations are too cumbersome and complex for ordinary laboratory use. Thus it is easy to see that hexane is a nonpolar solvent and that water is a strongly polar solvent, but compounds that fall between these extremes are harder to judge and are evaluated empirically.

A precise measure of the polarity of a molecule is given by its *dipole moment,* a function of its internal charge separation. A derived measure of the dipole moment is the *dielectric constant,* which may be determined

Table 7–2. Selected Physical Properties of Some Common Solvent System Components

Compound	Dielectric Constant, ϵ	Molar Volume, V	Compound	Dielectric Constant, ϵ	Molar Volume, V
Isooctane	1.9	165.0	α-Picoline	9.8	98.0
Hexane	1.9	130.6	1,2-Dichloroethane	10.4	78.7
Heptane	1.9	146.5	Pyridine	12.3	80.9
Cyclohexane	2.0	108.0	*t*-Butanol	12.5	98.0
Carbon tetrachloride	2.2	96.8	Isoamyl alcohol	14.7	108.4
1,4-Dioxane	2.2	85.3	Isobutanol	15.8	92.0
Benzene	2.3	88.9	*n*-Butanol	17.0	91.5
Triethylamine	2.4	139.4	Isopropanol	18.3	76.5
Toluene	2.4	106.4	Methyl ethyl ketone	18.5	89.6
n-Butyl ether	2.8	169.3	Methoxyethanol	19.9	79.8
Butyric acid	3.0	91.9	*n*-Propanol	20.0	74.6
Trichloroethylene	3.4	88.2	Acetone	21.0	73.7
Diethylamine	3.8	103.5	Ethanol (absol.)	24.0	58.4
Isopropyl ether	3.9	140.7	Methanol (absol.)	33.0	44.5
Diethyl ether	4.3	104.0	Nitrobenzene	34.8	102.2
Chloroform	4.8	80.5	Dimethylformamide	36.7	77.3
N,N-Dimethylaniline	4.9	126.8	Acetonitrile	37.5	52.2
Propylamine	5.3	82.2	Ethylene glycol	37.7	55.9
Ethyl acetate	6.0	97.7	Dimethyl sulfoxide	47.0	71.0
Acetic acid	6.2	57.3	Formic acid	58.0	37.7
Methyl acetate	6.7	79.8	Water	80.0	18.0
Aniline	6.9	91.1	Formamide	84.0	39.8
Ethyl formate	7.1	80.8	NH_4OH (15.1 M) (sp. gr. = 0.88)	—	66.2
Tetrahydrofuran	7.6	81.1			
Dichloromethane	9.6	64.1	HCl (11.7 M) (sp. gr. = 1.19)	—	85.5
Quinoline	9.0	118.5			

by capacitance measurements. Equation 7–11 defines the capacitance of a parallel-plate condenser as:

$$C = \frac{q}{v} \tag{7–11}$$

where C is the capacitance, q = quantity of energy transferred from one plate to the other at a given frequency, and v = the potential difference between the plates. The entire system is assumed to operate in a vacuum. The capacitance of a condenser is increased if the vacuum between the plates is filled with some nonconducting substance. The dielectric constant, ϵ, is defined by the equation

$$\epsilon = \frac{C_x}{C_{vac}} \tag{7–12}$$

where C_x is the observed capacitance when the system is filled with the substance, x. In simple terms, the dielectric constant is a measure of intrinsic charge asymmetry, or the asymmetry induced in a charge field such as is presented by an adsorbent surface.

Knowledge of the dielectric constant makes it possible to rank precisely a large variety of solvents, as shown in Table 7–2. Moreover, it is possible to use this information in mixed solvent systems to relate separation results, in terms of R_f values, to exact solvent system composition. This has been demonstrated with considerable elegance by D. M. Pierce. He constructed two sequences of solvent systems, each involving fourteen different solvent mixtures. One sequence was acidic; the other was basic. He next selected three basic and three acidic synthetic pharmaceutical substances. Each of the basic pharmaceuticals was tested in each of the basic solvent systems; then similar studies were performed on the acidic pharmaceuticals in the acidic solvent systems. The collected R_f values were plotted against the calculated dielectric constants of the solvent systems, where the ϵ values ranged from about 2 to about 10. A good correlation was noted (the correlation coefficients ranged from 0.68 to 0.87), lending support to the hypothesis that the relative mobility of the solute is indeed proportional to the dielectric constant of the solvent system. From his data, it does not appear that Pierce made any corrections for nonideal solvent behavior (volume changes on mixing). Perhaps the correlation would have been better if such corrections had been made. In addition, only a small number of solutes were tested, and the tests were run only on silica gel thin layers. Further and wider experience will be required to evaluate the merit of Pierce's hypothesis.

Pierce calculated the ϵ values of mixed solvent systems according to the equation of Amirjahed and Blake, which states that

$$\epsilon_{\mathrm{m}} = \frac{\Sigma X_i(V_i + 2\rho_i)}{\Sigma X_i(V_i - \rho_i)} \qquad (7\text{–}13)$$

where ϵ_{m} = the dielectric constant of the mixed solvent system; X_i = the mole fraction of the component i, V_i = the molar volume, in milliliters, of the component; and ρ_i = the molar polarization, given by:

$$\rho_i = \left| \frac{\epsilon_i - 1}{\epsilon_i + 2} \right| V_i \qquad (7\text{–}14)$$

where the absolute value of the quantity within the vertical bars is taken.

In addition to demonstrating a correlation between dielectric constant and solute mobility, Pierce's work permits a second important conclusion. Since the value of ϵ varies significantly with even slight changes in solvent proportions, reproducible work can be expected only if mixed solvent systems are made up with considerable care. Each component should be measured in a clean, dry container, of a size consonant with the quantity to be measured.

TYPES OF PARTITION CHROMATOGRAPHY

Paper Chromatography (PC)

Paper serves as a support medium for the solvent system, which generally involves an aqueous component and some other material partially miscible with aqueous solutions. The forces that induce solute retardation may involve solute adsorption to the paper, but more typically solutes are partitioned between the more mobile solvent component and a less mobile solvent. Ionic effects usually play only a very small role in the separations possible.

Thin-Layer (Planar) Chromatography (TLC)

A thin layer (0.25–2.0 mm) of virtually any finely divided substance is deposited on an inert backing. The backing material may be glass, aluminum, or cellulose butyrate–acetate foil. For practical purposes, thin layers may be regarded as "infinitely thin" columns. The methods are simple and require little expensive apparatus. TLC is ordinarily performed on an analytical scale; however, with thicker layers it may be scaled upward to handle milligram quantities of solute and so may also be preparative. The retarding forces may involve adsorption, ion exchange, gel filtration, or partition. In other words, nearly any of the separating systems thus far devised can be used in thin-layer form, which is one of its great virtues.

Reverse-Phase Chromatography (RPC)

If an inert support, such as powdered silica, has some lipophilic substance covalently bonded to its surface, then solutes in a mixture may be adsorbed or partitioned on that surface as if it composed the entire mass of the particles. Aliphatic carboxylic acids with chain lengths of 8–18 carbon atoms can be bonded in this manner, and the modified silica therefore presents to a solute mixture a graded hydrophobicity. This is the meaning of *phase reversal*—the conversion of a hydrophilic to a hydrophobic surface. Sometimes it suffices to coat a powder with some hydrocarbon oil that will not be removed by the solvent system. Solvent systems invariably contain one component that is miscible with the fatty component as well as with the aqueous components.

Gas-Liquid Chromatography (GLC)

In their early work, Martin and Synge predicted on theoretical grounds that chromatographic separations should be as feasible in the vapor phase as in the liquid. Their early efforts to demonstrate this were frustrated by lack of a suitable detector for solutes in the vapor state, but with the help of an engineer-physicist named Lovelock, they generated first the argon diode and then the flame detector. This development overcame the problem of detector sensitivity, and GLC became a burgeoning field.

GLC requires a column that contains some very high-boiling liquid, coated either on a finely divided inert support or on the inner wall of the column itself. The term "liquid" must be interpreted loosely; typical materials are represented by stopcock greases, by polyoxyethylene esters, or by modified silicone resins. The column, the detector, and some sort of sample injector are all maintained at some controlled temperature well above the boiling point of any solute in the sample. The sample itself is rapidly vaporized by a flash heater in the injection system, and the vapors are driven through the system by some inert gas. The solute species then partition themselves between the flowing inert gas and the high-boiling liquid, and are thereby retarded.

There are problems and limitations in GLC applications. Proteins, for example, are virtually excluded from analysis by GLC since they cannot tolerate the elevated temperatures, frequently up to 250°C. Even much smaller molecules—sugars, some steroids, amino acids, and others—will be destroyed unless they are derivatized prior to GLC. Amino groups and hydroxyl groups can be acetylated or silylated, carbonyl groups can be converted to phenylhydrazones, and carboxyl groups can be esterified, but all the derivatizations are tedious and expensive, and the modifying reactions may not go to completion except under vigorous conditions. Much effort has been devoted to development of suitable derivatizing reagents as a necessary precursor to GLC. In spite of these drawbacks, GLC has become a powerful tool, particularly in the field of fatty acid and steroid analysis, where there had long been a dearth of other methods. Carbohydrate chemists have also made extensive use of GLC, especially in combination with a mass spectrometer as detector. More recently, the preeminence of GLC has been challenged by HPLC, which avoids some of the above-mentioned problems.

High-Performance Liquid Chromatography (HPLC)

Resin beads for liquid chromatography range from 150 to 300 μm (microns) in diameter. Typical powder for GLC ranges from 100 to 125 μm in diameter. Typical particles for HPLC have diameters of 5–10 μm. It is largely this decrease in particle size that gives HPLC its greater resolving power. The exposed surface is larger, and the void volume of the column (i.e., the volume that is free of packing) is also less. Consequently, HPLC can do as much, or more, in the way of retarding solutes, as did GLC, without the nuisance of carrier gases, high temperatures, or the need for derivatization. Proteins and other sensitive molecules can be subjected to HPLC, which is also a great advantage. Many different types of detectors can be employed, ranging from visible or ultraviolet spectrophotometers or fluorometers to polarographic or electrolytic detectors.

HPLC is sometimes translated differently, as high-pressure liquid chromatography. Because the packings are so fine, they offer a very great resistance to liquid flow. To move fluid through the columns at practical rates, it is necessary to pump the solvent through the system at pressures up to 2000 pounds per square inch (psi). Since the free end of the column is at atmospheric pressure, this means that the pressure drop across the column is quite large. The columns must be made of specially treated metal to withstand the strain, and their inner surfaces must be finely polished. Solvents and samples must be degassed and passed through very fine filters before admission to the column. Degassing is needed to remove dissolved air that could cause bubble formation in the system, interrupting smooth flow by forming void spaces in the packing. Filtration is required to prevent plugging of the fine interconnecting tubing or of the column itself. The very high sensitivity of HPLC also requires that all solvents, including water, be carefully purified to avoid spurious peaks in the output recording.

In spite of these scrupulous requirements, and in spite of the rather high cost of the apparatus, HPLC is the most rapidly growing version of chromatography on the current scene. On balance, its advantages offset its disadvantages. A constantly growing variety of column packings is being marketed, and the literature of new or refined methods is growing even more rapidly. HPLC methods can be performed in the *isocratic* mode, which means that the solvent system has a constant composition, or in the *gradient* mode, which means that the solvent composition can be changed according to some program. Any material that can be separated by GLC can be separated by HPLC, and HPLC will handle proteins, which GLC cannot.

Field Flow Fractionation (FFF)

A rather recent development, sponsored by Calvin Giddings at the University of Utah, has promise of great generality. In essence, the method

Figure 7–4. General plan of a field flow fractionation system. The field in the detailed insert is a crossflow. Zones A and B are compressed to different degrees, depending on differences in diffusivity. Zone B, with average thickness Z_b, will see faster flow lines than zone A, with thickness Z_a, and will thus appear to be less retarded than zone A. The typical thickness of the migration channel, w is 100–500 μm. (Reproduced by courtesy of Drs. J. C. Giddings and K. D. Caldwell, Dept. of Chemistry, University of Utah, Salt Lake City.)

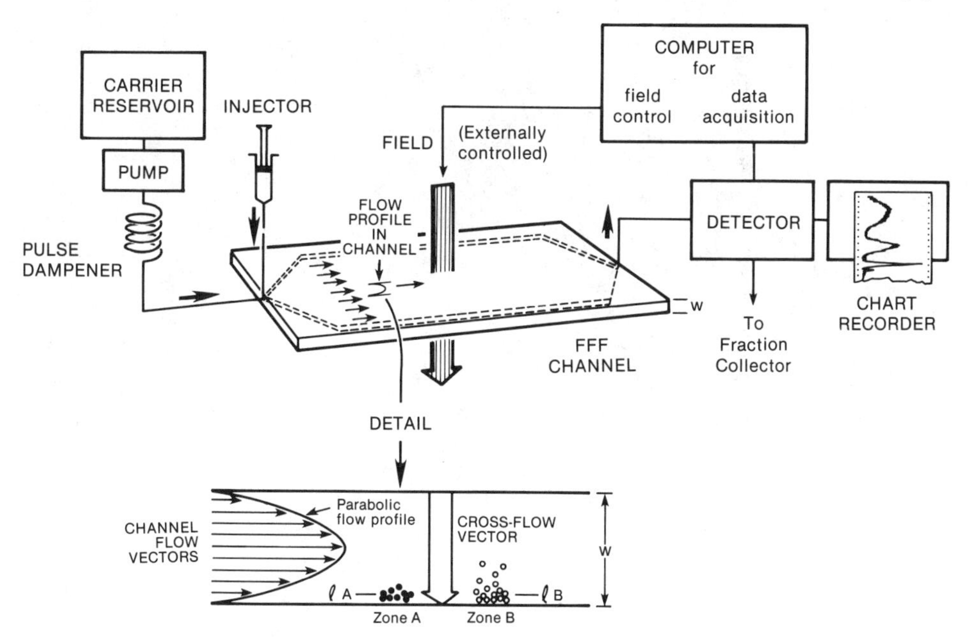

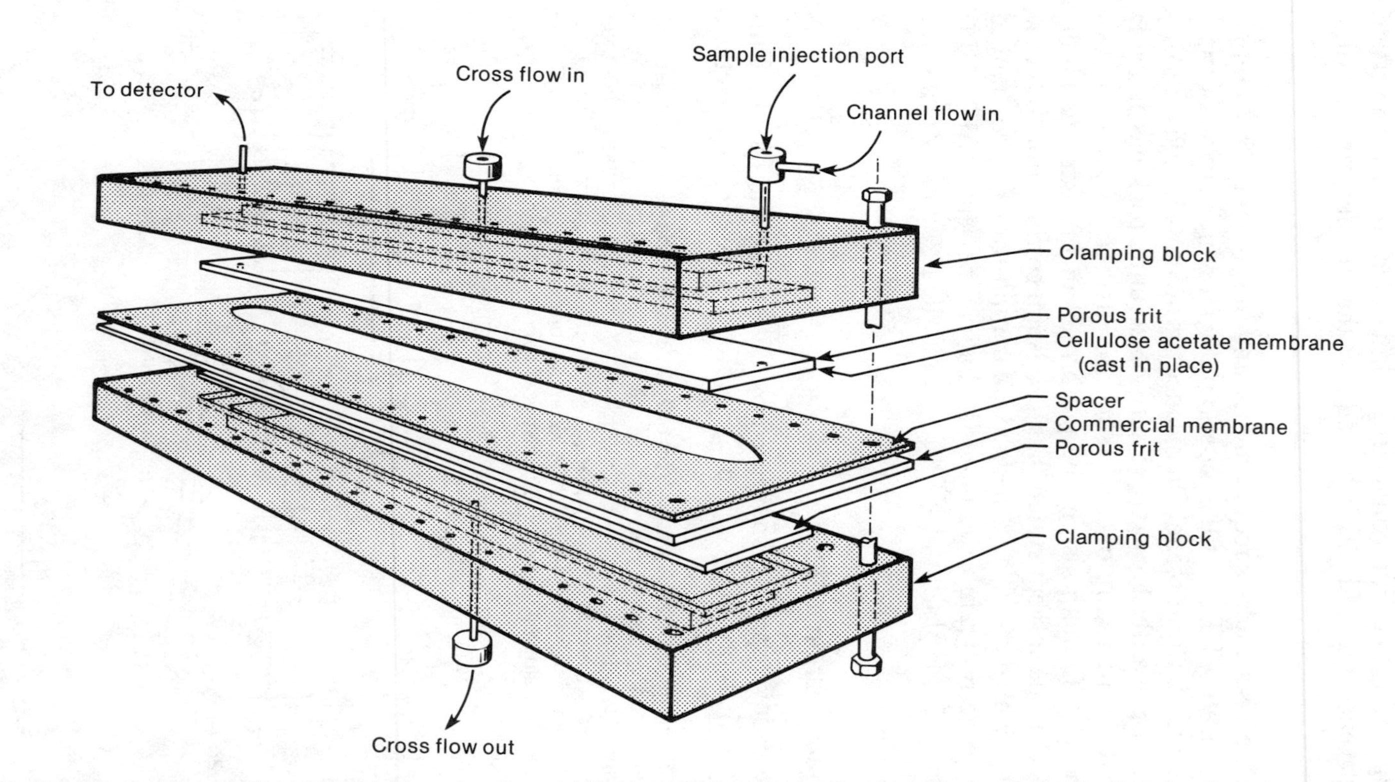

Figure 7–5. Exploded view of a field flow fractionation channel. This particular design, like the unit shown in Fig. 7–4, involves a crossflow of liquid. Other models have been built that involve thermal, electrical, or gravitational fields applied at right angles to the primary direction of flow. In the example shown, the applied crossflow causes solutes to accumulate at the bottom membrane (labeled "commercial membrane") which selectively permits the constituents of the buffer carrier to pass through the system. The clamping blocks may be constructed of glass, metal, or plastic. (Reproduced by courtesy of Drs. J. C. Giddings and K. D. Caldwell.)

involves a flow of solutes through a narrow channel in one direction, with the simultaneous application of another force field at right angles to the direction of flow. Typically, the other force fields are centrifugal, electrical, or thermal fields. Solutes or particles subjected to both fields describe a curved trajectory through the channel and can be collected through an exit port(s). These methods are still being developed, but Giddings has demonstrated their ability to separate intact cells from mixed cell populations, and they have been used for the purification of virus particles, red blood cells, nucleic acids, and colloidal suspensions. FFF methods may use packed or open channels and a variety of detectors (see Figs. 7–4 and 7–5). For further details, see references listed at the end of this chapter.

ION-EXCHANGE CHROMATOGRAPHY

Basic Principles

Ion-exchange resins may be generalized by the notation:

$$R-X_n \rightleftharpoons R^{+n} + nX^+$$

where R is a polymeric matrix bearing some ionizable group, X. As noted earlier, X may be acidic or basic, but for simplicity the present discussion will be illustrated only with carboxylic acid resins. For our purposes, such a resin may be looked upon as a polycarboxylic acid, the anionic and undissociated forms of which are insoluble. Because R is in the form of macroscopic beads, any charges on the bead surfaces are rigidly fixed, and only the soluble counter ions are free to mix and to exchange with ions of like sign in the surrounding medium. Thus:

$$\text{(R)}-COO^-,H^+ + \overset{+}{N}H_3R',X^- \rightleftharpoons \text{(R)}-COO^-,\overset{+}{N}H_3R' + H^+,X^-$$

$$\text{NaOH} \downarrow\uparrow \text{HCL}$$

$$\text{(R)}-COO^-,Na^+ + \overset{+}{N}H_3R',X^- \rightleftharpoons \text{(R)}-COO^-,\overset{+}{N}H_3R + Na^+,X^-$$

Several features of these equations are worth noting:

1. Resin carboxylates can be associated with a variety of counter ions; in the above scheme, two possible choices, H^+ and Na^+, are shown, but one has great freedom to select among others. Resins are frequently sold with the markings "hydrogen form," "ammonium form," etc. As the scheme shows, these forms are readily interconverted.
2. When some solute, say lysylglycine, is percolated through an ion-exchange resin bed *at the proper pH*, exchange of the soluble ions will occur until an equilibrium state is reached. In addition to an effect of pH, an effect due to ionic strength may also be noted.

3. For dilute solutions one can write

$$\frac{[M_1^+]_r\,[M_2^+]_s}{[M_1^+]_s\,[M_2^+]_r} = K_{eq} \tag{7-15}$$

where M_1^+ and M_2^+ are exchangeable ion species, $[M_1^+]$ and $[M_2^+]$ are their respective concentrations, and subscripts "r" and "s" indicate species bound to the resin and species free in solution, respectively. It is the difference in K_{eq} among different pairs of ions that make separations possible.
4. Like other chromatographic media, ion-exchange resins have a finite, saturable capacity, usually expressed as microequivalents (μeq) of exchangeable ions per gram of resin. There is, however, an additional complication. To limit swelling in aqueous media, these resins are cross-linked. More highly cross-linked resins suffer less swelling so that a larger quantity of exchangeable groups can be packed into a column. Whether or not all of the capacity is available to a solute depends on the solute size, since the degree of cross-linking affects the pore size of the resin particles.
5. K_{eq} in Eq. 7–15 may also be taken as the distribution coefficient of the solute ion. In other words, $K_{eq} = K_d$, which is a measure of the separability of two ions: the more K_d differs from 1, the greater the separation that can be achieved. When $K_d > 1.2$–1.3, reasonable separations can be accomplished, but they may not be complete. When $K_d > 1.8$–2.0, complete separation should be possible.

Some Practical Considerations

It would be tedious to present a complete theoretical analysis of ion-exchange chromatographic behavior for the following practical reasons. Although the mechanism by which ion-exchange resins retard solutes differs from the adsorptive mechanisms already discussed, the net effect of the two processes is essentially the same. Ion exchange can thus be considered as a special case of adsorption, conditioned by the facts that only charged particles can be retarded, and then only if the charge is opposite to that on the resin. The process is modulated by the ionic strength, since $\Gamma/2$ affects the ionization process in any electrolyte solution. The reader should understand that, with minor exceptions, the principles of adsorption and ion-exchange chromatography are easily transposable from one type of separation to the other.

Applications

The single most widely used application of ion-exchange chromatography is the analysis of amino acid mixtures obtained by hydrolysis of proteins under controlled conditions. This method was elaborated and highly developed by Moore, Spackman, and Stein. Several commercial versions of automated hardware have put this method at the disposal of scientists everywhere. With the more recent versions of amino acid analyzers, the total analysis can be performed in about 2 hours. Usually two separate resin columns are employed, and buffers of different pH and ionic strength are required. The system is also programed for several temperatures. The method has good precision for samples containing about 5 pmol of each amino acid. Here, successful results are achieved with the use of very fine resin beads, carefully selected for uniformity of size, and operation of flow rates of about 1 ml/min, which requires operating pressures of about 150–250 psi. There are some single-column versions of amino acid analyzers, but they have not been as widely adopted. The references at the end of the chapter provide further details about the equipment and results obtained with it.

EXCLUSION CHROMATOGRAPHY

Basic Principles

Some sources refer to this as molecular sieve or gel filtration chromatography. Regardless of which name is applied, the fundamental principle is simple. Exclusion media may be prepared from organic or inorganic materials, fabricated in a way that produces multitudinous fine channels or pores through the mass of the material. With organic media, the pores are produced by cross-linking of a polymer. With inorganic materials the pores are produced by an etching or solution process that removes part of the mass; this is the method by which porous glass beads are produced.

When a slurry of an exclusion medium in buffer is poured as a column bed, the total column volume, V_t, may be regarded as containing three compartments:

$$V_t = V_o + V_m + V_i \qquad (7\text{–}16)$$

where V_o is the volume external to the medium, V_m is the volume of the gel itself, and V_i is the volume included within the gel. A solute species will distribute between V_o and V_i with a distribution coefficient, K_d. We can at once write that

$$V_e = V_o + K_d V_i \qquad (7\text{–}17)$$

where V_e is the elution volume of the solute. If we define the retention, R, as V_o/V_e, then Eq. 7–16 becomes:

$$R = \frac{V_o}{V_o + K_d V_i} \qquad (7\text{–}18)$$

which is identical in form with Eq. 7–8.

Each of the volume terms in Eqs. 7–16, 7–17, and 7–18 can be experimentally measured. V_o is determined by passing through the bed some material of large enough molecular weight that it is effectively excluded from the interior compartment. A number of materials are available for this purpose. A commonly used material is Blue Dextran, a polysaccharide to which a blue dye, Cibacron Blue, has been covalently attached. It has a molecular weight of $\sim 2 \times 10^6$. V_i can be determined by passing through a smaller molecular species, known to distribute equally in V_i and V_o; adenine, lithium urate, or dinitrophenyl aspartate may be used for this purpose. V_t is determined directly from the dimensions of the gel column. V_m is what is left; frequently, V_m is ignored because it is such a small part of the total volume.

Use in Molecular Weight Estimations

In gel permeation chromatography, there is a close and obvious relation between elution volume and the size of a solute. For many solutes, molecular size parallels molecular weight. A reasonably good empirical description of these relations is given by:

$$V_e = A + B \log \mathrm{MW}_s \tag{7–19}$$

where A and B are constants for a given column, packed with a given medium. When V_e is plotted against log MW_s, a straight line is obtained, at least over some useful range of molecular weights characteristic of the exclusion gel. Equation 7–19 provides chemists with a rapid and rather reliable means of estimating molecular weights *provided that markers of known mass are used to calibrate the column.* It is only by means of the markers that the values of A and B can be determined for an individual case.

Other Applications

Exclusion chromatography has another important use. In the course of protein isolation or of metabolite synthesis, the desired product may be contaminated by salts or similar smaller molecular species. It is possible to desalt a product by passage through an exclusion gel column that just excludes the desired product but freely admits the contaminant species. The product will elute in the void volume while the contaminant is retarded in the pores of the gel. This method is quick, gentle, and eminently practical.

Properties of Exclusion Gel Media

Several exclusion gel media are now on the market. Some are based on polyacrylamide, some on cross-linked dextran polymers, and some on agarose. Some useful data on these media are given on Table 7–3. Note carefully that there is no simple relation between the manufacturers' designations and the operational characteristics of the products.

Table 7–3. Some Properties of Exclusion Gels

Trade Designation	Molecular Weight Fractionation Range	Water Regain (ml/g gel)	Bed Volume (ml/g gel)
Sephadex®[a] G-10	<700	1.0 ± 0.1	2–3
G-15	<1500	1.5 ± 0.2	2.5–3.5
G-25	1,000–5,000	2.5 ± 0.2	4–6
G-50	1,500–30,000	5.0 ± 0.3	9–11
G-75	3,000–70,000	7.5 ± 0.5	12–15
G-100	4,000–150,000	10.0 ± 1.0	15–20
G-150	5,000–400,000	15.0 ± 1.5	20–30
G-200	5,000–800,000	20.0 ± 2.0	30–40
Bio-Gel®[b] P-2	200–2600	1.5	4
P-4	500–4000	2.4	6
P-6	1,000–5,000	3.7	9
P-10	5,000–17,000	4.5	12
P-30	20,000–50,000	5.7	15
P-60	30,000–70,000	7.2	20
P-100	40,000–100,000	7.5	20
P-200	80,000–300,000	14.7	35
P-300	100,000–400,000	18.0	40
Bio-Gel®[c] A-0.5m	<500,000		
A-1.5m	<1,500,000		
A-5m	<5,000,000		
A-15m	<15,000,000		
A-50m	<15,000,000		
A-150m	<150,000,000		

[a]Data are provided by Pharmacia Fine Chemicals. Sephadex is a dextran polymer, cross-linked with epichlorohydrin. Water regain and bed volume data are based on the dry gel. Swelling of the dry gels is best accomplished by heating for 1–5 hours on a boiling water bath.

[b]Data are provided by Bio-Rad Laboratories. Bio-Gels P are polyacrylamide gels. All water regain data are cited with ±10% limits. Swelling of dry gels is best done on a water bath at room temperature. Swelling of P-60 or larger gels is quite slow. See manufacturer's instructions for further details.

[c]Data are provided by Bio-Rad Laboratories. Lower exclusion limits are not well defined. These agarose gels should never be dried out. Values cited are upper exclusion limits.

As noted in Table 7–3, agarose gels should never be dried out; they are sold as suspensions in dilute salt solutions. Polyacrylamide and dextran gels are sold in the dry state and require extensive hydration before use. The data on water regain and on bed volume per gram of gel are stated in terms of the dry gels. This information is helpful in determining how large a column may be required to prepare a gel bed of known hydrated volume. Glass media do not swell, and therefore no comparable data are given for porous glass preparations. Dextran and agarose gels are subject to bacterial decomposition, and wet preparations of these media should be protected by addition of 0.1–0.2% of NaN_3 whenever these columns are stored. The NaN_3 should be washed out before the column is used for separations or desalting operations; this compound is poisonous and, in the form of a heavy metal salt such as lead azide, potentially explosive. *Take care* not to dry down azide solutions but dispose of them using a large volume of water.

AFFINITY CHROMATOGRAPHY

In the most general sense, all chromatography depends on some measure of affinity between a solute and some structural feature of a chromatographic medium, be it a hydrogen bond-forming site, an ionized group, or a pore. However, the affinities involved are generally of a low order and are not at all specific. The term *affinity chromatography* is reserved for a special group of situations where the affinity is quite high, as may be the specificity. Suppose, for example, that one wished to isolate the protein known as haptoglobin from blood serum. It is known that haptoglobin forms a very tight complex with hemoglobin; indeed, that is the normal biological function of haptoglobin. If one could bind hemoglobin to the surface of some inert support, then percolate a diluted serum sample through a bed of the hemoglobin-containing medium, the haptoglobin should be tightly bound; all of the other proteins would wash through the bed. This is the fundamental basis of affinity chromatography. In this example, both the affinity and the specificity are high, and the separation is virtually unique. A second example can be described as follows. Suppose one wished to isolate lactate dehydrogenase from a tissue preparation. It is well known that this enzyme has a high affinity for NAD^+; thus NAD^+ could be bound to an inert support, the tissue extract could be poured through the bed, and, in theory, the lactate dehydrogenase should be bound with high affinity. Indeed, this is exactly what does happen, but the extract probably contains many dehydrogenases with nearly the same high affinity for NAD^+, and the isolated product would by no means be pure. This is a case in which the affinity is high but the specificity is low. In short, affinity chromatography may (but need not) give a singular result, depending on the affinity of a *bound ligand* for one or more than one solute in the crude mixture.

Success in affinity chromatography requires careful attention to three factors:

1. *Proper choice of the inert medium.* It must be mechanically and chemically as stable as possible. It should be as "open" (porous) as possible to present the largest possible surface for ligand binding, and it should not be adversely affected by the chemistry required for ligand binding. Agarose and polyacrylamide gels have both been used for this purpose.
2. *Selection of the most appropriate ligand for the isolation in question.* As noted above, the best results are obtained when the ligand reacts with only a single component of the mixture to be separated. This is not always possible, but it is certainly the goal. The best ligand must also be reasonably stable under the experimental conditions used.
3. *Binding of ligand with minimal loss of affinity.* The bond formed between the ligand and the support must not involve any atoms or functional groups that contribute significantly to the affinity. If, for example, a particular amino acid residue were essential to ligand binding, it would be fruitless to involve that group in attaching the ligand

to the support. Steric hindrance of the affinity binding must also be avoided. In separating large, bulky solutes it is frequently necessary to interpose a chain of carbon atoms between the ligand and the support. This alkyl chain, containing 4–15 carbon atoms, is commonly known as a *spacer arm.* How these are introduced is explained below.

Activation of Agarose-Based Gels

Agarose (M in the scheme below) reacts with cyanogen bromide in the presence of hydroxide ion (pH 11) to form a very reactive *imine* (half-life of about 15 min at 4°C). This imine can add an *amine* to give a mixture of products, an iminocarbonate (I in the scheme below) and an isourea (II). Although the exact reaction mechanism is not entirely clear, the isourea appears to be the predominant product. H_2N-R can be a monofunctional amine, a diamine, or an ω-NH_2 carboxylic acid. In theory, one might suspect that use of a diamine would lead to extensive cross-linking reactions, but in practice the great excess of the amine appears to prevent aberrant reactions, and the major product is the isourea (II).

$$\text{M}(\text{OH})_2 + \text{OH}^- + \text{CNBr} \longrightarrow \text{M}(\text{O})_2\text{C{=}NH (imine)} + \text{H}_2\text{O} + \text{Br}^-$$

$$\text{M}(\text{O})_2\text{C{=}NH} + \text{H}_2\text{NR} \nearrow \text{M}(\text{O})_2\text{C{=}NR} + \text{NH}_3 \quad \text{(I)}$$

$$\text{M}(\text{O})_2\text{C{=}NH} + \text{H}_2\text{NR} \searrow \text{M}(\text{OH})\text{–OC(=NH)–NHR} \quad \text{(II)}$$

In II, R is a spacer arm of desired chain length. If the arm has an -NH_2 group at its free end, a ligand may be connected to this group via some -COOH group of the ligand. Alternatively, if the arm terminates in a -COOH group, a ligand may be attached via some amino group of the ligand. The condensing reagent is in most cases a soluble carbodiimide. Thus, a ligand is attached to the agarose gel with the spacer arm between. Many variations on this basic reaction pattern have been employed, and full details are given in the references at the end of this chapter. One word of warning is necessary. *Cyanogen bromide is extremely toxic; it should be handled only in a good hood!* Commercial suppliers of "activated agarose" proceed in a slightly different manner. They utilize an ω-NH_2-R-COOH as a spacer arm and prepare from it an *N*-hydroxysuccinimide ester by a carbodiimide coupling reaction. This is a stable product. The succinimide moiety readily exchanges with amino groups of proteins or other ligands. Although these commercial products are fairly expensive, they do avoid the hazards of working with CNBr. This reaction pattern can be generalized as follows:

$$\text{(M)}-R-COOH + R'-N{=}C{=}N-R'' \longrightarrow R'-N(H)-C(=N-R'')-O-C(=O)-R-\text{(M)}$$

N-hydroxysuccinimide

$$\text{(M)}-R-C(=O)-ON(COCH_2CH_2CO) + R'-NH-C(=O)-NH-R'' + R'-N{=}C{=}N-R'' + H_2O$$

succinimide ester — a substituted urea

$$\xrightarrow{L-NH_2} \text{(M)}-R-C(=O)-ON(H)-L$$

In the context used here, R represents the free end of a spacer arm already attached to a gel medium (M). See the next section for further details on the addition of the spacer arm to either agarose or acrylamide gels of desired porosities.

Activation of Acrylamide-Based Gels

The amide groups of polyacrylamide gels can be modified by carefully controlled treatment based on partial hydrolysis or aminolysis. Two amines frequently employed for this purpose are hydrazine hydrate and anhydrous ethylenediamine. The products generated are hydrazides, substituted amides with a free amino terminus, or carboxylates, as shown in the scheme below.

$$\text{Matrix}-\overset{\overset{\displaystyle O}{\|}}{C}-NH_2 + NH_2NH_2 \xrightarrow{90°C} NH_3 + \text{Matrix}-\overset{\overset{\displaystyle O}{\|}}{C}-NHNH_2$$

$$\text{Matrix}-\overset{\overset{\displaystyle O}{\|}}{C}-NH_2 + H_2NCH_2CH_2NH_2 \xrightarrow{47\text{–}50°C} NH_3 + \text{Matrix}-\overset{\overset{\displaystyle O}{\|}}{C}-NHCH_2CH_2NH$$

$$\text{Matrix}-\overset{\overset{\displaystyle O}{\|}}{C}-NH_2 + OH^- \xrightarrow[\text{buffer)}]{\text{(bicarbonate}} NH_3 + \text{Matrix}-\overset{\overset{\displaystyle O}{\|}}{C}O^-$$

A variety of spacer arms can then be added and ligands attached using the general methods already described. As was true with agarose, commercial products are available with spacer arms attached and with free ends made ready for ligand binding.

It is important that the bonds connecting the spacer to the matrix and to the ligand should be chemically stable. The examples given above are amide bonds, which are acceptable. An even more stable bond that has been employed is an ether linkage. It is introduced by the use of butane-diol diglycidyl ether, as outlined below:

$$HOCH_2-\overset{\overset{\displaystyle OH}{|}}{CH}-CH_2-O-(CH_2)_4-O-CH_2-\overset{\overset{\displaystyle OH}{|}}{CH}-CH_2OH$$

$$\Big\downarrow \; NaBH_4,\; OH^-$$

$$[\overset{\frown O \frown}{CH_2-CH}-CH_2-O-(CH_2)_4-O-CH_2-\overset{\frown O \frown}{CH-CH_2}]$$

$$\Big\downarrow \; \text{Sepharose}$$

$$\overset{\frown O \frown}{CH_2-CH}-CH_2-O-(CH_2)_4-O-CH_2-\overset{\overset{\displaystyle OH}{|}}{CH}-CH_2-O{\sim}\text{Sepharose}$$

$$\Big\downarrow \; H_2N-CH_2-CH_2-NH_2$$

$$H_2N-(CH_2)_2-\overset{\overset{\displaystyle H}{|}}{N}-CH_2-\overset{\overset{\displaystyle OH}{|}}{CH}-CH_2-O-(CH_2)_4-O-$$

$$CH_2-\overset{\overset{\displaystyle OH}{|}}{CH}-CH_2-O{\sim}\text{Sepharose}$$

This linkage has been successfully used, for example, in the purification of plasma sex steroid binding proteins (SBP), by Mickelson et al. (1978).

Useful methods exist by which almost any ligand can be fixed to one or another of the typical inert supports. This allows preparation of affinity media for almost any desired purpose. The life of an affinity medium depends on the intrinsic stability of the ligand and its possible destruction by enzymic attack from components of mixtures to be separated. The shelf life, or refrigerator life, of an affinity medium may be several months up to a year. Such packings may be cycled as many as fifty times before the capacity is seriously diminished. Thorough washings between cycles is obviously important if normal lifetime is to be expected.

Recovery of Product from Affinity Media

In other forms of chromatography, continued percolation of solvent usually will gradually remove all of the bound substances, but affinity chromatography involves tight as well as quite specific binding. Continued washing will not, as a rule, elute the affined substance. In fact, one usually washes the bed until all but the affined substance has been removed. Three general methods for product recovery are used in most instances:

1. One can increase the ionic strength of the medium, which tends to weaken ionic forces that may hold the affined substance to the bound ligand.
2. One can shift the pH of the medium, with comparable effects; sometimes both the ionic strength and the pH are simultaneously altered.
3. One can percolate through the bed some substance that has a greater affinity for the ligand than the material to be recovered.

Obviously, the first two procedures are preferred since they pose less of a problem with respect to column regeneration and cleanup. Special tricks are sometimes useful. For example, when lactate dehydrogenase is bound to an affinity column where the ligand is an analog of NAD^+, washing with NADH can be used as an eluting ligand for the enzyme, since NADH binds more tightly to the enzyme than does NAD^+. Once free of the packing, the enzyme can be dialyzed or otherwise separated from the eluting reagent (see Experiment 9).

SUPERCRITICAL FLUID CHROMATOGRAPHY (SFC)

If some gases are subjected to high pressure and to temperatures exceeding their critical temperature, they assume a condition that has been considered a fourth state of matter. With CO_2, for example, the pressure used would exceed 1100 psi and the temperature a moderate 31°C. The CO_2 now has the capacity to dissolve organic compounds, like a ligand, although the viscosity is low, as for a gas.

The use of such fluids as the mobile phase in chromatography was first suggested by Klesper et al. (1962); they have since been used industrially for large-scale extractions, but their use in analysis is just emerging. In

SFC, nonisocratic elution is effected by an increase in the pressure and thereby an increase in the density of the fluid; as a result, the solute solubility in the fluid increases. Instrumentation for SFC is understandably somewhat different from that for GC or HPLC; the high pressure must be maintained up to the column-detector interface, where the pressure drops to ambient. The column material that is used can be cross-linked methylpolysiloxane bonded to a fused-silica capillary coiled column. Recent extensions of the analytical uses of EFC have been facilitated by instrumental improvements and availability of information about the solvent behavior of supercritical fluids. Articles by Lee and Markides (1987) and by Greibokk et al. (1987) update the current status of SFC.

SOLUTE DETECTION IN CHROMATOGRAPHY

Very little has thus far been said about detection of materials emerging from a chromatographic bed. Detectors for this purpose must operate on very small samples and frequently at quite low concentrations. Thus, a typical GLC or HPLC sample might contain ten to fourteen different fatty acid methyl esters in a total volume of 10 μl. The elution volume of each component might range from 1 to 5 ml of carrier gas (or liquid), so the actual concentration of a single ester in the eluate stream is small. Similar magnitudes may be encountered in other forms of chromatography. It is perfectly correct to say that the development of chromatography as a whole has closely followed the development of adequate detectors.

For many purposes spectrophotometry and spectrophotofluorometry are adequate means of detection. No new principles are involved here, but considerable ingenuity has been expended to make the light path as large as possible while keeping the detector volume very small. Flow-through cells have been designed with light paths of 5 mm and total volumes as low as 10 μl. These are quite acceptable, but hard to keep free of bubbles.

For GLC, which may operate at temperatures up to 225°C, spectrophotometry is rarely used. One of the early detectors that found wide application was the *argon diode* of Lovelock and Martin. When argon is bombarded by β radiation from ^{90}Sr, the argon molecules are activated and become metastable. The metastable state has a very short half-life. If, before they return to the ground state, the metastable argon molecules collide with a solute vapor, the solutes are ionized and can serve to carry a current between two charged electrodes. The argon diode is very sensitive and can detect as little as 2×10^{-12} mol of most organic materials. Unfortunately, this detector has a limited dynamic range and is rendered almost totally "blind" by water vapor, a disturbing drawback for many biological applications.

A later development was the flame detector, still in very common use. A small flame, fueled by hydrogen and compressed air, serves as an energy source sufficient to ionize solute molecules fed into the burner base. The body of the burner and a surrounding ring electrode are properly charged, and once again ions flow between them to generate a current proportional to solute concentration. Flame detectors are at least an order of magnitude less sensitive than argon diodes, but have a greater dynamic range and are totally insensitive to water vapor (a product of hydrogen combustion).

Other detectors have been devised, but are not as generally employed for lack of sensitivity or because they are easily fouled by one circumstance or another. Examples include hot-wire detectors, which depend on the heat capacity of solute vapors; dielectric constant detectors, which depend on polar properties of solutes; and radiation detectors, which depend on the presence of a radioisotope. A few detectors, quasi-specific for halogens or phosphorus, have come into use for pesticide analysis and similar specialized problems.

A variety of detectors for liquid chromatography, independent of radiant energy, have also been developed. A general-purpose detector measures change in refractive index as solute flows through a special cell, but its sensitivity is low. Further, it is sensitive not only to changes in solute concentration but also to changes in salt or solvent composition, since refractive index is a colligative property.

Much interest has been aroused by a special-purpose detector based on voltammetry. Peter Kissinger and his students have pioneered the development of these systems and have carefully evaluated factors such as electrode composition, pH, and the electroactivity of the solvent, among others. The basic premise of these systems is that certain substances are reducible at properly polarized electrodes. Electroactive substances like epinephrine, norepinephrine, and their metabolites are among those easily reduced and are also of considerable biological interest. Furthermore, these substances have, at the concentrations commonly encountered, few other properties that could allow for easy detection.

Faraday's law tells us that any oxidation–reduction system involves a certain quantity of electricity per mole of substance oxidized (or reduced):

$$Q = n\mathcal{F}N \tag{7–20}$$

where Q is the quantity of current in coulombs, n is the number of electrons transferred, $\mathcal{F}$ is Faraday's constant, and N is the moles of product formed. The current flow is given by:

$$i = \frac{dQ}{dt} = n\mathcal{F}\frac{dN}{dT} \tag{7–21}$$

With modern instruments, it is easy to measure currents of 10^{-9} amperes (amp); this corresponds to product formation at a rate of 10^{-14} equivalents per second. Although the actual current efficiency may be a little less than the theoretical efficiency, voltammetric detectors are sensitive and rugged.

On-Line Derivatization

All of the above systems operate on solutes without any modification. Much or all of the solute can be recovered after it passes through detector systems of the sorts already described. It is possible to proceed in a different manner, forming more readily detectable derivatives of solutes immediately as they emerge from a separation system. The classic example of this is the amino acid analyzer of Stein, Moore, and Spackman, already mentioned. Here the derivatizing reaction was based on ninhydrin. Other on-line reactions have been employed, in which an eluate stream is mixed with a reagent, then passed through some photometric device. A typical example is the reaction of *o*-phthalaldehyde in the presence of amino acids or amines at about pH 10. The product is a substituted isoindole that is strongly fluorescent. The reaction is shown below:

$$C_6H_4(CHO)_2 + HOCH_2CH_2SH + RNH_2 \longrightarrow \text{(isoindole, } SCH_2CH_2OH \text{, } N{-}R) + 2H_2O$$

On-line derivatization reactions have become increasingly important as automated adjuncts to microprocessor-controlled analytical systems. Reagent addition, component mixing time, and temperature of the reaction can be preprogramed for readout by standard photometric devices virtually without operator intervention; even automatic sample injection is becoming increasingly practical.

Gas-Liquid Chromatography–Mass Spectrometry (GLC–MS) Combinations: MS as a Detector

The most elegant (and most expensive!) detector for gas chromatography is a mass spectrometer, in which a solute vapor is first ionized, then separated from its congeners in terms of its charge/mass ratio. This provides an absolute physical characterization of the ion. During the ionization process, a certain fraction of the ions are fragmented, and the pattern of fragments often provides substantial additional information concerning the structure of the original solute. Until recently, dilution of the solute by carrier gas has posed some problems, but newer vapor separators, operating at extremely low pressures (10^{-7} torr) have largely eliminated this problem, and GC–MS instruments are now available in several models. Even the least of these can cover the mass range up to 1000, and some of the more sophisticated systems can go considerably higher.

WHAT CHROMATOGRAPHY DOES NOT DO

It is easy to be lulled into a false sense of security by chromatography, especially when a reasonably Gaussian peak, more or less well separated from other solute peaks, is obtained. Chromatography does *not* guarantee that the material collected in a single peak or fraction is homogeneous from a molecular point of view. It is not at all uncommon to find that a material carefully isolated by chromatography will, in electrophoresis, generate more than a single band or spot, or may produce more than one peak when examined by some different chromatographic modality. The reasons for this can be inferred from Martin's definition, quoted at the beginning of this chapter. The essence of the chromatographic process is differential retardation by one means or other. It is clear that more than one substance may have exactly the same ionic charge; these could not be resolved by ion-exchange procedures. Two different substances may be adsorbed just as tightly on a TLC plate; these might have different structures but identical boiling points. Such substances could not be resolved by a single GLC column. Even affinity chromatography, with its high specificity, may not give absolute separations. Thus, Cibacron Blue affinity columns will bind many NAD-dependent dehydrogenases, as is easily demonstrated. Like all other separation sciences, chromatography is powerful but not all-powerful. This goes to the heart of our concept of purity. The more methods are improved, the more one must realize that absolute purity is a seldom realized goal.

REFERENCES

Amirjahed, A. K., and Blake, M. I. Relationship of composition of nonaqueous binary solvent systems and dielectric constant. *J. Pharm. Sci. 63:*81 (1974).

Barth, H. G., Barber, W. E., Lachmüller, C. H., Majors, R. E., and Regnier, F. E. Column liquid chromatography. *Anal. Chem. 58:*211R (1986).

Brockmann, H. Chromatography of colourless substances and the relation between constitution and adsorption affinity. *Disc. Faraday Soc. 7:*58 (1949).

Clement, R. E., Onuska, F. I., Yang, F. J., Eiceman, G. A., and Hill, H. H., Jr. Gas Chromatography. *Anal. Chem. 58:*321R (1986).

Cuatrecasas, P. Protein purification by affinity chromatography. *J. Biol. Chem. 245:*3059–3065 (1970).

Giddings, J. C., Myers, M. N., Caldwell, K. D., and Fisher, S. R. Analysis of biological macromolecules and particles by field-flow fractionation. *Methods Biochem. Anal. 26:*79–136 (1980).

Greibokk, T., Berg, B. E., Blilie, A. L., Doehl, J., Farbrot, A., and Lundanes, E. Techniques and applications in supercritical fluid chromatography. *J. Chromatogr. 394:*429–441 (1987).

Heftman, E. *Chromatography,* 3rd ed. Van Nostrand-Reinhold, New York (1975).

Inman, J. K., and Dintzis, H. M. The derivatization of cross-linked polyacrylamide beads. Controlled introduction of functional groups for the preparation of special-purpose, biochemical absorbents. *Biochemistry 8:*4074–4082 (1969).

Janca, J. *Field-Flow Fractionation: Analysis of Macromolecules and Particles.* Marcel Dekker, New York (1987).

Karger, B., Snyder, L. R., and Horvath, C. *An Introduction to Separation Science.* Wiley-Interscience, New York (1973).

Klesper, E., Corwin, A. H., and Turner, D. A. High pressure gas chromatography above critical temperatures. *J. Org. Chem. 27:*700–701 (1962).

Krstulovic, A. M. *Reversed-Phase High-Performance Liquid Chromatography: Theory, Practice, and Biomedical Applications.* John Wiley and Sons, New York (1982).

Later, D. W. Supercritical fluid chromatography. *Am. Lab. 18*(8):108 (1986).

Lee, M. L., and Markides, K. E. Chromatography with supercritical fluids. *Science 235:*1342–1347 (1987).

Levy, E. J., Lurcott, S., Oneill, S., Yocklovich, S., Cohen, R., Pfeiffer, K., Wampler, T. P., and Liebron, S. A. Supercritical fluid chromatography. *Am. Lab. 19*(8):66 (1987).

Mayer, S. W., and Tompkins, E. R. Ion exchange as a separation method. IV. A theoretical analysis of the column separation process. *J. Am. Chem. Soc. 69:*2866–2874 (1947).

Mikelson, K. E., Tellerm, D. C., and Petra, P. H. Characterization of the sex steroid binding protein of human pregnancy serum. Improvements in the purification procedure. *Biochemistry 17:*1409–1415 (1978).

Moore, S., Spackman, D. H., and Stein, W. H. Chromatography of amino acids on sulfonated polystyrene resins. *Anal. Chem. 30:*1185–1189 (1958).

Perry, J. A., Haag, K. W., and Glunz, L. J. Programmed multiple development in thin layer chromatography. *J. Chromatogr. Sci. 11:*447–451 (1973).

Pierce, D. M. Selection of solvents for thin-layer chromatography by means of a simple ranking system based on dielectric constants. *Xenobiotica 11:*857 (1981).

Porath, J. Gel filtration of proteins, peptides and amino acids. *Biochim. Biophys. Acta 39:*193–207 (1960).

Snyder, L. J. *Principles of Adsorption Chromatography.* Marcel Dekker, New York (1968).

Snyder, L. R., and Kirkland, L. *Introduction to Modern Liquid Chromatography.* John Wiley and Sons, New York (1979).

Spackman, D. H., Stein, W. H., and Moore, S. Automatic recording apparatus for use in chromatography of amino acids. *Anal. Chem. 30:*1190–1206 (1958).

Stahl, E. *Thin-Layer Chromatography,* 2nd ed. Springer Verlag, New York (1969). [Some useful theoretical discussion was deleted from the second edition; for this reason the first edition is still useful.]

Thompson, S. T., Cass, K. H., and Stellwagen, E. Blue dextran–Sepharose: An affinity column for the dinucleotide fold in proteins. *Proc. Natl. Acad. Sci. U.S.A. 72:*669–672 (1975).

Trayer, I. P. Affinity chromatography. In *Techniques in Protein and Enzyme Biochemistry,* Part 1, Vol. B1/I. Elsevier, Amsterdam (1978).

Wilson, J. E. Applications of blue dextran and Cibacron Blue F3GA in purification and structural studies of nucleotide-requiring enzymes. *Biochem. Biophys. Res. Commun. 72:*816–823 (1976).

CHAPTER EIGHT

The Centrifuge and Centrifugal Separations

Centrifugal separations are designed to separate or fractionate mixtures by application of a large force field in place of gravity. Originally, centrifugal separations were applied to solids more dense than the medium in which they were suspended. The method was next extended to particles less dense than the suspending medium and ultimately was extended to include substances dissolved in another medium. This steady march of applications was made possible by improved technology that resulted in the development of centrifuges of ever higher speed. Today, centrifuges are divided into three major categories, depending on the maximum speed of which they are capable. Low-speed equipment can operate to a maximum of about 5×10^3 revolutions per minute (rpm), high-speed instruments operate to a maximum of about 2×10^4 rpm, and ultracentrifuges operate to a maximum of 10^5 rpm.

CENTRIFUGE COMPONENTS

Stripped of nonessentials, a centrifuge involves only two components. The first is a drive mechanism, the source of rotary motion. The second is a rotor, around the periphery of which are located sample holders for the materials to be fractionated. Large amounts of energy must be expended in order to drive the rotor at the necessary speeds, and the strain on the substance of the rotor may be enormous. For this reason, the rotors are massively built and are precisely balanced around their axis of rotation. Kinematic considerations require that introduction of the samples not disturb the balance. The simplest way to accomplish this is to place samples of equal mass at opposite ends of the rotor diameter. Most rotors therefore hold an *even* number of samples, although a few special-purpose rotors accept an *odd* number of samples equally spaced around the rotor circumference. Special balances are available to ensure that the weights of a pair of samples, plus the weights of their holders, are the same within a few tenths of a gram.

Low-speed centrifuges usually are started by applying power and stopped by shutting off the power and the rotor mount. Some may incorporate timers, mechanical or dynamic braking, or similar auxiliary items, but many do not even include a calibrated speed control. High-speed centrifuges are more elaborate; they almost always include a timer, a dynamic brake, and a calibrated speed control. One can purchase several rotors of different capacities. To remove frictional heat generated by the rotor speed, they usually have mechanically refrigerated rotor chambers, but, like the low-speed equipment, they operate at atmospheric pressure. Ultracentrifuges can reach their upper speed limits because geared-up drive mechanisms are used in place of direct drives, and because the rotor is operated in a vacuum. Recently, there has been a revived interest in an old idea: cutting carefully designed turbine slots in the bottom of a small rotor. By passing a jet of compressed air across the turbine slots, the rotor is made to revolve at very high speeds, balanced and supported on the cushion of air. These instruments easily reach speeds of $>10^5$ rpm in a very short time because the rotors are small and of low mass. The sample capacity is correspondingly small, and they are of use only in specialized applications.

THEORY OF THE CENTRIFUGE

Centrifugal force, as defined by Isaac Newton, is expressed by the tendency of any rotating object to move away from its center of rotation. Thus, when a stone tied to a string is whirled about one's head, centrifugal force is what keeps the string taut.

In a vacuum, the centrifugal force (F_c) is given by:

$$F_c = \omega^2 rm \tag{8-1}$$

where m is the mass of the object or particle, ω is its angular velocity, and r is the radius of rotation. In any medium other than a vacuum, centrifugal force is opposed by two others, the buoyant force and the force of viscous drag. The effects of these forces will be discussed shortly.

The angular velocity, ω, is measured by the angle swept through per second. Because one complete revolution sweeps through 2π radians, one can relate ω to rotor speed, expressed as rpm, by the equation:

$$\omega = \frac{2\pi \text{ (rpm)}}{60} \tag{8-2}$$

From Eqs. 8–1 and 8–2, we can quickly learn how impressive are the forces involved in centrifugation. Suppose one had a centrifuge rotor operating at 50,000 rpm, and carrying a pair of tubes. The tops of the tubes are located 6.0 cm, and the bottoms 10.0 cm, from the center of rotation. The total mass of each tube plus its content is known to be 10.0 g. What force must the bottom of the tubes withstand? The force may be calculated as approximately 2.7416×10^9 g·cm/sec^2. This is equivalent to 2.7416×10^4 kg·m/sec^2. The units are those for Newtons; since 1 Newton corresponds to a force of about 0.102 k_g, the force at the bottom of the centrifuge tube is equivalent to a mass of about 2742 kg! It is not surprising, therefore, that high-speed rotors are robustly built; they need to withstand tremendous strains. In spite of their heavy construction, high-speed rotors show signs of metal fatigue with continued use. To avoid catastrophic accidents, it is necessary to derate their safe maximum speeds on a scheduled basis, usually 10% of initial maximum speed per 1000 hours of use.

Equation 8–1 represents only the centrifugal force acting on a sedimenting particle. A more complete representation of the behavior of the particle would allow for buoyant and viscous forces as well. We can then write $F_c = (\rho_p - \rho_m)V\omega^2 x$, where ρ_p and ρ_m are the densities of the particle and the medium, and V is the volume of the particle and hence of the medium it displaces.

The expression $(\rho_p - \rho_m)V$ substitutes for m and corrects for buoyancy; it is, in fact, the particle mass in the suspending medium. Opposing the centrifugal force are frictional or viscous forces; Eq. 8–3 represents the balance of all forces acting on the particle.

$$(\rho_p - \rho_m)V\omega^2 x = k\frac{dx}{dt} \qquad (8\text{–}3)$$

where k is a factor related to the viscosity of the medium, and dx/dt is the distance the particle moved down the length of the centrifuge tube per unit time.

Equation 8–3 can be solved for dx/dt in terms of the other parameters, to give:

$$\frac{dx}{dt} = \frac{(\rho_p - \rho_m)V\omega^2 x}{k} \qquad (8\text{–}4)$$

We can now define S, the *sedimentation coefficient,* as:

$$S = \frac{dx/dt}{\omega^2 x}$$

or

$$S = \frac{(\rho_p - \rho_m)V}{k} \qquad (8\text{–}5)$$

The dimensions of S are seconds, and a value of S equal to 10^{-13} seconds has been named the Svedberg, in honor of the Swedish physical chemist who built the first ultracentrifuge. Note that S depends entirely on the molecular properties of the system being separated, and not at all on any property of the centrifuge itself. The value of S depends on the viscosity of the medium, and thus on the temperature. Convention has it that S values reported in the literature are reported as $S_{w,20}$, meaning that the measurement is corrected to the viscosity of water (w) at 20°C. Correction of any experimental value to the conventional value is made by the equation:

$$S_{w,20} = \left[S_{s,T} \right] \frac{\eta_{s,T}(1 - \bar{\nu}\rho)_{w,20}}{\eta_{w,20}(1 - \bar{\nu}\rho)_{s,T}} \tag{8–6}$$

where η is the viscosity under the conditions shown by subscripts and $\bar{\nu}$ is the partial molar volume under the same conditions. Thus, $S_{s,T}$ is the value of S in some solution, s, at a temperature of T °C.

Svedberg and Pedersen showed that S can be related to the molecular weight of biopolymers. Their equation, here stated without proof, is:

$$S = \frac{M(1 - \bar{\nu}\rho)D}{RT} \tag{8–7}$$

where D is the diffusion coefficient of the polymer, M is its molecular weight, ρ is the polymer density, R is the universal gas constant, and T is the absolute temperature. This development gave great importance to analytical ultracentrifugation and provided a new way of determining the molecular weights of very large molecules. At the time, there were not many other reliable procedures available. However, the hardware is very expensive, and the procedure is tedious and quite time consuming. With the development of high-quality exclusion gels and of electrophoresis in stabilized media, the ultracentrifugal method has been largely supplanted. Interested readers should consult the references at the end of the chapter for further details.

Other uses of S remain important and eminently practical. Consider the following example: A sample of isolated hepatocytes was suspended in a solution of 0.25 mol/L sucrose containing 25 mmol/L of KCl. The cells were homogenized gently in a Potter–Elvejhem homogenizer, and the mixture was transferred to centrifuge tubes. The various subcellular organelles (Table 8–1) could be separated in a state of reasonable homogeneity essentially as described below. How knowledge of S values proved to be useful in this effort is discussed is the remainder of this section.

Recall from the definition of the Svedberg coefficient that:

$$S = \frac{dx/dt}{\omega^2 x}$$

Thus one can replace dx/dt by its equivalent, distance moved per unit time, to give:

$$\frac{\text{distance moved}}{\text{unit time}} = S\omega^2 x \tag{8–8}$$

On a purely theoretical basis, this equation implies the following:

1. Assuming a medium of initially uniform density, and assuming the centrifuge is driven with sufficient speed, the subcellular fractions with

Table 8–1. Sedimentation Characteristics of Subcellular Fractions

Subcellular Fraction	Range of S Values[a]	Density (g/ml)
Soluble proteins	2–25	~1.30
RNA, mixed	4–50	
DNA	≤100	~1.5–1.6
Ribosomal subunits	30–60	—
Mitoribosomes	55–60	
Ribosomes	70–80	1.6–1.75
Polysomes	100–400	—
Microsomes	$0.5\text{–}15.0 \times 10^3$	1.16
Lysosomes	$1\text{–}2 \times 10^4$	1.21
Peroxisomes	$2.0\text{–}2.2 \times 10^4$	1.23
Mitochondria	$2\text{–}7 \times 10^4$	1.19
Plasma membrane vesicles	$10^2\text{–}10^5$	1.16
Nuclei	$4\text{–}10 \times 10^6$	—

[a]In aqueous media containing 0.25 mol/L sucrose and 25 mmol/L KCl at 4°C.

Sources: G. D. Birnie, ed., *Subcellular Components: Preparation and Fractionation,* 2nd ed., University Park Press, Baltimore (1972); H. A. Sober, ed., *Handbook of Biochemistry,* 2nd ed., Chemical Rubber Publishing Co., Cleveland (1968); O. M. Griffith, ed., *Techniques of Preparative, Zonal, and Continuous Flow Ultracentrifugation,* 3rd ed., Beckman Instrument Co., Spinco Division, Palo Alto, Calif. (1979).

the largest S values will sediment (or reach the bottom of the tube) fastest.

2. The conclusion in (1) applies equally well to single, soluble molecular species and to more organized structures such as organelles.
3. Because the range of S values shows some overlap, it is not always possible to separate completely one kind of organelle from the others in a single differential centrifugal step.

These theoretical implications are completely supported by experiment. If one takes a homogenate of the sort described above, the nuclei can be more or less completely sedimented at $800 \times g$ in about 15 min when the homogenate is placed in a 50-ml centrifuge tube with its center at about 8 cm from the axis of rotation. Next, the mitochondria can be sedimented at $8500 \times g$ in about 15 min, but there will be considerable contamination by peroxisomes and modest contamination by lysosomes, as judged by electron microscopy of the pellet. To collect ribosomes, one might have to centrifuge at $100{,}000 \times g$ for more than an hour, and the initial preparation will probably not be totally free of other organelles.

It is interesting that peroxisomes have been recognized as true organelles only during the past thirty years or so. Prior to that time, they were identified as the "light" mitochondrial fraction. More recent studies have clearly shown that peroxisomal structure, properties, and functions are quite distinct from those of mitochondria. The demonstration was made possible largely by improved methods of centrifugal separations.

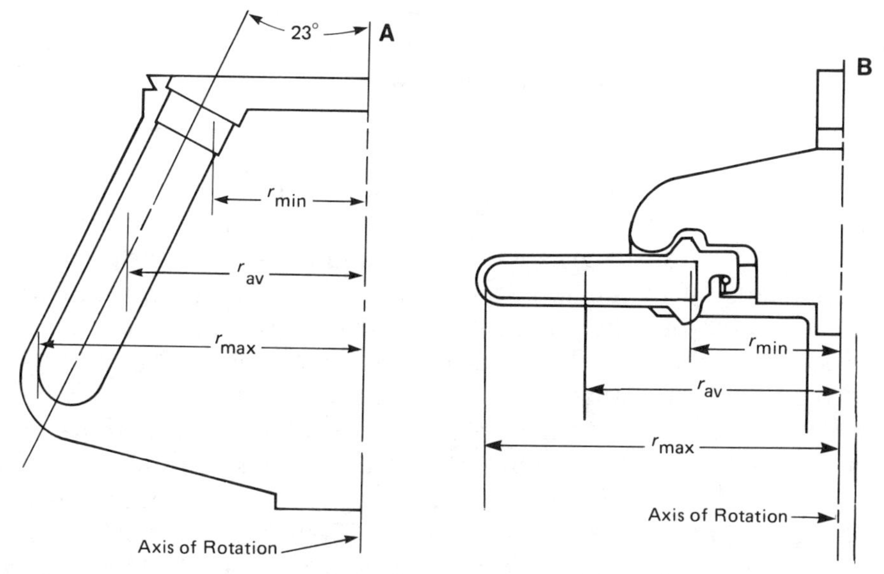

Figure 8–1. Sectional views of fixed-angle (A) and swinging-bucket (B) centrifuge rotors. See text for details. (Reproduced by courtesy of Spinco Division of Beckman Instruments, Inc.)

TYPES OF CENTRIFUGE ROTORS

All centrifuge rotors can be classified into three major categories: *fixed-angle, swinging-bucket,* and *zonal.* Rotors in the first two classes are in common use for small-scale preparative and for analytical purposes, but zonal rotors are restricted almost entirely to large-scale preparative purposes. Discussion of zonal rotors is deferred to the latter part of this chapter.

Analytical-Scale Separations: Fixed-Angle and Swinging-Bucket Rotors

Description and Use. Fixed-angle rotors have a tapered section, and the cavities into which the sample tubes are placed are precisely bored out of the mass of the rotor at some angle to the axis of rotation, as shown in Fig. 8–1A. The individual sample tubes need not be capped since the rotor is closed by a single, circular cover which can be tightly fastened to the rotor top. A variant form of fixed-angle rotor holds its tubes parallel to the axis of rotation; these are the so-called vertical rotors. Regardless of the angle, all of these fall into the first category since the tubes are held at some angle *independent of the rotor speed.*

Swinging-bucket rotors are designed differently. The sample carriers are external to the mass of the rotor itself and are held to the underside of the rotor by hooks or by pins. Each carrier has a wide flange built into its top, just below the point of attachment. Cut into the rotor edge are yokes which engage these flanges as the rotor comes up to speed; the free ends of the carriers thus swing upward and outward until they lie in a plane at right angles to the axis of rotation. As a result, little of the stress of the applied centrifugal force is borne by the suspending hooks or pins. A sectional view of a carrier locked into its yoke is shown in Fig. 8–1B.

Sample carriers are very carefully matched for mass, and they are always numbered. Corresponding numbers are stamped into the rotor; in this way each carrier can always be loaded in the same position. This precaution ensures that dynamic balance is maintained, provided that the carriers are properly loaded with equal masses of sample. Before starting up a swinging-bucket rotor, one should make a careful check of the carrier positioning; otherwise vibrations due to imbalance could cause the rotor to fly off the drive, with great damage to the centrifuge. Each of the carriers is closed with a threaded cap, since there is no way in which a single rotor cover could be employed, and it is important that the caps be firmly seated. In general, swinging-bucket rotors have a larger exposed surface than do fixed-angle rotors of comparable capacity. As a result, they generate more turbulence in the surrounding air as the rotor revolves. Unless adequate refrigeration is provided, the turbulent pattern of air flow rapidly causes considerable sample heating. Even with adequate refrigeration, when swinging-bucket rotors are driven at speeds much greater than 10,000 rpm, the rise in sample temperature at atmospheric pressure is so great that many sensitive samples would be damaged when centrifuged for longer than a few minutes. The only satisfactory solution is to reduce friction by operation of the rotor in a vacuum.

Relative Merits. In a fixed-angle rotor, any particle suspended in the bulk of the medium is subjected to a lateral force, F, depending on the rotor speed and the distance of the particle from the axis of rotation. The path through the liquid is relatively short, since the tube diameter is small as compared to its length. When the particle strikes the side of the tube, it is driven downward by a force, F', where $F' = F \sin\theta$ and θ is the angle of inclination with respect to the rotor axis (see Fig. 8–1A).

Vertical tube or zero angle rotors were introduced in the mid-1970s and have proven quite useful. The system to be centrifuged is established in tubes as for other rotor types. When centrifugation is started, the separation zones shift so that at full speed they are established along the *short* axis of the tubes. Since there is a shorter distance for the components in the tubes to travel, separation requires less time. When the centrifuge is turned off, and as the rotor slows, the components reorient themselves. When the tubes can be removed, the separation vector is again parallel to the long axis of the centrifuge tube. There is surprisingly little mixing during this reorientation. To anticipate our discussion below, when density gradients are used in such rotors, the gradient also reorients successfully during the course of centrifugation.

In swinging-bucket rotors, the particle is subject to a constant force, F, but the path length is increased so that it is now equal to the depth of the suspension; thus the sedimentation time will be increased accordingly.

For practical purposes, there is little difference in the efficiency of the

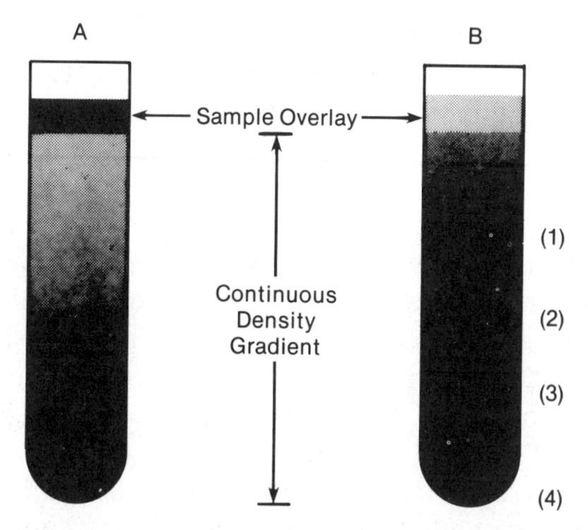

Figure 8–2. Isopycnic separation in a continuous density gradient. The gradient is represented by the continuous half-tone in both A and B. In A, at the beginning of the separation, the sample is overlaid in the gradient, as shown by the darker band at the top of the tube. In B, after the separation is complete, bands 1 through 4 have descended through the gradient until each indicated fraction has reached a gradient density equivalent to the particle density—the isopycnic point. Further movement of each fraction is prevented by the buoyant force, which equals the centrifugal force at the isopycnic point. Note that for optimal results, the sample suspension is usually applied in a medium of lesser density than any part of the gradient.

two categories of rotors as long as the suspending medium has a uniform density. In view of the fact that fixed-angle rotors are less expensive to produce and easier to maintain, they have become very popular for all but certain specialized applications. Fixed-angle rotors are also easier to use, and, as noted above, they tend to generate less heat as a result of their smooth profiles. Depending on the model, they usually have larger capacities than swinging-bucket rotors of the same mass.

DENSITY GRADIENTS AND ISOPYCNIC CENTRIFUGATION

The discussion thus far has centered on systems about which two assumptions have been made: (1) the suspension is initially homogeneous; that is, its physical properties are the same throughout the volume of the centrifuge tube; and (2) the density of the suspending medium is less than the density of the suspended matter. These assumptions indicate that if the suspension is centrifuged long enough and hard enough, all of the suspended matter will be pelleted to the bottom of the tube.

Alternatively, the centrifuge tube can first be nearly filled with a medium of graded density, so that—from the top of the tube to the bottom—the density might range from 1.02 to 1.26 or higher. The sample suspension is then carefully overlaid on top of this gradient. Initially, the tube will appear as in Fig. 8–2A, in which the increasing density of the suspending medium is indicated by the increased tonal density. On application of centrifugal force at some predetermined value, particles contained in the sample overlay will move downward through the gradient until they encounter a density equivalent to their own. That level in the tube is known as the *isopycnic point* for those particles; there they will stop because the buoyant force will just balance the centrifugal force. If no volume element of the gradient is sufficiently dense to buoy up some fraction of the suspended matter, that fraction will reach the bottom of the tube, provided the centrifugal force is applied long enough. This is the case with material identified as fraction 4 in Fig. 8–2B. Fractions 1, 2, and 3 reach their isopycnic points within the higher reaches of the gradient, so they are stopped as shown in Fig. 8–2B. If the tube pictured in Fig. 8–2B is viewed by transmitted light, each of the fractions will be visible, with more or less clear spaces in between. The exact locations of the bands will depend on the S values of the suspended particles, their densities under the prevailing experimental conditions, the nature of the gradient, and of course the relative centrifugal force applied. It is important to understand that once a fraction of the particulate matter has reached its isopycnic point, a longer duration of centrifugation will not significantly affect the result.

Recovery of Fractions from Gradient Separations

The simplest means of recovering material from gradient separations is to use a Pasteur pipet fitted with a suction bulb, sucking off successive fractions from the top of the open tube. This method requires some skill and daring since it is all too easy to stir up and disperse the fractions.

More precise recovery can be accomplished with the aid of a tube punch and pump device, several of which are commercially available. These can be used only with light-walled, disposable centrifuge tubes, usually made of cellulose acetate. The bottom of the tube is penetrated by a sharpened needle, through which very dense media can be forced by a peristaltic pump. The top of the tube is capped by a cover through which a small tube passes. As the dense solution is delivered into the bottom of the centrifuge tube, it causes liquid to pass through the upper tube to a fraction collector. To avoid undesirable mixing, the pumping solution must be distinctly denser than the densest portion of the gradient, and the rate of sample extrusion must be slow. Some punch/pump units are designed to operate in the reverse mode from that described here, that is, light fluid is added at the top of tube and heavy fractions are drained from the bottom; however, as a rule these are less functional because of difficulty in clean fraction recovery.

Materials Suitable for Gradient Preparation

To be useful in forming density gradients, a material should have the following properties:

1. *High water solubility,* to give the greatest possible range of densities to the solutions.

2. *High purity and moderate cost,* since these solutions may range up to concentrations of several moles per liter.
3. *Little or no toxicity to cells or organelles,* because if recovered fractions are to be useful, they must not be poisoned by the process of separation.
4. *Little or no absorbance in the ultraviolet (especially from 260 nm to 365 nm), and colorless in the visible range,* to allow protein and/or coenzyme assays by the usual methods, without overwhelming interference from the gradient materials.
5. *High stability and freedom from microbial attack,* to allow stock solutions to be made and kept for reasonable periods of time.
6. *Lowest possible viscosity,* to minimize forces of viscous drag when one is working with dense solutions.
7. *Moderate to high molecular weight,* to minimize osmotic effects on cells and membrane-bound organelles.

When one takes into account all of the above requirements, the number of materials that have a useful combination of properties is actually quite small, and none in general use is of universal applicability. The common gradient solutes are not good buffers; where required, buffers, antioxidants, and chelating agents must be added separately, as in solutions for differential centrifugation. The merits and demerits of the more common gradient solutes are listed below.

Cesium Chloride (MW = 168). Favorable solubility characteristics make possible solutions of high density (Fig. 8–3A) and low viscosity. The viscosity is too low to allow plotting on the scale used in Fig. 8–3B. However, the ionic nature of the salt and its low molecular weight also cause intolerable osmotic effects when cells or organelles are to be separated. Cesium chloride is therefore used almost exclusively for separations of RNA and DNA. Cesium chloride solutions generate a gradient by simple centrifugation; starting with a solution of uniform density, gradients from 1.099 to 1.90 g/cm^3 can be obtained.

A second problem is the high cost of the salt; another is contamination of the commercial material with substances that absorb in the ultraviolet. The latter substances must be removed by tedious recrystallization procedures. A specially purified grade of CsCl is sold at about twice the cost of the usual grade. For this reason, CsCl is frequently recovered from gradient separations.

Hypaque (MW = 636). Hypaque (3,5-diacetamido-2,4,6-triiodobenzoic acid, sodium salt) was originally developed as a radiographic contrast medium, and is still employed for this purpose in clinical studies. The iodination also makes a solution of this material very dense. A number of related compounds, differing in their acyl groups, are also available. All of these give gradients of moderate density and low viscosity (see Fig. 8–3A,B). Materials are usually available in sterile solutions, which may be diluted as needed.

The major drawbacks to Hypaque are its relatively large osmotic effect and its high concentration of Na^+; these make it unacceptable for some

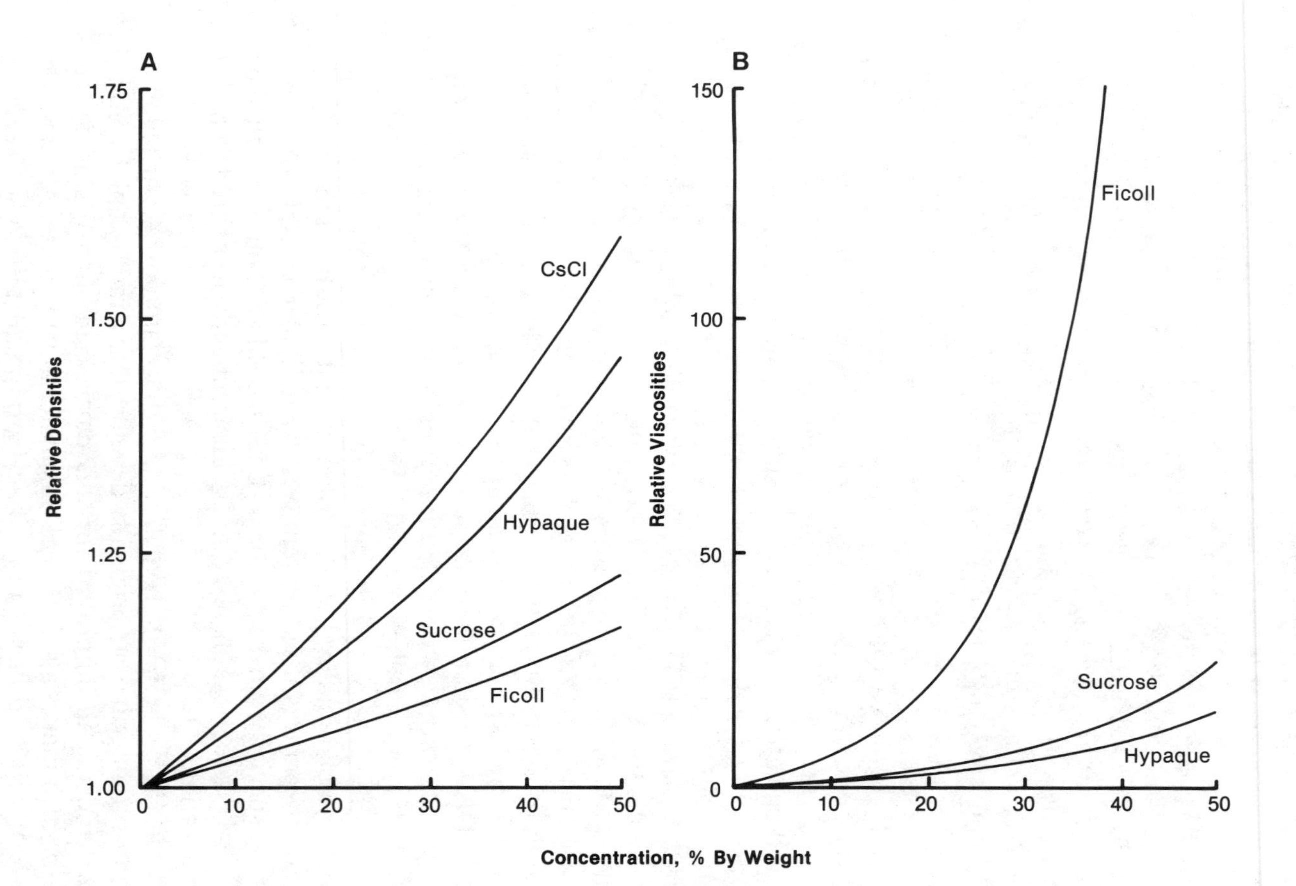

Figure 8–3. Density (A) and viscosity (B) of some common gradient solutes. The densities are expressed relative to water at 20°C, and the viscosities relative to water at the same temperature.

purposes. It is also quite expensive. Hypaque has been mixed with other gradient solutes for blood cell fractionation. Thus, when mixed with Ficoll®, it can cleanly separate granulocytes and lymphocytes from whole blood. Hypaque is not routinely used for subcellular fractionations.

Sucrose (MW = 342). Although the maximum density that can be attained with sucrose is about 1.29, the low cost of the material and its modest viscosity make it the most commonly used gradient solute. However, sucrose can be troublesome in certain applications because it has an osmotic effect, which is a function of its molecular weight and its ability to penetrate some biological membranes; it may thus cause shrinkage of some organelles by osmotic withdrawal of water from them.

Ordinary (table grade) sucrose cannot be used for making gradients; a specially purified reagent grade is required because table sugar is treated with calcium sulfate to keep it from lumping, and the high concentration of Ca^{2+} introduced into gradient solutions would be disastrous to mitochondria, for example. One must also keep in mind that sucrose solutions at a concentration of 0.25 mol/L support microbiological growth. Older solutions frequently become contaminated with toxic substances generated by molds or bacteria. Solutions that are the slightest bit turbid or cloudy should be discarded promptly. Stronger solutions ($\geq$1 mol/L) are not as likely to become contaminated, probably because of their osmotic powers.

Ficoll® (MW = ~400,000). Originally developed as a plasma substitute for transfusion purposes, Ficoll (a polymer of sucrose and epichlorohydrin) is a fairly stable nonionic polymer, sold as a freeze-dried preparation. It has become a useful gradient solute because of its very low osmotic effects even though its solutions are quite viscous, and even though it is more difficult to dissolve in high concentration than is sucrose. The density of its solutions is at best modest compared to other gradient solutes, but its advantages in cellular separations offset these difficulties. It is also compatible with other solutes in mixed gradients, as noted above. Ficoll is also quite expensive, which tends to limit its uses to essential cases. As received, Ficoll is contaminated with small, ionic material, and preparations should be extensively dialyzed with good stirring to rid the polymer of impurities.

Ludox. When first marketed, Ludox (a colloidal silica preparation) showed great promise because of the fairly high densities of the suspension. Subsequent experience indicated that sorptive effects were damaging to many cells and organelles. Ionic contamination was also a problem. For these reasons the product has not been widely adopted as a general-purpose gradient material.

Percoll®. The promising results possible with colloidal silica stimulated efforts to modify the silica surface to minimize the problems noted. By chemically bonding polyvinylpyrrolidone, the sorptive effects were largely eliminated without sacrificing any of the advantages of silica dispersions. Thus Percoll, a polyvinylpyrrolidone-coated colloidal silica, can form gradients with density ranges from 1.0 to 1.3 g/ml, is nontoxic to cells, and has a very low viscosity. Its osmotic effects are also very low and independent of concentration, so gradients are isosmotic. Depending on the solution in which the silica is dispersed (NaCl at 0.15 mol/L, or water), the silica particles swell or shrink slightly, but the size range runs from 25 to 35 nm.

Ludox or Percoll® suspensions provide for self-generating gradients by simple centrifugation. An initially uniform suspension, when centrifuged at forces $>10{,}000 \times g$, redistributes itself to form a gradient, the steepness of which changes with time of centrifugation, the geometry of the rotor, and the initial concentration of the suspension. This can be an advantage, but only if the conditions of separation are precisely controlled from run to run. For further details, the manufacturer's brochure (Pharmacia Fine Chemicals, AB) should be consulted.

Gradient Preparation and Calibration

The simplest form of a gradient is the *discontinuous* or *step gradient,* made by adding successive layers of solution to a centrifuge tube. One must begin, of course, with the heaviest solution or layer. By adding the successive layers slowly, it is possible to build a gradient "layer-cake fashion," with little or no mixing at the interface of each added layer. This is illustrated in Fig. 8–4.

There are some very real advantages of discontinuous over continuous gradients. First, they are simple to prepare, requiring only a series of stock solutions. Second, the height of each layer can be adjusted to provide good spacing between adjacent densities, allowing good separation between fractions that collect at the selected interfaces and facilitating recovery of the fractions. Third, when layer 6 (Fig. 8–4) is made distinctly heavier

Figure 8–4. A discontinuous sucrose gradient. Layer 1 represents the sample, dispersed in 0.25 mol/L of sucrose. The successive layers (reading downward) represent layers of increasing sucrose concentration. Layer 6, for example, might be 2 mol/L, layer 5 might be 1.2 mol/L, layer 4 might be 0.95 mol/L, layer 3 might be 0.6 mol/L, and layer 2 might be 0.45 mol/L. Note also that the volumes of the layers need not be identical.

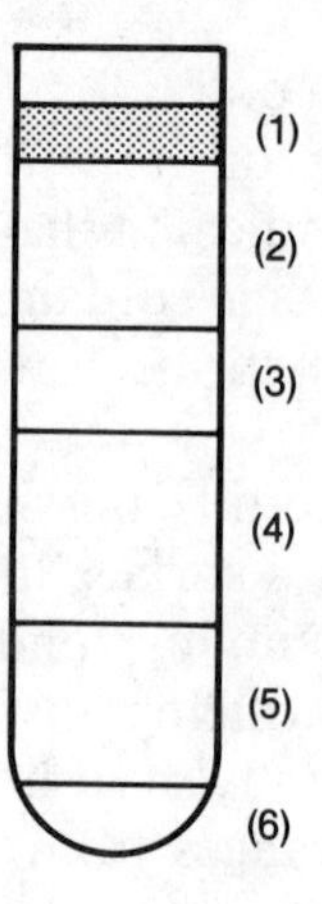

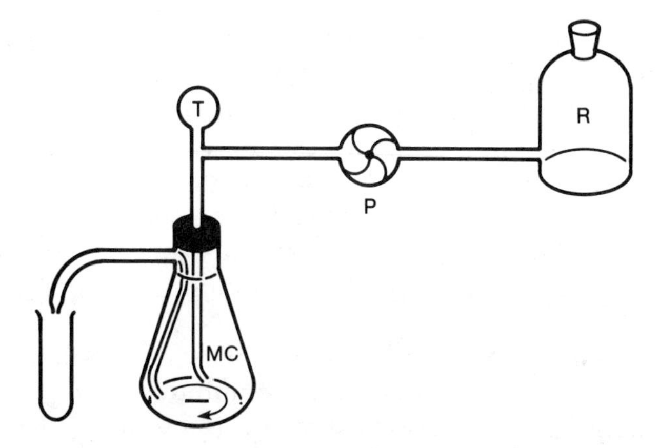

Figure 8–5. Anderson's gradient mixer. R is the open reservoir bottle for the heavy solution, P is a small peristaltic pump to propel the heavy solution into the mixing chamber, MC. T is a trap to collect bubbles that may be produced by action of the pump. A magnetic stirring bar is shown in the bottom of the mixing chamber, and the outlet is shown dipping into a centrifuge tube that will hold the gradient solution.

than any of the anticipated fractions, it provides a "pillow" through which no suspended matter will pass, so that punch/pump assemblies will collect without loss of any material that sediments at the interface of layers 5 and 6. Provided the characteristic densities of the suspended material are known even approximately, it is easy to tailor the gradient to the need at hand.

Continuous gradients may be produced by continuous addition of a heavy solution to a fixed volume of a lighter solution, contained in a closed, stirred mixing chamber, the outflow of which delivers to a centrifuge tube. Such a gradient mixer is shown schematically in Fig. 8–5.

If the heavy solution is not too viscous, and if the reservoir is supported well above the mixing chamber, the pump may be omitted; however, usually some source of pressure other than gravity is necessary. Stirring should be vigorous to ensure good mixing in the chamber, and the output solution should be allowed to run down the side of the centrifuge tube. This design, credited to N. G. Anderson, has the great virtue that it can be assembled out of ordinary glassware. More elaborate designs have been marketed, but they offer very little advantage over the assemblage shown in Fig. 8–5. Note that the fundamental principle is identical with that used in mixing gradients for chromatography.

One can estimate the form of the gradient mathematically if certain parameters of the system are known. If one represents the concentration emerging from the mixer by C_t, the corresponding volume withdrawn by V_t, the volume of the mixing chamber by V_m, and the concentrations of the denser and lighter solutions by C_D and C_L, respectively, then the concentration of the emerging solution at any time is given by:

$$C_t = (C_D - C_L)e^{-V_t/V_m} \quad (8\text{–}9)$$

The steepness of the gradient is clearly a function of the difference in concentration of the dense and light solutions, other things being equal. Because the volume of a given centrifuge tube is finite and usually not more than 50 ml, one must sometimes adjust the volume of the mixing chamber to fit the circumstances; however, that is easy to do with this simple design, which therefore has the additional advantage of flexibility. From the known volume of the centrifuge tube and its cross section, it is possible to estimate the density of the contained solution at any given height in the tube.

Precise Gradient Calibration. Most gradients, discontinuous or continuous, are ordinarily made from stock solutions of the gradient solute. These must be made with analytical care if the results are to be reliable and consistent from experiment to experiment. Dilutions of very concentrated solutions must also be made with care and mixed especially well to avoid errors due to their high viscosity. In diluting solutions such as 2 mol/L of sucrose, for example, the pipetting error may be very large because of the slow drainage of pipets. It is therefore a good idea to have a number of stock solutions available, and to avoid handling solutions more concentrated than are actually required for the work at hand.

The ultimate method for calibration is to weigh a known volume of a solution and thus to determine its density. Small pycnometers are available for this purpose, but the same problems mentioned in connection with pipetting apply here also. It is difficult to adjust the volume of a thick, viscous solution to an exact mark in a pycnometer, as in a volumetric flask. Small errors in volume may represent significant errors in weight. Besides, pycnometry is very tedious if many solutions are to be measured.

A second calibration procedure depends on measurement of the refractive index of a solution, since the refractive index is a measure of the number of particles per unit volume and so can be directly related to the concentration. The Abbe refractometer is ordinarily used and requires only a drop of solution. Tables are readily available relating the index to the concentration for virtually all of the materials commonly used in gradient centrifugation. Refractometry is rapid and quite sensitive, properties that enhance its use.

A third method of gradient calibration depends on the availability of polymeric microspheres of known density. These are commercially available and are known as Density Marker Beads. Ten different bead types are produced, each dyed a different color. Between them, they cover the density range 1.017–1.142. Samples of these beads are usually added to a gradient containing no sample, but otherwise identical to the experimental gradients. The final position of each color of beads in the tube marks a known density position. These preparations are expensive, but much less so than a refractometer, and they fill a need where refractometers are not to be had. With care, the beads can be recovered, washed, and used again. The manufacturer points out, correctly, that the actual buoyant density of the beads will vary slightly with the osmolarity of the medium, and with its ionic strength, but for close approximations these variations are not too limiting. Further details can be obtained in the manufacturer's brochure describing the product.

PREPARATIVE-SCALE SEPARATIONS: USE OF ZONAL ROTORS

Zonal rotors were first developed by N. G. Anderson, of the Oak Ridge National Laboratory. Subsequent efforts of Anderson's group gave rise to a number of models, differing among themselves, but all depending on the same basic principles. We shall describe only one of the possible models here. In essence, a zonal rotor is a hollow bowl of up to several liters in volume, useful for large-scale separations. Connections are provided so that material may be pumped into or out of the rotor while it is revolving; this accounts for the high cost of zonal rotors and explains why some modifications must be made to the remainder of the centrifuge.

Figure 8–6 shows a cross section of a typical zonal rotor. In the upper left (1), a gradient is generated in the rotor bowl while it is spinning at low speed, for example, about 2000 rpm. The gradient shown is discontinuous, but this is only for pictorial convenience; in fact, the usual operating mode employs a continuous gradient. The rotor is loaded by pumping solution from the outer edge of the bowl toward the center. In the upper right (2), sample is loaded from the center by a piston-driven pump or syringe. The rotor is then driven up to speed (20,000–30,000 rpm), and the various components suspended in the sample are distributed throughout the gradient according to their densities, as shown in the lower left (3). When equilibrium has been achieved, the rotor is unloaded at low speed by pumping dense solution into the bowl, this time from its periphery (4). The samples emerging from the bowl may be monitored by absorbance or similar measurements, and the product distributed through a fraction collector. As diagramed in Fig. 8–6, right-hand panel (B-29 Core), the core of some rotors may be changed to one that permits the introduction of samples and recovery of fractions from the edge and also the center of the zonal rotor.

The salient differences between zonal rotors and the others described previously are as follows:

1. No centrifuge tubes are used with zonal rotors, and nearly the entire volume of the rotor is available for holding the gradient and the sample suspension.
2. Zonal rotors are always loaded and unloaded under dynamic conditions; this is almost never true of tube-type rotors. Dynamic loading and unloading is necessary to avoid the disturbing effects of Coriolis forces (the same forces that make the water swirl in a sink or bathtub drain).

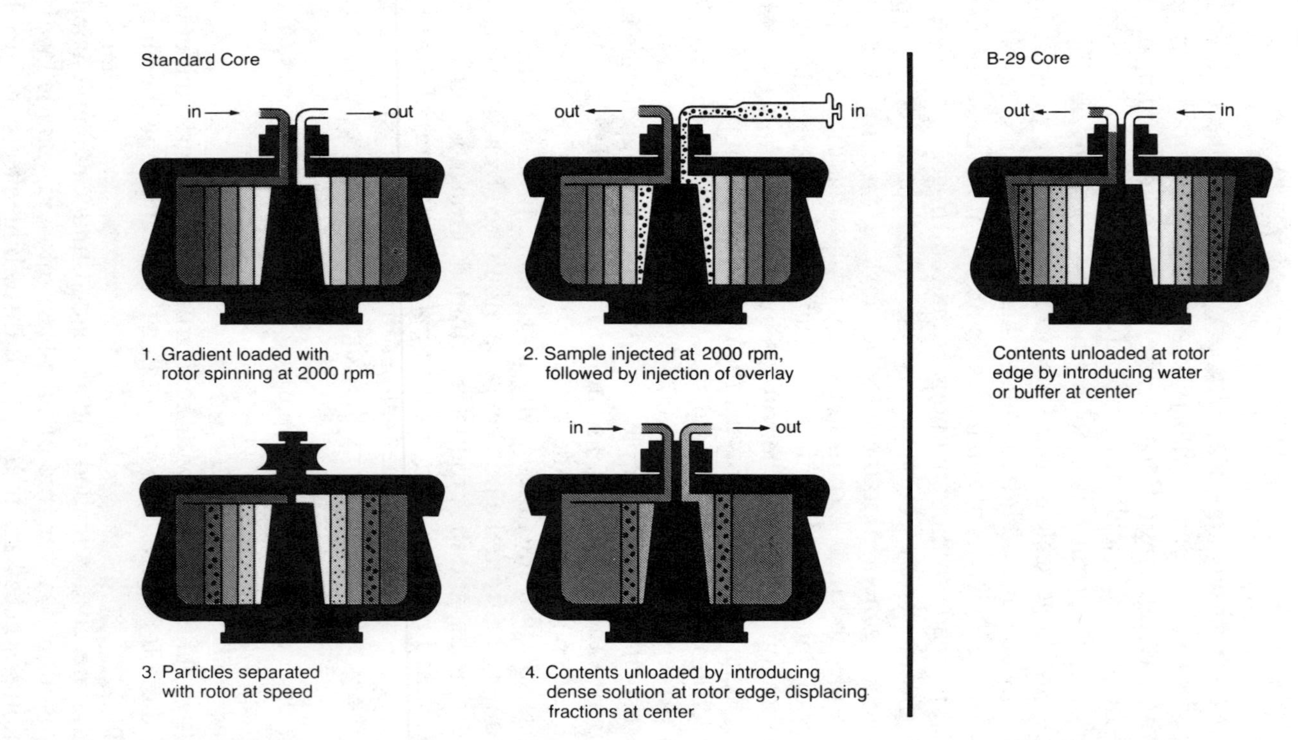

Figure 8–6. Stages in loading, separating, and unloading of a zonal rotor. The four views show how the gradient is loaded into the rotor under dynamic conditions, how the sample is loaded at low speed and separated at full speed, and how the separated components are then pumped out of the rotor at low speed. For further details, see the text. (Courtesy of Spinco Division of Beckman Instruments, Inc.)

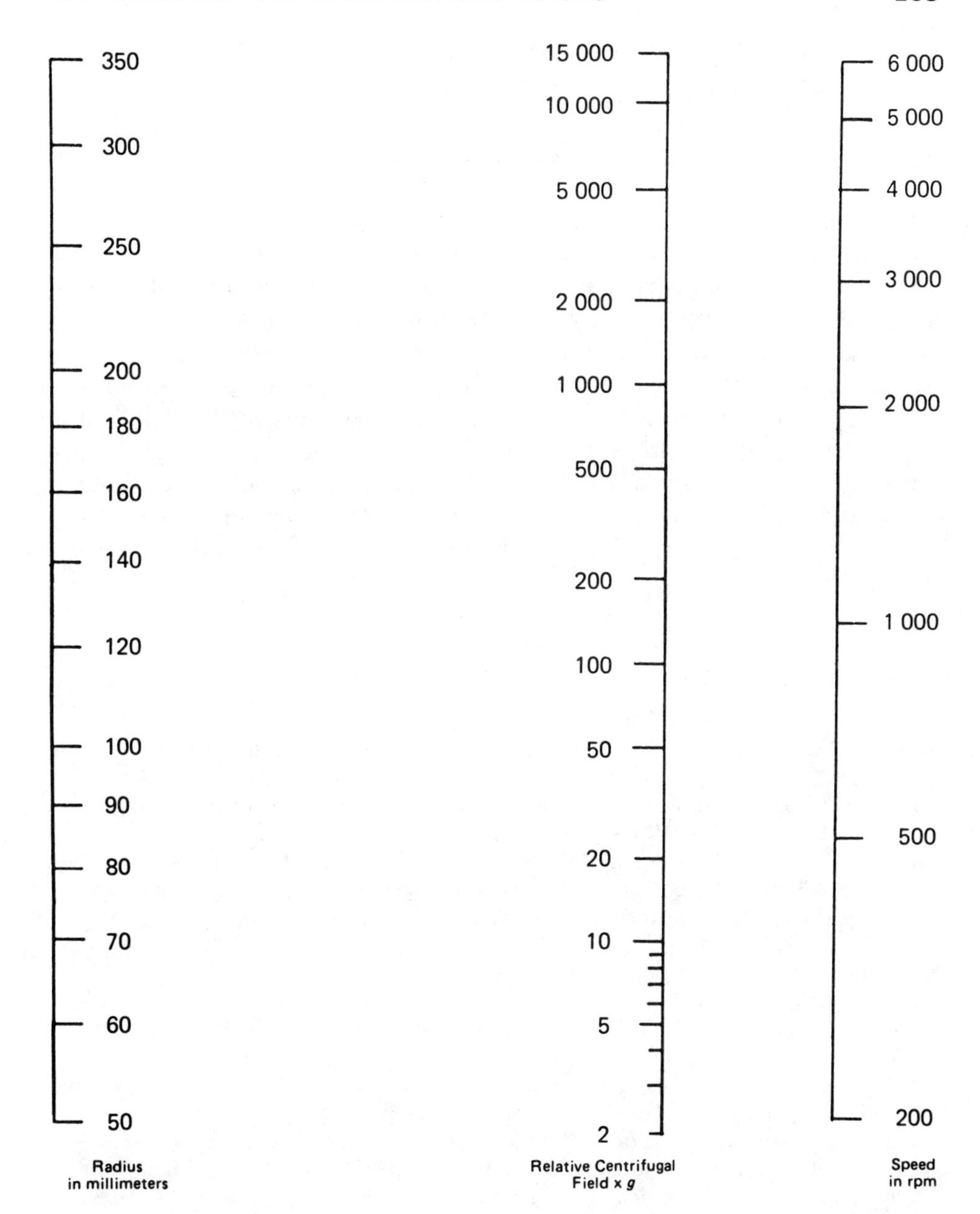

Figure 8–7. A nomogram for measurement of relative centrifugal force. Align a straightedge through known values in two columns; read the desired figure where the straightedge intersects the third column. (Reproduced by courtesy of Spinco Division of Beckman Instruments, Inc.)

3. Coriolis forces are minimal in tube-type rotors because the cross-sectional area of the mass of fluid in a single tube is less than the corresponding measurement in a zonal rotor.
4. Zonal rotors are designed for fairly large samples and are much less practical for small, analytical samples.

NOMOGRAMS: CONVENIENT MEASUREMENT OF RELATIVE CENTRIFUGAL FORCE

It is customary to describe the forces that operate in a centrifuge in terms of the relative centrifugal force, always with respect to the force of gravity. Thus, a force of 5000 *G* is a force equivalent to $5000 \times g$, where *g* is the force due to gravity. Instead of calculating *G* for each set of circumstances, it is convenient to construct a nomogram in terms of the radius of a rotor, its speed, and the value of *G*. A typical nomogram is shown in Fig. 8–7.

A word of caution is necessary in using this or other nomograms. Regardless of the rotor type involved, one can measure a maximum radius (at the extremity of the suspension from the axis of rotation), a minimum radius (at the other extremity of the suspension), or an average radius (midway between extremities). Forces existing in the suspension will obviously differ with the effective radius. When one estimates centrifugal force, it is usually the average value that is considered, unless otherwise stipulated.

REFERENCES

Anderson, N. G. The zonal ultracentrifuge: a new instrument for fractionating mixtures of particles. *J. Phys. Chem. 66:*1984–1989 (1962).

Anderson, N. G. Analytical techniques for cell fractions: a simple gradient-forming apparatus. *Anal. Biochem. 21:*259–265 (1967).

Beaufay, H., Jacques, P., Baudhuin, P., Sellinger, O. Z., Berthet, J., and deDuve, C. Resolution of mitochondrial fractions from rat liver into three distinct populations of cytoplasmic particles by means of density equilibration in various gradients. *Biochem. J. 92:*184–205 (1964).

Birnie, G. D. *Sub-Cellular Components: Preparation and Fractionation.* University Park Press, Baltimore (1972).

Cline, G. B., and Ryel, R. B. Zonal centrifugation. *Methods in Enzymol. 22:*168–204 (1971).

De Duve, C. *Separation and characterization of subcellular particles.* Harvey Lectures Series No. 59, pp. 49–87. Academic Press, New York (1965).

De Duve, C. Tissue fractionation: past and present. *J. Cell Biol. 50:*20D–55D (1971).

Gmelig-Meyling, F., and Waldmann, T. A. Separation of human blood monocytes and lymphocytes on a continuous Percoll gradient. *J. Immunol. Methods 33:*1–9 (1980).

Gray, G. W. The ultracentrifuge. *Sci. Am. 184:*42–51 (1951).

Rickwood, D. An assessment of vertical rotors. *Anal. Biochem. 122:*33–40 (1982).

Schachman, H. K. *Ultracentrifugation in Biochemistry.* Academic Press, New York (1959).

Svedberg, T., and Pederson, K. O. *The Ultracentrifuge.* Oxford University Press, New York (1940).

CHAPTER NINE

Recovery of Biopolymers from Solution by Freeze-Drying (Lyophilization) and Other Processes

Many proteins are sensitive to denaturation by a variety of environmental factors. They may be precipitated from solution by changes in ionic strength, by pH adjustment, and by addition of some organic solvents or solutes. Whether or not these procedures result in denaturation frequently depends on the temperature and one or more of the factors listed above. Whenever possible, it is safer to avoid methods that depend upon addition of salts or organic solutes (urea, guanidine, etc.) and to recover the desired solute by the mildest possible means, especially when it is necessary to collect a product in the pure, dry state.

FREEZE-DRYING (LYOPHILIZATION)

Freeze-drying depends on removal of solvent by the process of sublimation; it begins with a frozen solution and transfers solvent molecule from the solid to the vapor state without passing through the liquid phase. Because the solution is kept in the frozen state, there is no possibility of foaming, which often leads to surface denaturation when unfrozen solutions are evaporated at reduced pressure. It is particularly suited to sensitive biopolymers, since most of these show improved stability at low temperature.

In its early days (the 1930s), freeze-drying was also known as lyophilization, and the machines used are still known as lyophilizers. However, the major journals and similar authorities prefer to describe the process and related hardware as freeze-drying, and as freeze-dryers. The preferred term is clearly more descriptive.

Although most commonly used for strictly aqueous solutions treated at very low pressures, neither of these requirements is essential. The process can be applied to solutions in mixed solvents, such as ethanol–water or acetone–water. The sole requirement is that the freezing point of the system be above the most convenient low-temperature bath available. That

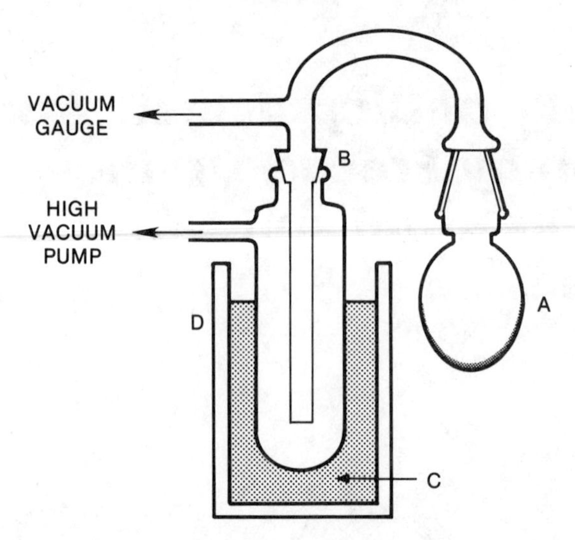

Figure 9–1. Essential components of a freeze-drying apparatus. The sample flask (A) is shown with the sample frozen in a thin layer over the interior surface. It is connected to the vapor trap (B) by a standard taper joint and then to a special rubber adapter. The trap is kept cold by a dry ice–acetone slush (C), which is protected against premature warm-up by the Dewar flask (D). Larger instruments provide for the connection of more than one flask, and most commercial units replace the freezing bath by mechanical refrigeration units.

temperature is commonly taken as −76°C, the temperature of a dry ice-acetone slush. Meryman showed, first on theoretical grounds and then on an experimental basis, that freeze-drying could proceed at atmospheric pressure as well as at very low pressure; although the demonstration was correct, most situations pose limits of time and sample volume which make operation at atmospheric pressure totally impractical.

Essential Components of a Freeze-Dryer

The components of a freeze-dryer are connected together as shown in Fig. 9–1. The sample flask (A) should have a volume at least four to five times the volume of the sample to be dried. This is connected to a vapor trap (B), which can be kept at about −70°C by means of a surrounding dry ice-acetone slush (C) contained in a Dewar flask (D). A Dewar flask is double-walled, the insides of the walls are silvered to reduce heat exchange, and the space between the walls is evacuated to reduce heat transfer even further. Mechanically refrigerated traps are also widely available and most convenient. Also shown in the figure is a vacuum pump capable of reducing the pressure in the closed system to about 10–30 μm (1 μm = 10^{-3} mm Hg) and a vacuum gauge capable of accurately reading pressures in the same range.

The usual vacuum pumps are mechanical, although mercury vapor pumps may be employed. Mechanical pumps are sealed with low-vapor-pressure oils, for good lubrication and as a secondary vacuum seal. The mechanical parts are fitted and polished to very close tolerances in order to operate in the necessary vacuum range. As a result, they are easily damaged when exposed to the corrosive effects of volatile acids or bases. These should *never* be drawn into mechanical pumps without special auxiliary traps to collect corrosive vapors before they reach the finely machined surfaces of the moving parts.

Theory of the Mechanical Pump

A cross section of a typical mechanical pump is shown in Fig. 9–2. The outer casing contains a cylindrical cavity fitted with a eccentric rotor. Fitted against the rotor, and passing through the center of the outer case top is a spring-loaded metal vane. The pump inlet and outlet tubes are shown at the upper left and right, respectively, of the top.

The inner cavity is divided into two chambers by the spring-loaded vane and the point of contact between the rotor and the case. As the rotor is driven in the counterclockwise direction, the volume of the inlet chamber increases and the pressure within it drops accordingly. This pressure drop draws into the chamber a portion of any molecules in the vapor contained in some connected apparatus, as shown by the small arrows. At the same time, the volume of the outlet chamber decreases; thus the pressure in it rises and vapor molecules are exhausted through the outlet tube, which is fitted with a check-valve. Properly maintained mechanical pumps can produce pressures of 10^{-4} mm Hg, but the speed with which this pressure can be achieved depends on the volume of the closed system, its freedom from leaks, and the pump capacity. Pump capacity is usually defined in terms of the number of liters per minute of air at 760 mm Hg which the pump can move. Single-stage mechanical pumps provide capacities of about 25 L/min, although larger, double-stage pumps are often rated at 100 L/min.

Theory of the Vapor Pump

Vapor pumps, like the common water aspirator, operate on Bernouilli's principle. A cross section of a typical water aspirator is shown in Fig. 9–

Figure 9–2. Cross section of a mechanical pump. See the text for a description.

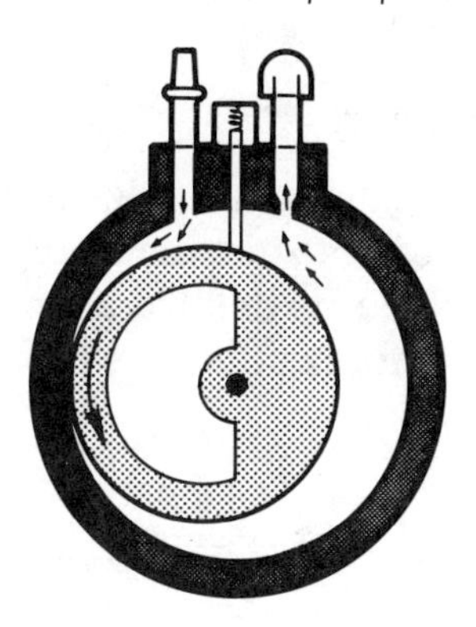

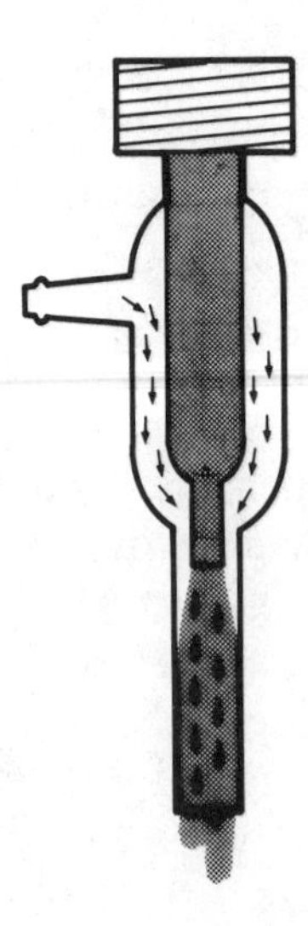

Figure 9–3. Cross section of a water aspirator. See the text for a description.

3. A jet of water emerges under considerable pressure from a narrow orifice enclosed in a tube of slightly larger diameter. The rush of the water stream entraps vapor molecules from the space around and just above the jet, thereby lowering the pressure in whatever vessel is connected to the side arm. The movement of vapor molecules is shown in the figure by the small arrows.

Similar devices, fitted with boilers and condensers, may be driven by mercury or oil vapor instead of a water stream. The ultimate pressure that can be reached is limited by the vapor pressure of the operating fluid; at any given temperature, the pump cannot reduce the pressure below that value. Thus, on a warm summer's day, water aspirators cannot reduce the pressure to $<$30 mm Hg, although when the water is cold ($\sim$10°C) the ultimate pressure may be 9.5 mm Hg. Because mercury and many pump oils have low vapor pressures at or near room temperature, they can provide ultimate low pressures of about 10^{-8} mm Hg. As compared to mechanical pumps, vapor pumps have low capacities, and they are frequently "backed up" by mechanical pumps.

Mean Free Paths of Vapor Molecules at Reduced Pressures

Because of their kinetic energies, molecules are constantly moving and colliding with one another or with the enclosing surfaces. The mean free path, L, of molecules in a vapor is defined as the average distance a molecule travels per unit time before colliding with another. From the kinetic theory of gases, one can show that:

$$L = \frac{kT}{\sqrt{2}\pi\sigma^2 P} \quad (9\text{–}1)$$

where k is Boltzmann's constant, T is the absolute temperature, P is the pressure, and σ is the distance separating the centers of two molecules at their closest approach. If one idealizes water vapor molecules as inelastic spheres, σ can be interpreted as the diameter of such a sphere. For water, the value of σ is about 3.85 Å. From the formula for L given above, one can calculate that the mean free path of a water molecule in the vapor phase at 0°C is about 100 Å if the pressure is 1 atm, whereas at a pressure of 1 μm (1 μm = $\sim 1.3 \times 10^{-6}$ atm), the mean free path has increased to about 4.5 cm. To be efficient, lyophilizers must be constructed with tubing of the largest practical diameter, and distances between heat sources and heat sink should be as short as possible. The same rule applies to connections between the sample flask(s) and the rest of the device.

Pressure Measurements at High Vacuum

McLeod Gauge. Simple Torricellian manometers are not ordinarily used for reading pressures in the range of 10–30 μm because they cannot be read with sufficient accuracy. The most frequently used instrument is known as a McLeod gauge, a cross section of which is shown in Fig. 9–4. It consists of four tubes or "arms." The arm marked M in the figure is the measuring arm; it is joined to the others by a diagonal, lower tube, but its top is closed off with a seal that is as flat as possible. The diameter of the

Figure 9–4. Diagram of a McLeod vacuum gauge. The volume of the measuring arm (M) and of the attached bulb, determines the working range of the gauge. The reading scale, which is nonlinear, is attached to the side of M. The leveling arm (L), the connecting arm (C), and the reservoir arm (R) are all connected above the maximum mercury level, shown by the stippled area. The arrow shows the direction of rotation of the gauge about its central pivot. Connection is made through a tube extending at right angles to the plane of the figure and to the rear.

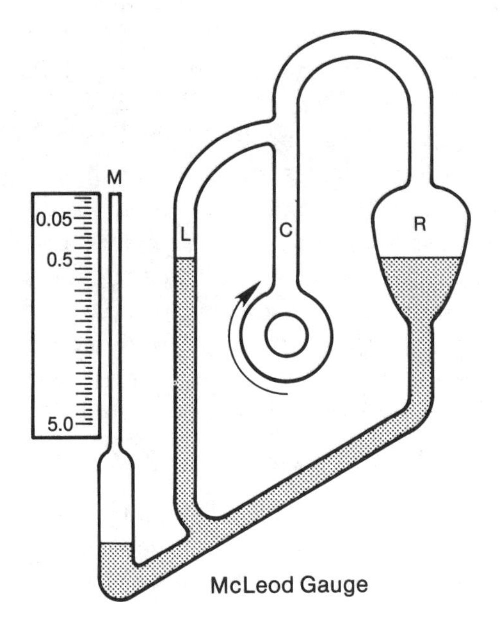

measuring arm is also made considerably smaller than the others. Affixed to the bottom of the measuring arm is a small bulb of somewhat larger diameter; the volume of the arm and the bulb must be precisely known for calibration purposes. This is done by the manufacturer, who scribes the attached measuring scale accordingly. Just to the right of the measuring arm is the leveling arm, L. In use, the level of mercury in M and L is made the same to ensure that the pressure in both is identical. To the right of L is the connecting arm, C. Projecting at right angles to the plane of the figure is the connecting tube, which is therefore seen end-on in the drawing. The entire instrument is arranged to pivot around the connecting tube, as is shown by the arrow in the figure. On the right side is the reservoir arm, R, which contains a portion of the mercury with which the gauge is filled. Note that L, C, and R are always at equal pressure, since they are connected above the mercury.

The McLeod gauge is an application of the Gay–Lussac law, which states that:

$$PV = nRT \tag{9–2}$$

R, of course, is the gas constant (in liter-atmospheres), V is the fixed volume of the measuring arm and bulb, n is the number of moles, and T may be taken as substantially constant over any series of measurements (or it may be measured and proper corrections made). As material is pumped out of the system, P must decrease.

To use a McLeod gauge, it is *slowly* rotated in a clockwise direction, allowing mercury to run into the reservoir bulb until the bottom of the measuring arm and bulb is open to the leveling arm. This allows the pressure in L and M to equalize. The gauge is then *slowly* rotated counterclockwise until the height of the mercury in L and M is the same. This maneuver has trapped a volume of gas in M; by making certain that the heights of the mercury columns in M and L are the same, one has established that the pressure in M is that of the system to which it is connected. If that pressure lies within the working range of the gauge (<5 mm Hg), it can be read directly. Note that there is no theoretical basis for the 5-mm Hg working limit. Rather, it depends on the design of the glassware and the necessity for pivoting about the connecting tube. In the design discussed here, gauges with a higher working limit would be quite fragile and so large as to be unwieldy. Note further that rotary movements of the gauge should always be *slow* and *deliberate.* Mercury has a specific gravity of 13.6, so a moving mass of mercury has considerable momentum. A rapidly moving column of mercury might slam into the closed end of the measuring arm with sufficient force to break the expensive glassware (counterclockwise gauge movement) to say nothing of endangering sample recovery. A rapidly descending column could be "broken" or interrupted, leaving spaces between intervals of mercury.

Pirani Gauge. In larger, commercial installations, for example, where large quantities of penicillin or antisera are being prepared, McLeod gauges are sometimes replaced by Pirani gauges. These are thermionic

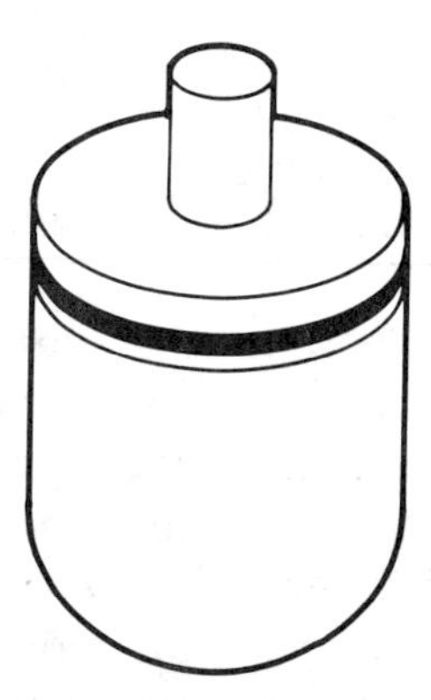

Figure 9–5. A two-piece sample flask. The compressible rubber O-ring is held in a groove, cast in the glass, where the two pieces meet. The ring is shown as a dark band in the drawing.

devices, consisting of a heated filament and a flat plate electrode placed closely and parallel to the filament. The filament is given a negative charge and the plate a positive charge, connected in series with some current-measuring device. The heated filament emits electrons which are carried to the plate by ionizing vapor molecules that lie between the electrodes. The current flow is therefore a measure of the pressure in the system. Pirani gauges can measure very low pressures (about 10^{-8} mm Hg) but they are fragile and require considerable hardware in the form of power supplies and current detectors. For these reasons the McLeod gauge remains the most common means of measuring pressure in laboratory-scale freeze-dryers.

Sample Flasks

Although ordinary round-bottomed flasks can be used for sample flasks, it is sometimes inconvenient to recover the dried product because the length of the flask neck interferes with the scraping-off of the adherent product with a spatula. A number of other styles have been designed. Some of these are fitted with the male half of a standard joint, rather than the normal female half. The reverse joint allows for lubrication with some grease, but one should keep this grease from contact with the dried material. Some sample flasks are pear-shaped, with short necks, to allow easier contact of a spatula with the inner surface. Another development is the two-piece flask, shown in Fig. 9–5. This has the advantage of a very wide mouth, since the bowl of the flask has more or less straight sides. The top is held in place by a compressible rubber O-ring, which also provides a tight seal. Whatever style of flask is used, it should be carefully inspected for severe scratching. Badly scratched flasks should be avoided because the surface imperfections may set up internal strains that could cause the flask to implode under high vacuum.

Starting Up a Freeze-Dryer

Freeze-dryers operate by transferring solvent water from sample flask(s) to the freezing trap, which has a finite capacity. Before connecting a sample to the machine, it is important to determine that the cold trap has sufficient remaining capacity to accept whatever volume of water is to be removed. If the total trap capacity is known, it is a good idea to log sample volumes as they are dried, so that a running tally of collected water may be kept. One must also be sure that the trap has been cooled to −50° to −75°C. In mechanically refrigerated machines this is easily done by checking the thermometer which manufacturers ordinarily build in. In machines cooled by dry ice–acetone mixtures, the cooling time is usually very short. Lastly, a check should be run on the vacuum that can be achieved without load, to be sure that all connections are tight before samples are committed.

Preparation of the Sample(s): Shell Freezing

An auxiliary dry ice-acetone bath is prepared to freeze the sample(s) on the inner wall of the sample flask(s). This bath should be prepared with careful attention to the following advice.

1. The dry ice should be finely divided. This is done by placing the solid mass of dry ice in some toweling or sacking and crushing it with a hammer, or else by means of an ice mill. The rate of cooling is proportional to the surface, not to the mass, of the ice. The only purpose of the acetone is to transfer heat promptly and efficiently.
2. The crushed dry ice should be transferred to a flat, open vessel; acetone is then added *slowly and carefully* until the free liquid surface is about 2 or 3 cm above the surface of the crushed ice. There will be an initially violent ebullition of CO_2 and acetone vapor as the mixture cools down, and there may be some splashing of the mixture over the top of the dish. This is the point at which care is required. The early, violent bubbling will soon cease as the mixture becomes increasingly cold. **CAUTION: Never put acetone into the dish first and then add dry ice. This error will almost certainly cause violent splashing and increase the danger of being burned.**
3. The bath is ready when the contents appear slightly sirupy in consistency, and when the rate of bubbling is very low. The bath is now at a temperature of −75°C. More dry ice may be added from time to time; because the bath is now so cold, no further violent ebullition will be noted when additional ice is added.

Transfer each sample to an appropriate flask (Figs. 9–5 and 9–6) and fit the latter with a temporary handle. As shown in Fig. 9–6, this may be a spare half of a standard taper joint, or it may be a length of glass tubing passed through a one-hole stopper. The only purpose of the handle is to permit safe manipulation of the flask while the sample is being frozen. If

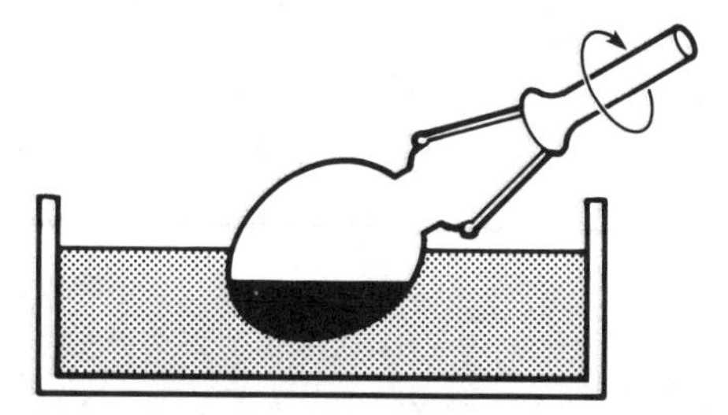

Figure 9–6. Auxiliary freezing bath for sample preparation. A dry ice-acetone slush is prepared in an ordinary metal basin. The sample flask, fitted with a temporary handle, is immersed in the freezing bath. The handle is used to twirl the flask between the fingers so that the sample freezes in as thin a layer as possible, over the inner surface of the flask. It should apper as in Fig. 9–1.

the handle is made of a standard joint, take care that the surfaces between the joint parts are not wet, or this film of moisture may also freeze. Even when they are dry, keep the ground area of the joint well above the bath surface. If the joint gets too cold, it may be difficult to remove the handle when the sample is solidified.

Immerse the bulb of the flask into the freezing bath while rapidly rotating, or twirling, the handle between the fingers. The idea is to cause the liquid sample to freeze over as much of the inner wall as possible. Continue the twirling motion until all of the sample is frozen. As the solid cools, it may crack in several places, with popping noises, but this can be ignored as long as all of the sample is solid. Remove the handle and connect the flask to the apparatus. A film of frost usually appears on the outside of the flask(s).

Connecting the Sample Flask(s)

Commercial, mechanically refrigerated machines often use a stainless-steel freezing chamber of 20–30 L in volume. The tops are usually of glass or plastic, allowing the user to inspect the volume of accumulated ice. Around the periphery of the steel chamber are as many as twenty to thirty sample ports, each with a built-in valve and a free end to which the sample flasks are attached. This allows the sample to be connected or disconnected without total loss of interior vacuum. Simpler, all-glass freeze-dryers often have one to four ports and depend on standard taper joints for connections.

Users will note a significant change in the sound of mechanical pumps, depending on the internal pressure of the system. When the pressure is high, the pump will make a "burbling" or "popping" sound, which quickly diminishes as the pressure drops. As vapor is being removed from the frozen sample(s), the outer surface of the flask(s) becomes covered with a thickening coat of frost that remains until all of the water is removed from the frozen sample. When the sample(s) are dry, this frost coating will melt and disappear, signaling that the necessary work has been completed. The pressure may then be returned to atmospheric and the flasks removed from the machine.

Recovery of the Dried Sample(s)

Freeze-dried samples frequently carry a surface charge of static electricity. When one inserts a metal spatula into the mass of dried material, the sample may cling to the spatula or may be repelled, depending on the sign of the charge. It is therefore a good idea to cover a lyophilized sample loosely and allow it to stand overnight, or until the static charge is dissipated. Alternatively, one may use a gunlike static charge dissipator, which can be triggered to discharge a pulse of electricity at whatever is closely targeted.

Theory of the Freeze-Drying Process

Transfer of molecules from the solid to the vapor phase occurs only at the interface. The energy needed to drive this process comes from heat transfer through the walls of the sample flask(s) and through the layers of ice, inside and out. Depending on the heat transfer rates and on the pumping speed, the actual temperature of a flask connected to the apparatus may rise or fall with respect to the bath in which it was prepared.

The Clausius–Clapeyron equation states that vapor pressure varies with temperature as:

$$\frac{dP}{dT} = \frac{\Delta H_{\text{sub}}}{T(V_{\text{v}} - V_{\text{s}})} \qquad (9\text{–}3)$$

where ΔH_{sub} is the heat of sublimation, V_{v} is the specific volume of the vapor, and V_{s} of the solid (both V_{v} and V_{s} are taken at absolute temperature, T, when the pressure is equal to the vapor pressure). Unfortunately, ΔH_{sub} varies somewhat with temperature, and it is not easy to measure T in a freeze-drying flask. Where experiments have been done, the relation between temperature and the vapor pressure of water takes the form shown in Fig. 9–7. At temperatures below 0°C, the vapor pressure drops sharply; if water is to be removed efficiently, the pump must produce pressures well below the vapor pressure, which explains why freeze-drying is practical only at low pressures.

It is difficult to find precise data for ΔH_{sub} in the range of temperatures from 0° to −50°C, but a reasonable approximation is that 700 cal are required to convert 1 g of ice to water vapor at the same temperature. If the pumping speed is too low, excess heat uptake will cause the sample to melt as its temperature rises. If the pumping speed is too great, the flask will suffer a drop in temperature, slowing sublimation. This tends to be self-correcting, because then the temperature will rise and vapor removal rate will again increase. The highest working temperature that will keep the contents of sample flasks in the frozen state is the best temperature.

Caution. Solutions of high ionic strength must be avoided since their freezing points are too low for the sample to remain frozen as a result of sublimation in a vacuum.

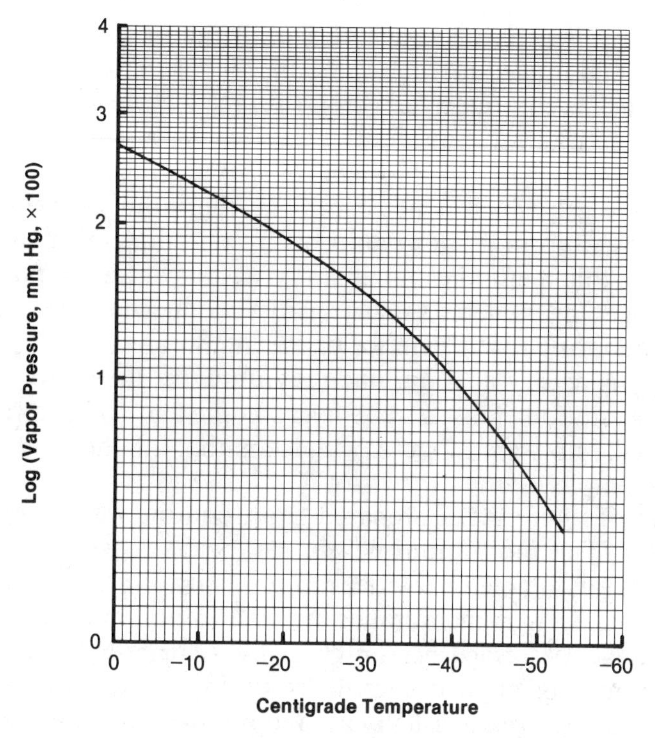

Figure 9–7. Plot of the logarithm of the vapor pressure as a function of temperature. For convenience in plotting, the vapor pressure was multiplied by 100 before taking the logarithms.

DRYING BY ULTRAFILTRATION

Dialysis membranes are freely permeable to water. The cutoff limits for larger molecules vary with the manufacturing process and the nature of the membrane. The limits are not always sharp, and are commonly expressed in terms of molecular weights of substances that will just permeate the membrane (e.g., MW = 3000, or 10,000, etc). Dialysis membranes are ordinarily used for purposes of concentration, but they may also be used for final drying if one is not too critical about removal of the last traces of water. Several operational procedures are possible, as set forth below.

1. Properly washed dialysis tubing is used, by knotting the ends, to form a sack filled with the sample. This can be hung from a clamp, and a small fan used to blow air over the surface. The stream of air causes evaporation, which continues until nearly all of the water is removed. The sack is cut open, and the highly concentrated or even semisolid or dry residue is collected. This process is called *pervaporation.* A major problem is that the membrane also dries, whereupon it becomes brittle and cracks easily. As a result, it is sometimes difficult to recover the product free of membrane fragments. This process is best suited to rough work (i.e., not requiring great precision).

2. The sample is tied into a sack of dialysis tubing as in (1) and the sack is then immersed into solid, high-molecular-weight dextran (Ficoll®, Bio-Gel®). The sack is completely covered with the dextran, which is too large to pass into the interior, but avidly binds water as it passes through the membrane. This process, although effective, consumes considerable amounts of the gel, which is not cheap. Furthermore, as the gel picks up water, it tends to form a sticky gum before it completely dissolves. Like procedure (1), this method is better suited to concentration purposes than to final drying. One must also guard against product contamination by the sticky gum adhering to the exterior of the membrane.

Instead of using dry dextran, one may immerse the sample into a 30–40% dextran solution. This is quite hygroscopic and still has considerable capacity to take up water. One advantage is that a buffered dextran solution may be employed.

3. Instead of using sacks prepared from continuous tubing, one may employ discs of ultrafiltration membrane, held in special filter devices. These discs are available in a wider range of cutoff values than is the continuous tubing. Similar membranes are sold as hollow fibers which have a greatly extended surface per unit weight. A pressure of up to several atmospheres may be applied over the solution to speed up the dialysis of water; a tank of nitrogen is usually used to provide pressure. The discs and filters may be reused, a great advantage in terms of cost. Pressure dialysis methods tend to be fast and efficient, and do not require bulky equipment. Final drying may not be as complete as by lyophilization, but the methods are widely used for small- and large-scale concentration.

Unlike freeze-drying, where everything contained in the sample flask ends up in the residue, dialysis methods tend to lose buffer salts and similar small molecular species. One must be careful to distinguish between drying and precipitation phenomena due to ionic-strength effects.

REFERENCES

Daniels, F., and Alberty, R. A. *Physical Chemistry,* 2nd ed. John Wiley and Sons, New York (1961).

Flosdorf, E. W., and Mudd, S. Procedures and apparatus for preservation in the "lyophile" form of serum and other biological substances. *J. Immunol. 29:*389–425 (1935).

MacKenzie, A. P. Physico-chemical basis for the freeze-drying process. In *Freeze-Drying of Biological Products Developments in Biological Standardization,* Vol. 36 (V. J. Cabasso, and R. H. Regamey, eds.), pp. 51–67. S. Karger, Basel (1977).

Meryman, H. T. Sublimation freeze-drying without vacuum, *Science 130:*628–629 (1959).

Rey, L. *Theoretical and Industrial Aspects of Lyophilization.* Hermann, Paris (1964).

CHAPTER TEN

The Basis of Immunochemical Methods

Immunochemical methods are increasingly important tools of modern biochemical research. Their power stems from their sensitivity and specificity; typical immunochemical procedures function best in the mass range from 0.1 to 100 ng; thus immunochemical methods are truly micromethods.

Immunochemical methods also have *resolving power,* that is, the ability to separate individual components from very complex mixtures, for which both high sensitivity and high specificity are essential.

In immunochemical reactions, the specificity depends on the "fit" among the reaction components; the principle is analogous to that of enzyme reactions, where the "fit" of a substrate to the active site of an enzyme helps determine whether catalysis occurs and at what rate. The high specificity of an immunochemical reaction is due to the very large number of *immunochemical determinants* or *epitopes,* which define the specificity. In addition to the number of epitopes, their arrangement within the architecture of a molecule may have some effect on the reaction system.

The sensitivity of an analytical method, other things being equal, depends on the actual operation employed to estimate or determine an end point. Well-designed immunochemical methods may result in products with net charges that are quite different from the charges on the reaction components. As a result, the products frequently are much less soluble than the reactants. The products may bind quite differently to charged surfaces, and thus they can readily be separated and recovered. The sensitivity of end point detection may be amplified in one of several ways. A reaction component can be tagged (1) by introduction of a radioactive atom, such as ^{125}I; (2) by covalent binding of some totally unrelated species, such as fluorescein; or (3) by covalent binding of some totally unrelated enzyme which itself can be determined with high sensitivity (see ELISA, under Enzyme-Linked Immunoassays, below). Because the intrinsic sensitivity of the immunochemical assay is further enhanced by the sensitivity of the fluorometric, spectrophotometric, or liquid scintillation technology, the overall sensitivity of the amplified system is very high indeed. In simpler cases, of course, one could proceed by staining the gels or by using a reliable chemical method for determination of the precipitated protein.

THE FUNDAMENTAL IMMUNOCHEMICAL REACTION

The basic reaction that we will use can be written as:

$$Ag + Ab \rightleftharpoons Ag \cdot Ab$$

where Ag represents an *antigen* and Ab represents an *antibody*. The symbol Ag·Ab represents a noncovalent complex formed from the reactants. Frequently, but not always, the Ag·Ab complex is insoluble, and then is known as a *precipitin*. Hence, the antigen–antibody reactions that lead to formation of a flocculent or solid product are sometimes identified as *precipitin reactions*. Representation of the product as Ag·Ab is probably incorrect; from careful experimentation it appears that $Ag_x \cdot Ab_y$ would be a more accurate representation of what actually occurs. Furthermore, either x or y may have nonintegral values; since $Ag_x \cdot Ab_y$ does not show constant stoichiometry, it cannot be regarded as a true compound.

We can classify antigen–antibody interactions as a subset of ligand binding interactions, similar in many respects to the interaction between biotin and avidin (see Expt. 4). Mathematical methods for studying immunochemical systems are similar to those employed in general ligand-binding systems. The important distinction between immunochemical systems and generalized ligand binding systems is two fold. Broadly speaking, generalized ligands include compounds with a wide range of molecular weights. Examples of low-molecular-weight ligands include biotin and palmitic acid (a good ligand for serum albumin). Insulin, with a fairly high molecular weight, may be regarded as a ligand for its specific receptor on plasma membranes. Antigens, on the other hand, are of rather high molecular weight. Most antigens are proteins (or large polypeptides) although some are non-protein-containing complex polysaccharides.

The second distinction lies in the ligand receptors of immunochemical systems, the antibodies. All antibodies are globulins that are produced as part of the immune response by cells of the reticuloendothelial system, and are then delivered to the serum. Therefore antibodies are also known as *immunoglobulins*. In humans and some other animals, five distinct classes of immunoglobulins (Ig) can be detected in the blood serum and in such bodily secretions as saliva, tears, nasal mucus, etc. The physiologic purpose of immunoglobulins is to distinguish between "self" and "non self" by combining with and/or promoting inactivation or destruction of molecular species that arise outside the host organism. Thus, if one repeatedly injects a rabbit with purified human serum albumin ("immunizes" the rabbit), within a week or so the rabbit's blood will contain a specific subset of IgG which, when added to a dilute solution of human albumin, gives a positive precipitin reaction. This particular set of IgG would be known as rabbit anti-human albumin, or rabbit anti-hA.

There are several profound consequences of these observations. Rabbits, in the normal course of their lives, do not encounter human albumin; the production of anti-hA occurs *only* as a direct consequence of the challenge of immunization. One may therefore speak of "raising" antibodies to a given antigen of one's choice, in the host animal of one's choice. It is this characteristic of the immune response that makes immunochemical methods extremely versatile. Suppose that one has a sample of anti-hA and mixes it with an albumin preparation from some nonhuman primate. Whether or not a precipitin reaction will occur depends entirely on the degree of homology between human albumin and the primate albumin. An antibody that gives a positive response with two (or more) closely related proteins is said to *cross-react* with those proteins. One can extend such observations through larger segments of the Animal Kingdom; from such experiments one can construct a kind of biochemical "family tree" based entirely on the known properties of proteins. Such geneologies have been established for a small number of proteins, most notably serum albumin and cytochrome c. When coupled with amino acid sequence studies, immunochemical methods of this kind provide amazing insights into phylogeny.

ANTIBODIES

Properties and Structure

As noted above, immunoglobulins can be divided into five major classes. Table 10–1 lists some of the properties of each class, as found in humans. The molecular weights of the Ig classes cover quite a range of values. The smallest species, IgG, can cross the placental membranes, whereas the larger species cannot. All of the classes contain carbohydrate, but its relative content varies widely from class to class. Serum levels of the Ig classes are surprisingly disparate, probably in relation to their respective biological functions, an aspect that we shall not address here. When a rabbit is injected subcutaneously, intramuscularly, or intraperitoneally with a foreign protein, the usual response is the generation of antibodies of the IgG class. So it is with synthetic antigens also.

Striking as these differences are, the similarities in Ig structures are even more striking. All immunoglobulins are tetramers, composed of two

Table 10–1. Properties of Immunoglobulins

	IgA	IgD	IgE	IgG	IgM
MW $\times 10^{-3}$	160	185	200	150	900
Carbohydrate content, %	8	13	12	3	12
Serum Conc., mg/ml	1–4	0–0.4	$<10^{-5}$	8–16	0.5–2.0
Light chains	$\kappa+\lambda$	$\kappa+\lambda$	$\kappa+\lambda$	$\kappa+\lambda$	$\kappa+\lambda$
Heavy chains	α	δ	ϵ	γ	μ

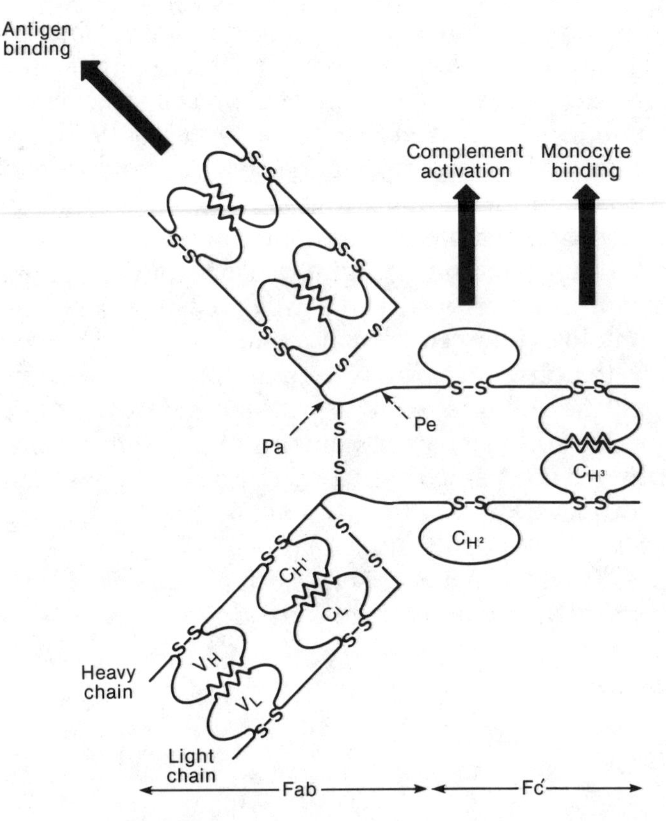

Figure 10–1. The general structure of immunoglobulins, showing the disulfide bonds that form closed loops in the light and heavy chains as well as the disulfide bonds that hold the chains together. The antigen-binding function is localized to the amino-terminal portions of the light and heavy chains, whereas activation of complement and binding to monocytes (other important functions of immunoglobulins) are localized to other portions of the Y-shaped tetramer. The cleavage sites of the proteolytic enzymes papain (Pa) and pepsin (Pe) are shown by the broken arrows. The variable and constant regions of the light (V_L, C_L) and heavy chains (V_H, C_H) are shown, for convenience, only on the lower branch of the schema. (Adapted by permission from I. Roitt, *Essential Immunology*, 5th ed. Blackwell Scientific, London, 1985.)

heavy (H) chains characteristic of each class and two light (L) chains (or perhaps three) which occur in all Ig classes. The H chains are usually identified by Greek letters which specify the separate classes (e.g., α, γ, μ), whereas the L chains are identified as either κ or λ. Thus, the molecular structure of an IgA might be represented by ($\alpha_2\kappa_2$ $\alpha_2\lambda_2$) with the two tetramers held together by disulfide bonds. IgM forms a pentamer of tetrameric units, with the tetramers again held together by disulfide bonds. Each tetrameric unit—say, $\gamma_2\kappa_2$ or $\gamma_2\lambda_2$—is held together by numerous disulfide bonds, as shown in Fig. 10–1.

It is clear from Fig. 10–1 that disulfide bonds join the L chains to the H chains and also join the H chains together, roughly at the centers of the H chains. In IgM and IgA, additional disulfide bonds form, closer to the carboxy-terminus of the H chains, allowing for dimerization or pentamerization of the basic tetrameric units, as already mentioned. In any event, the nonpolymerized (monomeric) species, such as IgG, has a Y-shaped structure, as determined by x-ray crystallographic studies, and thus is drawn accordingly. The area about the center of the Y is denoted the *hinge region;* it is also the site at which one or more disulfide bonds join the H chains, as well as being the site of attachment of the carbohydrate.

The H and L chains also contain numerous internal disulfide bonds along their lengths. These cause the formation of closed loops of 50–70 amino acid residues in length. Each of these closed loops comprises a separate protein domain and obviously affects the folding pattern of the immunoglobulin. So consistent is this folding pattern that it has become known as the *immunoglobulin fold.* Space-filling models show that the L chains are helically wrapped about the adjacent portions of the H chains, and that the distal parts of the H chains are wrapped about each other.

Early studies of immunoglobulins were designed to determine which part of the Ig structure was responsible for antigen binding. The proteolytic enzyme, papain (Pa in Fig. 10–1; for convenience and clarity, the site of papain cleavage is shown on only one branch of the Y), cleaves an Ig molecule into two fragments, roughly in the middle of the molecule. The fragment derived from the amino-terminal segment of the Ig is knows as *Fab,* since it retains all of the antigen-binding capacity of the Ig from which it was obtained. The remainder of the molecule, which does not have this capacity, can be crystallized; hence it is known as *Fc.* A second proteolytic enzyme, pepsin (Pe in Fig. 10–1), cleaves at a slightly different point, generating fragments known as $F(ab')_2$ and Fc. $F(ab')_2$ still has a disulfide bond, and thus is represented as a dimer; it still has the antigen-binding capacity of the original Ig. Enzyme cleavage of antibody thus established that the antigen-binding properties of Ig molecules rest entirely on the association of L chains with at least the amino-terminal half of the H chains. The carboxy-terminal moiety of the H chains has specific binding properties of its own, but these are secondary to this discussion in that they do not involve antigen-antibody interactions.

Analytical data and sequence studies show that the H and L chains contain regions of considerable constancy of amino acid composition, as well as regions in which the amino acid composition is quite variable. The organization of these regions is shown in Fig. 10–2, where, for simplicity, V_L, C_L, V_H, C_H^1, C_H^2, etc. are indicated only on the upper branch of the Y-shaped molecule. The numbers on the lower branch of Fig. 10–2 are the numbers (*not* positions) of amino acid residues in the respective regions. Within the variable regions, certain stretches of the primary sequence are especially prone to alteration; such subregions are known as *hypervariable regions.* They are indicated in Fig. 10–2 by the symbol HV, together with the numbered locations within the amino acid sequence. It is currently believed that the hypervariable regions are of special importance in "fine-tuning" the structure of an Ig to provide for the demonstrated specificity toward one antigen out of many. It is not at all clear that specificity depends only on alterations in the hypervariable regions, but the actual

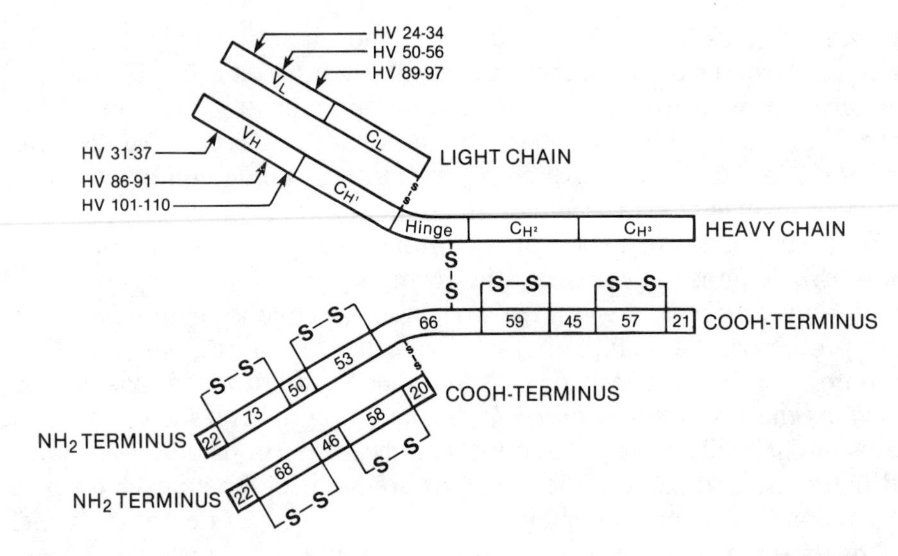

Figure 10–2. A representation of immunoglobulin structure showing the *numbers* of amino acid residues incorporated into the closed loops and the segments between them. The hypervariable regions are indicated by the symbol HV; the numbers after the symbol specify the *positions* of greatest variability. (Adapted by permission from C. Milstein, and J. Svarti, *Prog. Immunol.* *1*:33, 1971.)

number of completely specific antibodies that could be generated by such alterations alone is very large indeed. The diversity and specificity of antibody production depends on assembly of numerous gene segments to from a complete Ig gene (see Chapter 11).

Antibody Production and the Immune Response

Injection of antigen triggers a two-stage response in the host. The first, or *primary* stage, requires from 5 to 15 days. Although the concentration of circulating antibody rises significantly during the primary period, little IgG is found in the mixture. The *secondary* stage, which is induced by repeated injection of the antigen ("booster shots"), results in a significantly faster production of antibodies so that the concentration of IgG greatly increases. Proper immunization of an animal frequently requires as many as five injections, in order to maximize production of serum IgG.

The dual response can be explained as follows: The primary response is initiated by small circulating lymphocytes. These either originate in or are in some way processed by the thymus; hence they are known as T cells. Once activated by antigen, T cells produce little antibody by themselves, but, in ways that are still uncertain, they interact with and stimulate the production of antibody by plasma cells, which are clones of cells produced in mammalian bone marrow (B cells) and are exported to the blood and the lymph nodes. The high titer of circulating antibodies noted in the secondary stage of the immune reaction is due largely to antibody production by the B-derived cells. The memory of the immunologic challenge may be transitory, or it may persist through the life of the host. If antibody production begins to diminish, it frequently can be invigorated by an additional booster shot. Thus, a properly immunized rabbit or guinea pig becomes a valuable source for additional antibody raising and harvesting.

The strength of the immune response is conditioned by the size, the frequency, and the site(s) of the immunizing injections. The physical form of the antigen, which modulates the rate of its absorption by the host, may also be influential. Thus, a stronger response is usually obtained when the antigen is injected as an emulsion rather than as a solution. Emulsifiers may incorporate *adjuvants,* or mixtures of inert irritants, which stimulate the immune response by induction of mild inflammation at the injection site(s). A final factor in determining the strength of the immune response is the host itself; certain species are more responsive than others to a given antigen, and even within a species certain individuals have more pronounced responses than others.

Isolation and Purification of Antibodies

In the simplest case, purification may be limited to collection of serum from the immunized host. Many immunoassays function quite well even in the presence of the other serum proteins. In more discriminating assays, it is possible to remove the major protein, serum albumin, by fractionation with ammonium sulfate. By modification of the salting-out procedure, it is also possible to remove α and β globulins, leaving a concentrate of γ globulins. Purification to this stage is simple and straightforward, but to go beyond this and to isolate a single, homogeneous antibody is considerably more complicated.Four purification procedures are described in the following subsections.

Affinity Chromatography. In affinity chromatography (see also Chapter 7), the antigen against which the antibody is directed may be covalently bound to an inert support such as agarose beads, which are then packed in a column. When a γ globulin concentrate containing the antibody to be purified is passed through the column, the antibody is tightly bound and the unbound proteins may be washed through the column. The antibody may be recovered from the column by a sharp drop in the pH or by a large increase in the salt concentration. The products that are recovered are very nearly homogeneous; thus this is one of the best purification methods known. If, for any reasons, only a small quantity of antigen is available, this method may require an extravagantly large amount of antigen.

Adsorption Procedures. This could be described as a kind of inverse affinity chromatography. Suppose one were studying the in vitro translation of an isolated mRNA message, perhaps in a noncellular wheat germ translating system (see Chapter 11). It is clear that the translating system would contain many proteins, from which isolation of the *purified* message product would indeed be difficult.

One could prepare an affinity column by covalently binding to agarose beads the mixture of proteins found in the translating system, *less* the added message. If a rabbit is next immunized with proteins from the translating system *plus* the added message—a mixture of antigens that presumably contains the translated product of the added message—the rabbit will produce antibodies against all of the protein components. If the mixed antibodies are then passed through the column, only the antibody of the message product will be found in the eluate; all of the other antibodies will have been bound to the column packing or, as the immunologists describe it, they will have been *adsorbed.* This is adsorption of a very special and specific kind. Other, more traditional adsorbents—charcoal, powdered glass, etc.—have been used for similar purposes, but they do not operate as well because they are not specific. Immunologic adsorption methods are costly and somewhat difficult to carry out, but are capable of giving high resolution.

***Staphylococcus aureus* Protein A.** Certain strains of *Staphylococcus aureus* produce a substance known as protein A. Some classes of IgG from some species are bound very strongly by protein A. Rabbit IgG is especially well bound, whereas goat IgG binds much less strongly. The site of the IgG responsible for binding to protein A is presumed to be located in the Fc region, and it appears that the carbohydrate moiety may be partly responsible for the binding. Protein A has been isolated and is now commercially available bound to agarose beads. It provides a different kind of affinity chromatography for antibody purification but, clearly, is not always a feasible approach.

Ion-Exchange Chromatography. DEAE-cellulose (see Chapter 7) has a high affinity for γ globulins, but its capacity is not great. It may be applied directly to whole serum or to material previously treated with ammonium sulfate to reduce the concentration of extraneous proteins. In either case, the DEAE-cellulose may be employed batchwise or as a column packing. Batch operations are more suitable for large-scale work, but are intrinsically less efficient than columns. It may be necessary to repeat the treatment several times in order to attain a desired degree of purification. The higher efficiency of column operations is sometimes offset by their lower capacity.

ANTIGENS

Properties and Structure

An antigen must have three properties:

1. It must contain a polypeptide or polysaccharide of fairly high molecular weight. Glycopeptides or proteins and lipopolysaccharides or lipoproteins, as well as nucleoproteins, are effective antigens from the standpoint of their molecular weights.
2. When injected into an animal that recognizes it as foreign, an antigen must stimulate production of antibodies directed against it.

3. It must show a specific reaction when mixed with the antibody in vitro. Ideally, a precipitin will be formed in the mixture.

Antigens exhibit these properties because they contain certain structural features that have been variously identified as *haptenic groupings, immunologic or antigenic determinants,* or *epitopes.* In a broad sense, these terms have nearly identical functional significance, but there are some distinctions that must be kept in mind. The phrase "immunologic or antigenic determinants" is ordinarily used to describe that subset of possible epitopes that are natural structural elements of a protein, etc., that cause it to be antigenic. The term, *hapten,* on the other hand, is ordinarily reserved for small molecules or ligands that, when coupled to some weakly antigenic or nonantigenic protein, cause it to become an effective antigen.

Haptens, by themselves, are nonantigenic. Injection of haptens does not lead to antibody formation. However, antibodies directed against hapten-containing antigens will react positively with that antigen *and/or* with the free hapten.

Epitopes, whether natural immunologic determinants or deliberately added haptenic groups, ordinarily have one or both of two structural features: (1) a center or cluster of charges, as in arsenilic or trinitrophenyl sulfonic acids, or even lysine; and (2) the presence of rings, such as the oxygen-linked rings of certain polysaccharides. Even thyroxine or certain steroids can act as haptens, which makes it possible to generate antibodies against these small hormones for use in immunoassays of the hormones. Immunoglobulins themselves can act as antigens, providing the basis for double-antibody immunoassays in which, for example, very small amounts of an antigen–antibody complex formed with a rabbit (r)IgG is detected by means of a goat-derived anti-rIgG.

The great power of immunochemical methods stems largely from the almost limitless ability to design antigens for a particular purpose. One needs only a suitable hapten, a suitable carrier protein, and a method for covalently joining the two. The number of suitable carrier proteins, which should be available in high purity and at low cost, is limited. In the early days, ovalbumin and casein were most popular; more recently, greater use has been made of serum albumin from the cow, pig, rabbit, sheep, or human. In general, one would not attempt to immunize a rabbit with an antigen based on rabbit albumin, since the hapten introduced is only one of many potential kinds of epitopes that might exist in the carrier structure. The carrier, as well as the hapten, should be as foreign as possible in order to elicit a strong immune response. One reason serum albumin is such an outstanding carrier protein is the relative lack of structure conservation between species, which translates into antigenicity. Insulin, on the other hand, has a high degree of structure conservation, which allows human diabetics to be treated with insulin obtained from the cow or the pig. Only very rarely does a human develop antibodies against the foreign hormone. Conservation appears to be the rule for other protein hormones as well.

"Valency" and "Combining Power"

It was noted earlier that antigen-antibody complexes, when thrown out of solution as precipitins, did not exhibit constant stoichiometry. Yet the literature of immunology contains many references to "valence" and "combining power" of antigens with respect to antibodies. These terms are put in quotation marks because, as the chemistry of the immune reaction became better understood, it also became clear that the sense of these terms was not the same as when they were applied to nonimmunologic systems. As used by immunologists, "valence" does not refer directly to ions, nor is "combining power" used to describe the behavior of 1.008 g H^+ ions. It would perhaps be better if these terms were altogether abandoned in favor of equivalence of antigens and antibodies.

Detailed chemical studies of the precipitin reaction began about twenty-five years ago. From the early work of Heidelberger and Kendall, extended by the later work of Bordet, Marrack, and others, we know that antigens contain numerous epitopes, not all of which are identical in nature. To form a precipitin, an antigen must contain as least two epitopes (a *complete antigen*) whereby it may serve as a cross-linking adduct, of a non-covalent type, to join two antibody molecules. Such a complex could become large enough to become insoluble. An antigen that contained only one epitope (an *incomplete antigen*) could not serve as a cross-linking adduct and the probability that an insoluble complex would be formed is very low. As a rule of thumb, the number of epitopes in a typical protein is approximately given by the expression, MW/10,000. Of this total number, some epitopes are latent or cryptic because they are buried deep within the structure of the protein as it folds in space; only those epitopes that are exposed at the surface are functional in antigen–antibody binding. Evidence for this hypothesis comes from examination of cleavage products of antigens, produced by controlled proteolysis. As antibody structure was elucidated, the Y shape of Ig species suggested that there were two binding sites per molecule, each of which could bind to some recognized feature(s) of an antigen. These concepts are summarized schematically in Fig. 10–3. The antigen-binding sites are shown as open circles (or triangles) at the ends of the Y-shaped molecules. Epitopes are shown as filled circles (or triangles), placed arbitrarily at the ends of heavy lines which represent antigens. In actual antigen molecules, of course, the functional epitopes would be scattered in some random manner over the exposed surface.

Figure 10–3A(1) shows the reaction of an incomplete antigen mixed with an antibody in nonequivalent proportions. A complex would form, but it would not produce a precipitin because cross-linking could not occur. Even when mixed in equivalent proportions, as in Fig. 10–3A(2), the complex formed could not give rise to a precipitin, for the very same reason.

Figure 10–3B depicts the situation for the simplest possible complete antigen, containing only two epitopes which are assumed to be different.

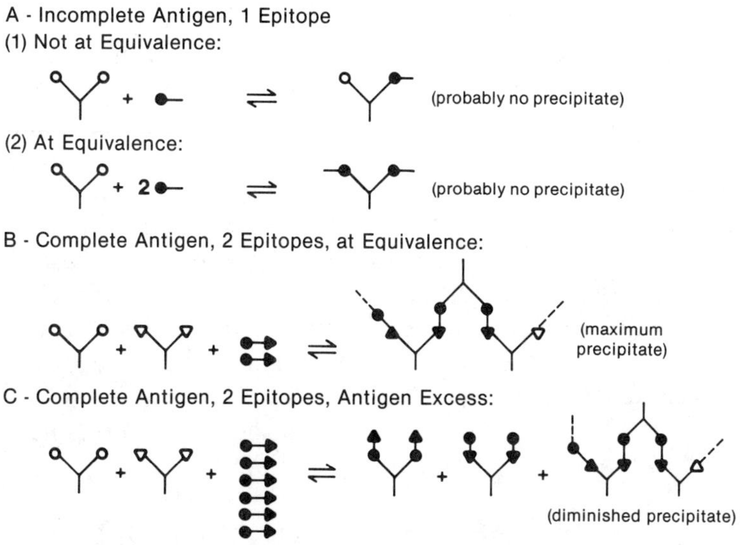

A - Incomplete Antigen, 1 Epitope
(1) Not at Equivalence:
(probably no precipitate)
(2) At Equivalence:
(probably no precipitate)
B - Complete Antigen, 2 Epitopes, at Equivalence:
(maximum precipitate)
C - Complete Antigen, 2 Epitopes, Antigen Excess:
(diminished precipitate)

Figure 10–3. A schematic representation of antigen–antibody interactions. Nonidentical antigenic epitopes are shown by closed circles or triangles, and corresponding antigen-binding sites on the Y-shaped antibodies are shown by open circles or triangles. Reactions A(1) and A(2) illustrate incomplete antigens, in nonequivalent or equivalent proportions, respectively. Because cross-linking is not possible in either A(1) or A(2), formation of a precipitate is unlikely. The complexes simply cannot grow to sufficient size. Reaction B illustrates a complete antigen mixed with an antibody at equivalence; here the reactants combine to form a large, three-dimensional complex which has a high probability of forming a fairly stable precipitate, even though some unsatisfied binding sites persist at the extremities of the complex as it extends in space. Reaction C shows what happens when there is an excess of antigen. This free antigen competes for free antibody with the rapidly associating and dissociating complex. As a result, the quantity of precipitate is diminished, as compared to the situation in B, where antigen and antibody are at equivalence. A similar argument can be constructed for antibody excess. (The effects of extraneous substances, although real, are omitted from this diagram for simplicity.)

These are represented by a closed circle and a closed triangle. *Note that two nonidentical antibody species are required to neutralize this antigen, but that these antibodies cannot be operationally distinguished by the precipitin reaction.* At equivalence, the antigen will combine with these antibodies to form a large aggregate; there is a very high probability that a precipitin will form. Since most antigens contain more than two epitopes, the high probability of precipitin formation becomes a virtual certainty if the reactants are in equivalent proportions. It follows also that the number of nonidentical but indistinguishable antibodies that will react with a particular antigen must increase as the number of functional epitopes increases.

Figure 10–3C depicts the case when the theoretical antigen with only two epitopes is present in significant excess. Because immunochemical reagents are expensive and tedious to prepare, one always attempts to work at or near the equivalence point. There is, however, a more compelling reason for using equivalent proportions. Recall that the formation of an antigen–antibody complex is reversible. Then, even if a complex large enough to form a precipitin is momentarily formed, it could likely be degraded by the slower equilibrium which would ultimately prevail because of the antigen excess free in the solution. Any antigen or antibody molecules exposed at the surface of the complex could associate with or dissociate from the complex in a random manner. Once dissociated from the complex, the odds are statistically greater that an antibody molecule would encounter a free antigen molecule before it encountered the complex. Hence, the efficiency of precipitin formation under these conditions would be lower than if the interacting species were at equivalent proportions. A formally similar situation would exist in the presence of excess antibody. For these reasons, antigen and antibody preparations must always be titrated against each other, usually by empiric serial dilution methods, to establish the equivalence point of a given precipitin reaction system and to maximize precipitin formation.

CURRENT AND NOVEL APPLICATIONS OF IMMUNOASSAYS

Immunossays are special cases of ligand binding assays. The precipitin reaction has been the most widely exploited of the immunochemical reaction types. Its earliest applications were based on simple solution chemistry; a solution of an antibody was carefully overlaid on an antigen solution, and the presence or absence of a precipitin at the interface of the solutions was noted after some suitable time. From there, technology progressed to diffusion or electrophoresis through gels of agarose or polyacrylamide, thus avoiding the problem of convective disturbance. These methods are still widely used, with end point and/or quantitation performed by densitometry of suitably stained preparations or nephelometry of unstained preparations (see Experiment 11).

As an alternative procedure, a precipitin product may be generated in free solution, collected by centrifugation or filtration, and determined by a radioactive or fluorescent ligand. These methods are in common practice as *direct* procedures, in which one or the other immunochemical component must be tagged. Because of the expense and difficulty of labeling antibodies, *indirect* or *competitive* assays have become increasingly popular. Here the antibody is raised against some hapten, perhaps a steroid or similar species. To carry out the determination, antibody is added to the unknown sample which has been mixed with a known amount of radiolabeled hapten at a known specific activity. The endogenous and the labeled hapten compete for binding sites on the antibody. After a suitable period to allow for equilibration of the system, a second antibody, raised against the first, is added to form a precipitin, which is collected and counted. Note carefully that the fundamental principle is the competition between labeled and unlabeled *hapten;* the addition of the second antibody is merely a convenient and sensitive way to collect the product complexes specifically. Many assays of this type are used for determination of steroids, peptides, and similar small molecules.

An alternative approach is to bind antibodies covalently to beads of inert supports. This very much simplifies collection of antigen–antibody complexes. In other instances, antibodies are covalently attached to the inner surfaces of small glass or plastic tubes. In a competitive binding assay, for example, one need only incubate the reaction mixture, drain and rinse the tubes, then transfer them to an automatic scintillation counter. No filtration or centrifugation is required. Although such beads or tubes are expensive and can be used only once, they are economically justified if large numbers of the same assay are being performed on a routine basis.

Two-dimensional electrophoresis is widely used to separate and/or identify individual proteins in complex mixtures. To pick one out of many hundreds of molecular species is a complex task, but it is a task for which the specificity of immunochemical methods is ideally suited. Incorporation of a radioactive label is not always practical, and the number of suitable isotopes is not large. An alternative is fluorescence tagging of the antibody. Fluorescein isothiocyanate (FITC) is an excellent fluorophore (i.e., is highly fluorescent), and the conditions of labeling are quite mild. One performs the electrophoretic separation in the usual way in a slab of agarose or polyacrylamide gel. The gel is then overlaid by a thin sheet of nitrocellulose paper moistened with buffer. By diffusion or, better, electrophoresis, the separated proteins are transferred from the gel to the nitrocellulose paper in what amounts to the third dimension. Because the distance involved is small and because the proteins are rather tightly bound to the paper, the transfer is fairly rapid and loss of resolution by diffusion is fairly small. The nitrocellulose paper is then dipped into a solution of the antibody tagged with fluorescein, then briefly rinsed. When observed under ultraviolet light, only the spot where the antigen–antibody complex was deposited will fluoresce ("light up"). This procedure of transfer of proteins from gel to nitrocellulose paper is an example of a "Western" transfer. If desired, one can recover the protein sought, after cutting out the spot, by adjusting the pH or the ionic strength.

Fluorescent antibody methods have also been used to localize certain proteins on the surfaces of plasma membranes or organelles. Localization of flagellar antigens of bacteria and of some capsular antigens has been perfected by immunochemical procedures of this kind. In general, antibodies are too large to penetrate biological membranes, so their use is largely restricted to substances in contact with the external surfaces of cells.

Many isolated organelles are too small to be studied with the light microscope. In these instances, the fluorescent ligand can be replaced by an electron-dense ligand, such as ferritin. Binding of ferritin-tagged antibodies can be studied with the electron microscope.

ENZYME-LINKED IMMUNOASSAYS

Enzyme-linked immunoassay (EIA) is a term applied to a variety of related procedures used to enhance the sensitivity of an immunoassay. In general terms, instead of simply observing the precipitin lines in immu-

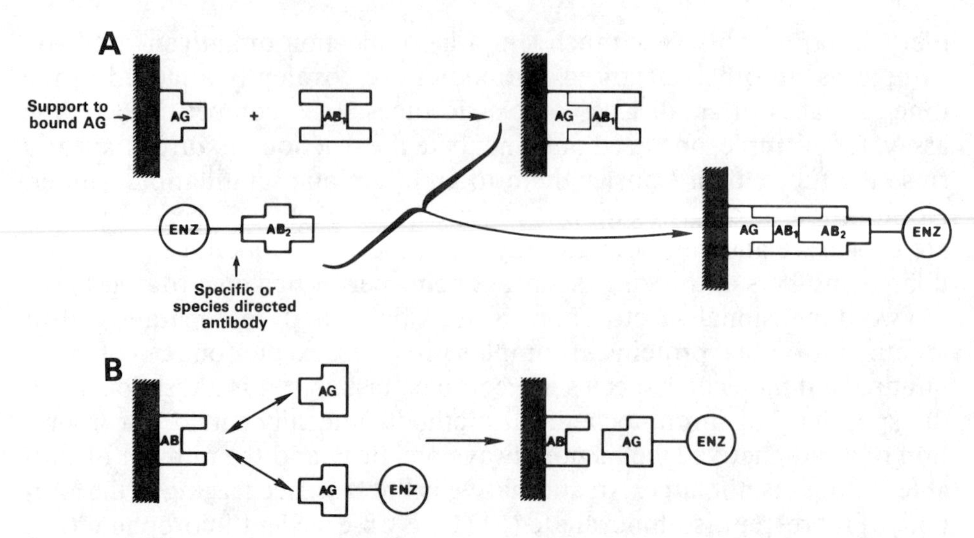

Figure 10–4. General schematic for ELISA procedure: (A) noncompetitive; (B) competitive. AB_1 is antibody to antigen; AG is antigen; AB_2 is antibody specific for the animal species in which AB_1 was raised; ENZ is horseradish peroxidase, alkaline phosphatase, or another enzyme.

nodiffusion systems, or separating the antigen–antibody complex by centrifugation, one uses an antibody that has been covalently linked to an enzyme. This antibody–enzyme reagent is then reacted with the antigen. The presence of this antigen–antibody–enzyme complex is then detected by supplying the appropriate substrate for the enzyme and measuring the amount or concentration of product. The enzymes most frequently employed are alkaline phosphatase, horseradish peroxidase, glucose oxidase, and β-galactosidase.

There are two general approaches to EIAs: The "separation-free" (homogeneous) system relies on the discovery that when an antibody raised against an antigen binds to an enzyme labeled with that same antigen, that enzyme's catalytic activity is modulated (inhibited or activated). We will not describe this approach further. The second approach is the "separation-required" or heterogeneous technique. It may be further subdivided into sequential (noncompetitive) and competitive methods. The sequential, heterogeneous immunoassay method, also known as the enzyme-linked-immunosorbent assay (ELISA), is most widely used.

ELISA (Noncompetitive EIA)

Figure 10–4 is a schematic representation of systems in which an antigen is bound to a supporting surface. The antigen against which the first antibody was raised—in, for example, a goat—is added to form an immobilized antigen–antibody complex. A second antibody, which was raised against goat proteins (i.e., is species specific), and which also bears a covalently bound enzyme, is next added to form a surface-bound ternary complex. The appropriate enzyme substrate is added and the product formed is assayed.

The supporting surface upon which the antigen–antibody reaction is carried out and to which the antibody, or the antigen, must bind may be of latex, nylon, Sepharose, cellulose, or polystyrene. Trays or affinity columns—or "dipsticks" for rapid assay—may be used. If a clear polystyrene tray with multiple wells is used and if the enzyme product absorbs light in the visible range, the intensity of color may be read in recording spectrophotometers especially constructed to scan each well in the tray. Such instruments are commercially available. The general procedures are to soak the tray-wells in the first antibody solution and rinse off the excess, or to apply the antigen first as in Fig. 10–4A. In some laboratories, a dilute albumin solution is used as a wash to minimize nonspecific binding. Adsorption of the interactants is about 80% complete in 3 hours, with maximum adsorption accomplished after 36 hours at 37°C. The degree of adsorption increases linearly with concentration up to about 100 ng/cm^2 (Pesce et al., 1977).

Enzyme Conjugation. There are a number of reagents that can be used to bind the enzyme covalently to the second antibody, for example:

1. Glutaraldehyde, which links amino groups of the enzyme with those of the antibody
2. Maleimides
   a. *N,N′-o*-Phenylenedimaleimide, which reacts with -SH groups (see Ishikawa et al., 1978), and
   b. *m*-Maleimidobenzoyl-*N*-hydroxysuccinimide ester, which reacts in the following manner:

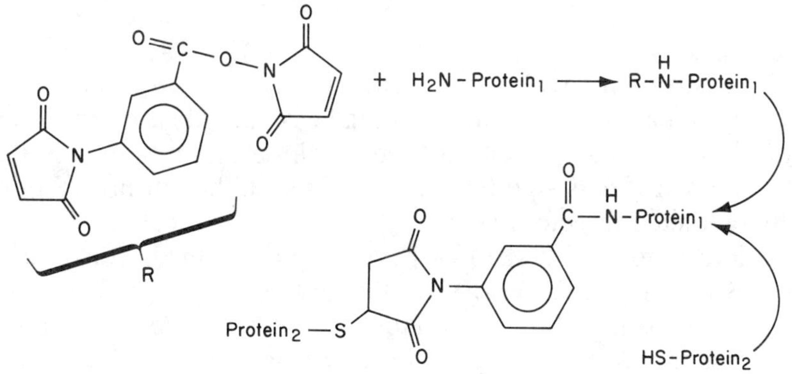

Biotin–Avidin Variation (BAELISA). In a modification of the double-antibody technique, the strong affinity between avidin and biotin is capitalized upon. The biotin–avidin ELISA, or BAELISA procedure is outlined in Fig. 10–5.

In the above sequence the reaction well could be in a polypropylene tray and the antigen could be a purified preparation of chicken lactate dehydrogenase (LDH; see Expt. 9). The antibody to the LDH would be raised

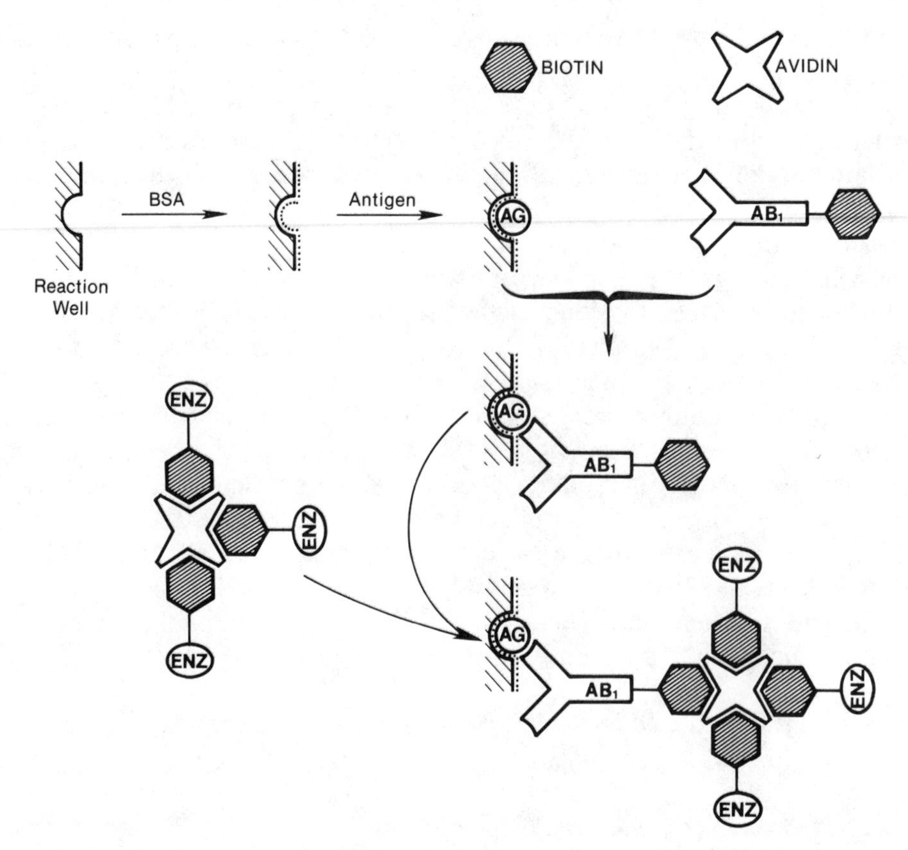

Figure 10–5. Biotin-avidin variation of ELISA (BAELISA). Application of bovine serum albumin (BSA) in the first step is optional. AG is antigen (e.g., LDH from Expt. 9); AB_1 is rabbit antibody 1, raised against the antigen; ENZ is the reporter enzyme (e.g., alkaline phosphatase or peroxidase). The figure depicts the maximal interaction of biotinylated-enzyme (3 for each AB_1), a level not usually achieved.

in a rabbit and then covalently attached to biotin. The reporter enzyme, for example, alkaline phosphatase, would be covalently attached to avidin. As a result of the quadrivalency of the avidin–biotin interaction, the sensitivity of BAELISA is significantly increased, although not to the theoretically predicted threefold level.

Several variations of BAELISA have been developed. For example, in bridged BAELISA (BRBAELISA), the antibody and the enzyme are labeled with biotin, and native unlabeled avidin is used. The antibody is reacted with its adsorbed antigen, and then washed; and the avidin is subsequently added. After incubation and washing, biotin-labeled enzyme is added, and its activity is eventually determined. In another variation, biotin-labeled antibody and enzyme-linked avidin are added to adsorbed antigen.

Bergmeyer et al. (1986) describe a number of applications of ELISA including its use with antigen-secreting cells, antibody-secreting cells, monoclonal antibody screening, red blood cell autoantibodies, anti-sperm antibodies, herpes simplex virus, and human Tcell leukemia–lymphoma/

lymphadenopathy-associated virus (HTLV-III/LAV). It has been used in studies of gene and cell mapping and insulin receptor interaction, among other systems.

This brief glimpse at recent developments exemplifies the ways in which antigen–antibody interactions are used in problems of separation, identification, and quantitation. The appended references provide entry into a rapidly growing literature. In addition manufacturers' brochures can be consulted for the most current information on applications of interest.

REFERENCES

Amzel, L. M., and Poljak, R. J. Three-dimensional structure of immunoglobulins. *Annu. Rev. Biochem. 48:*961–997 (1979).

Bergmeyer, H. U., Bergmeyer, J., and Grassl, M., eds. *Methods of Enzymatic Analysis,* Vol. 10, 3rd ed. VCH,Weinheim (1986).

Catt, K. J., and Tregear, G. W. *Science 158:*1570–1572 (1969).

Clausen, J. Immunochemical techniques for the identification and estimation of macromolecules. In *Laboratory Techniques in Biochemistry and Molecular Biology* (T. S. Work and E. Work, eds)., American Elsevier, New York (1970).

Cooper, T. G. *The Tools of Biochemistry,* pp. 274–277; 304–306. John Wiley, New York (1977). [In this experiment avidin is used as an antigen to raise antibodies in a rabbit. The antibody is harvested and then reacted with avidin to which radioactive biotin is firmly, although noncovalently, attached. Radioactivity in the antigen–antibody precipitate is measured. This is *not* an EIA.]

Crowle, A. J. *Immunodiffusion,* 2nd ed. Academic Press, New York (1973).

Davies, D. R., Padian, E. A., and Segal, D. M. Three-dimensional structure of immunoglobulins. *Annu. Rev. Biochem. 44:*639–667 (1976).

Ishikawa, E., Yamada, Y., Hamaguchi, Y., Yoshitake, S., Shiami, K., Ota, T., Yamamoto, Y., and Tanaka, K. Enzyme-labeling with maleimides and its application to the immunoassay of peptide hormones. In *Enzyme-Labelled Immunoassays of Hormones and Drugs* (S.B. Pal, ed.). Walter de Gruyter, New York (1978).

Ngo, T. T., and Lenhoff, H. M., eds. *Enzyme-Mediated Immunoassay.* Plenum, New York (1985).

Pal, S. B., ed. *Enzyme-Labelled Immunoassay of Hormones and Drugs.* Walter de Gruyter, Berlin (1978).

Pesce, A. J., Ford, D. J., Garzulus, M. Z., and Pollak, V. E. *Biochim. Biophys. Acta 492:*399–407 (1977).

Roitt, I. *Essential Immunology,* 5th ed. Blackwell Scientific Publications, London (1985).

Schrader, W. T., and O'Malley, B. W. *Laboratory Methods Manual for Hormone Action and Molecular Endocrinology,* 6th ed. Houston Biological Associates, Houston, TX (1982).

Sternberger, L.A., *Immunocytochemistry.* Prentice-Hall, Englewood Cliffs, NJ (1974).

Van Vunakis, H., and Langone, J. J., eds. *Immunochemical Methods,* Parts A, B, and C, *Methods Enzymol. 70* (1980), *73* (1981), and *74* (1981).

CHAPTER ELEVEN

Introductory Aspects of Molecular Biology

Molecular biology is the study of biological phenomena in terms of molecular structures, interactions, and interconversions. The ultimate goal of molecular biology is to understand in strictly chemical terms all of the myriad reactions that can and do occur in living cells; it therefore represents a fusion of the classical disciplines of biochemistry and cell biology.

This chapter deals with the elementary aspects of "molecular biology" in sufficient depth to provide a background for certain experiments in Section II of this book.

Although its philosophical origins may be traced back to the work of Mendel and Darwin, molecular biology as it is understood today is a new science, scarcely more than thirty years old. It is fruitless to argue about the exact beginnings of the new science; most would agree that it stemmed largely from the work of McLeod, Avery, Watson, and Crick, to name a few of the pioneers. In the ensuing years, biological science, especially genetics, has been transformed. Through the use of natural or induced mutations, molecular biologists have learned much concerning the operation and regulation of genetic systems. Proceeding from that understanding, a more or less elaborate technology has developed whereby it is possible to add a foreign gene (that is, a gene not characteristic of a given cell line or species), or to modify in a known way some preexisting gene. This kind of research and development, now being intensively pursued, gave rise to terms such as "genetic engineering" or the less flamboyant "recombinant DNA technology." In the face of these developments, the once somewhat distinct lines between microbiology, biochemistry, and physiology have become even less clear. In effect, all modern biologists and modern clinicians must understand the unit operations of the new technology.

Fundamental to the idea of molecular biology is the concept that DNA and RNA (deoxyribonucleic acid and ribonucleic acid, respectively) are *informational macromolecules.* Within its sequence of purine and pyrimidine bases, DNA contains all of the information necessary to specify the primary sequences of all the catalytic, regulatory, and structural proteins contained within an organism. However, not all of the information contained in the plan needs to be read at once, and in some cells part of the information may never be read. This implies that expression of some of the coded information is repressed and that it can be read out (and operated upon) only when derepressors allow that to happen. Alternatively, expression of part of the information may require a positive stimulus, such as hormonal activation at a suitable stage of development. DNA is the substance of genes; no cell can, under normal circumstances, produce any protein for which the information is not precoded in the DNA. Furthermore, as stem cells become specialized, some parts of the information may become permanently sequestered.

DNA does not participate directly in the synthesis of proteins. Instead, it is transcribed into several species of RNA. In this sense, messenger RNA (mRNA) serves to define the order in which amino acids are to be assembled to produce a certain protein. The amino acids must first be activated and attached to one of a smaller species of RNA, transfer RNA (tRNA), each of which bears a recognition site (a triplet of bases) that matches some site(s) along the mRNA. Translation of data from the mRNA to a peptide sequence is catalyzed by ribosomes, organelles that contain ribosomal RNA (rRNA) along with other protein factors.

The flow of information described above, in which DNA is first transcribed to RNA, which is then translated into proteins or peptides, is nearly universal, but a few exceptions are known. These come from the realm of viruses and phages. Instances are known in which viral RNA acts as a template for the production of DNA in infected cells. This transfer of information is catalyzed by an enzyme known as *reverse transcriptase,* which acts as a DNA polymerase. It operates on the template provided by the single-stranded viral RNA to produce a single-stranded complementary DNA (cDNA). The latter is then converted to the usual double-stranded form by the synthetic machinery of the infected cell. This step is necessary because the mechanisms of transcription appear to require double-stranded DNA in the intact cell. Although some protein synthesis can be brought about by single-stranded DNA species in vitro, there is no good evidence for such a process in vivo.

Another exception to the general scheme occurs in the case of some small, simple RNA viruses or phages. Among these are the polio virus and the phage known as Qβ. When Qβ infects a cell, the infecting, single-stranded RNA is replicated by formation of a cRNA, not a DNA. This process is catalyzed by an enzyme known as *Qβ replicase,* a tetrameric enzyme. Only one of the four subunits comes from the phage; the others are generated by the infected cell, from information contained in the viral RNA.

All of the above possibilities can be summarized by the familiar triangular diagram shown in Fig. 11–1. The solid lines represent the usual flow of information. The straight broken line on the left represents the action of reverse transcriptase, the dotted curve at the lower left represents the effect of Qβ replicase, whereas the dotted line on the right represents the

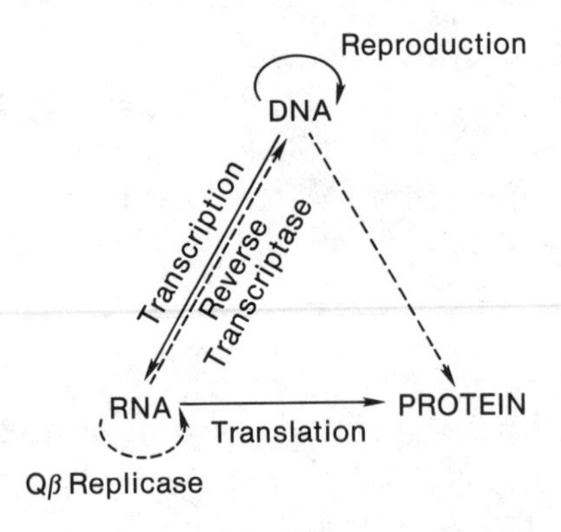

Figure 11–1. Possible pathways of information flow from DNA to protein. The normal flow of information is mediated by RNA, as shown by the solid lines. The dashed lines indicate several possible variations that have been observed in special cases.

in vitro synthesis of protein from single-stranded DNA. The variant cases represented by the dotted lines are uncommon but not unimportant. RNA viruses can cause some significant human disease. In the laboratory, availability of reverse transcriptase and Qβ replicase has enriched the technology of molecular biology.

NUCLEIC ACID STRUCTURE AND PROPERTIES

Deoxyribonucleic Acid (DNA)

DNA consists of a double helix, each strand of which is a polymer of β-D-deoxyribonucleosides joined by phosphodiester linkages. The phosphodiester bonds are formed between 5′ and 3′ carbons of adjacent deoxysugar moieties. Attached to the C^1 carbon of each deoxysugar is either a purine or a pyrimidine base, and it is the sequence of these bases that confers upon the DNA its function as a coding device. The purines, adenine (A) and guanine (G), are found in both DNA and RNA. The pyrimidines found in DNA are usually thymine (T) and cytosine (C), whereas in RNA thymine is replaced by uracil (U). The structures of these bases are shown below. Many of the bases in a DNA may be methylated at one or more C or N atoms. Methylation occurs after assembly of the polymer and is catalyzed by specific methylases. Methylation of bacterial DNA inhibits, or restricts, attack by endonucleases originating in viruses or other bacteria. However, in the natural battle for survival, cells have developed endonucleases that can attack methylated as well as nonmethylated bases; indeed, some of these so-called restriction endonucleases have recognition sites that must include methylated bases. Restriction enzymes have become important and powerful tools in fragmenting or modifying DNA species, as we shall see later on.

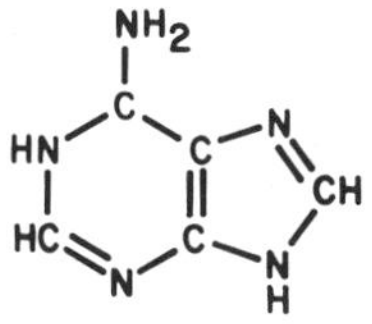

Adenine (A)

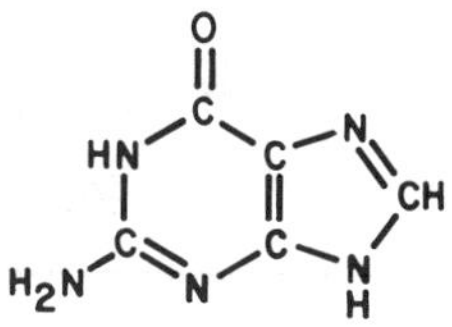

Guanine (G)

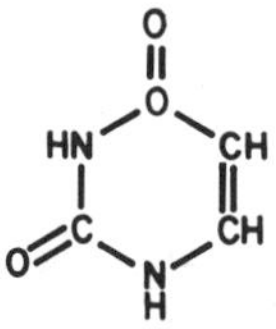

Uracil (U)

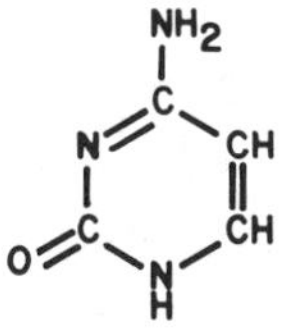

Cytosine (C)

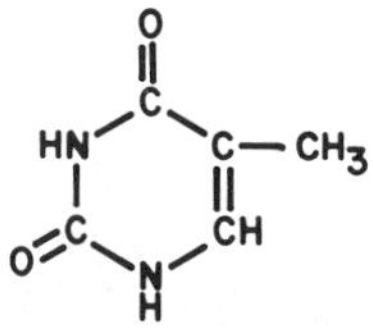

Thymine (T)

The two helical DNA strands are known to be *antiparallel;* in other words, the 5′ end of one strand is adjacent to the 3′ end of the other. The strands wrap around each other, as shown in the diagram. The strands are

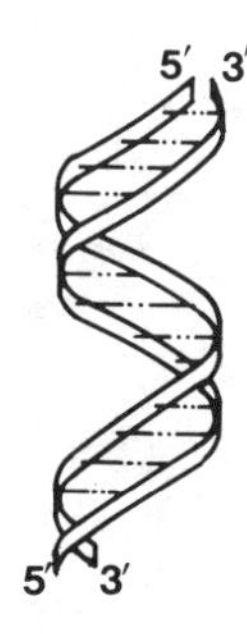

complementary, so that a purine base located on one strand lies close to and points quite directly at a pyrimidine base on the other. In the diagram, this relationship is represented by the broken lines between the two strands. The dimensions of the helix are such that for each complete turn there exist ten base pairings at which hydrogen bonds form between a purine base on one strand and a pyrimidine base on the other. Two hydrogen bonds can form between T and A, but three can form between C and G. The complementarity of the H bonding is a function of the distances between bondable groups in the two chains and is illustrated in Fig. 11–2. Although hydrogen bonds are not strong ($\Delta G^0 \simeq 5$ kcal/mol) there may be many thousands of such bonds in a DNA molecule and they play a considerable role in maintaining the double helical structure. Close stacking of the purine and pyrimidine bases along a given strand also contributes to stability. Purines and pyrimidines are fairly flat molecules, and the aggregate of the van der Waal's forces between them can be considerable, even though the attractive force between any two rings is quite modest.

In sum, the considerable stability of the double helix arises from the very large number of interactions, none of which are, individually, of great

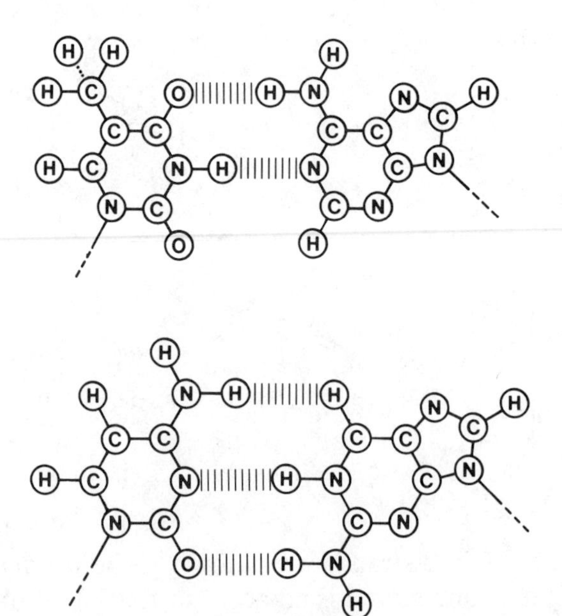

Figure 11–2. Hydrogen bonding and DNA structure. The upper model shows the two hydrogen bonds between thymine and adenine in antiparallel and complementary strands of the double helix. The lower model shows the three hydrogen bonds between cytosine and guanine. The broken (dashed) lines indicate attachments of the bases to the deoxyribose backbones of the helices. By complementarity, which always opposes a purine and a pyrimidine or vice versa, the distance between the backbones is always precisely filled. Two purines, or two pyrimidines, would not meet the requirement for precise space filling.

strength. It is therefore not surprising that the double helix is readily denatured (so it can be separated into its separate strands) by relatively mild means. One such means involves heating to temperatures of about 65°C. This brings about considerable strand separation which can be followed by examination of the ultraviolet absorption spectrum; as the strands separate, the absorbance at 260 nm increases; this phenomenon is called *hyperchromism.* The temperature at which the absorbance change is half-maximal has been defined as the "melting" temperature, or T_m, for a given DNA species. In comparing a number of DNA molecules, the exact pH and ionic strength of the buffer must be controlled, but even then it is clear that the value of T_m is related to the base composition of DNA. As the percentage of G + C increases, the double helix becomes more stable, and T_m rises. Slow cooling, or *annealing,* allows the denatured strands to return to the double-helical state, but rapid cooling, or *quenching* may prevent the proper configuration from being reached. The melting and annealing properties of DNA are useful because they can be put to use in studying the homology of two related DNA species.

Ribonucleic Acid (RNA)

The most obvious differences between DNA and RNA include the substitution of deoxyribose by ribose and of thymine by uracil, so that an RNA backbone can be described as a polymer of β-D-ribosides joined by phosphodiester linkages between the 3′ carbon of one sugar unit and the 5′ carbon of the next. As in DNA, the purine and pyrimidine bases are attached to the C^1 carbon atoms of the sugars.

Some less obvious differences also exist. For one, RNA does not, ordinarily, exist as a double helix. Typical RNA species are single-stranded helical structures and they generally have considerably lower molecular weights than DNA, reflecting the different biological functions of RNAs. Hydrogen bonding does exist in RNA species, but it depends on the peculiar dispersion of bases that allows purines and pyrimidines, some distance apart, to come into suitable apposition. This is particularly apparent in the case of the tRNAs, where three or more loops are formed, as shown in Fig. 11–3. As deduced from crystallographic data on the three-dimensional structure, tRNAs are actually L-shaped (or "boomerang" shaped) as a result of folding of the noncoding arms of the molecule. As a class, mRNAs exist in an extended form, free of loops that could impede read-

Figure 11–3. Complete sequence and folding pattern of yeast alanine tRNA. The four major bases are represented by U* (5,6-dihydrouridine), ψ (pseudouridine, or 5′-ribosyluracil), G^m (1-methylguanosine), and I^m (1-methylinosine). For simplicity, the sites of hydrogen bonding are shown by single lines between the paired bases. The 3′ end of all tRNA species terminates in the —C—C—A sequence to which the activated amino acid, in this case alanine, is joined by the action of an aminoacyl-tRNA synthetase (AAS). The anticodon loop contains a triplet, here I—G—C, that is complementary to a codon in an mRNA, here C—C—G; hence the anticodon loop confers specificity in peptide assembly on a message. The TψC loop is involved in ribosomal binding of the aminoacyl tRNA, and the AAS binding loop facilitates binding of the tRNA to the activating synthetase. The tRNAs for other amino acids may be larger, and the loops may be of different sizes. In particular, the U* loop, here only four bases long, may be larger. Many of the tRNAs have been crystallized.

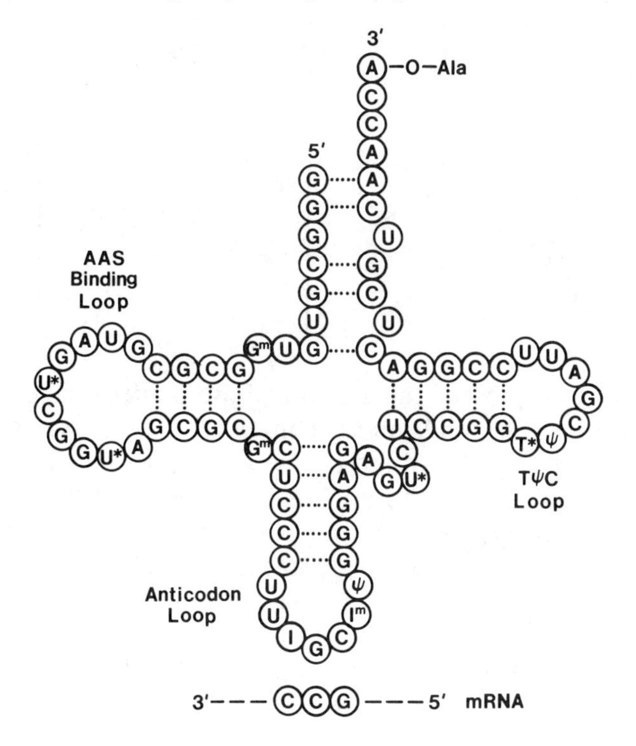

out of the message. Sometimes adventitious loops do form in isolated mRNAs, lowering the efficiency of readout. This problem can be overcome by melting and quenching the mRNA to destroy the unwanted secondary structure.

Figure 11–3 shows also that tRNAs contain some unusual or modified bases not found in DNA. One of these is 5,6-dihydrouridine. A number of ribosylated bases, 5-ribosyluracil or pseudouridine (represented by the symbol, ψ) and 5-ribosylthymidine (T*), have been found. A number of methylated bases, including 1-methylinosine, 1-methylguanosine, and 2,2-dimethylguanosine, are also known.

Ribosomal RNAs (rRNAs) are the largest RNA species known, having molecular weights much greater than that of either tRNA or mRNA. Much less detailed information about the structure of rRNA is available, and we shall not pursue the matter of rRNA properties any further. The function of rRNA is clear; in association with other proteins, with an energy source, and in response to various kinds of regulatory agents, the rRNAs form the biosynthetic machinery by which proteins are produced. They do this by assembly of aminoacyl-tRNAs according to directions prescribed by an mRNA.

"SHORT-HAND" NOTATIONS FOR THE NUCLEIC ACIDS

Several "short-hand" notations have been devised to simplify description and discussion of nucleic acid molecules. One such scheme is shown below:

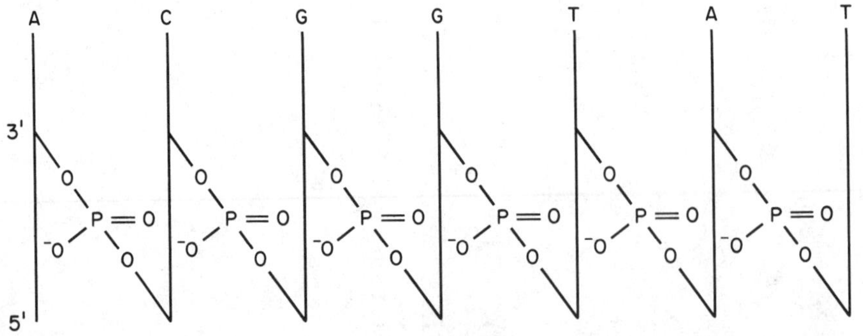

The vertical lines represent the sugar—ribose or deoxyribose. Which of these sugars is meant depends on the context of the discussion. The letters at the tops of the lines refer to the purine or pyrimidine bases. The phosphodiester bonds are shown, running between the adjacent sugar moieties. The important conventions of this notation are that the free 5′-OH group is always drawn at the left, just as the free amino-terminal group of a peptide is always drawn at the left, and the base sequence is always read from left to right.

Two other even simpler schemes are in use. The first of these omits all structural detail, represents the 3′→5′ phosphodiester linkages simply by the letter "p," and represents the bases by the single-letter convention already noted. Terminal phosphate groups are also indicated by the letter "p," written to the *left* when they are attached to a 3′ carbon and to the *right* when attached to a 5′ carbon. Thus, adenosine-3′-monophosphate (3′-AMP) would be written as pA, adenosine-5′-triphosphate (5′-ATP) would be written as Appp, etc. The oligonucleotide shown in the scheme above would be written as ApCpGpGpTpApT. The simplest scheme of all abbreviates a nucleic acid by merely listing the order of bases, with the free 5′-OH group on the left—for example, ACGGTAT, applying the convention to the above example. Where terminal phosphate groups are present, they must be indicated by the letter "p." The use of these conventions vastly simplifies discussion of nucleic acids, their derivatives, and their precursors.

DNA Restriction Mapping and Sequence Analysis

ENZYMES USED AS TOOLS IN NUCLEIC ACID MODIFICATION AND ANALYSIS

Base Sequence-Nonspecific Enzymes

Many enzymes that degrade nucleic acids have been isolated and purified from animal, vegetable, and bacterial sources. Some of these enzymes have become important tools in structural studies and in the preparation of deliberately modified molecular species of DNA and RNA. The substrate requirements of nucleases are diverse, ranging from a fairly low order or a quite high order of specificity. Nucleases can be classified as *exonucleases* or *endonucleases;* exonucleases attack phosphodiester bonds only at the termini of nucleopolymers, whereas endonucleases attack similar bonds in the interior regions of such polymers. Further subdivision can be made on the basis of the contained sugar; thus one speaks of ribonucleases and deoxyribonucleases.

Some nucleases act only on single-stranded substrates, whereas others act on double-stranded species. As we will show, still other enzymes can act on either type of polymer, depending on the experimental circumstances. Since the nucleases act to break (or make) phosphodiester bonds, it is important to understand that nucleases may leave the phosphoric acid residue attached to a 5′ or a 3′ linkage, depending on the properties of the enzyme. This is illustrated by the following scheme:

$$\text{pApTpGpGpA} \xrightarrow[5'\rightarrow 3' \text{ nuclease}]{} \text{pApTp} + \text{GpGpA} \quad (11\text{-}1)$$

$$\text{pApTpGpGpA} \xrightarrow[3'\rightarrow 5' \text{ nuclease}]{} \text{pApT} + \text{pGpGpA} \quad (11\text{-}2)$$

In reaction 11–1, the enzyme cleaves on the 5′ side of the phosphate, resulting in a new 3′-phosphorylated end. In reaction 11–2, the situation is exactly reversed.

Many areas of biopolymer chemistry have been aided by the use of enzymes as probes. In protein chemistry, for example, pepsin, trypsin, and chymotrypsin have played an important part. They have proved useful because they cleave substrates into fragments at a limited number of sites. Pepsin preferentially cleaves peptide bonds in which both amino acid residues are aromatic, whereas trypsin cleaves bonds involving the carboxyl groups of basic amino acids, and so forth. It must be kept in mind that the utility of the enzymes depends not only on the specificity of the enzymes but also on the diverse properties of the amino acids themselves. In dealing with the informational macromolecules, DNA and RNA, the situation is somewhat different. Two distinctions should be kept in mind. First, there is much greater monotony in the structures of the deoxyribonucleotides than in the structures of the amino acids for which they code. (Similarly, there is greater monotony in the symbols of the Morse code than in the letters of the alphabet for which they code.) This fact imposes a different kind of rule on the specificity of enzymes that act on nucleic acids. Second, in the case of proteins, no single enzyme is known that can ligate (or join together) separate peptides. However, as we shall show, oligodeoxynucleotides can be ligated through terminal phosphodiester bonds by any of a number of single enzymes, operating on a template of a complementary nature. Presumably, the reason for this stems from the sensitivity of nucleic acids to damage by ever-present physical and chemical agencies, and the consequent need for repair processes. Whereas damaged proteins are generally discarded and degraded, damaged DNA is more frequently repaired.

Table 11–1 lists a number of important enzymes used in manipulation of the nucleic acids, along with the most common sources (given in the left-hand column in parentheses) and a brief description of some key properties. Nucleases are metalloenyzmes; the majority of them are Mg^{2+} dependent, but a few employ Cu^{2+}, Co^{2+}, Mn^{2+}, or Zn^{2+}. In some instances, changing the metal ligand by chemical procedures results in a change in enzyme action. For example, DNase I works with either Mg^{2+} or Mn^{2+}, although the mix of products depends on which ligand is employed. Note also that the nuclease known as Bal 31 shows a preference for single-stranded DNA when it contains Co^{2+}, and for double-stranded DNA when it contains Cu^{2+}.

DNA polymerase I (frequently abbreviated as DNA pol I) is described in Table 11–1 as an exonuclease possessing three separate activities. After mild proteolytic digestion with subtilisin, the 3′→5′ exonuclease and the polymerase activities are retained, but the 5′→3′ exonuclease activity is lost. The digestion product, comprising the bulk of the original polymerase, is known as the *Klenow fragment.* The polymerase activity requires a primer with a free 3′-OH group and serves to add dNTP units complementary to a template. A very similar polymerase is produced by T_4

Table 11–1. Properties of Some Sequence-Independent Enzymes Important in DNA Technology

Enzyme (Source)	Substrate Species	Required Ion	Reaction Catalyzed	Comments
			Exonucleases	
Exonuclease III	ds	Mg^{2+}	performs stepwise 3′→5′ removal of 5′-mononucleotides if 3′-OH end is free	also has 3′-phosphatase activity
Bal 31 exonuclease (*Brevibact. albidum*)	ds	Cu^{2+}	removes a mixture of mono- and oligonucleotides from both ends of ds DNA	DNA ends may be blunt or staggered
	ss	Ca^{2+}	same as above, but acts on ss DNA only	can be completely inactivated by EGTA
Nuclease S_1 (*Aspergillus oryzae*)	ss	Zn^{2+}	yields 5′-mono- or oligonucleotides; may cleave duplexes at nicks or small gaps	works on RNA also, but with lower activity; often used with ExoIII to make deletions in DNA strands
λ exonuclease (T_4-infected *E. coli*)	ds	Mg^{2+}	substrate must have a terminal 5′-P; releases 5′-mononucleotides in stepwise fashion	used for removal of protruding ends, and prepreparation for sequencing by dideoxy method of Sanger
DNA polymerase I	ds	Mg^{2+}	has *three* distinct activities:[a]	
(*E. coli*)			1. 5′→3 polymerase activity	needs ss template and a primer with a free 3′-OH
			2. 5′→3′ exonuclease activity	degrades ds DNA from a free 5′-end; mono- and oligonucleotides are produced
			3. 3′→ 5′ exonuclease activity	attacks ds or ss DNA from a free 3′-OH end; this action is blocked by 5′→3′ polymerase activity
Klenow fragment (obtained by treating DNA Pol I with subtilisin)	ds	Mg^{2+}	loses 5′→3′ exonuclease activity but retains: 1. 5′→3′-polymerase activity	adds dNTP to fill out the free 3′-OH ends of a primer on a ss template.
	ds, ss		2. 3′→5′-exonuclease activity	activity is like that of intact DNA Pol I.
T_4 DNA polymerase (T_4-infected *E. coli*)	ds, ss	Mg^{2+}	has same activity as *E. coli* Pol I.	its exonuclease activity is much higher than that of DNA Pol I.

Table 11–1. Properties of Some Sequence-Independent Enzymes Important in DNA Technology (*Continued*)

Enzyme (Source)	Substrate Species	Required Ion	Reaction Catalyzed	Comments
			Endonucleases	
Deoxyribonuclease I (bovine pancreas)	ds, ss	Mg^{2+}	with Mg^{2+}, DNase I randomly attacks each strand or a single strand.	with Mg^{2+}, produces a mixture of 5′-mono- and oligonucleotides.
		Mn^{2+}	with Mn^{2+}, its points of attack are very nearly the same in each strand.	with Mn^{2+}, fragments are more nearly blunt-ended than with Mg^2
			Ligases and polymerases	
T_4 DNA ligase (T_4-infected *E. coli*)	ds	Mg^{2+}	1. forms phosphodiester bonds between ds DNA with overlapping complementary (cohesive) ends having 3′-OH and 5′-P termini	used in repair of "nicked" DNA.
			2. joins blunt-ended segments of ds DNA meeting the above requirements.	works at fairly high concentrations of DNA segments.
RNA-dependent DNA polymerase (myeloblastoma virus, avian), a/k/a reverse transcriptase	ds	Mg^{2+}	has 5′→3′ polymerase activity, producing a DNA strand complementary to primer RNA; although DNA also can be a primer, it is not usually employed as such in recombinant DNA technology.	used to prepare cDNA from some messenger RNA; also has 3′→5′ exoribonuclease activity.

Terminal transferase (calf thymus)	ds, ss	Mg^{2+}	adds deoxynucleotides to the 3′-OH end of ss DNA, or ds DNA with protruding ends.	good for ^{32}P labeling of the ends of DNA strands.
	ds	Co^{2+}	works on blunt-ended strands if Mg^{2+} is replaced by Co^{2+}.	works primarily with ds DNA.
T_4 RNA ligase (T_4-infected *E. coli*)	ss		covalently joins 5′-DNA *or* -RNA to 3′-OH ends of ss DNA *or* ss RNA.	used to increase the efficiency of T_4 DNA ligase.
PolyA polymerase *(E. coli)*	ss	Mg^{2+} and Mn^{2+}	adds pApApA . . . to ss RNA species at free 3′-OH ends, for "tailing."	used to prepare RNA for cloning experiments.
			Kinases	
T_4 polynucleotide kinase (T_4-infected *E. coli*)		Mg^{2+}	transfers a γ-P from ATP to the free 5′-OH end of ss or ds DNA *or* RNA.	needs protection by a thiol reagent, usually DTT; with [γ-^{32}P]ATP is often used to label 5′ termini in Maxam–Gilbert sequencing studies; is also used in reverse to remove such a label where required.
			Alkaline Phosphatases	
Calf intestinal alkaline phosphatase	ss, ds		removes P_i from 5′ end of either ss or ds DNA *or* RNA.	P_i removal may be done prior to ^{32}P labeling; used for sequencing studies or to prevent self-ligation in the presence of other enzymes in the system.

Note: Sequence-specific endonucleases (restriction endonucleases) are described in the next section of text and are listed in Table 11–2.

Abbreviations: ds, double-stranded; ss, single-stranded; dNTP, deoxynucleotide triphosphate; DTT, dithiothreitol; EGTA, ethyleneglycolbis (β-aminoethyl ether) *N,N′*-tetraacetic acid.

[a]Clearly, the action of *E. coli* DNA pol I depends on the circumstances of use. An important application is in labeling of cloned DNA by "nick" translation processes.

phage-infected cells; the phage product is a more potent exonuclease than the *E. coli* enzyme.

General and rather nonspecific endonucleases typified by DNase I (and RNase I) are normal digestive enzymes of higher species, a common source of which is bovine pancreas. DNase acts more or less at random on either strand of DNA. As noted in Table 11–1, the sites of attack are usually not identical in each of the strands, but become more nearly so when the enzyme is manipulated to contain Mn^{2+} instead of Mg^{2+}.

As already mentioned, ligases serve to join the 5′ end of one DNA segment to the 3′ end of another, under conditions listed in Table 11–1. The T_4 phage-derived enzyme is an important tool in recombinant DNA technology. It works best on segments that have overlapping ends but can be made to operate on blunt-ended fragments as well.

Reverse transcriptase is an RNA-dependent DNA polymerase that acts to produce single-stranded DNA complementary to the RNA template. Like the other DNA polymerases, it also has 3′→5′ and 5′→3′ exonuclease activities, but these appear to be diminished when it is acting as a polymerase. Because the single-stranded DNA is an exact complement of the RNA, the product is often defined as a cDNA. It is readily converted to double-stranded material by action of DNA pol I. Reverse transcriptase is a very powerful tool because it provides a means for generating the DNA complementary to an isolated RNA message. When that DNA is introduced into some cell line, and incorporated into its genetic apparatus, the cloned cells continue to produce the product of the message thereafter. This is the concept embodied by the straight, broken arrow on the left-hand side of the triangle in Fig. 11–1.

Terminal transferase, like some of the other enzymes already discussed, shows a difference in affinity for, and activity toward, single- or double-stranded DNA, depending on the metal ligand involved. It acts to add dNTP to the free 3′ end(s). It is widely used for labeling the end(s) of a DNA strand with ^{32}P-labeled nucleotides in preparation for other studies. It is important to note that terminal transferase does not require a template. This enzyme shows a marked preference for addition at termini occupied by guanine. This activity can also be found in differentiating lymphocytes during antibody gene recombination.

T_4 RNA ligase appears in phage-infected *E. coli* cells, along with the T_4 DNA ligase already mentioned. There is a curious and poorly understood synergy between the RNA and the DNA ligases. Recall that the DNA ligase catalyzes blunt-end ligation of double-stranded DNA. The RNA ligase does not catalyze this reaction, but it markedly stimulates (six to seven fold) the activity of the DNA ligase, especially at low concentrations of the latter. Single-stranded DNA is also a substrate for the RNA ligase, strange as that may seem.

Polynucleotide kinase is a phosphorylating enzyme that transfers a γ-^{32}P moiety from labeled ATP to a free 5′-OH end group of DNA or RNA. Its major laboratory use is to tag the end of a molecule without altering the base sequence or composition. This is generally done in preparation for base sequence analysis, as we shall see later on. If the molecule under study already bears a 5′-P, it can first be removed by treatment with an alkaline phosphatase; then the radiolabel can be put in place through action of this kinase.

Sequence-Specific Endonucleases: Restriction Enzymes

The enzymes listed in Table 11–1 act quite independently of the actual base sequence of a DNA (or RNA). Yet the information content of the DNA rests altogether on sequential ordering of the bases, so it is clear that much value would be attributed to enzymes that recognized some specific sequence of bases, even though that sequence might be very short with respect to the total sequence length under examination. Many bacteria have evolved one or more endonucleases designed to restrict the effects of phage infection by cleavage of the phage genome at specific sites. These endonucleases are collectively known as *restriction enzymes.* More than 200 such enzymes have been described, many of which have been purified and marketed on a commercial basis. A partial listing of available restriction enzymes is presented in Table 11–2. The individual enzymes are identified by acronyms related to the organisms from which the endonuclease is obtained. Because more than one enzyme has been isolated from certain species, the acronym is followed, where needed, by a Roman numeral indicating the order of isolation priority. Thus, EcoRI is the first restriction enzyme isolated from *E. coli,* MboII is the second restriction enzyme isolated from *Moraxella bovis,* and so forth.

Restriction enzymes have some important characteristics that clearly set them apart from other nucleases.

1. Restriction enzymes are specific to the strain of organism from which they come. Each strain may produce several restriction activities as well as corresponding methylase activities. The latter activities involve transfer of methyl groups from *S*-adenosylmethionine (SAM) to specific bases in order to protect the cell's DNA against its own restriction enzymes. Careful study of restriction enzymes and their function demonstrates that three different structural classes can be defined. The classification depends in part on whether or not the endonuclease and methylase activities occur as separate proteins or as subunits of a more aggregated system, and on a requirement for ATP, but we shall merely note these differences and not pursue them further.
2. For each restriction enzyme there exists a precisely defined recognition sequence, as shown in the second column of Table 11–2. The number of bases in the recognition sequences ranges from four (AluI, HaeIII, etc.) to six (HpaI, SmaI, etc.) in the instances shown; not shown, but also known, are examples of eight-base recognition sequences. One (or sometimes more than one) of the bases may be methylated (Sau 3A, DpnI). An interesting corollary is that methylation of the other bases in the sequences may block cleavage by the enzyme.
3. Although the sequences composing the recognition sites are most fre-

TABLE 11–2. Some Commonly Used Restriction Endonucleases

Enzyme	Recognition Sequence	Number of Cleavage Sites				
		ϕX174	λ	Ad2	SV40	pBR322
Alu I	AG↓CT	24	143	158	34	16
Ava I	C↓PyCGPuG	1	8	40	0	1
Ava II	G↓GCC	1	35	>73	6	8
Bal I	TGG↓CCA	0	18	17	0	1
Bam HI	G↓GATCC	0	5	3	1	1
Bcl I	T↓GATCA	0	8	5	1	0
Bgl I	GCC$(N)_4$↓NGGC	0	29	20	1	3
Bgl II	A↓GATCT	0	6	11	0	0
Bst EII	G↓GTNACC	0	13	10	0	0
Cto I	GCG↓C	18	215	>375	2	31
Dde I	C↓TNAG	14	>104	>97	20	8
Dpn I	G^{Me}A↓TC		(cleaves only methylated DNA)			
Eco RI	G↓AATTC	0	5	5	1	1
Eco RII	↓CCGG	2	>71	>136	17	6
Hae II	PuGCGC↓Py	8	>48	>76	1	11
Hae III	GG↓CC	11	>149	>216	18	22
Hha I	GCG↓C	18	>215	>375	2	31
Hinc II	GTPyPuAC	13	35	>25	7	2
Hind III	A↓AGCTT	0	7	12	6	1
Hinf I	G↓ANTC	21	>148	>72	10	10
Hpa I	GTT↓AAC	3	14	6	4	0
Hpa II	C↓CGG	5	>328	>171	1	26
Kpn I	GGTAC↓C	0	2	8	1	0
Mbo I	↓GATC	0	>116	>87	8	22
Mbo II	GAAGA$(N)_8$↓	11	>130	>113	16	11
Nci I	$CC\binom{C}{G}GG$	1	>114	>97	0	10
Pst I	CTGCA↓G	1	28	30	2	1
Pvu II	CAG↓CTG	0	15	24	3	1
Sal I	G↓TCGAC	0	2	3	0	1
Sau 3A I	↓GATC	0	116	87	8	22
Dam	↓G^{Me}ATC	0	116	87	8	22
Sau 96 I	G↓GNCC	2	>74	>164	11	15
Sma I	CCC↓GGG	0	3	12	0	0
Sph I	GCATG↓C	0	6	8	2	1
Sst I	GAGCT↓C	0	2	16	0	0
Sst II	CCGC↓GG	1	4	>33	0	0
Taq I	T↓CGA	10	121	50	1	7
Tha I	CG>CG	14	157	303	0	23
Xba I	T↓CTAGA	0	1	5	0	0
Xho I	C↓TCGAG	1	1	6	0	0
Xor II	CGAT↓CG	0	3	7	0	1

Note: ϕ174, phage 174; λ, lambda phage; Ad2, adenovirus 2; SV40, simian virus 40; pBR322, plasmid BR322. In the recognition sequences column, N indicates that the enzyme will accept any base; Pu, a purine only; Py, a pyrimidine only.

Source: *Reproduced by permission of Bethesda Research Laboratories, Life Technologies, Inc., Gaithersburg, MD. BRL is a commerical source for all of the enzymes listed in this table.

quently defined exactly, this is not always the case, as Table 11–2 shows. A few enzymes will recognize and accept any base (specified in the table by N) at a particular position and either a purine (Pu) or a pyrimidine (Py) at some other. Nevertheless, these modest exceptions do not diminish the impression that restriction enzymes have evolved to require a very high sense of order in their substrates.

4. One must carefully distinguish between the recognition site and the actual site of nucleolytic cleavage. In some instances, the cleavage site is symmetrically located within the recognition sequence (AluI, HaeIII, etc.) while in others it is not (EcoRI, XbaI, etc.). In still others, the cleavage site is located at the 5′ end of the recognition sequence (EcoRII, Sau 3A, etc.). Still more complicated cases are known. MboII has a recognition sequence of five bases, but the point of cleavage is eight bases beyond the 3′ end of the sequence. Similarly, the case of BglI is complicated by an apparent split in the defined recognition sequence; the site of cleavage lies between the defined regions, which are joined by an undefined region of any four bases.

All of these situations can be summed up by the generalization that there is no apparent rule governing the lengths of recognition sequences or the relation between them and the cleavage sites. Each enzyme must be characterized on an individual basis; nevertheless, these enzymes are of immeasurable value in probing, or mapping, the sequences of DNA derived from phages, plasmids, and other genomic structures. Through their agency it is also possible to make deliberate modifications of DNA at known sites and so to observe the effects of such modifications on protein translation products.

THE PACKAGING OF GENETIC INFORMATION: EXTRACHROMOSOMAL ELEMENTS

Two instances are known in which significant parts of the total genetic information are packaged outside the nuclear chromosomes. These are the mitochondrion and the plasmid.

Mitochondria

Mitochondria are organelles that contain a small, circular DNA of their own, along with a corresponding set of mitochondrial ribosomes and tRNA species. These do not mingle with the remainder of the genetic apparatus, presumably because they cannot cross the barrier of the mitochondrial membranes. Indeed, the similarity of the mitochondrial genetic system to that of bacteria has given rise to the speculation that mitochondria are present-day residua of some primordial invasion of more primitive cell types. Speculation aside, it is clear that the mitochondrial apparatus contains relatively little information. It codes primarily for several subunits of essential mitochondrial proteins and for mitochondrial RNA.

It is also known that the mitochondrial genetic code differs slightly from that employed by the cytoplasmic mRNA. The entire mitochondrial apparatus operates separately but in concert with the nuclear apparatus; that is, when the nuclear chromosomes cease production of their complement of mitochondrial protein subunits, such as those involved in formation of the F_1 ATPase*, then the mitochondrial apparatus also shuts down, and vice versa. Exactly how this is accomplished remains unclear.

Plasmids

In many kinds of bacteria a second kind of extrachromosomal packaging occurs. Plasmids are double-stranded, circular DNA species containing 1–200 $\times$ 10^3 kilobases (kb). In general, plasmids contain genes coding for "maleness" or for resistance to certain antibiotics, among others. Plasmids can replicate independently of the major genomic circle, under either of two distinct sets of controls. Some plasmids occur only in a small number (1–10) of copies per cell and are said to be under *stringent control;* others may occur in much greater number (up to 200 copies per cell) and are said to be under *relaxed control.* When protein synthesis is shut down (by chloramphenicol treatment, for example), production of plasmids under relaxed control continues and thus their copy number increases. This is not true of plasmids under stringent control nor of the major genomic circle itself. Thus it is important to distinguish between the two forms of control. Chloramphenicol treatment allows one to manipulate a bacterial culture so as to increase the yield of plasmids under relaxed control; clearly, they are the better of the two types as laboratory tools. Because the size of typical plasmids is significantly different from the size of the major genomic circle, it is possible to separate the two DNA forms by differential centrifugation using density gradients, thus providing a good means of harvesting plasmids.

It is important to realize that although some genetic information may be packaged in the genomic circle and some in the plasmids, both packages form intrinsic parts of the bacterial genetic bank. Depending on the circumstances, there may be ready exchange of information between the two kinds of packages in accordance with the principles of genetic recombination. This fact, coupled with the ability to increase the copy number of certain plasmids, is the basis of their importance in purposeful manipulation of bacterial DNA. The exact base sequence has been determined for a number of plasmids, along with "maps" showing how each is fragmented by restriction enzymes.

A plasmid widely used in the recent past is identified as pBR322; a reference to the published sequence of 4362 nucleosides is given at the end of this chapter. A restriction map of pBR322 is reproduced as Fig. 11–4.

*The F_1 ATPase is a large complex molecule that in submitochondrial vesicles catalyzes the slow hydrolysis of ATP to ADP + P_i in the presence of Mg^{2+}. In its normal function in intact mitochondria, F_1 catalyzes the reverse reaction—the synthesis of ATP, which accompanies electron transport.

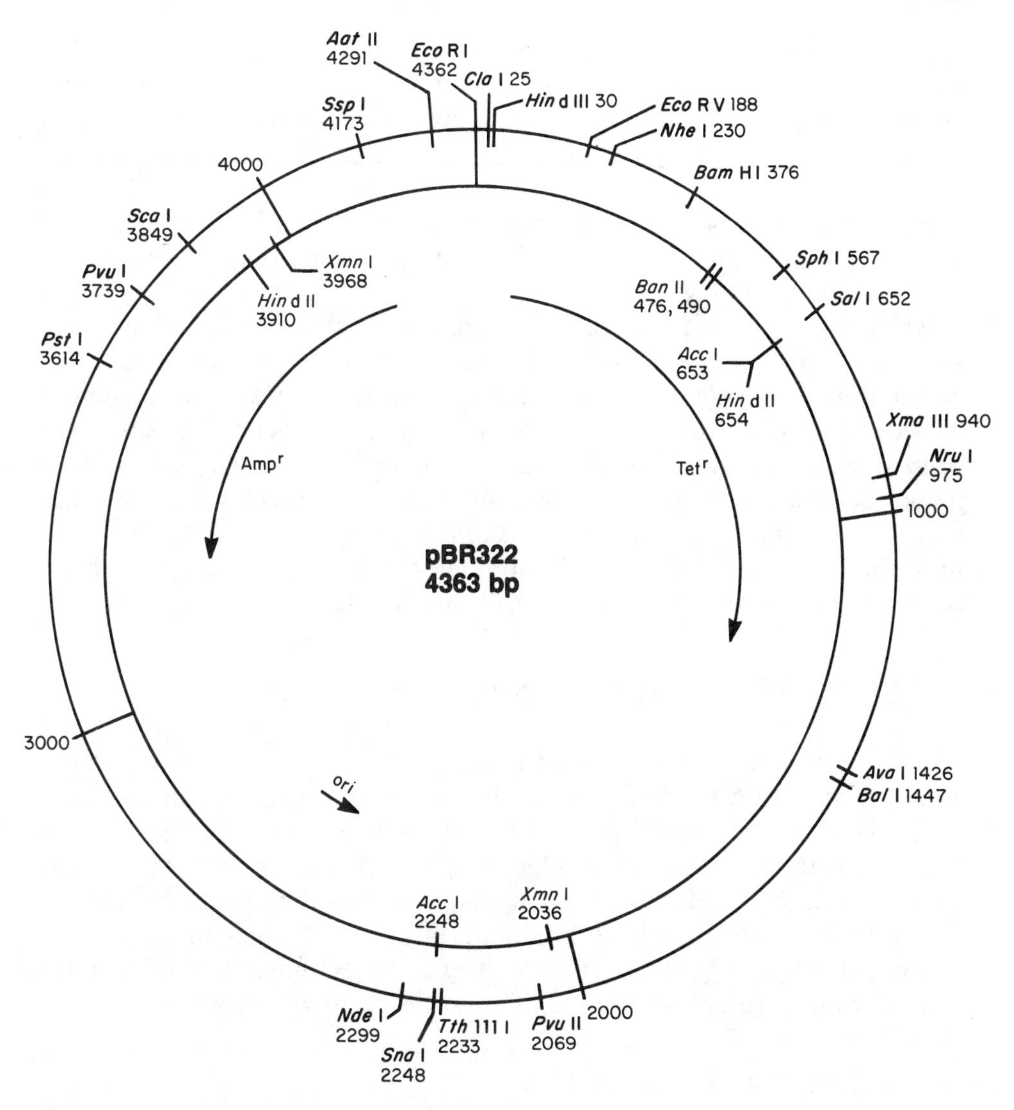

Figure 11-4. Restriction map of pBR322. Restriction endonucleases that cleave once are shown in bold type. (Reproduced by permission of Bethesda Research Laboratories, Life Technologies, Inc.)

There are several reasons for the popularity of pBR322. The first is its modest size, which minimizes the risk of mechanical damage by shearing during its isolation. The second is that this plasmid grows under relaxed control, increasing yields from cell populations of modest size. Third, pBR322 contains genes that confer resistance to two antibiotics, ampicillin and tetracycline. A culture of wild-type *E. coli* will die when exposed to these antibiotics, especially if both are added to the culture simultaneously. However, any cells that contain pBR322 will survive antibiotic treatment; they can be separated and recovered after drug treatment.

Other plasmid vectors that provide significant advantages over pBR322 are now in use. For example the pUC series, which are small molecules (~2.7 kb), grow to a high copy number. They contain an ampicillin-resistant gene, the same replication origin as pBR322, plus some polylinker sites identical to those in M13 (see below). Also commercially available are specially synthesized plasmids that are likewise small (~2.9 kb), and contain multiple cloning sites (e.g., SP6, T7, and T3), RNA polymerase promotion sites, as well as an f1* origin of replication for single-stranded phage production. Such vectors make it possible to carry out several protocols using the same plasmids and enormously simplifies various operations in biotechnology.

Isolated samples of the plasmids can be modified by introduction of still some other gene, and the modified plasmid can be reintroduced into a cell population. Any cells that then survive drug treatment must have contained the plasmid resistance factors *plus* the added gene. The surviving cells then become the basis of cloned colonies that may be exploited for laboratory purposes. Although one cannot always be certain that the extraneous gene will be expressed in terms of RNA species and of protein products, that is certainly the goal. Some methods to improve the likelihood of expression will be discussed later in this chapter.

USE OF RESTRICTION ENZYMES IN PLASMID MAPPING

Mapping of an unknown plasmid involves four steps: (1) cutting the circular plasmid into linear fragments by one or more restriction enzymes; (2) separation and visualization of the individual fragments; (3) determination of the relative molecular weights of the fragments, usually expressed in terms of base pair lengths; and (4) assembly of the data to give a logical ordering of the fragments from which their linkage in the native plasmid can be deduced. In concept, the technique is similar to the use of limited proteolysis in studying the structure of proteins.

Restriction Enzyme Digestion of Plasmids

The specificity of restriction enzymes has already been described, yet even fairly small plasmids may contain more than one cleavage site at which a particular enzyme can act. Although the use of enzymes that make single cuts somewhat simplifies the ultimate data analysis, common experience indicates that enzymes frequently attack plasmids at more than one site; fortunately this poses no serious limitation. As plasmid size increases, the chances of finding enzymes that make single cuts decrease on a purely probabilistic basis.

Restriction enzymes are sensitive to oxidative and thermal denaturation; they should be stored at −20°C in dilute glycerol and should be protected with mercaptoethanol or a similar sulfhydryl reagent. Commercial

*f1 phage belongs to the filamentous phases, Ff (i.e., F-specific filamentous), distinguished by their ability to adsorb to the F pilus present on cells carrying a sex factor. Its genome is a single-stranded circle of DNA. Another member of the Ff phages, the M13 phage, is discussed in a later section.

samples should not be used if there is any suspicion that they have thawed during shipment. The enzymes are prepared from selected strains of microorganisms; although highly purified, they are not totally free of extraneous, nonspecific nucleases. This contamination will, if digestion times are not carefully controlled, result in unwanted digestion of fragments generated by restriction enzyme action. Recall that most nucleases are metalloenzymes; consequently, digestion can be stopped quite effectively by addition of EDTA or EGTA,* protecting the restriction fragments against further breakdown. This is important because the use of restriction enzymes for mapping is based on the hypothesis that the total plasmid structure is retained in the fragments. To the extent that a few bases may be lost here and there, an error is introduced into the final mapping analysis.

For plasmids of modest size (4000–5000 bp), useful data can be generated by as few as three restriction enzymes, although more precise analysis does require a larger number of enzymes. Thus, Sutcliffe generated the currently accepted structural map of pBR322 with seven enzymes (his elegant paper is cited in the bibliography), each with a unique site of action, as shown in Table 11–2.

Separation and Visualization of Plasmid Fragments

Many of the major fragments will be fairly large and can be separated by differential gradient centrifugation; however, the smaller fragments will probably not be well resolved by this method. Gel electrophoresis is a faster, less expensive, and less tedious procedure. The gel medium is a highly refined agarose, preferred over acrylamide for its better mechanical properties at low concentrations. A 2% agarose gel has a working range of 0.1–3 kb; a 0.6% gel, a range of 1–20 kb, and a 0.3% gel, a range of 5–50 kb. Flat slabs are preferred over cylindrical gels so that several samples and standards can be simultaneously separated under identical conditions.

Although solutions of nucleic acids show a strong absorbance in the region of 260 nm, this does not provide a practical basis for gel analysis. The structure of the gel itself causes too much energy scattering and nonspecific absorption. Instead, use is made of the characteristic binding between nucleic acids and certain fluorescent dyestuffs. The best known of these dyes is *ethidium bromide,* an aromatic and fairly planar molecular structure. Ethidium bromide is an example of a type of compound known as *intercalating* species; this stacks, or intercalates, between the adjacent bases of nucleic acids. (Other intercalating species include the antibiotics bleomycin and actinomycin.) The fluorescence of intercalated ethidium bromide is more intense than that of the dye free in solution; thus the separated bands of "stained" nucleic acid are brighter than the background when viewed under ultraviolet radiation. The forces that bind eth-

*EDTA, ethylenediaminetetraacetic acid or edetic acid; EGTA, ethyleneglycolbis (β-aminoethyl ether)*N,N'*-tetraacetic acid.

idium bromide to nucleic acids appear to be largely nonpolar, even though the dye contains a quaternary nitrogen atom that confers ionic properties on the molecule.

Determination of Fragment Molecular Weights

Molecular weight standards may be obtained commercially or prepared in the laboratory. Standards ranging from <100 to >5000 base pairs are available. The actual molecular weights of the standards are less consequential for plasmid analysis than are the base pair lengths, since the latter property is the one commonly employed in constructing the plasmid map. However, as will be shown shortly, restriction enzyme digestion may result in the complication of two distinct fragments with identical base pair lengths. Since these cannot easily be resolved by electrophoresis, some additional characterization is required. This could be accomplished, for example, by sequence analysis of the overlapping fragments.

Construction of the Restriction Map

Generation of the map depends on three implicit hypotheses. These are that (1) the recognition sequences of each restriction enzyme in a mixture is unique; (2) in a mixture, the action of any one enzyme is independent of the others; and (3) the map must account for the total known base length of the plasmid; that is, no significant loss of bases has occurred due to extraneous nuclease action. The problem then reduces to one of numerical analysis. How the analysis is performed can best be shown by some simple examples.

Example 1. You are given a plasmid known to contain 4500 base pairs. You are also informed that the three restriction enzymes EcoRI, PvuII, and PstI each makes a single cut in the plasmid. The plasmid is digested first with all three possible pairs of the enzymes taken two at a time, then with the three enzymes taken together. The digestion mixtures, along with suitable standards, are examined by agarose gel electrophoresis in the presence of ethidium bromide. A schematic representation of the finished gel is shown in Fig. 11–5, and, for convenience, the fragment lengths measured from the gel are reproduced in the table below.

EcoRI + PvuII	PstI + EcoRI	PstI + PvuII	PstI + PvuII + EcoRI
2600	4050	2350	2150 (1)
1900	450	2150	1900 (2)
			450 (3)
4500	4500	4500	4500

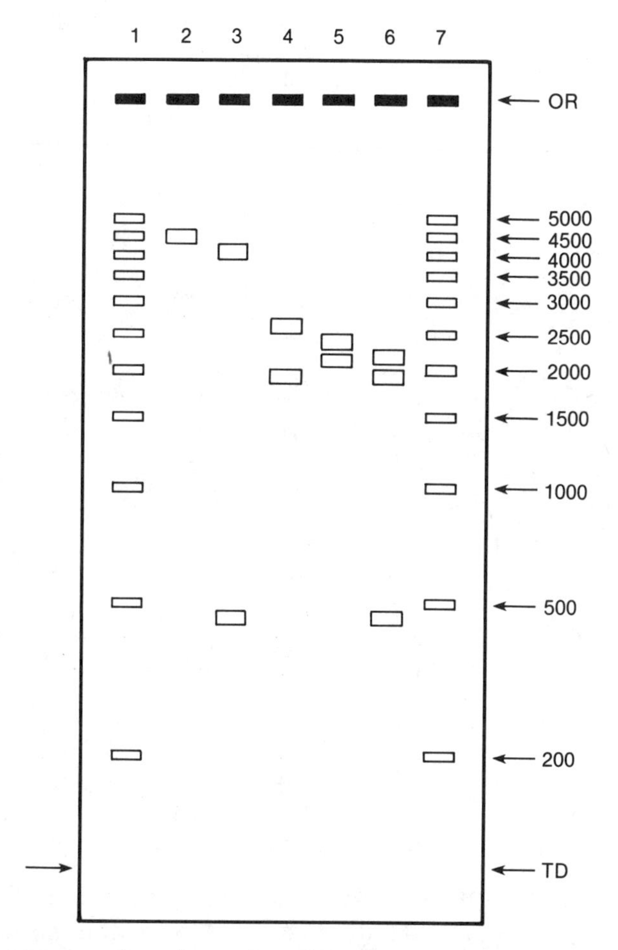

Figure 11–5. Diagram of an electrophoretic separation of ethidium bromide-DNA complexes after restriction enzyme digestion of a hypothetical plasmid containing 4500 base pairs. Lane numbers are identified across the top of the figure. Material was applied to sample wells, shown by the blackened rectangles near the top of the slab. This is the origin of the migration and is so indicated by the arrow marked OR, to the right of the figure. Lanes 1 and 7 were loaded with an identical mixture of standards, the base pair lengths of which are shown by the arrows to the right of the figure. Lane 2 was loaded with material resulting from EcoRI digestion. Lane 3 was loaded with material from digestion with EcoRI + PstI; lane 4 with material from EcoRI + PvuII digestion; and lane 5, with material from PstI + PvuII digestion. Lane 6 contained material from a digest with all three enzymes present.

When observed under ultraviolet light, the intercalated ethidium bromide fluoresced strongly while the remainder of the gel surface remained relatively dark. In this monochrome representation, the bright fluorescence is indicated by the rectangles. The tracking dye is indicated by the arrow marked TD, near the lower right of the gel.

Generation of the map means the determination of the order in which fragments (1), (2), and (3) must be assembled to express the composition of the plasmid. The intact plasmid can be represented by a circle; from the problem statement, it follows that the circumference of the circle represents a length of 4500 base pairs. The problem also states that each enzyme makes a single cut in the plasmid, a fact that could be represented by drawing a single arrow at the 12 o'clock position. However, this is such a trivial

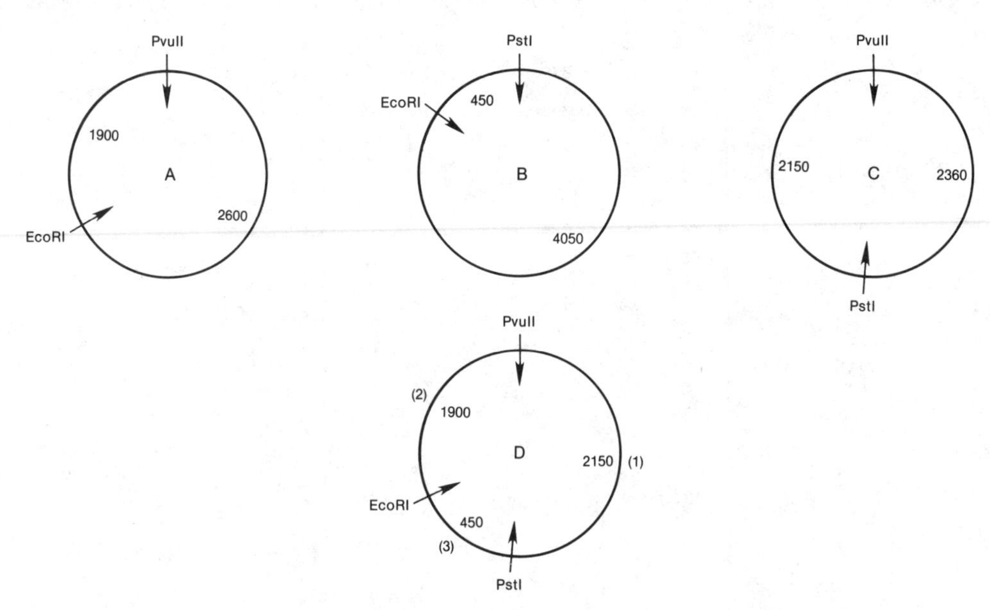

Figure 11–6. Preparation of restriction enzyme map of a plasmid. The assumptions are that each of the three indicated enzymes acts at a single cleavage site in the plasmid of 4500-base pair length. Circles A, B, and C represent digestions with the indicated enzyme pairs; the cleavage sites are indicated by the arrows. Circle D is a similar representation of digestion in the presence of all three enzymes taken at once. The numbers in parentheses around the periphery of circle D describe the proper ordering, or mapping, of the fragments. See the discussion of example 1 in the text for further details.

case that circles with single arrows have been omitted from Fig. 11–6. The figure does show the results of digestions with the possible pairs of enzymes (circles A, B, and C). Note that the arrows are positioned around the circumference in proportion to the base pair lengths of the recovered fragments. The sums of the fragment lengths indicate that all of the base pairs have been recovered; that is, there was no significant loss of material due to extraneous nuclease activity. Furthermore, the differences in fragment lengths show that the enzymes did act on their recognition sites independently of one another. From circle B of the figure, it is clear that there must be an interval (base pair length) of 450 pairs between PstI and EcoRI cleavage sites. From circle A it is also clear that a corresponding interval of 1900 pairs must exist between the PvuII and EcoRI cleavage sites. Since that same interval is retained in circle D, it follows that the site of PstI cleavage must lie between the PvuII and EcoRI cleavage sites, when going in a clockwise direction. This is exactly what is shown in circle D, the final map of the plasmid in terms of the three enzymes noted above. The assumption that each enzyme acts at only a single site was made to simplify the argument; the logic is identical and the procedure is only slightly more complicated for multiple sites, as shown in the next example.

Example 2. You are given a plasmid of 4500 base pairs in length; each of the restriction enzymes EcoRI, PstI, and PvuII makes two cuts in this plasmid. Proceed by determining the base pair lengths of fragments generated by each enzyme alone, by the enzymes acting in all possible pairs, and by all three enzymes acting at once. The data generated by gel electrophoresis are summarized by the numbers that appear inside the circles of Fig. 11–7. One notes quickly, in this second idealized case, that base recovery was complete and that the enzymes acted independently of one another. What remains is to analyze the data and demonstrate that the map shown in circle G is correct.

Circles A, B, and C represent the effect of each enzyme acting alone. The circles can be regarded as clock faces, where the arrows indicate the ends of the hands. If one were to take circle B and superimpose it on circle A, clockwise rotation of B about the fixed centers by approximately "2 hours worth" would generate circle D. Similarly, superposition of circle B on

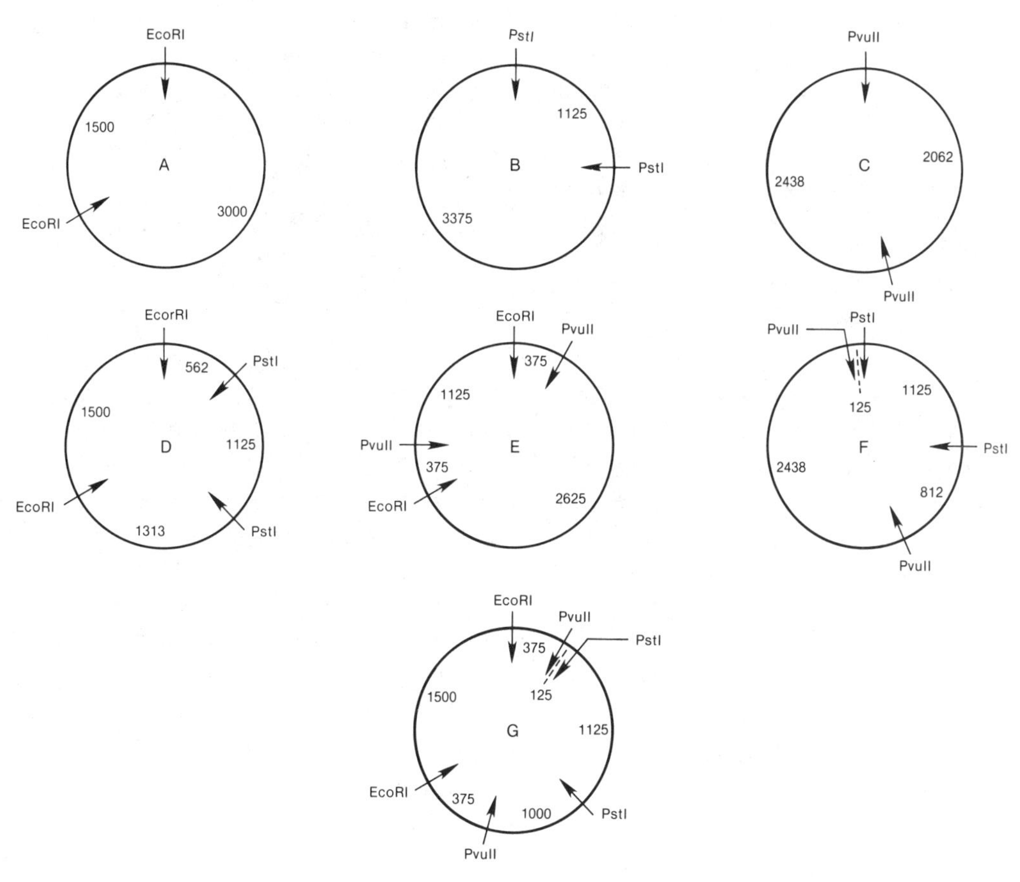

Figure 11–7. Preparation of a restriction map of a plasmid, assuming two sites of attack for each enzyme. The original plasmid contains 4500 base pairs and can be digested by the three enzymes shown. Cirlces A, B, and C show the fragments generated when each enzyme acts alone. Circles D, E, and F show the fragments generated by the indicated pairs of enzymes, and circle G shows the fragments generated by all three enzymes taken at once. Circle G therefore represents the desired restriction map of the plasmid, since it is evident that no bases were lost in the separate digest. See example 2 in the text for further details.

circle C, with counterclockwise rotation of about "minus 1 hour's worth" would generate circle F. Superposition of circles D and F would generate circle G, the final restriction map. The clock analogy emphasizes two of the hypotheses already noted; taken one at a time, the recognition sites of restriction enzymes are unique (however, the problem of *isochizomers* raises a real need for caution when enzyme mixtures are employed; see below) and the enzymes do act independently of each other. Circles D and F each give rise to sets of fragments that are distinct. No two fragments in either set have the same base pair length. The situation with circle E is different. Two fragments do have identical base pair lengths. There is a very high probability that they would not have identical structures, but electrophoretic separations would not resolve the differences. Note that the identical length of these small fragments persists in circle G, the final restriction map. This duplication is, of course, required by the logic of the analysis.

These oversimplified problems were carefully designed to illustrate the fundamental theory of plasmid mapping by restriction enzymes, but the problem in the real world is rarely so tidy. We turn now to some of the shortcomings of mapping and some of the problems that must be addressed. Most plasmids are cut not once, but many times by the majority of restriction enzymes. For example, Sutcliffe has shown that pBR322 is cut at 31 sites by the enzyme, HhaI. At least four other enzymes make 22 cuts in the same plasmid (see Fig. 11–4). These observations indicate the potential complexities of restriction enzyme mapping, because with an increased number of fragments there is an increased probability that more than one fragment will have some given base pair length. From direct observation of agarose gels, it is sometimes possible to detect duplication of base pair lengths in one or more bands by virtue of their more intense fluorescence. The increased concentration of duplicated base pair lengths increases the likelihood of diffusion and so leads to band spreading. However, these phenomena are not always easy to interpret in a reliable way, and some more precise means of detecting multiple fragments of identical base pair length is necessary. The only reliable method is that of sequence analysis, in which each fragment is taken apart, base by base, for final structural proof. Sequence analysis of DNA fragments is comparable to the technique of peptide dissection, amino acid by amino acid, in proof of protein or peptide structures, although the chemistry involved is necessarily of a quite different kind.

Matters may also be complicated by the problem of isoschizomeric restriction enzymes. Two restriction enzymes are said to be *isoschizomers* if the recognition site of one is contained within the recognition site of the other. As an example, the restriction enzyme MboI (and also Sau 3A) recognizes the sequence shown below, where the arrows indicate points of cleavage:

↓GATC
CTAG↑

The enzyme BamHI recognizes the sequence:

G↓GATC C
C CTAG↑G

It is clear that the tetranucleotide sequence recognized by MboI falls within the hexanucleotide sequence recognized by BamHI. Were these two enzymes to be used simultaneously in a restriction mapping experiment, the results would certainly be confusing. Once again, the product fragments probably would not be resolved clearly by simple gel electrophoresis, but the result would also depend on the *total length* of the fragments that were so similar. The best solution is to be mindful of the problem and, wherever possible, to avoid the use of enzymes that make up isoschizomeric pairs. (As a useful exercise, how many isoschizomeric pairs can you find among the enzymes listed in Table 11–2?)

DNA SEQUENCING METHODS: TWO STRATEGIES

Two quite different strategies have been developed for sequencing DNA. The first is largely the work of Sanger's laboratory at the MRC Laboratories of Cambridge University, whereas the second is largely the work of Maxam and Gilbert of Harvard University. Sanger's group actually developed several methods; although these differ in details, they all depend on interruption of chain lengthening of some primer while it is bound to a template in the presence of DNA polymerase. Chain lengthening also requires the addition of all four nucleotides: dATP, dGTP, dTTP, and dCTP. If any one of the four is omitted, chain growth will cease whenever that nucleotide is demanded by the template sequence. In a variation on this approach, Sanger's group has employed dideoxynucleotides (ddNTP), each of which blocks incorporation of the corresponding dNTP. By omitting in turn each of the four dNTPs (or by adding in turn each of the ddNTPs) to a series of otherwise identical incubation mixtures, one can generate a family of variously extended oligonucleotides, still hydrogen-bonded to the template. After urea denaturation, the newly generated single strands are collected and examined by electrophoresis on polyacrylamide gels, as outlined earlier. By comparison of the fragment lengths, the sequence of the template can be written down over some considerable portion of its length. Further details of how this is done will be explored after a brief consideration of the strategy developed by Maxam and Gilbert.

The approach taken by Maxam and Gilbert is based on the structural chemistry of DNA. It has been demonstrated previously that methylation of purines (at N^3 and N^7 for adenine, and N^7 for guanine) weakens the glycosidic bond between the purine and the deoxyribose in a nucleotide; heating methylated, purine-containing nucleotides at neutral pH causes breakage of the glycosidic bond. Subsequent heating with 0.1 M NaOH cleaves the bonds between the depurinated deoxyribose and the adjacent phosphate groups. Purine methylation is easily accomplished with the inexpensive reagent, dimethyl sulfate. By adjustment of the experimental conditions, it is possible to enhance cleavage at A as opposed to G, providing reasonable specificity of attack. It had also been shown that hydrazine caused ring opening of the pyrimidines, after which the base, piperidine, displaced the products of ring opening from the deoxyribose to which they had been attached. Slight modifications of the experimental conditions suppress reaction of thymine with hydrazine, allowing distinction between the pyrimidines. Thus, with only two simple and inexpensive reagents employed in slightly different circumstances, it is possible to identify selectively each of the common bases that occur in DNA by electrophoresis of chemically generated fragments.

To summarize the two strategies, the Sanger methods depend on interrupting growth of a polynucleotide chain, either by addition of a chain-lengthening inhibitor (ddNTP) or by omission of a required nucleotide (dNTP). A primer, a template, and DNA polymerase are also essential. The sequence to be determined is, of course, that of the template. The Maxam and Gilbert method works in the reverse direction. One starts with a preformed polynucleotide, then subjects samples of it to attack by reagents that preferentially attack at sites of one base or another. Ordinarily, in their procedure, the 5′ end of the polynucleotide is tagged with ^{32}P or, more recently, with an ^{35}S-tagged nucleotide; the electrophoretic separations are then examined by autoradiography. Note that neither the Sanger nor the Maxam and Gilbert strategy can be directly applied to intact DNA; an intact DNA is simply too long. Instead, sequence analyses are performed on fragments 20–250 base pairs in length, produced by use of restriction enzymes or similar means. The smaller fragments are better able to withstand the required manipulations without physical damage.

Sanger's Dideoxy Chain-Terminating Method: Further Details

Space limitations prevent detailed discussion of all the Sanger methods, and the dideoxy method has been selected for further analysis by virtue of its simplicity and ability to cope with fairly long sequences in a single stage.

The 2′,3′-dideoxynucleotides (commercially available) inhibit chain growth because they cannot form phosphodiester bonds. If a primer and template, mixed with the Klenow fragment, are incubated with the four essential dNTPs *plus* one ddNTP, a family of fragments will be generated. Each of these will have identical 5′ ends (which will be the same as the 5′ end of the primer) and each will terminate at some specific base, depending on the ddNTP added. Suppose, for example, that ddTTP was added. Chain growth will continue until a T is required by the template sequence (at an A location in the template). At that point, the polymerase may add a ddTTP, in which case chain growth will be terminated. Or, the polymerase might add a dTTP (since both dTTP and ddTTP are present), and

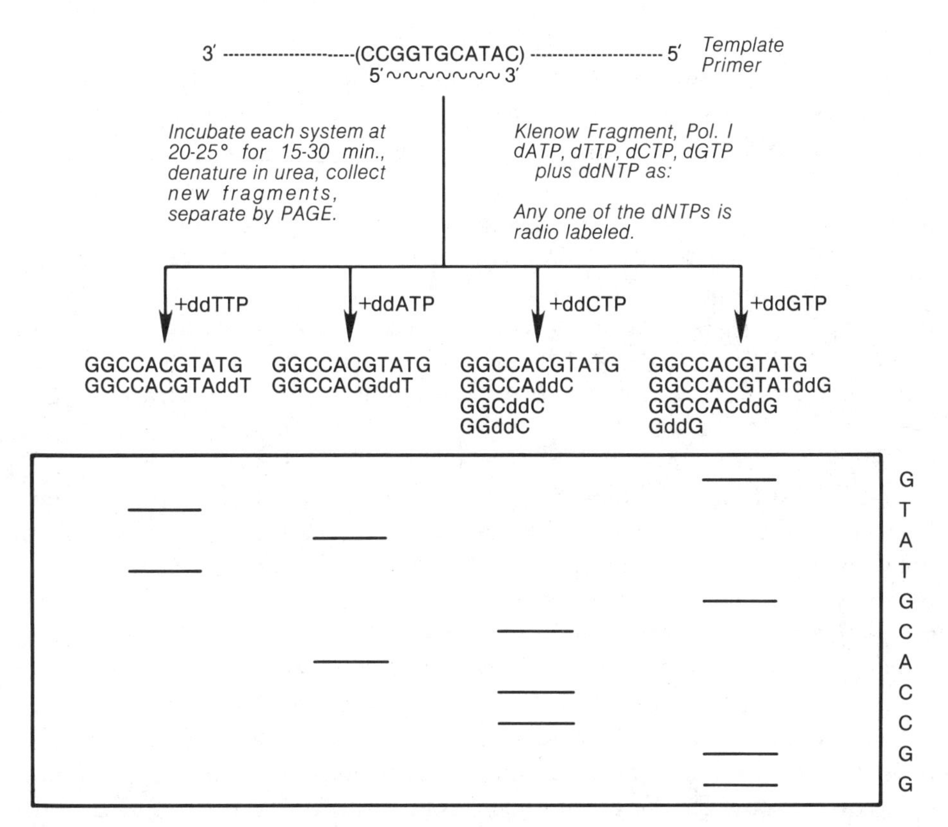

Figure 11–8. A schematic of the dideoxy (dd) sequencing procedure. After a small primer is annealed to a template, samples are incubated with a Klenow preparation of DNA polymerase I and a mixture of the four deoxynucleotides. To each of four aliquots, one or another of the dideoxy derivatives is also added. After a suitable incubation period, the products are denatured with urea, cleaned up by gel filtration, and then separated by electrophoresis on a single polyacrylamide gel. The autoradiogram is sketched in the lower part of the figure. It is clear that fragments differing by only a single nucleotide can be separated. The sequence can be read from the gel, as shown to the right of the diagram.

chain growth will continue in the 5′→3′ direction at least until the next T is called for by the template. At that point, the possibilities of chain termination or elongation again present themselves. A dideoxy sequence study requires that four separate incubations be run in parallel. They must be identical in composition except for the difference in ddNTP added. The four product mixtures are denatured with urea, and the variously terminated products are simultaneously examined on a single polyacrylamide gel slab. After autoradiography the sequence can be read from the positions of the dark bands on the film. Frequently, it is possible to read 15–200 nucleotides from the 3′ end of the primer, with reasonable accuracy, from a single 40-cm gel.

Figure 11–8 is a simplified scheme illustrating how the dideoxy analysis is performed. Shown at the very top of the figure is a template containing a short run of bases for which a complementary primer is available. The primer is indicated by the "squiggly" line just below the template. The bases shown in parentheses along the length of the template are, initially, not known. They are shown here only to emphasize the logic of the procedure. The template, to which the primer has been annealed, is then incubated in four separate samples, to each of which a different ddNTP was added as shown. After incubation, the nucleotides produced are denatured, separated by electrophoresis, and subjected to autoradiography. The rectangle in the lower part of the figure shows how the autoradiogram might appear. In the actual experiment, the added nucleotides would be extensions of the primer, but to simplify the figure, all primer sequences have been omitted. The reader must keep in mind that the primer remains attached to the 5′ ends of the newly synthesized fragments, but omission of the primer from the diagram does not affect the logical analysis.

The largest fragment in each lane of the gel results from minimal interference with chain growth, whereas the smallest fragment results from maximal interference. Accordingly, the bands line up in terms of the number of nucleotides incorporated into each fragment. One therefore reads the gel from the smallest fragment (at the lower right-hand corner) to the largest fragment (in the upper right-hand corner). In between, one moves from lane to lane to determine the base that occupies a given position in the unknown sequence. A correct reading of the sequence determined by the gel is shown along its right-hand edge. The lowermost of the symbols clearly represents the 5′ end of the sequence. If this reading is added to the 3′ end of the primer shown at the top of the figure, it immediately becomes clear that it is precisely the complement of the unknown sequence of the template, shown in parentheses.

Maxam and Gilbert's Specific-Base Cleavage Method: Further Details

This procedure is directly applicable to any single- or double-stranded DNA species of as many as 500 base pairs in length, provided the sample has a free 5′-OH group. Where that is not the case, existing 5′-P groups can be removed by treatment with alkaline phosphatase (see Table 11–1). The 5′-OH can then be labeled with [γ-^{32}P]ATP or, in a modified technique, with an ^{35}S-labeled nucleotide in the presence of polynucleotide kinase. It is important that the labeled ATP be of high specific activity since only a single ^{32}P is introduced into each strand and since final measurements are dependent on autoradiography of electrophoretic gels, as in the Sanger methods.

The rationale of the procedure can be illustrated by the following example. Imagine a DNA fragment represented by:

*pGpCpTpGpCpTpApGpGpTpGpCpCpGpApCpG . . . (Compound I)

The asterisk shows that the 5′ end has been labeled with ^{32}P. Compound I may be single- or double-stranded; for simplicity, only one strand is shown. Neglecting the chemistry for the moment, imagine next that it is possible to cleave this extended fragment; first at each of the positions occupied by a G, then at each of the positions occupied by a C, and so forth. A different pattern of subfragments would result in each case; here we will show only those resulting from cleavages at positions occupied by G. The mixture of subfragments would include:

*p
*pGpCpTp
*pGpCpTpGpCpTpAp
*pGpCpTpGpCpTpApGp
*pGpCpTpGpCpTpApGpGpTp
*pGpCpTpGpCpTpApGpGpTpGpCpCp
*pGpCpTpGpCpTpApGpGpTpGpCpCpGpAp
*pGpCpTpGpCpTpApGpGpTpGpCpCpGpApCpG . . .

Other products would also be formed, but they would not be labeled and so they would not appear on an autoradiogram of the final electrophoretic gel in which the sets of fragments had been separated by their lengths. Such an autoradiogram would be read from the bottom upward, since the smallest fragment would indicate the 5′ end of the sequence of compound I. If the chemical attack at the site of each base is indeed specific, the resulting sequence reading should be unambiguous. Before examining the autoradiogram, we will briefly discuss the chemistry of the degradation methods and their specificity.

Chemistry of Cleavage at Purine-Bearing Sites. A scheme of the reaction pathways that break DNA at G sites is shown in Fig. 11–9. A very similar pathway exists for A sites, except that the adenine ring may also be methylated at N^3, as shown by the broken arrow in the top, left-hand structure of the figure. Treatment with dimethyl sulfate and base results in ring opening of the imidazole portion of the purine, as shown at the upper right of the figure. Treatment with piperidine then displaces the ring-opened A or G residue and at the same time breaks the 3′ and 5′ phosphate bonds to the associated sugar. As a result, the DNA has been cleaved into two fragments at the point of attack.

Because G reacts fivefold faster with dimethyl sulfate than does A, the scheme depicted in Fig. 11–9 is known as a strong G/weak A, or G>A, cleavage. An autoradiogram of these fragments has a pattern of light and dark bands, reflecting the difference in reaction rates of the two bases. By heating the methylated DNA in mildly acid medium, rather than at neutrality, a strong A/weak G, or A>G, cleavage results, reversing the pattern of banding in the autoradiogram. Taken in combination, these methods provide a fairly reliable means of locating purine bases in a sequence.

Chemistry of Cleavage at Pyrimidine-Bearing Sites. Hydrazine attacks pyrimidine rings, leading to the formation of new pyrazolone structures as shown in Fig. 11–10. Continued reaction with hydrazine splits out the pyr-

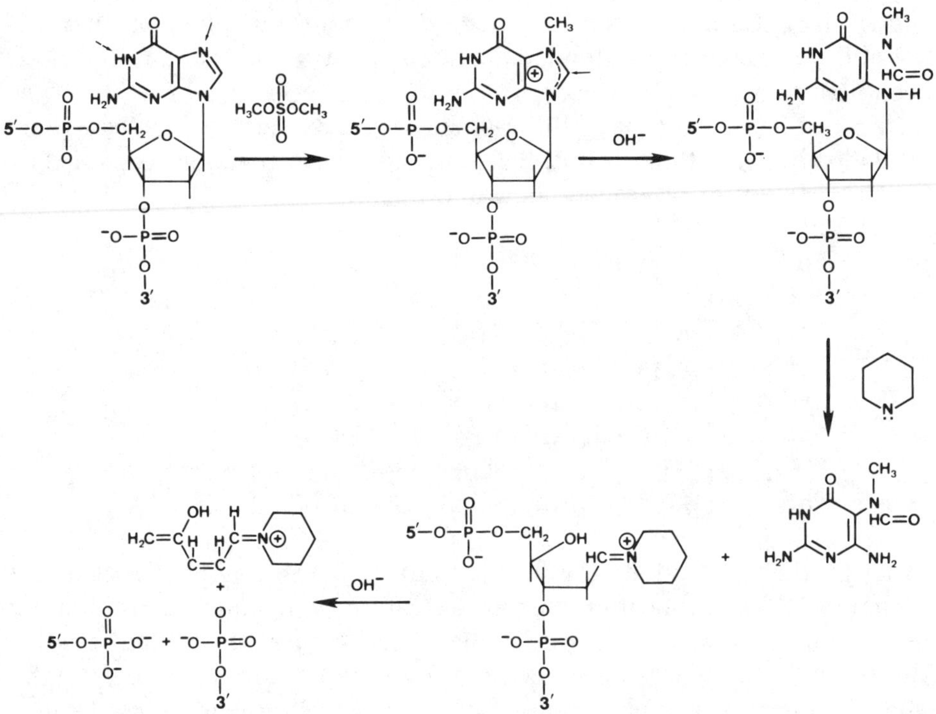

Figure 11–9. Cleavage at guanine sites by sequential reactions with dimethyl sulfate and piperidine. Alkylation at N^7 fixes a positive (+) charge on a portion of the imidazole ring of guanine, as shown in the top, central structure. When base attacks C^8, the imidazole ring is opened (top, right). Piperidine then displaces the ring-opened methylguanine and catalyzes β elimination of the phosphates from the deoxyribose. The sugar ring is also opened (bottom, center), possibly to form a piperidone (bottom, left). The broken arrow at N^3 (top, left) shows another site of attack by dimethyl sulfate, usually observed at A residues. Methylation at N^3 by itself does not result in ring cleavage, nor does it significantly affect the remainder of the reaction pathways shown. Note also that the phosphates remain at the 3′ and 5′ ends of the fragmented DNA after cleavage at the given guanine sites.

azolone, opens the deoxyribose ring, and produces a sugar hydrazone. Treatment with piperidine forms a piperidone and cleaves the 3′ and 5′ phosphate bonds, resulting in cleavage of the DNA strand. As with the purines, the reaction described here is not absolutely specific. The scheme shown is best described as a T+C cleavage. Specificity for C site cleavage can be accomplished by replacing water in the reaction mixture with 5 M NaCl. Salt addition markedly depresses reactivity of T with the hydrazine. It has also been observed that the efficiency of these specific reactions is critically dependent on the concentration and purity of the hydrazine reagent.

Reading the Autoradiogram. Figure 11–11 is a schematic of the autoradiogram that would be produced by analysis of compound I, defined earlier. Each gel lane was generated by the indicated cleavage system. When a band appears at the same level in C as in C+T, it should be read as a C. When it appears in C+T but not in C, it should be read as a T. The two left-hand columns are independent of each other and are read as the stronger band of the pair. The reader can quickly verify that the sequence represented is identical to the sequence given earlier for compound I.

Problems with the Method. As noted earlier, reliability of this method is very much a function of the experimental conditions and the quality of the reagents. In addition, both dimethyl sulfate and hydrazine are quite toxic. All operations in which they are employed should be performed in a good fume hood. The additional hazard of the high specific activity required in the ATP should also be noted. Use of ^{35}S-labeled nucleotides reduces the radiation danger and, in fact, yields sharper, less diffuse bands in autoradiographs. It is, however, more difficult to monitor the various steps on the laboratory bench top with a Geiger counter because of the lower energy of ^{35}S compared to ^{32}P.

In addition to the four systems described above, Maxam and Gilbert have outlined eight additional variants developed in an effort to make the sites of cleavage more nearly base-specific and more clear-cut. References to this work appear at the end of this chapter.

Figure 11–10. Cleavage at thymine sites by sequential reactions with hydrazine and piperidine. Hydrazine attacks pyrimidines at C^4 and C^6, opening the ring (top, left); then it cyclizes with C^4, C^5, and C^6 to form a new, 5-membered pyrazolone ring (top, center). Further reaction with hydrazine releases the pyrazolone and leaves a ureide derivative of the deoxysugar (top, right). Still further attack may result in formation of a hydrazone, shown at the lower right. Alternatively, piperidine reacts with the ureide derivative to form piperidones (bottom, center), and also catalzyes β elimination of the phosphates (bottom, left). This part of the reaction sequence is comparable to the mechanism that characterizes cleavage at purine sites. Note that the phosphates remain attached, at the 3′ and 5′ positions, to the resulting new DNA fragments.

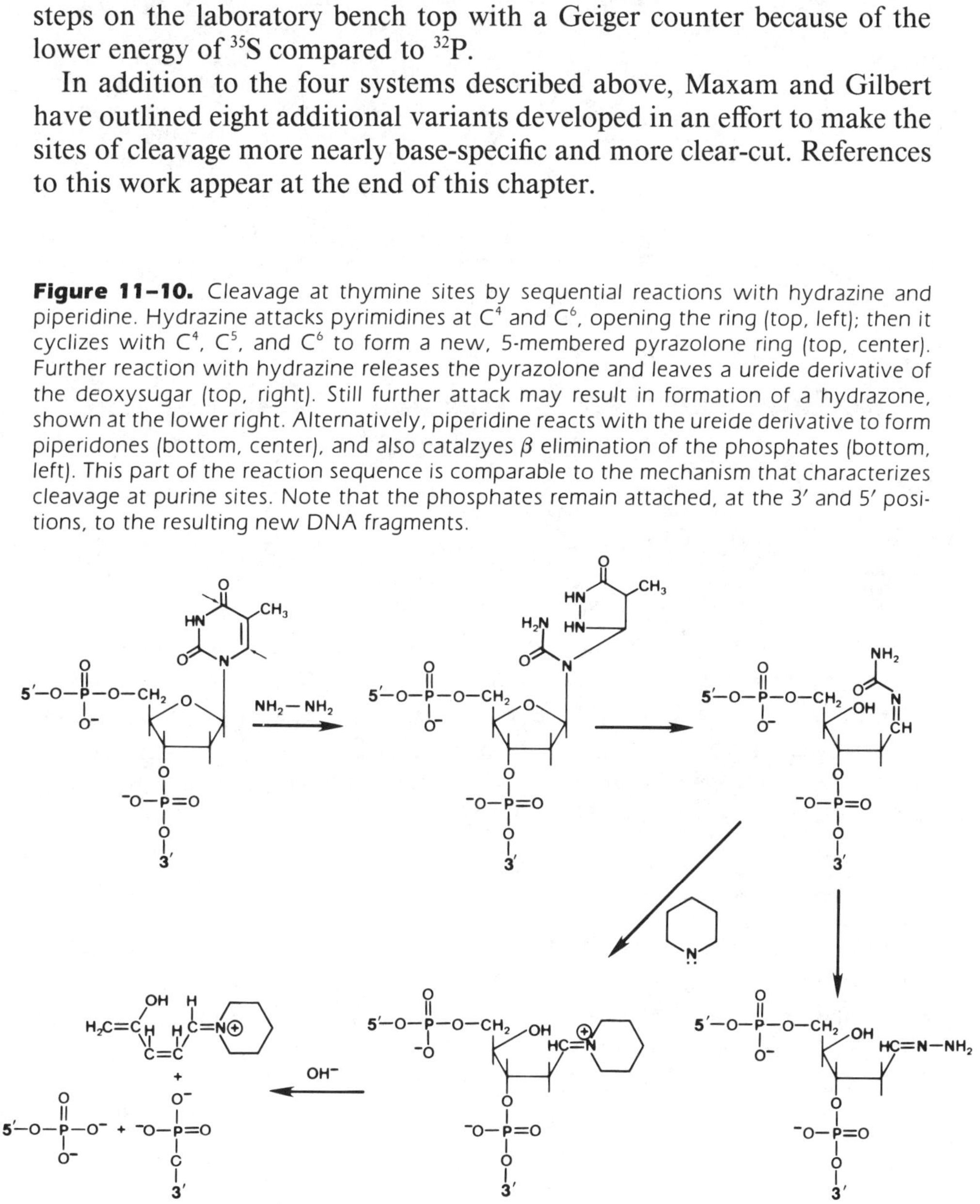

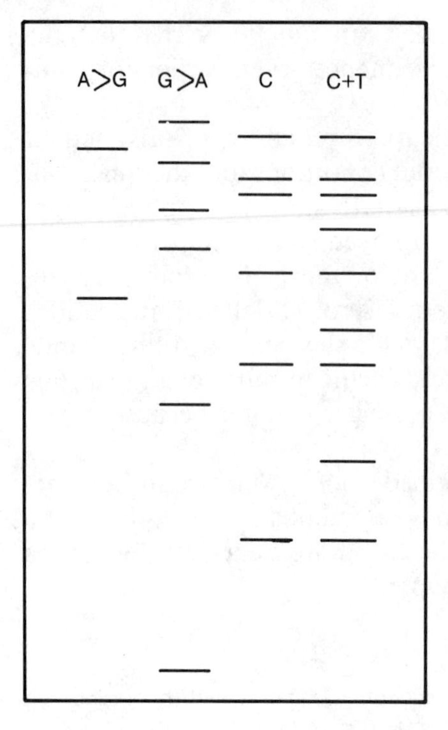

Figure 11–11. Schematic of an autoradiogram generated by a Maxam and Gilbert sequence study of the DNA fragment identified in the text as compound I. The lanes are labeled to indicate the specific cuts that account for the bands shown. The strong adenine (A > G) and strong guanine (G > A) cleavages are shown at the left-hand side of the gel. Bands in these lanes do not occur at the same levels. Bands in the C and the C + T lanes may occur at the same level, signifying that the cut was made at a C position. Bands that appear only in the T + C lane are therefore due to T. The sequence is read from the bottom of the gel to the top.

Recombinant DNA Technology: Principles and Operations

USE OF RESTRICTION ENZYMES IN GENETIC RECOMBINATION IN VITRO

In vivo genetic recombination may result in multiple alterations in DNA sequences, due to the many possibilities for crossing over. Controlled, in vitro recombination allows deliberate modifications of a DNA molecule, either by addition or by deletion of short or long polynucleotide sequences. Thus a second and important application of restriction enzymes is in recombinant DNA technology.

Generation of Blunt and Sticky (Cohesive) Ends

All restriction enzymes can be divided into two classes with respect to the termini they generate. As shown in Table 11–2, EcoRI and PvuII each has a distinctive recognition sequence composed of six base pairs. The scission point of EcoRI falls within its recognition sequence, but it is off center. The scission point of PvuII also falls within its recognition sequence, and lies at the center of the sequence. These observations are detailed as shown:

$$\text{EcoRI:}\ \begin{array}{l} 5'\ldots \text{G}{\downarrow}\text{AATTC}\ldots 3' \\ 3'\ldots \text{CTTAA}{\uparrow}\text{G}\ldots 5' \end{array} \rightarrow \begin{array}{l} 5'\ldots \text{G} \\ 3'\ldots \text{CTTAA} \end{array} + \begin{array}{r} \text{AATTC}\ldots 3' \\ \text{G}\ldots 5' \end{array}$$

$$\text{PvuII:}\ \begin{array}{l} 5'\ldots \text{CAG}{\downarrow}\text{CTG}\ldots 3' \\ 3'\ldots \text{GTC}{\uparrow}\text{GAC}\ldots 5' \end{array} \rightarrow \begin{array}{l} 5'\ldots \text{CAG} \\ 3'\ldots \text{GTC} \end{array} + \begin{array}{l} \text{CTG}\ldots 3' \\ \text{GAC}\ldots 5' \end{array}$$

The products of the PvuII reaction are said to have *blunt ends;* each base in proximity to the new ends is hydrogen-bonded to its complement, and thus there are no unfulfilled possibilities for additional hydrogen bond formation between the new ends and any other DNA. This is sometimes described by the statement that there are no *free tails* in the product. With EcoRI, the products are said to have *sticky ends* or *cohesive ends* because, on each strand, there is (are) one (or more) base(s) that do not have hydrogen bonding partners. In this case, there are free tails. One can draw similar conclusions for all restriction enzymes. Sticky ends have the great advantage that they will readily anneal with any other molecule that has a complementary sticky end. After annealing, the strands may be joined together by DNA ligase. For example, an EcoRI-directed scission would yield a product with one terminus designated on the left as:

$$\begin{array}{l} \text{Terminus: } 5'\ldots \underline{\text{AATT}}\text{C}\ldots 3' \\ \qquad\qquad\qquad\qquad\ \text{G}\ldots 5' \\ \qquad\qquad\qquad\qquad\ \text{Complement: } 5'\ldots \text{CGA}\ldots \text{GGCC}\ldots 3' \\ \qquad\qquad\qquad\qquad\qquad\qquad\qquad\quad 3'\ldots \text{GCT}\ldots \text{CCGG}\underline{\text{TTAA}}\ldots 5' \end{array}$$

A complementary species, indicated by the underscored bases in the sequence on the right, will readily anneal with the EcoRI-derived terminus shown on the left, after which the two segments can be ligated to form a new, single DNA species. The energy source that drives the reaction comes from ATP plus Mg^{2+}.

Suppose that one had a plasmid in which a single cut had been made by PvuII, generating blunt ends, and that one wished to join to one end a DNA fragment derived from some other source by the action of EcoRI. The obvious problem is that a blunt end cannot be annealed to an overlapping end. Some change must be made in either the plasmid or the EcoRI-derived material. Only one among the various possibilities, in which the blunt ends of the plasmid DNA are modified, will be described.

Blunt-ended ligation is catalyzed by T_4-DNA ligase, and is greatly enhanced if the DNA ligase contains RNA ligase, as was noted earlier. The energy source for this ligase reaction is also ATP plus Mg^{2+}. The double-stranded DNA segments to be joined must be present in fairly high concentration and must have 5′-P and 3′-OH groups; these groups are easily introduced by means of methods already described. One can prepare (or purchase) what is known as an *EcoRI linker sequence.* This is a small, blunt-ended oligonucleotide that contains an EcoRI recognition sequence. One commercially available linker sequence of this type has the structure:

5′ . . . GG | AATT   CC . . . 3′
3′ . . . CC   TTAA | GG . . . 5′

The dashed lines are added to emphasize the similarity between the sequence as prepared and the typical terminal structure of an EcoRI digest. Once the linker sequence has been attached to the opened plasmid by blunt-ended ligation, it can be treated with EcoRI to provide a terminus compatible with, and suitable for annealing with, the EcoRI-derived material to be inserted. Linker sequences are now commercially available for many of the common restriction enzymes, making blunt-ended ligation a practical and useful procedure.

INTRODUCTION OF FOREIGN DNA INTO INTACT CELLS

The ability to modify DNA structure is a powerful tool for advancing molecular biology, but this tool could not be fully exploited until methods were developed for introduction of carefully tailored DNA into cells, wherein the modified molecules could find biological expression. In this section, we examine how cells, which normally resist invasion by foreign or xenobiotic molecules, can be made to accept foreign DNA species, incorporating such species, either as plasmids or as open, linear structures, into their own genetic apparatus.

Specialized proteinaceous structures, including the base plate and tail fibers, allow phage particles to infect bacterial cells, literally by injection of the phage DNA through a perforation of the host cells' plasma membranes. Recall also that in bacterial sexual conjugation, plasmids are exchanged by passage through the hollow center of a protoplasmic bridge between the mating cells. In both of the above cases there is a clearly demonstrable pathway for DNA movement. Twenty or more years ago, it had been shown that free (i.e., phenol-extracted) DNA could be taken up by coliform cells in the presence of a "helper phage," suggesting that changes in the plasma membranes of the infected cells extended over a greater area than the exact sites of phage attachment. It was also noted that DNA uptake was more efficient if the free DNA had a cohesive end complementary to the end of the phage DNA. Methods dependent on helper phage were clumsy and inefficient because they introduced the phage DNA as well as the free DNA, constituting a potential source of ambiguity in the observed results.

In 1970, Mandel and Higa showed that free DNA could be passed into intact cells without any helper phage, in the presence of excess Ca^{2+} under ice-cold incubation. They developed a procedure that is still widely used today. Later, Graham and Van der Eb showed that the DNA of human adenovirus 5 could be taken up by monolayered cultures of human KB cells if the DNA, in a low-P_i buffer, was first treated with a 100 mM $CaCl_2$ solution. Apparently, the DNA was adsorbed to the precipitate of calcium phosphate formed. In their experiments, no necessity for low-temperature incubation was noted. Calcium phosphate adsorption has since been applied to other cell lines and to other viruses, and so this method also seems to be a general one. Today, free plasmid uptake by bacteria is generally induced by some version of the Mandel and Higa procedure.

Exactly how Ca^{2+} promoted the passage of DNA across plasma membranes of cells is not clear. Experiments reveal that the effect is not permanent, and that Ca-treated cells remain receptive to either circular or linear DNA for only a limited time after treatment. For certain cells (yeasts, bacteria) that have a heavy cell wall, it is necessary to prepare protoplasts before introduction of DNA is attempted. A number of enzymes or enzyme mixtures have been employed to remove cell walls, based on lysozyme or Glusulase® treatment with or without added agents such as penicillin. Glusulase® is a commercial name for a mixture of β-glucuronidase and sulfatase prepared from snails.

CELL–CELL FUSION: HYBRIDOMA PRODUCTION

In addition to passing free DNA into a cell, it is possible to fuse two unlike cell types together, forming hybrid cells, or *hybridomas.* Because these contain the total DNA of both precursor cell lines, they are of considerable interest. A number of methods have been developed for production of the "manufactured" fused cell lines.

Cell Fusion Induced by Viruses

One of the earliest methods used to achieve cell fusion depended on the effect of certain viruses. Of these, the best known is Sendai virus, named after the Japanese city where it was first isolated during an influenza-like epidemic. Later studies showed that this virus is very "sticky," adhering firmly to cell surfaces and causing adjacent cell membranes to fuse together. This property persists even after the pathogenicity of the virus is destroyed by controlled ultraviolet irradiation. Inactivated Sendai virus became a useful tool in cell fusion experiments, but it has since been replaced by simpler methods.

Cell Fusion Induced by Simple Chemicals

Cells can also be fused by treatment with polyethylene glycol (PEG) 1500, under appropriate conditions. A typical example is the work of Oi et al. (1983), who fused bacterial protoplasts with cultured lymphoid cells so that the bacterial plasmids transformed the lymphoid cells. As the with $CaCl_2$ treatment mentioned earlier, no precise mechanism for this transformation has been established. This reagent also causes membranes to fuse at points of contact, and the membranes appear to disintegrate where they touch, allowing mixing of the total contents of the fused cells. The lymphoid cells used by Oi et al. were derived from a myeloma tumor that reproduces repeatedly in culture; in common parlance, such cell lines are said to be "immortal." When hybridized with, say, an antibody-producing cell line from the spleen or other reticuloendothelial tissue, the product hybridoma becomes a more or less permanent source of *monoclonal antibody.* A given lymphoid cell produces only a single antibody species during its life, as do its cloned descendents. When such clones are fused with tumor cells that reproduce indefinitely, the production of antibody is perpetuated.

Cell Fusion Induced by Physical Means

Recently, attention has been called to hybridoma production by a process of electric field-induced cell fusion. This work was pioneered and developed by Ulrich Zimmermann, and several commercial versions of the required instrumentation are now available. The method is intrinsically simple and fast. Under some conditions, depending largely on the sizes of cells to be fused, it is possible to observe the process under the light microscope.

Most finely divided matter, when suspended in aqueous electrolytes, demonstrates a net surface charge. Thus, *E. coli* cells show a negative charge and move toward the anode in an electrophoretic cell to which a direct current field is applied. Similar observations have been made on erythrocytes, on oil droplets, and even on tiny air bubbles. Cells can also be suspended in low-ionic-strength media such as isosmotic solutions of sucrose or mannitol. Because these solutions have very poor conductances, they are described as *dielectric media.* When a source of high-frequency, high-voltage alternating current is applied to a cell suspended in a dielectric medium, and when the electrodes are of approximately the same area, a significant dipole moment will be induced in the cell. However, as shown in Fig. 11–12, the cell will suffer very little movement because of the rapidity with which the field reverses. This is shown in Fig. 11–12A. The figure also shows that the lines of force across the field are essentially uniform, since the electrodes are about the same size. If one electrode is made much smaller than the other, as shown in Fig. 11–12B, then the field becomes inhomogeneous, as shown by the converging lines, and the cell(s) will move toward the smaller electrode. This is true, regardless of the actual charge on the electrode, as shown in Fig. 11–12C, because the induced dipole also changes in polarity (note that in C and D, the dipole charge signs have been omitted). Ultimately, a cell will contact the smaller electrode, where it sticks. When more than one cell is suspended, they will line up at the smaller electrode, as shown in Fig. 11–12D; under the microscope the adherent cells have the appearance of a string of pearls. The dielectric force causes some flattening of the cells so that an initial point of contact becomes a small plane of contact. A better depiction of this phenomenon is provided by Fig. 11–13A, representing a higher-power view of two cells in contact in the dielectric field. The plane of contact has

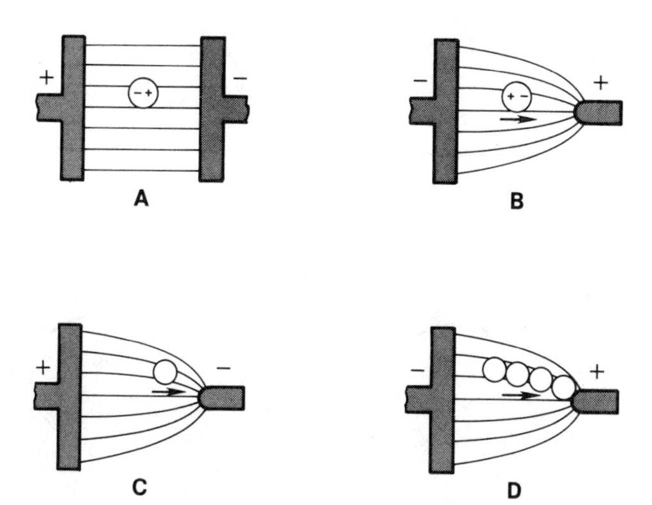

Figure 11–12. Effect of alternating current fields on a cell suspended in a dielectric medium. (A) A cell is suspended in a uniform field; there is little or no movement. (B) The electrodes are shaped to produce a nonuniform field, shown by the converging lines of force. The nonuniform field results in dielectrophoresis, and the cell moves toward the smaller electrode. (C) The cell continues to move. Note that the field has alternated, and that the dipole signs within the cell have been omitted for clarity. (D) The first cell has reached the smaller electrode, and others have moved behind it to contact the first cell, giving rise to the "string of pearls" appearance. The cells contact each other; the point of contact becomes a plane of contact as the membranes alter their shapes under the influence of the electric field. In this state, the cells are prepared for the brief direct current pulse that will drastically alter membrane permeability and thereby permit cell–cell fusion. To prevent harmful electrolysis, the cells are suspended in a sucrose or mannitol solution, containing very low concentrations of ionizable materials. The frequency of the alternating field ranges from 1 to 10 kHz, at a field strength of 200 volts/cm. The gap between the electrode ranges from 50 to 200 μm, depending on the cell size.

been established, but the individual nuclei are distinctly separate and the visible extranuclear structure can still be clearly associated with the separate cells, as indicated by the small, scattered lines in each.

The alternating current is briefly interrupted, and a single very short pulse of direct current is applied to the electrodes, causing drastic disorganization of the membranes across the plane of contact. Under the microscope, this transformation is observed as a "blurring" of the membranes (see Fig. 11–13B) that is soon followed by a breakdown of the contiguous membranes into small vesicles (Fig. 11–13C). The nuclei move together and begin to fuse, and there is evident protoplasmic streaming that results in cytoplasmic mixing. Within 5–15 min, the interfacial vesicles gradually disappear and the fused nuclei begin to assume a spherical shape, as does the plasma membrane of the newly formed hybridoma (Fig. 11–13D). Thus, within a fairly short span of time, the individuality of the precursor cells has been submerged in the generation of the hybridoma.

The situation shown in Fig. 11–13 assumes that only two cells were fused; if, however, the initial string of pearls arrangement contains more than two cells, it is possible to cause them all to fuse into a single giant

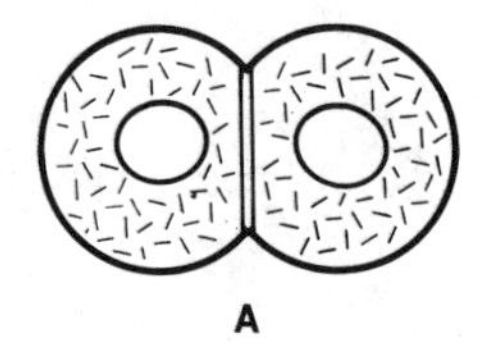

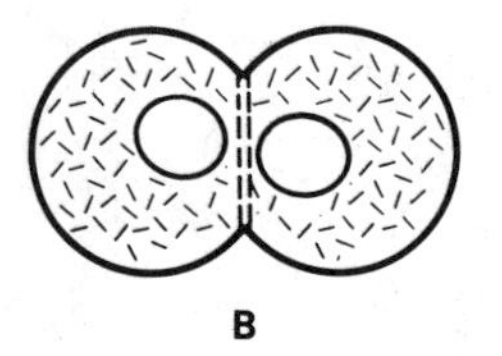

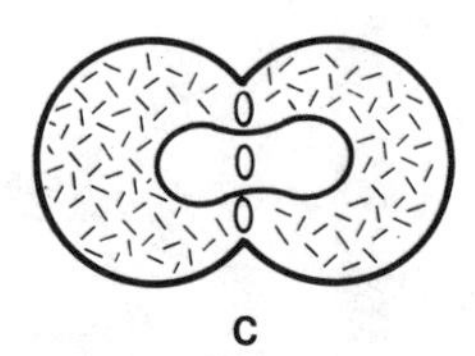

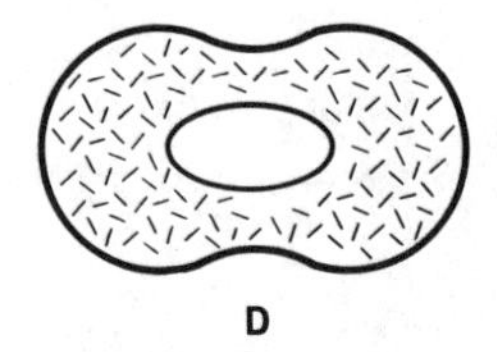

Figure 11–13. Steps in obliteration of membranes in contact after a brief, high-intensity direct current pulse. Cells in the "string of pearls" configuration have a common, flattened plane of contact (A). Application of a direct current pulse of very short time base and high intensity breaks down the regular array of phospholipids in the plasma membranes, causing pores or holes to appear, as shown by the broken lines in B. Within a minute or so, the disorganized membrane fragments form visible vesicles, as shown in C. The intact parts of the cell membranes begin to assume a more ovoid appearance. The small vesicles shortly disappear, as shown in D. The boundaries of the cells begin to merge, and ultimately become spheroidal as the internal contents of both begin to mix. At this stage, the cells have effectively been fused. The entire process takes less than 15 min and can be continuously observed through a microscope. If only two cells are collected at the smaller electrode, then only these two will fuse. If the "string of pearls" contains more than two cells, they all will fuse, giving rise to multinucleate hybridomas. Multiple fusion can be controlled by the spacing between electrodes and by the cell population.

cell. With care, twenty or more cells can be fused into a single entity. Irrespective of the number of cells, the final fusion product must be aspirated from the dielectric medium by micromanipulators and transferred to other media for growth and reproduction. Thus far, the method has been applied to fusion or protoplasts prepared from yeasts and higher plant cells; and murine as well as human hybridomas have also been made. Liposomes have been fused to each other and also to a variety of cells. It is this broad scope of possible applications that makes the field-induced fusion method one of considerable promise.

The field-fusion method is not without its drawbacks. Many variables—electrode size and spacing; choice and composition of the dielectric medium; frequency, voltage, and pulse length of the alternating current; intensity and duration of the direct current fusion pulse; and cell population—need optimizing.

SOME TECHNICAL PROBLEMS AND THEIR ORIGINS

The discussion just concluded demonstrates that recombinant DNA species can be prepared in several ways. Regardless of the method used, the yield is likely to be small in terms of the number of cells recovered. It is

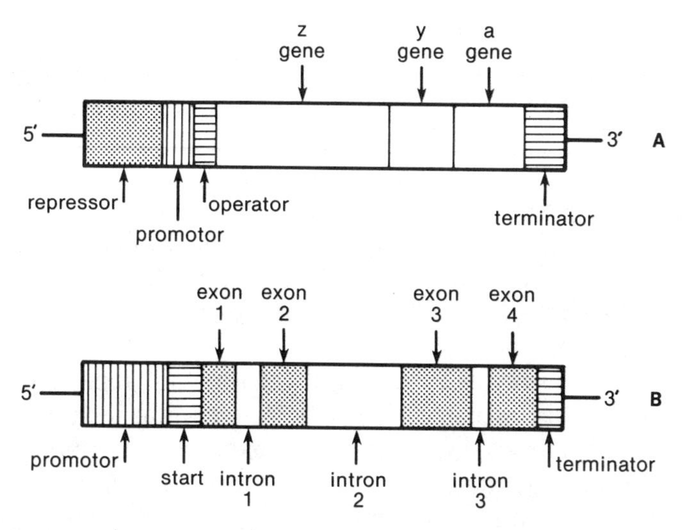

Figure 11–14. Differences in organization of prokaryotic (A) and eukaryotic (B) genes. (A) The organization of the *lac* operon of *E. coli;* each region of the operon is identified. Note that the three genes are continuous, and recall that the operon serves as a single transcription unit. (B) A hypothetical eukaryotic gene. Note that the single gene is interrupted; the shaded segments identified as exons constitute portions of the message, whereas the clear spaces, identified as introns (or intervening sequences) are removed during mRNA processing. Tissue-specific changes in splicing pathways sometimes allow an intron for one gene to serve as an exon for a related gene. This occurs, for example, in amylase genes of the liver and salivary glands of the mouse. Other examples of varied splicing patterns are known. For clarity, the relative base pair lengths in A and B are not drawn to scale.

necessary to "select out" the transformed* cells from the others and then to increase the transformed population to useful numbers of cells. One must then determine that the desired modification did, in fact, take place and that some undesirable modification did not. Depending on the growth rates of the cells employed, this may take considerable time; even with the more efficient methods of DNA modification and with less than fastidious cells, such an experiment may take as long as several weeks to complete.

The Complexity of Eukaryotic Genes

Further complications arise from differences in the internal architecture of prokaryotic and eukaryotic genes. Bacterial genes are commonly arranged in *operons.* An operon is a single transcriptional DNA unit that defines a number of enzymes, all of which are closely associated with some metabolic pathway. Operons include DNA subsegments identified as repressor, promotor, and operator regions, as shown in Fig. 11–14A. These are reg-

*Genetic modifications are variably defined, depending on the source of the modifying DNA and on the cell type into which it is introduced. When phage DNA is integrated into a prokaryotic cell, the process is termed *transduction.* When nonphage DNA is integrated into a bacterial genome, the process is known as *transformation.* The latter term, as applied to eukaryotic cells, refers to their conversion to a state of unrestrained growth, as in tumors. *Transfection* of eukaryotic cells is the acquisition of new genetic markers by incorporation of DNA, usually from other eukaryotic sources.

ulatory regions, capable of binding metabolites of various kinds. Many operons exist in a state of repression (i.e., in which some repressor substance is bound to the repressor region) and so are not read out until derepressed. A classic example is the *lac* operon, best known in *E. coli.* This operon codes for a lactose permease that promotes lactose uptake by the bacteria, and the *a* gene codes for a thiogalactoside transacetylase, whose function is not altogether clear. Other known operons are still more complicated and are subject to finer control by means of corepressor and/or attenuators that modify reading out of the genes in accord with the momentary needs of the cells.

Eukaryotic genes may carry operator regions, but repressor regions are not as common as in prokaryotic genes. The coding segments of eukaryotic genes are frequently interrupted by extraneous segments of nucleotides (intervening sequences, or *introns*), dispersed between information-bearing segments (*exons*), as shown in Fig. 11–14B. The purpose(s) of these organizational patterns of eukaryotic genes cannot be explicitly stated, but their consequences are clear. When eukaryotic genes are modified by recombinant technology to remove all of the introns, some of these recombinant forms still function in a reasonably normal manner, but others do not. Introns may also serve other purposes not yet discovered. In any event, the intron–exon mix makes the task of elucidating eukaryotic structure more difficult than the corresponding studies of prokaryotic genes. This can be exemplified by a closer comparison of relative gene lengths in prokaryotes and eukaryotes.

The *lac* operon of *E. coli* contains 6277 base pairs. Of this length, 1040 base pairs make up the repressor region, the next 122 base pairs comprise the promotor-operator region, leaving 5115 base pairs to express the three functional enzymes. The *z* gene (3510 base pairs) cannot code for a peptide of more than 1170 amino acid residues, the *y* gene (780 base pairs) cannot code for a peptide longer than 260 amino acid residues, and the *a* gene product cannot be longer than 275 amino acid residues. Thus, each gene is actually specified by a relatively short length of DNA, out of the $\sim 4 \times 10^6$ base pairs of the genome. It follows that there is a fair probability that cleavage of the *E. coli* genome by any of a number of restriction enzymes will provide fragments in which one or more of the *lac* genes will remain intact and useful for recombinant experimentation.

Consider now a different case. The mammalian enzyme, xanthine oxidase, is a dimeric metalloprotein with a mass of about 256,000 daltons. It contains Mo(VI) and nonheme iron. Even after subtracting the masses of the non-amino-acid components, and allowing for the dimeric structure of the protein, each monomeric peptide would have a mass of about 127,500 daltons. On the average, an amino acid residue represents a contribution of 100 mass units, and each amino acid requires a triplet codon for its specification. It follows that the length of DNA needed to code for a xanthine oxidase monomer could not be less than 3825 base pairs. Given the existence of introns and exons, the actual length might well be twice that value, or 7650 base pairs. At least in theory, we see that specification of this single mammalian enzyme monomer could require as much in the way of base pair length as did the entire *lac* operon! This argument explains why, at least up to the present time, progress in elucidation of gene structure for larger, eukaryotic proteins has been slow. It is true that the xanthine oxidase gene, even granting all of the above speculations about its possible length, is a very small portion of the chromosome in which it resides, but the statistical probability of cutting into its structure is increased, relative to the *lac* genes, by its greater length.

GENETIC RECOMBINATION IN VIVO: THE IMMUNE SYSTEM

An entirely different order of complexity is exhibited in the genetics of the immune system. An organism may be exposed during its lifetime to a potentially enormous number of antigens, against which it must quickly develop antibodies. In an earlier chapter, we saw that an immunoglobulin, or antibody, shows a very high specificity toward its antigen; in many instances the antigen–antibody reaction is unique, which is why antibodies are so valuable as analytical tools. The question before us now is how an organism can generate as many as 10^6 or more specific immunoglobulins, each the product of a gene.

Recall that immunoglobulins are tetramers, each composed of two heavy (H) chains and two (or three, in the case of IgM) light (L) chains. The H chains are identified by Greek letters that specify the Ig class, of which there are eight, whereas the L chains are identified by the letters κ or λ, one or the other of which is found in virtually all Ig classes. Both L and H chains can be further characterized as containing *constant* and *variable* regions (as shown in Fig. 10–2). In any animal species, the constant regions of any one class of L chains have a virtually fixed composition. Similarly, the constant regions of the H chains in any one class have a fixed composition, but the two fixed regions (H and L) are not identical. The variable regions, V_H, and V_L, show a great diversity in their compositions, independent of the Ig class. They are the regions responsible for antigen binding and thus for the specific antigen–antibody interaction. The simplest unifying explanation for these observations can be summarized as follows:

1. There is a family of genetic elements that codes for the κ light chains, another that codes for the λ light chains, and a third that codes for the various types of heavy chains (μ, δ, α, etc.).
2. Within each family there are separate genetic elements that code for the various V and C regions of both light and heavy chains.
3. There are additional genetic elements, the J gene segments (in H and L chains) and the D genetic elements (in H chains only), that serve to join, or connect, the V and C elements that make up the final, complete Ig gene.

In other words, the genes that code for a given Ig are themselves connected systems of independent (unlinked) genetic elements, selected from a finite, preexisting population. Therefore, each Ig gene must be the result of extensive recombination.

The initial skepticism concerning this hypothesis arose in part out of the question of antibody diversity; it was not believed that the hypothesis as stated could account for the very large number of possible antibodies. Subsequent research has eliminated all skepticism on the following bases:

1. The exact number of V_κ genes is not known with certainty, but a good estimate of that number lies between 100 and 300.
2. Similar estimates for the number of V_H genes fall between 70 and 200.
3. At least four J_κ and a like number of J_H genes are known.
4. An additional source of diversity was discovered by demonstration of the D (for diversity) genetic elements involved in formation of H chain genes. The best estimate is that there may be 10–100 D gene segments.

Combinatorial laws (see Chapter 12) can now be applied to the problem of fitting the above hypotheses to the known facts of antibody diversity. If one assumes only 300 V_κ gene segments and 4 J gene segments, these could give rise to 1200 distinct products. If one further assumes only 100 V_H, as few as 10 D, and 4 J_H segments, these could combine to form 4000 distinct products. Taken together, these conservative estimates show that one can account for about 4.8×10^6 specific antibodies, generated by random recombinations of a small number of genetic elements. When one takes into account certain other sources of diversity, especially the high rate of mutations found in Ig genes, it is not unreasonable to conclude that as many as 10^9 or 10^{10} distinct antibodies could be produced! It is unlikely that any single organism would ever be called upon to exercise so many options in antibody construction.

Antibody Specificity: How the Options Are Exercised

It is believed that all small, immature lymphocytes act as random generators of IgM and/or IgD antibodies. These antibodies are exported to the exterior of the cell, where they remain attached to the lymphocyte surface by their C_H termini. The V_H and V_L regions therefore extend into the space surrounding the cell. When challenged by an antigen, the small B lymphocyte whose surface bears the IgM or IgD that best fits the antigen is triggered, by means yet unknown, to further development and maturation involving three easily demonstrable kinds of change. First, the cell grows in size and begins to reproduce. Second, the triggered cell (and all its progeny) becomes committed to production of antibody directed against the triggering antigen, and to that antibody alone. Thus, the process of maturation is one of clonal generation and contributes (memory cells) to long-term immunity. Third, in the course of maturation the surface-bound IgM and IgD species are largely replaced by secreted Ig species, such as IgG, IgA, etc., that fulfill the immune functions at sites other than the immediate surface of the generating lymphocyte. This shift to production of free, secreted antibodies in addition to surface-bound antibodies is known as *class-switching.* Class-switching is mediated by, or at least involves, some two out of eight widely separated noncoding sequences of about 2000 nucleotides each. These fall between the $V_H/D/J_H$ segments and the genetic element coding for C_α, C_γ, etc. No other overt function has been found for these noncoding sequences.

There is much yet to be learned about the precise mechanisms governing the genetics of the immune system, but this brief overview establishes several key concepts:

1. The genes that code for Ig production are generated by extensive recombinations of a relatively small number of gene segments. From approximately 10^3 known genetic elements, it is possible to produce between 10^6 and 10^9 (or more) different antibodies.
2. Lymphocytic maturation is associated with commitment to production of but a single antibody species per cell, implying that a particular program or recombination has become fixed in that cell and in its progeny. If such a programed lymphocyte can be hybridized with a continuously reproducing tumor cell, the hybridoma that results becomes a continuous source of a monoclonal antibody; monoclonal antibodies are now very important tools in many areas of research.
3. Recognition of a foreign antigen is initiated by an in vivo DNA recombination and is ultimately expressed through the medium of immunoglobulins, but other regions of Ig molecules are responsible for complement fixation and for binding to monocytes. It is clear that the genetic reshuffling occurs without loss of a truly remarkable set of properties characteristic of Ig molecules in general.

Applications of Recombinant DNA Technology and Related Techniques

So far we have outlined the major principles, or unit operations, upon which recombinant DNA technology rests. We have emphasized its great dependence on bacterial or cultured cells, on plasmids or viruses. At this point, it seems fitting to examine some typical applications of these principles toward the achievement of specific goals. The stories of these applications are of interest in their own right, as examples of well-designed research plans. Moreover, one of the two examples we shall discuss has resulted in the development of a valuable tool, now commerically available, that is finding a widening role in DNA sequencing studies by the dideoxy method. Finally, our discussion of the two stories will permit at least brief mention of several other techniques not previously discussed.

TAILORING A VIRAL CLONING VEHICLE: THE M13 PHAGE

The Nature of Phage M13

M13 is a small, single-stranded DNA phage containing about 6.4 kb. This length is divided into ten genes and an intergenic space (IS region). The latter is 507 nucleotides in length and of undetermined function. The mature virion takes the shape of a supercoiled filament, the physical length of which is ordinarily about 100 times its diameter. The outer surface of the filament is "shingled" with a large number of fairly small coat protein subunits. In addition, one end of the filament carries several copies of a somewhat larger "pilot" protein, important in the adsorption of the phage to *E. coli* cells. The amino acid sequences of the major coat protein, coded by phage gene 8, and of the pilot protein, coded by phage gene 3, are known.

Curiously, M13 can infect only F^+ coliform cells—that is, only those cells bearing the "maleness" factor in the form of an F pilus. The phage attaches at or near the free end of the pilus, as can be demonstrated by electron microscopy, but just how the pilus serves to promote viral infection is not known with any certainty.

During penetration of the coliform plasma membrane, the phage coat proteins are neither discarded nor totally destroyed; instead, a significant quantity of the coat materials is sequestered in the inner leaflet of the plasma membrane and is used later to coat some of the progeny virions. The pilot protein, on the other hand, remains with the advancing tip of the viral DNA as it penetrates to the interior of the cell. It is assumed that the pilot protein facilitates DNA transport across the lipid bilayer of the membrane, since at least a part of the pilot protein remains associated with the inner membrane surface during replication of the phage genome. The remainder of the pilot protein, which travels with the penetrating DNA, fulfills yet another function, as described below.

Replicative Cycle

There are three distinct stages in the replication of M13:

In the *first stage,* the single-stranded infectious form (SS) must be converted to a double-stranded replicative form (RF). For this conversion, the phage depends almost entirely on the host's replication machinery. However, the process must be triggered by the presence of at least part of the pilot protein. Phage mutants with missing or defective gene 3 products cannot perform the SS→RF conversion.

Production of the second DNA strand is catalyzed by an RNA polymerase that generates a short RNA segment serving as a primer for the subsequent extension by DNA polymerase I. Proof that RNA polymerase is required for replication comes from experiments performed in the presence of rifampicin, a known inhibitor of RNA polymerase. Replication cannot proceed in the presence of rifampicin.

The *second stage* involves multiplication of the RF; this depends largely on the host's machinery also, but there is strong evidence that this part of the replicative cycle must be initiated by the protein product of phage gene 2. Many copies of the viral genome are produced as circular duplexes in a fairly short time. Since M13 is a temperate virus, it does not kill the host cells, even though highly active phage reproduction may slow the growth of the host cells.

The *third stage* involves disassembly of the RF and the packaging of the SS form for export. The product of phage gene 5 is a binding protein with a high affinity for single-stranded DNA; it appears to be necessary for the disassembly process. Note, however, that this binding protein is distinct and not associated with any of the coat proteins. Presumably, the binding protein does not leave the cell, although its exact fate is uncertain. Depending on the total virions produced, new coat proteins are synthesized so that, when added to the coat materials sequestered earlier, each new virion is properly dressed.

Sensitivity to Restriction Enzymes

One of the characteristics of M13 is its limited sensitivity to the common restriction enzymes. For example, the wild-type phage is not attacked by EcoRI, Hsu, or SmaI. It is attacked by HindIII at from 2 to 20 sites, depending on experimental conditions. At least one of the HindII sites falls within the intergenic region, where it does not inactivate any of the essential genetic components of the phage. This relative resistance to restriction enzymes undoubtedly plays some protective role for the phage in its normal situation, and it certainly is advantageous to biologists, as we shall soon see.

Properties of M13 Favoring Its Use as a Vector

When Joachim Messing and his collaborators were searching for a new vector, they likely reasoned as follows:

1. M13 is one of the so-called filamentous phages, described earlier; addition of cloned DNA would not grossly alter the form of the phage, but would merely extend the length of the filament. This would make it easy to package long or short segments of foreign DNA.
2. Wild type M13 is a phage of moderate size. Its total base length compares favorably with that of the plasmid, pBR322. Furthermore, the internal structure of M13 is quite simple, being arranged in a small number of genes and an identifiable intergenic region. All of the genes are essential for phage reproduction; therefore, any modifications have to be localized to the intergenic region.
3. M13 is a temperate phage; that is, it does not kill infected cells. It rapidly produces a large number of progeny phage with yields reported as up to 1 g per 10 L of cultured cells. These good yields more than offset the limitation that only F^+ cells can be infected with it.

4. The fact that two of the ten cleavage sites sensitive to the restriction enzyme BsuI fall within the intergenic region should allow insertion of a moderate length of extraneous DNA with no effect on any of the phage genes. Ideally, the DNA insert should have some distinguishing property that would allow its ready detection as a screening device.
5. It was believed that, if the manipulations described above proved successful, then the modified phage product would constitute a useful vector. If the new vector carried one or more restriction enzyme recognition sites (e.g., an EcoRI site) not found in the phage itself, this would be most useful for cloning additional DNA into the new vector. This, of course, was the ultimate aim of the project.
6. Plasmids like pBR322 had already proved useful for cloning purposes, but they could be employed only in the presence of selective pressures, such as antibiotic resistances. A new vector based on M13 would not depend on drug effects as a screen for successful transformation. Because one would no longer need to resort to testing for transformation by growth inhibition, results would be available more quickly.
7. Because vectors derived from M13 would also be single-stranded DNA, the same would be true of any extraneous DNA cloned into the new vehicle. Such chains would be ideal templates for sequencing studies by Sanger's dideoxy chain-termination method. By eliminating the need to separate duplex strands, sequencing studies would be expedited.

Design and Construction of M13mp1*

Messing's group first worked out conditions whereby the enzyme, BsuI, cut the replicative form of the phage genome at only a single site. By limiting the enzyme concentration and by carefully controlling time and temperature, they were able to separate linearized DNA of unit length, and to show that the cutting point fell within the nonessential portion of the intergenic region. The desired product was collected and characterized by agarose gel electrophoresis and by equilibrium centrifugation in a potassium iodide density gradient. BsuI recognizes the sequence:

$$\begin{matrix} -\mathrm{CC}{\downarrow}\mathrm{GG}- \\ -\mathrm{GG}{\uparrow}\mathrm{CC}- \end{matrix}$$

and, as shown by the arrows, generates blunt ends.

They then proceeded to implant a portion of the *lac* operon into a plasmid by ligation of a HindII fragment of the operon into a suitable carrier, pMG1106. Their purpose was to amplify the quantity of *lac* fragment by chloramphenicol treatment of the growing culture. The plasmids were iso-

*The designations, mp1, mp2, etc., refer to the first and second modifications of the wild-type M13. Current versions up to at least mp19 have been constructed, but we shall not explore all of the more sophisticated details since they do not change the fundamental hypotheses on which Messing's work has been based.

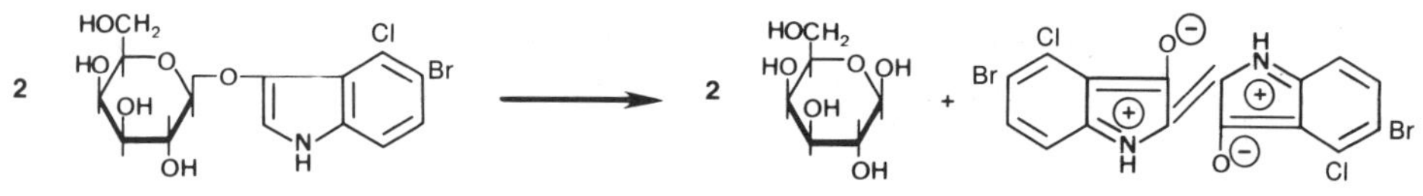

"X - Gal"

5-Bromo-4-chloro-3-indolyl-β-D-galactoside

Galactose

Halogenated Indigo (blue in color)

Figure 11–15. Color reaction of "X-Gal." Hydrolysis of the galactoside forms a substituted indigo, which has a deep blue color. See the text for further details.

lated and treated with HindII a second time. This treatment released, among others, the amplified *lac* fragment, which was recovered from the mixture by binding to *lac* repressor that had been fixed to nitrocellulose filters. After proper washing, the product was released from the filters with a solution of isopropyl thiogalactoside (IPTG), the allosteric effector of the *lac* repressor. The eluted DNA was then prepared for insertion into the RF of M13. Before proceeding further, we must note that the recognition sequence for HindII* is:

$$\begin{array}{c} -\text{GTPy}\downarrow\text{PuAC}- \\ -\text{CAPu}\uparrow\text{PyTG}- \end{array}$$

so that it also generates blunt ends. Accordingly, the linear M13 DNA previously obtained and the recovered *lac* operon fragment were mixed and subjected to blunt end ligation by T_4 DNA ligase. The product was designated as M13mp1 and was used to transform *E. coli* (strain 71-18) by the calcium chloride procedure.

To appreciate fully the elegance of these procedures, it is necessary to know something more about strain 71-18 of *E. coli.* These cells produce a defective β-galactosidase (lacking amino acid residues 11–41) and overproduce the *lac* repressor. On the other hand, the HindII *lac* fragment inserted into M-13mp1 includes the entire regulatory region of the operon *plus* the first 145 amino acid residues of the β-galactosidase. Consequently, M-13mp1 serves to complement the defect in the coliform cells and allows expression of the functional galactosidase, provided the intrinsic overproduction of repressor can be reduced or offset.

Derepression was readily accomplished by adding IPTG to the soft agar medium in which the cells were grown, and could be demonstrated by the simultaneous addition of a chromophoric galactoside sometimes identified as "X-Gal." This galactoside, developed and used earlier as a cytochemical stain for galactosidase, is more properly known as 5-bromo-4-chloro-3-indolyl-β-D-galactoside. When it is hydrolyzed, it forms the dye, indigo, readily identified by its blue color, as shown in Fig. 11–15. Thus, transformed colonies give rise to blue plaques, whereas untransformed

*Hind II is not currently commerically available. An isoschizomer of this enzyme, Hinc II, has the same intermediate base sequence recognition site.

cells give rise to colorless plaques. One can select blue plaques at a glance and subject them to further analysis, an obvious advantage.

Use of Phage M13 in Rapid DNA-Sequencing Studies

As a new cloning vector, M13mp1 could already be regarded as a success, but further developments made the use of modified M13 recombinants as vectors even more versatile. By deliberate selection of a mutation that changed a G:C pair to an A:T pair at position 13 of the *z* gene sequence, a unique EcoRI site was created in the RF of the modified phage. Insertion of a DNA fragment at this site resulted in a *decreased* complementation, so that now, in the presence of X-Gal, the recombinant plaques would be identified as the colorless ones. These would serve as excellent templates for sequencing studies, provided a primer could be developed that would hybridize with the template at just the right place. To greatly condense a considerable body of work, Messing and others first produced a 92-bp primer fragment by AluI and EcoRI digestion of M13mp2 RF DNA. Even more versatile cloning sites and a number of synthetic primers have been generated; these are commercially identified as mp7, mp8, and so forth. These new vehicles are now being employed for the rapid sequencing, by the dideoxy method, of DNA pieces ranging from 15 to 250 nucleotides in length. Where the unknown fragment inserted into the vector is much longer than 250 nucleotides, the transformed DNA tends to become unstable. In spite of this and some other modest limitations, the entire venture must be regarded as an outstandingly successful example of deliberate phage manipulation.

REGULATION OF SECRETORY PROTEIN BIOSYNTHESIS: HAPTOGLOBIN PROTEIN STUDIES

Figure 11–1 shows that the flow of biological information generally runs from DNA→RNA→protein, and up to now we have focused entirely on the highest level of this pathway, on DNA and its manipulation. The story to be related next addresses a quite different concern. In effect, it asks (and partly answers) the questions: Given a protein, what can we say about the RNA message(s) that coded for it? What can we determine about the nature and the location of controls on the protein biosynthetic process? In other words, can we work backward from an end product to reconstruct the kind of pathway exemplified in Fig. 11–1?

All living things must produce proteins. In the majority of cases the proteins are designed for use within cells, but in other cases one or more proteins are designed for export from the cells of their origin. Typical exported proteins include certain hormones, some toxins, a number of digestive enzymes, the blood plasma proteins, and a variety of binding or carrier proteins (some but not all carrier proteins circulate in the blood plasma). Haptoglobin is an example of the latter class, since it functions in mammals to bind hemoglobin liberated from circulating erythrocytes by trauma or inflammation. Removal of free, circulating hemoglobin is important to avoid kidney damage. The haptoglobin–hemoglobin complex is delivered to the liver, where part of the complex is excreted into the bile and the remainder is degraded by cells of the reticuloendothelial system. The total amount of haptoglobin leaving and entering the liver is about 0.5 g/day under normal circumstances. The actual concentration in the blood plasma ranges from 120 to 260 mg/dl, although active inflammatory processes may cause these figures to double. Clearly, there is a slow but constant production of haptoglobin, subject to considerable increase in the face of disease or injury. What controls the normal production, and what evokes the increase during disease are not clearly understood. Further, because haptoglobin must be exported from the hepatocytes to the circulation, it must share certain features that are typical of exported or secretory protein in general.

Haptoglobin and Hemoglobin: Comparison of Properties

Since haptoglobin serves as a binding protein for free hemoglobin, it is interesting to compare these proteins. Haptoglobin is a heterotetramer; its two α subunits ($M_r \simeq 9{,}500$) and its two glycosylated β subunits ($M_r \simeq 38{,}000$) are joined by interchain disulfide bonds. The total mass of haptoglobin is thus considerably larger than that of its ligand, hemoglobin ($M_r \simeq 64{,}500$). Hemoglobin is also a heterotetramer, but its subunits are held together by noncovalent forces.

It is well known that many mutant hemoglobins exist (over 200 are known), some of which are compatible with reasonably normal health. Shortly after birth there is a normal switch from HbF to HbA, representing a gradual shift in gene expression. So far as is known, there is no comparable shift in haptoglobin species. Although it is clear that different haptoglobin genotypes do exist in the general population, the number of forms found to date is quite small.

It has been established, both in humans and in some other species, that the genes for α and β globins are located on different chromosomes. By means of in vitro protein-synthesizing systems, using methods to be described later, it has also been shown that globin synthesis is inhibited in the absence of heme and that the inhibition can be released by addition of heme. The added heme turns on the globin genes by suppression of initiation inhibitors. The haptoglobin case is quite different. Recent evidence indicates that a single gene codes for the entire haptoglobin structure in the form of a preprohaptoglobin, and that subsequent processing of the primary transcript accounts for the formation of the α and β subunits as well as for the glycosylation of the latter. The demonstration of this pathway forms the basis of this research story. Although the work to be described clearly depends on the high affinity between haptoglobin and hemoglobin, it nevertheless exemplifies a general approach taken when working from a protein product back toward its nucleic acid progenitors.

Fundamentals of the Experimental Design

In the presence of some signal, a gene is transcribed to provide one or more copies of a message. After necessary processing and passage to the cytoplasm, the message may be read out and translated by more than one ribosome at a time. As each ribosome passes from the 5′ end of the message toward the 3′ end, the nascent peptide chains grow in length. Such a collection of ribosomes, all working at once on a single mRNA, is known as a *polysome* or *polyribosome.* Even though the peptides emerging from the individual ribosomes are incomplete, they may begin to fold in some pattern typical of the finished product while still attached to the biosynthetic machinery. In other words, immunologically recognizable epitopes may form before protein synthesis is completed. This would be the basis of a precipitin reaction between a suitable antibody and a polyribosome. By collection of the immunoprecipitate, one can recover the desired message from a mixture of possible messages in the system. After dissociation of the immunoprecipitate and separation of the captured message, the latter is added to a fresh protein-synthesizing system, substantially free of other messages. There, it can once again form polysomes and allow translation of its information in the form of a peptide or a protein.

After demonstrating competence of the message, one can go a step further. By addition of reverse transcriptase and the proper mixture of dNTPs, a single-stranded cDNA can be generated from the message. This cDNA can be hybridized with the original genome. If the hybridization is complete, one can argue that the recovered message sequence is a faithful mirror of the message coded in the DNA. If hybridization is incomplete—that is, if there are lengthy, hairpin loops or open areas—one can argue that some excisions or deletions must have occurred between synthesis of the primary transcript and the finished message. We turn next to the methods by which this design was executed.

Isolation and Purification of Rat Haptoglobin

To increase the concentration of plasma haptoglobin, inflammation was induced in rats by intramuscular injection of turpentine. At 24–48 hours after injection, the rats were bled and the plasma was collected. Haptoglobin was concentrated and purified by passage through a hemoglobin–Sepharose column. After washing, the haptoglobin was eluted with a solution of 6 mol/L of guanidine hydrochloride. Part of the product was retained intact; the remainder was treated with β-mercaptoethanol to liberate the α and β subunits. These were separated by exclusion chromatography using a column packed with agarose A-5M.

Preparation of Anti-Haptoglobin Immunoglobulins (IgG)

Rabbits were separately immunized against intact haptoglobin or isolated α and β subunits by multiple injections of the appropriate protein mixed with Freund's adjuvant. The rabbit sera were collected, and the IgG fraction was purified by affinity chromatography or columns of protein A–Sepharose. (Protein A is produced by the Cowan strain of *Staphylococcus aureus*. Protein A binds to the Fc region of some IgG classes and is useful in separating these from other immunoglobulins.) The anti-α and anti-β IgG species showed no cross-reactivity, and the anti-haptoglobin IgG showed only a single precipitin band when challenged by native haptoglobin or whole rat serum in adjacent wells.

Isolation of Hepatocyte mRNA from Turpentine-Inflamed Rats

Two precautions are important when one is isolating functional (intact) mRNAs. First, the method used should minimize nicking or breakage by physical means; and second, all traces of RNase activity must be removed. (RNase activity can be detected even in fingerprints!) It is important to work at low (0° to −10°C) temperature, to work carefully, and to work speedily.

Inflammation was induced by turpentine injection, as before. At 24–30 hours after injection, the rats were lightly anesthetized and the livers were perfused to remove blood. The livers were then excised and homogenized in a 4 mol/L solution of guanidinium thiocyanate (both ions of which are chaotropic) containing 0.1 mol/L of β-mercaptoethanol plus 0.5% Sarcosyl. The mercaptoethanol served to reduce disulfide bonds of protein in general, and especially of any RNase. The Sarcosyl was added to disrupt membrane structure. After differential centrifugation in sucrose, the total RNA was precipitated by ethanol in the cold. The redissolved pellet was further purified by centrifugation through a cesium chloride gradient to remove residual traces of DNA.

Fully processed mRNA contains a polyA tail. Processed mRNA can therefore be separated from total mRNA by passage through an affinity column packed with poly dT–Sepharose or polyU–cellulose, using buffers of low ionic strength for the binding step and of high ionic strength for elution of the product. In this way, the mature messages can be separated from heteronuclear (hn) RNA and rRNA. Note, however, that this procedure gives a mixture of all polyA-tailed mRNA species.

Preparation of Polysomes and Their Immunoprecipitation

Livers were removed from turpentine-inflamed rats as described above. The livers were quickly minced and homogenized, and a postmitochondrial supernatant was prepared by centrifugation. The polysomes were pelleted at 15,000 × *g*. The pellet was resuspended in an ice-cold HEPES buffer containing NaCl, 5 mmol/L of $MgCl_2$, and heparin (100 μg/ml). The heparin was required to minimize polysomal aggregation. A portion of the mRNA fraction obtained from the poly dT column is then added to this defined ribosomal system. After addition of the appropriate antibody, the mixture was incubated for an hour or more at 0°C, then pelleted by centrifugation. The washed and resuspended pellet was treated with SDS and extracted with phenol and chloroform. The ionic strength was increased by addition of NaCl to a final concentration of 0.1 mol/L, and the desired mRNA was precipitated by ethanol addition in the cold.

In Vitro Protein-Synthesizing Systems: Utilization of mRNA

Protein biosynthesis generally requires, in addition to the message, an energy source, initiation and elongation factors, a mixture of tRNAs and the synthetases that charge them, a mixture of the twenty common amino acids, and a quantity of competent ribosomes. A number of systems for protein biosynthesis have been developed, of which we shall describe two.

The Reticulocyte Lysate System. When rabbits are made anemic by daily intramuscular injection of phenylhydrazine for 5 or 6 days, they respond by releasing to their circulations large numbers of immature erythrocytes. The immature erythrocytes are known as *reticulocytes* because, when properly stained, a meshlike, basophilic network is revealed, running through the cytoplasm. Reticulocytes retain nuclei, mitochondria, ribosomes, and all the ingredients required for protein biosynthesis, and still produce hemoglobin as if they were going to become mature erythrocytes in the bone marrow. One of the disadvantages of this system is the high concentration of globin mRNA it contains, necessitating careful and controlled washing in high salt concentrations, along with controlled nuclease treatment, to remove most of the globin mRNA so that the system becomes dependent on the exogenous mRNA to be studied.

Lightly anesthetized rabbits are bled by cardiac puncture, and the blood is collected in heparinized syringes. The cellular elements are packed by centrifugation in the cold at 10,000 × *g* for 10 min; then the cells are washed three times in a modified saline solution (0.14 mol/L of NaCl, 5 mmol/L of $MgCl_2$, and 50 mmol/L of KCl). The washed cells are lysed by addition of an equal volume of a freshly prepared hypotonic solution (0.1 mmol/L of Na_2EDTA, pH 7.0, plus 1 mmol/L of dithiothreitol) with vigorous swirling. The mixture is again centrifuged in the cold at 10,000 × *g* for 20 min to pack cell debris, and the supernatant fluid is collected. The clear supernatant is the source of ribosomes, and initiation and elongation factors. When stored at −70°C, the supernatant fraction may be kept for a few months without significant loss of activity.

Ribosomes are prepared from the lysate by centrifugation at 95,000 × *g* for 2 hours; the resuspended polysomes are treated with 4 mol/L KCl to give a final concentration of 0.5 mol/L and allowed to stand in an ice bath for 10 min, during which time the contents of the vessel are frequently swirled to promote disaggregation of the polysomes. The free ribosomes are pelleted at 143,000 × *g* for 2 hours and finally resuspended in a solution of sucrose, Na_2EDTA, dithiothreitol, and $MgCl_2$. Contamination by the globin message may be still further reduced by repeating the high-salt wash, but only at the expense of a considerable loss of activity. Free ribosomes prepared in this way can be kept for some time if they are stored in liquid N_2.

A mixture of auxiliary factors (initiation factors, elongation factors, synthetases, etc.) is prepared from the postribosomal supernatant by first lowering the KCl concentration and then treating with DEAE-cellulose in a batch-type process. This step removes some inhibitory substances and an additional part of the globin mRNA. The supernatant from the DEAE-cellulose treatment is concentrated by ultrafiltration to about one-fifth of its original volume, then stored in small aliquots at liquid N_2 temperature. The stability of this mixture is improved by addition of creatine phosphate, ATP, and GTP, as well as by addition of small amounts of hemin. The presence of the energy source mixture and the hemin inhibits formation of inhibitory substances in the concentrated mixture.

To perform an in vitro protein synthesis, one mixes an aliquot of free ribosomes with an aliquot of the auxiliary factors. It is common to add additional creatine phosphate and some creatine kinase (CK), along with a mixture of the amino acids, one of which is radiolabeled at a high specific activity. Synthesis is initiated by additional of mRNA, and the total mixture incubated for an hour or more. For reasons that are not altogether clear, the optimum incubation temperature seems to vary with the mRNA. Similarly, the optimum concentrations of Mg^{2+} and K^{+}, both of which are essential ions, also vary with the mRNA. The reaction is stopped by addition of hot trichloroacetic acid (TCA). Any material insoluble in hot TCA is assumed to include the newly synthesized protein, the radioactivity of which is determined in the usual way. Other aliquots of the product mixture may be taken for immunoassay of the protein formed, or for electrophoresis. Because the actual quantity of protein produced is quite small, it is common to add some simple protein, such as bovine serum albumin, to the incubation mixture before precipitation with hot TCA, thereby increasing the bulk of insoluble material, and minimizing the chances of loss during the washing steps required to remove non-protein-bound radioactivity (see Experiment 21).

The Wheat Germ System. Viable wheat germ must be purchased from commercial sources, usually from a milling company that removes the germ prior to converting wheat to flour. Wheat germ is obviously a much tougher tissue than reticulocytes, and it must be disintegrated carefully by grinding with glass beads in a mortar and pestle, in the cold. Wheat germ, unlike reticulocytes, contains extracted lipid material which must be carefully and completely removed during processing. If the intact wheat germ is extracted with chloroform, the preparation seems to be more sensitive to an added mRNA, although the total activity of the system appears to be somewhat diminished. We describe here a preparation from nonextracted wheat germ, known as the "S-30" system, since it is based on a supernatant from a 30,000 $\times$ g centrifugation.

Equal volumes of wheat germ and acid-washed glass beads are ground at 4°C together with about 5 volumes of a homogenizing buffer (20 mmol/L of HEPES, pH 7.6, 0.1 mol/L of KCl, 1 mmol/L of $MgCl_2$, 2 mmol/L of

$CaCl_2$ and 6 mmol/L of mercaptoethanol) for 3 min. The homogenate is then centrifuged at 30,000 $\times$ g for 10 min. The surface lipid is removed completely and the supernatant is collected carefully to avoid the debris.

The supernatant fluid is then applied to a column of coarse Sephadex G-25, equilibrated with a gel filtration buffer. This is the same as the homogenizing buffer, except that the $CaCl_2$ is omitted. Using a fairly rapid flow rate (5–10 ml/min), the eluate is collected in 3-ml fractions as the tan-colored band approaches the column outlet. The pooled, highly turbid fractions are defined as the S-30 product. The S-30 product is taken up in a large, fine-tipped pipet, then allowed to fall as fine droplets into a stainless-steel beaker half-filled with liquid N_2. The solution freezes into small, spheroidal particles, each with a volume of 30–50 μl. Aliquots of the solid are distributed into small, prechilled tubes and stored at liquid N_2 temperature. Once thawed, they should not be refrozen; thus the dispensed aliquots should be designed to accommodate a single experiment.

Setting up an in vitro synthesis with the S-30 system is entirely comparable to using the reticulocyte lysate system. One needs to provide an energy source, a mixture of amino acids, a quantity of the thawed S-30 pellets, and a message. After incubation, the reaction can be stopped by addition of hot TCA, and thereafter the procedures for washing and counting are much the same.

Comparison of the Lysate and Wheat Germ Systems. The reticulocyte lysate system is expensive, because rabbits are expensive. Furthermore, it is not easy to remove all of the heavily contaminating globin message even when nuclease treatment is employed. As a result, electrophoretic gels of the peptide or protein product almost always show globin subunits along with the desired product(s). Offsetting this problem is the fact that the system is efficient; it can translate fairly large mRNAs, and it will release more than 90% of the peptide chain synthesized.

The wheat germ system begins with less expensive starting materials, and it is more humane in concept. It is easier to remove endogenous mRNA from wheat germ than from reticulocyte lysate, so electrophoretic gels of products tend to be cleaner. Unfortunately, the wheat germ systems as a class do not perform well with mRNAs for proteins larger than about 60,000 daltons, and even some smaller proteins are not readily released from the ribosomes.

Haptoglobin Products In Vivo and In Vitro

Returning to our main theme, the technology just described provides a means of comparing haptoglobin produced by the intact rat or by intact, isolated hepatocytes, with the product of a cell-free system. The protein obtained from intact rats or from isolated hepatocytes has been processed to generate fully mature haptoglobin. The cell-free product would be either *preprohaptoglobin,* the long, single-peptide precursor of haptoglobin, in which the α and β subunits are still covalently joined, together with any signal peptide that promotes export from the cells, but lacking the carbohydrate moiety characteristic of many exported proteins; or *prohaptoglobin,* material from which the signal peptide has been removed and in which core glycosylation has occurred, but where the α and β subunits remain covalently joined in a single long peptide. Experiments have shown that only preprohaptoglobin is obtained in the cell-free system.

The peptide of preprohaptoglobin is longer than can be accounted for by the lengths of the α and β subunits. Cleavage with cyanogen bromide resulted in small peptide fragments that were specifically immunoadsorbed with anti-α IgG or anti-β IgG. Partial sequencing analysis by the automated Edman method showed that the NH_2-terminus of the preprohaptoglobin contained a hydrophobic signal region of eighteen amino acid residues, directly followed by the α subunit. In the intact rat or in intact hepatocytes, cotranslational processing was shown to require two steps: First, the signal peptide region was removed by proteolytic attack in the endoplasmic reticulum. Second, core glycosylation of the β subunit occurred, giving rise to prohaptoglobin, in which the α and β subunits were still covalently linked. Prohaptoglobin then passed to the Golgi apparatus, where final modification of the asparagine-linked carbohydrate moiety was accomplished by sialylation, and where clipping of the long peptide occurred. Polymerization to the tetramer is also presumed to occur in the Golgi, although there is some evidence that some prohaptoglobin can be released to the blood as such, since incubation of prohaptoglobin in rat plasma or serum, or even in the sera of other animals, resulted in conversion of prohaptoglobin to haptoglobin.

Summary of the Haptoglobin Story

The salient features of the biosynthesis of haptoglobin, and of its regulation, can be summarized as follows:

1. Haptoglobin is a tetrameric glycoprotein and a normal component of mammalian blood plasma and serum. Production of haptoglobin can readily be increased by factors that induce inflammation, since inflammation results in erythrocyte breakdown and haptoglobin is required to bind free hemoglobin.
2. The mRNA for haptoglobin can be isolated by exposing homogenates from inflamed hepatocytes to antibodies directed against haptoglobin. These react with polysomes bearing nascent haptoglobin peptides. The collected polysomes can be disaggregated and the isolated mRNA recovered, now free of its congeners.
3. Several things can be done with the isolated message:
   a. It can be used, together with a reverse transcriptase, to produce a DNA, from which a duplex DNA can be formed. The duplex DNA can be inserted into a proper vector, then expressed in some host cell.
   b. It can be used, together with reverse transcriptase, to produce a cDNA, from which a duplex DNA can be formed. The duplex DNA

can be inserted into a proper vector, then expressed in some host cell.

c. The mRNA can be hybridized with hepatocyte DNA to search for introns and exons in the genome itself.

4. The primary transcript can be compared with the mature, fully processed haptoglobin to detect and characterize signal sequences, deletions, and other modifications such as glycosylation of one or both subunits.
5. Finally, this example shows that it is possible to reverse the usual direction of information flow; that is, one may regress from a protein to its origin in the nucleic acids, by combining classical biochemical techniques with advances in molecular biology.

SOLID-STATE SYNTHESIS OF OLIGONUCLEOTIDES

A major goal of chemistry is to demonstrate that a synthetic molecular species has the same properties (including biological activities) as some natural substance, thus closing the circle of structure proof. Furthermore, skillful chemists have sometimes been able to improve upon natural substances by creating analogs with more desirable properties, from one point of view or another. It is not surprising that a considerable research effort has been mounted to synthesize polynucleotides with specified based sequences. Even though it is not yet practical to synthesize the long nucleic acids typically specifying natural products, it is feasible to synthesize oligonucleotides of eighty or more bases in length. These short sequences have many uses. They may serve as primers for enzymic syntheses, or they may serve as probes for regions of complementarity in natural nucleic acids; they also may be used to modify existing nucleic acids by a variety of cloning procedures, or they may be joined to still other synthetic oligonucleotides to form longer and more interesting sequences. It was by just such a combination of synthetic and enzymic methods that Khorana's group determined the total sequence of the ϕX174 genome. This was regarded as the first synthetic gene and was the basis for awarding the Nobel Prize to Khorana. The methods then in use were classical, being based on reactions carried out in free solution. The yields were not always as high as might be desired, the manipulations were tedious, and it was necessary to purify carefully a great many intermediates.

The free solution methods were modified by covalent attachment of the first nucleoside in a desired sequence to the surface of some inert matrix. Polyacrylamide and silica have been used as matrices, with silica enjoying a slight edge in popularity. In subsequent steps, a second nucleoside, properly protected to block undesired side reactions, is coupled to the first by some analog of phosphoric acid, to form the $5' \rightarrow 3'$ phosphodiester bond. Additional nucleosides are added by repeating these steps; then the final product is cleaved from the matrix and the protecting groups are removed.

The logic is much the same whether one is working with deoxyribonucleosides or ribonucleosides; only the details differ.

There are some powerful advantages to solid-state syntheses, compared to syntheses in free solution. Immobilization means that all reaction intermediates are insoluble. They can be washed free of impurities and reagents by simple filtration; recrystallization is unnecessary. Because manipulative losses are reduced, the overall yields tend to be higher. Virtually the entire sequence of reactions is performed in a single container. This approach lends itself well to automation, and a number of machines are marketed for this purpose. One can now purchase properly blocked nucleosides already affixed to a matrix. The material is packed into a small reaction vessel that need only be mounted in the synthesizer to begin a programed production of a specified oligonucleotide. As a result of development and marketing of newer and more powerful condensing reagents, the total time required has been sharply reduced, and it is now possible to prepare a sequence of twenty bases (commonly known as a *20mer*) in about as many hours.

The exact details of the chemical procedures have been the subject of a number of recent reviews, several of which are cited in the references at the end of this chapter. We shall describe the procedures only in broad outline, and for simplicity deal only with oligodeoxyribonucleotides to avoid the complications attendant in proper blocking of the 2′-OH of ribonucleosides.

Preparation of Silica-Based Matrices

When fine particles of silica are briefly boiled with strong HCl, some of the bridge oxygen atoms at the surface of the silica are converted to free OH groups. The silica can then be refluxed with a toluene solution of γ-aminopropyltriethoxysilane, forming a covalent adduct with a free NH_2-terminus. Refluxing this product with succinic anhydride yields a hemiamide, and the product now has a free COOH-terminus. These details are shown in Fig. 11–16. The structure of the final adduct is shown, but it is easier to refer to it later on as "M⁓COOH."

Preparation of Blocked Nucleosides

The structure shown in the upper left of Fig. 11–17 represents a blocked nucleoside; that is, B represents any of the purine or pyrimidine bases encountered in DNA, suitably protected by acetylation, alkylation, or benzoylation at -OH or -NH_2 groups (other blocking reagents are also employed). It is necessary to block the 5′-OH prior to attachment of the nucleoside to M⁓COOH. A reagent commonly used for this purpose is dimethoxytrityl chloride (some workers use the monomethoxy compound). The prepared nucleoside is then condensed with the matrix by a carbodiimide condensation. The washed and dried condensate is then pre-

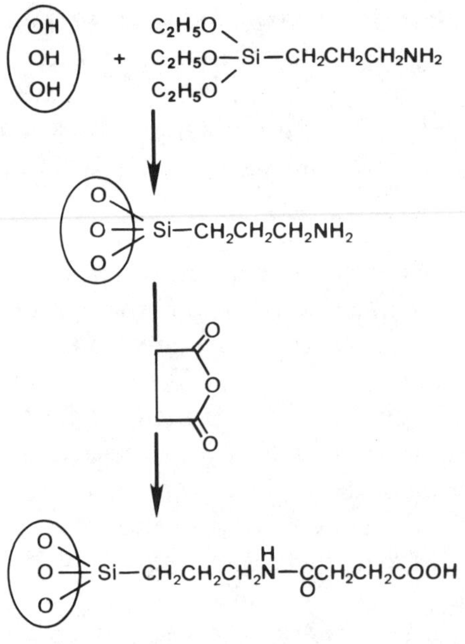

Figure 11–16. Preparation of a modified silica matrix for oligonucleotide synthesis. Hydrated silica gel particles (depicted in oval) are refluxed with γ-aminopropyltriethoxysilane. Ethanol is eliminated, and the amino side chain becomes covalently attached. This intermediate is then refluxed with succinic anhydride, forming the hemiamide shown at the bottom of the figure. The free carboxyl group is the point at which the first nucleoside in the desired sequence will be attached, by a carbodiimide condensation. The similarity of this procedure to those used in preparation of affinity packings is obvious. With care, it is possible to bind 100 μeq of binding sites per gram of silica gel.

pared for chain elongation by mild acid treatment to bring about detritylation.

The Choice of a Condensing Reagent

There are essentially two protocols used to generate the phosphodiester bond: the *modified triester approach* and the *phosphite triester approach.* The first method uses various substituted phosphochloramid*ates,* in which the phosphorus is trivalent since the compounds are derivatives of phosphonic acid. The phosphochloramid*ites* have come into use since Letsinger demonstrated that they react more rapidly and perhaps more completely than the phosphochloramidates. The phosphonodiesters formed by the condensation can be neatly oxidized to phosphodiesters by treatment with aqueous iodine solutions, as shown in Fig. 11–18. The nascent dinucleotide is then washed, dried, and detritylated in preparation for the next condensation step.

Cleavage of the Product from the Matrix

Cleavage of the oligonucleotide from the matrix is commonly done by heating with NH_4OH, although oximate ion has also been used for this purpose. Two problems can arise at this point; under the conditions of hydrolysis there may be some depurination, and destruction of some pyrimidines may also occur. For reasons that are not well understood, the likelihood of damage appears to increase with chain length.

Deblocking the Bases

The last step in the oligonucleotide synthesis is to remove the protecting or blocking groups that had to be added to the -OH and -NH_2 groups of the bases themselves. Good protecting groups are those that can be removed under conditions sufficiently mild to avoid damage to the structure of the nucleotide chain itself. Base-labile and acid-labile groups have been used, and the literature contains examples of rather sophisticated reagents. Newer solutions to the problem of better blocking reagents are being actively pursued in the hope of minimizing damage or rearrangements in the product nucleotides.

Possible Explanations for the Present Limitations on Chain Length

In theory, there is no reason why the present technology should not allow synthesis of much longer polynucleotides than can currently be achieved.

Figure 11–17. Binding of the blocked first nucleoside to the matrix. The 5′-OH is blocked by treatment with dimethoxytrityl (DMTr) chloride, as shown in the first reaction. Since the 3′-OH is now the only site of reaction, addition of dicyclohexyl carbodiimide (DCCD) to a mixture of the blocked nucleoside and the prepared matrix, M~COOH, splits out water and results in covalent binding of the nucleoside to the matrix, as shown in the second reaction. In preparation for the addition of a second nucleoside, the 5′-OH is detritylated by treatment with a mild acid.

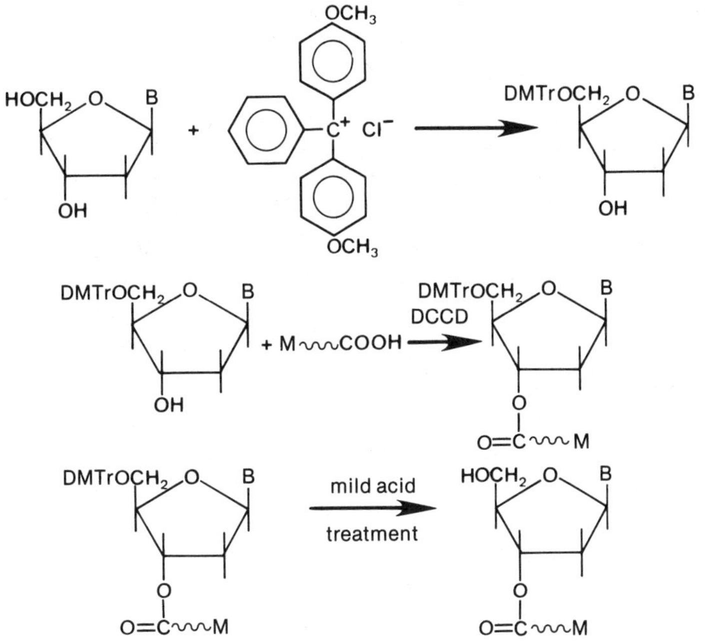

HOCH₂ O B / OH + (OCH₃-phenyl)₂(phenyl)C⁺ Cl⁻ → DMTrOCH₂ O B / OH

DMTrOCH₂ O B / OH + M~COOH —DCCD→ DMTrOCH₂ O B / O / O=C~M

DMTrOCH₂ O B / O / O=C~M —mild acid treatment→ HOCH₂ O B / O / O=C~M

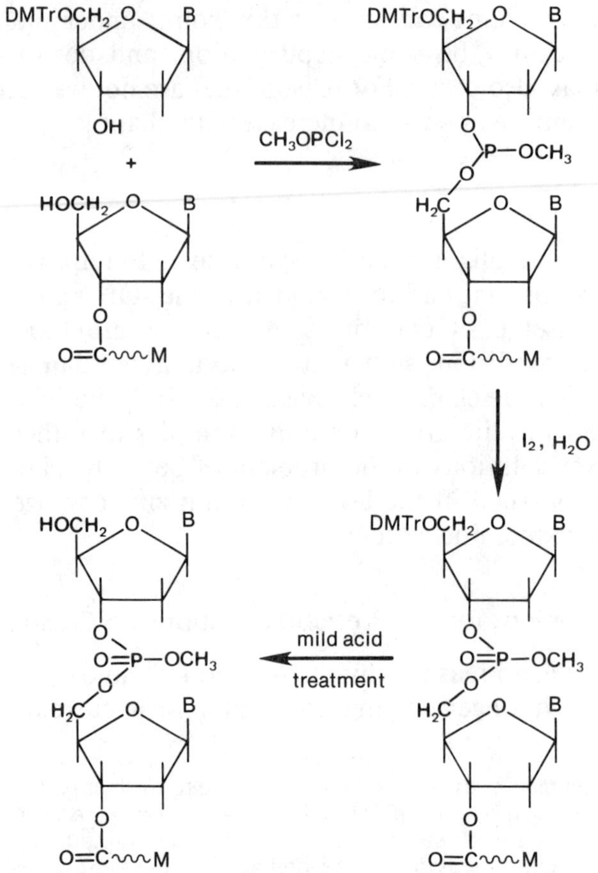

Figure 11–18. Coupling of blocked nucleosides by methyl dichlorophosphite. After coupling, the diphosphite ester bond is oxidized to a phosphodiester bond by aqueous iodine. Detritylation of the 5′-OH of the second nucleoside prepares the immobilized preparation for the next stage of condensation, and the cycle may be repeated to generate longer chains.

As noted above, the practical working limit is 50 to 100 bases in length. Even though the newer condensing agents are powerful and act quickly, some aberrant reaction products accumulate at each step. The individual reactions do not function with perfect efficiency. In the case of ribonucleotides particularly, unwanted rearrangements can occur. All of these problems are additive and increase with the number of repeated cycles in the synthesizer. After the chain has been synthesized, the scission from the matrix and removal of protecting groups further increase the likelihood of damage, as already noted. However, all of these problems are assumed to be technical and not a matter of theory. Continued refinements will, no doubt, resolve some of these problems. A kind of bootstrapping has been introduced, in which the synthesizer is fed not with single nucleosides, but with dinucleosides prepared externally. Although this ingenious technique allows a modest increase in realizable chain lengths, it does not address the problems associated with removal of product from the matrix or removal of the blocking groups. The use of glass beads with larger pore sizes, as well as more recent modifications of the chemical intermediates used (see Gait, 1984), have made possible the synthesis of oliognucleotides of up to 100 bases in length; even more improvements are being investigated.

REFERENCES

General Texts and Reviews

Abelson, J. RNA processing and the intervening sequence problem. *Annu. Rev. Biochem. 48*:1035–1069 (1979).

Alberts, B., Bray, D., Lewis, J., Raff, M., Roberts, K., and Watson, J. D. *Molecular Biology of the Cell.* Garland Publishing, New York (1983). [See especially Chaps. 8, 11, and 17.]

Denhardt, D. T. A comparison of the isometric and filamentous phages. In *The Single-Stranded DNA Phages* (D. T. Denhardt, D. Dressler, and D. S. Ray, eds.). Cold Spring Harbor Laboratory, Cold Spring Harbor, NY (1978).

Kornberg, A. *DNA Replication.* W. H. Freeman, San Francisco (1980).

Kreil, G. Transfer of proteins across membranes. *Annu. Rev. Biochem. 50:*317–348 (1981).

Lewin, B. *Genes.* John Wiley and Sons, New York (1983).

Nevins, J. R. Pathway of eukaryotic mRNA formation. *Annu. Rev. Biochem. 52:*441–446 (1983).

Presper, K. A., and Heath, E. C. Assembly, transfer and processing of carbohydrate side-chains of glycoproteins. In *The Enzymes* (H. Neurath, ed.), *16:*450–488 (1983).

Watson, J. D., Hopkins, N. H., Roberts, J. W., Steitz, J. A., and Weiner, A. M. *Molecular Biology of the Gene,* Vols. 1 and 2, 4th ed. Benjamin/Cummings, Menlo Park, CA (1987).

Watson, J. D., Tooze, J., and Kurtz, D. T. *Recombinant DNA: A Short Course.* W. H. Freeman, New York (1983).

Wu, R. DNA sequence analysis. *Annu. Rev. Biochem. 47:*608–631 (1978).

Yuan, R. Structure and mechanism of multifunctional restriction endonucleases. *Annu. Rev. Biochem. 50:*285–315 (1981).

Zubay, G. *Biochemistry.* Addison-Wesley Publishing, Reading, MA (1983). [See especially Chaps. 18–26.]

Laboratory Manuals

Davis, L. G., Ribner, M. D., and Battey, J. F. *Basic Methods in Molecular Biology.* Elsevier, New York (1986).

Hackett, P. B., Fuchs, J. A., and Messing, J. W. Introduction to *Recombinant DNA Techniques,* Benjamin/Cummings, Menlo Park, CA (1984).

Maniatis, T., Fritsch, E. F., and Sembrook, J. *Molecular Cloning.* Cold Spring Harbor Laboratory, Cold Spring Harbor, NY (1982).

Schleif, R. F., and Wensink, P. C. *Practical Methods in Molecular Biology.* Springer Verlag, New York (1981).

Plasmids and Genetic Recombination

Cohen, S. N., Chang, A.C.Y., and Hsu, L. Non-chromosomal antibiotic resistance in bacteria: genetic transformation of *E. coli,* by R-factor DNA. *Proc. Natl. Acad. Sci. U.S.A. 69:*2110–2114 (1972).

Sharp, P. A., Sugden, B., and Sambrook, J. Detection of two restriction endonuclease activities in *H. parainfluenzae* using analytical agarose–ethidium bromide electrophoresis. *Biochemistry 12:*3055–3063 (1973).

Sutcliffe, J. G. Nucleotide sequence of the ampicillin resistance gene of the *E. coli* plasmid, pBR322. *Proc. Natl. Acad. Sci. U.S.A. 75:*3737–3741 (1978).

Sutcliffe, J. G. pBR322 restriction map derived from the DNA sequence: accurate DNA size markers up to 4361 nucleotide pairs long. *Nucl. Acids Res. 5:*2721–2728 (1978).

Transfer of DNA

Caplan, A., Herrera-Estrella, D., Inze, D., Van Haute, E., Van Montagu, M., Schell, J., and Zambryski, P. Introduction of genetic material into plant cells. *Science 222:*815–821 (1983).

Chu, G., and Sharp, P. A. SV40 DNA transfection of cells in suspension: analysis of the efficiency of transcription and translation of T-antigen. *Gene 13:*197–202 (1981).

Graham, F. L., and Van der Eb, A. J. New technique for the assay of infectivity of the human adenovirus 5. *Virology 52:*456–467 (1973).

Mandel, M., and Higa, A. Calcium-dependent bacteriophage DNA infection. *J. Mol. Biol. 53:*159–162 (1970).

Gene Expression and Regulation

Dickson, R. C., Abelson, J., Barnes, W. M., and Reznikoff, W. S. Genetic regulation: the *lac* control region. *Science 187:*27–35 (1975).

Maniatis, T., Ptashne, M., Barrell, B. G., and Donelson, J. Sequence of a repressor binding site in the DNA of bacteriophage λ. *Nature 250:*394–397 (1974).

Oi, V. T., Morrison, S. L., Herzenberg, L. A., and Berg, P. Immunoglobulin gene expression in transformed lymphoid cells. *Proc. Natl. Acad. Sci. U.S.A. 80:*825–829 (1983).

Palmiter, R. D., Norstedt, G., Gelinas, R. E., Hammer, R. E., and Brinster, R. L. Metallothionine–human GH fusion genes stimulate growth of mice. *Science 222:*809–814 (1983).

Rice, D., and Baltimore, D., Regulated expression of an immunoglobulin κ Gene introduced into a mouse lymphoid cell line. *Proc. Natl. Acad. Sci. U.S.A. 79:*7862–7865 (1982).

Weil, P. A., Luse, D. S., Segall, S., and Roeder, R. G. Selective and accurate initiation of transcription at the Ad2 major late promoter in a soluble system dependent on purified RNA polymerase II and DNA. *Cell 18:*470–483 (1979).

Field-Induced Cell Fusion

Bischoff, R., Eisart, R. M., Schedel, I., Vienken, J., and Zimmerman, U. Human hybridoma cells produced by electro-fusion. *FEBS Lett. 147:*64–69 (1982).

Halfmann, J. H., Emeis, C. C., and Zimmerman, U. Electrofusion of haploid Saccharomyces yeast cells of identical mating type. *Arch. Microbiol. 134:*1–4 (1983).

Zimmerman, U. Electric field-mediated fusion and related electrical phenomena. *Biochem. Biophys. Acta 694:*227–284 (1982).

DNA Sequencing

Maxam, A. M., and Gilbert, W. A new method for sequencing DNA. *Proc. Natl. Acad. Sci. U.S.A. 74:*560–564 (1977).

Maxam, A. M., and Gilbert, W. Sequencing end-labeled DNA with base-specific chemical cleavages. *Methods Enzymol. 65:*499–560 (1980).

Sanger, F., Donelson, J. E., Coulson, A. R., Kössel, H., and Fischer, D. Use of DNA Polymerase I primed by a synthetic oligonucleotide to determine a nucleotide sequence in phage f1 DNA. *Proc. Natl. Acad. Sci. U.S.A. 70:*1209–1213 (1973).

Sanger, F., and Coulson, A. R. Rapid method for determining sequences in DNA by primed synthesis with DNA polymerase. *J. Mol. Biol. 94:*441–448 (1975).

The M13 Cloning System

Gronenborn, B., and Messing, J. Methylation of Single-Stranded DNA in vitro introduces new restriction endonuclease cleavage sites. *Nature 272*:375–377 (1978).

Heidecker, G., Messing, J., and Gronenborn, B. A versatile primer for DNA sequencing in the M-13mp2 cloning system, *Gene 10*:69–73 (1980).

Messing, J., Gronenborn, B., Müller-Hill, B., and Hopchneidor, P. H. Filamentous coliphage M-13 as a cloning vehicle: Insertion of a HindII fragment of the *lac* regulatory region in M-13 replicative form in vitro. *Proc. Natl. Acad. Sci. U.S.A. 74*:3642–3646 (1977).

Messing, J., Crea, R., and Seeberg, P. A system for shotgun DNA sequencing. *Nucl. Acids Res. 9*:309–321 (1981).

DNA Recombination in the Immune System

Adams, J. M. Organization and expression of immunoglobulin genes. *Immunol. Today 1*:10–17 (1980).

Early, P., Huang, H., Davis, M., Calame, K., and Hood, L. An immunoglobulin heavy chain variable region gene is generated from three segments of DNA: VH, D and JH. *Cell 19*:981–992 (1980).

Early, P., Rogers, J., Davis, M., Calame, K., Bond, M., Wall, R., and Hood, L. Two mRNAs can be produced from a single immunoglobulin μ by alternative RNA processing pathways. *Cell 20*:313–319 (1980).

Jerne, N. K. The immune system. *Sci. Am. 229*:52–60 (1973).

Leder, P. Genetics of antibody diversity. *Sci. Am. 246*:102–115 (1982).

Cell-Free Translation of mRNA

Hanley, J. M., Haugen, T. H., and Heath, E. C. Biosynthesis and processing of rat haptoglobin. *J. Biol. Chem. 258*:7858–7869 (1983).

Haugen, T. H., Hanley, J., and Heath, E. C. Haptoglobin, a novel mode of biosynthesis of a liver secretory glycoprotein. *J. Biol. Chem. 256*:1055–1057 (1981).

Pestka, S. Purification and manufacture of human interferons. *Sci. Am. 249*:37–43 (1983).

Ricciardi, R. P., Miller, J. S., and Roberts, S. E. Purification and mapping of specific mRNAs by hybridization–selection and cell-free translation. *Proc. Natl. Acad. Sci. U.S.A. 76*:4927–4931 (1979).

Chemical Synthesis of Oligonucleotides

Alvarado-Urbina, G., Sathe, G. M., Liu, W.-C., Gillan, M. F., Duck, P. D., Bender, R., and Ogilvie, K. K. Automated synthesis of gene fragments. *Science 214*:270–274 (1981).

Gait, M. J., ed. *Oligonucleotide Synthesis: A Practical Approach.* IRL Press, Oxford (1984).

Itakura, K., Rossi, J. J., and Wallace, R. B. Synthesis and use of oligonucleotides. *Annu. Rev. Biochem. 53*:323–356 (1984).

Narang, S. A. DNA Synthesis, *Tetrahedron 39*:3–22 (1983).

Reese, C. B. Chemical synthesis of oligo- and polynucleotides by the phosphotriester approach. *Tetrahedron 34*:3143–3179 (1978).

CHAPTER TWELVE

Practical Data Analysis by Elementary Statistics

Experimental data are always associated with an element of uncertainty. For example, we might measure the mass of a piece of brass and write down the following for repeated observations of the measurement: 1.000 g, 1.002 g, 1.007 g, and so on. A conscientious investigator, recognizing the fallibility of measurements, writes down the average of these values. An equally careful investigator might strike out the value, 1.007 g, on the grounds that it lay so far from the others as to be suspect of very gross error. Another investigator, although not certain, might write down the first of the measured values, perhaps because that was the value stamped on the piece. The intrinsic problem has nothing to do with balance sensitivity, weight calibration, or similar factors; it deals only with what we might call the *truth content* of any observation. In this connection, it might be pointed out that the only mass known with absolute certainty is the mass of the International Prototype; all other measures of mass are at best approximations, the truth content of which is not predicated by the observation itself. Although this example may seem trivial, it is crucial to the nature of observations and of hypotheses derived from them. A precise expression of this axion is given by the statement

$$O = O_0 + E \tag{12-1}$$

where O = any measured observation, O_0 = the accepted value, and E is the absolute value of the error, or uncertainty. Since $O_0 < O_0+E$, the truth content of $O < 1$; in general, the truth content will depend on the relative magnitudes of O_0 and E. Because observations may be used in many ways to formulate hypotheses, it is possible to accept a false hypothesis or to reject a true one. We wish to minimize both of these undesirable possibilities.

Statistics may be described as the quantitative study of mathematical probabilities as a basis for generalization from a given set of data. Application of statistical principles to a limited data set enables the investigator to extract the maximum amount of information at some predetermined level of reliability, thereby allowing the investigator to infer the "true" value which could be measured only with infinite resources of time and material. The structural elements of statistics are variables. If one has in hand *all possible values* of that variable, one has a *statistical population,* or *universe.* If one has less than all possible values, or has a limited *data set,* then one has a *statistical sample.* Statistical measures, such as the mean, the variance, or the standard deviation, are calculated from the data set(s), and statistical operations involve the examination and interpretation of such measures, according to the laws of mathematical probability, in support of a given hypothesis or hypotheses. A basic statistical hypothesis is known as the *null hypothesis.* One compares the calculated characteristics of a given set of variables with those of a known or theoretical set of variables. Tentatively, one acts on the hypothesis that the characteristics of the two sets are the same, so the difference between them is null (or zero). Acceptance or rejection of the hypothesis is conditioned by the details of the comparison. When the comparison reveals a character difference other than zero, the quantitative probability that the difference is due to chance can be determined. Applications of the null hypothesis constitute a major part of statistical methodology.

PROBABILITY FUNDAMENTALS

The variables most important in statistics are random variables, which means only that any numerical value of the variable may occur with an independent likelihood. In tossing a coin, it is clear that only a head or a tail can result, but we have no way of foretelling which will appear in a given toss. This is truly a random variable. In more general instances, the probabilities of specific outcomes are not equal, but they can be calculated by application of some basic definitions and rules. We shall now review some of these.

Theorem I: If an event can occur in N ways, each of which is equally likely, and if m ways among these are in some sense successful, then the probability of success is

$$p = m/N \tag{12–2}$$

Theorem II: If an event is certain to happen, then the probability of its happening must be 1, since any happening is by definition a favorable happening. If any event cannot happen, its probability must be zero, for no happening could be favorable from the definition.

Theorem III: If the probability of success in any case is p, then the probability of failure, q, must be given by

$$q = 1 - p \tag{12–3}$$

Theorem IV: If one thing can be done in m ways and another thing can be done in n ways, then both things can be done together or in succession in mn ways. It follows that if the probability of one event is P_1 and of the other is P_2, then the probability of both taken together or in succession is

P_1P_2. This formulation can be extended to any number of probabilities, as $P_1P_2P_3P_4 \ldots P_n$.

Theorem V: The number of arrangements of n objects taken r at a time $(n > r)$ is given by the expression

$$_nC_r = n!r!\,(n - r)! \tag{12–4}$$

where $n! = n(n - 1)(n - 2)(n - 3) \ldots 1$, etc.

Theorem VI: If the probability of an event in a single trial is p, then the probability of that event occurring *exactly* r times in n trials is given by the binomial law as

$$pr = {_nC_r}p^r(1 - p)^{n-r} = n!p^{r4}q^{n-r}/[r!(n - r)!] \tag{12–5}$$

Theorem VII: If the probability of an event in a single trial is p, then the probability of that event happening *at least* r times in n trials is given by the binomial law as

$$\tfrac{1}{2}r' = p^n_n C_1 p^{n-1}q + {_nC_2}p^{n-2}q^2 + \cdots + {_nC_{n-r}}p^r q^{n-r} \tag{12–6}$$

Theorem VIII: If the probability of an event in a single trial is p and if n trials are made, then the most probable number of successes is given by the greatest integer less than $p(n + 1)$. Since p is less than 1, the most probable number of successes is approximately np. This is the "most expected" number of successes.

We can illustrate these theorems by some simple examples.

Example 1. A coin and a die are tossed together. What is the chance that a head and a 4-spot will appear together? The probability of a head is 1/2 and the probability of a 4-spot is 1/6. Taken together, the probability is (1/2)(1/6) = 1/12.

Example 2. What is the probability that the ace will appear at least once in n throws of a die? The probability that the ace will *not* appear in n successive throws of the die is $(5/6)^n$, so the probability that it will appear at least once is given by $1 - (5/6)^n$.

Example 3. Let a beaker contain 20 red stoppers and 30 black stoppers, well mixed. What is the probability that (a) a red and a black stopper are drawn in succession, (b) a black and a red stopper are drawn in succession, and (c) three black stoppers are drawn in succession? There are 50 stoppers altogether. The probability asked for in (a) is given by theorem IV as p = (20/50) (30/48) = 12/49. This is identical with the probability requested in (b). In case (c), the probability is given by (30/50) (29/49) (28/48) = 29/140.

Example 4. A certain glucose-containing sample is known to give only a faintly positive test with Benedict's reagent. The chance is only 1/4 that analyst A will detect the glucose, but it is 2/3 that glucose will be found by analyst B. How are the prospects of a correct report improved by having both analysts test this sample? By theorem III, the chance that analyst A will *not* detect the glucose is 3/4; the chance that analyst B will *not* detect

Table 12–1. Probability of *n* Black Stoppers in Five Trials

n	p_n
0	0.0778
1	0.2592
2	0.3456*
3	0.2304
4	0.0768
5	0.0102

the glucose is 1/3. The chance that both analyst A and analyst B will not detect the glucose, according to theorem IV, is 1/4. Since the report will be positive unless both analysts miss the glucose, the prospect of a proper report is $1 - 1/4 = 3/4$.

Example 5. A beaker contains 4 red stoppers, 10 black stoppers, and 3 amber stoppers. What is the probability of drawing 4 black stoppers in 4 withdrawals from the beaker? By theorem V, the number of ways in which 4 black stoppers can be taken from the 10 present is ${}_{10}C_4$. Likewise, the number of ways 4 stoppers can be selected from the total of 17 present is ${}_{17}C_4$. The probability of the desired result is given by theorem I as ${}_{10}C_4/{}_{17}C_4$ and is equal to 3/34.

Example 6. What is the chance that the ace will appear exactly 4 times in 10 throws of a die? The chance of an ace in any given throw is 1/6. Recourse to theorem VII gives $p = {}_{10}C_4(1/6)^4(5/6)^6 = 0.0108$, or about 1 in 100.

Example 7. Consider a beaker containing 3 red and 2 black stoppers. Let a single stopper be withdrawn, its color noted, and the stopper replaced. If five drawings are made, what are the probabilities of drawing 0, 1, 2, 3, 4, or 5 black stoppers in the five successive trials? By reference to theorem VI, the probability that none of the stoppers is black can be calculated as $p_0 = {}_5C_0(2/5)^0(3/5)^5$. Similar calculations can be made for the remaining cases. Note also, in accord with theorem VIII, the "most expected" value should be $(n)(p)$. The probability of drawing a black stopper in any single trial is 0.4, so that $(n)(p) = (5)(0.4) = 2$. If one calculates all of the required probabilities according to theorem VI and tabulates the results, the data would resemble the display of Table 12–1. The maximum probability, marked by an asterisk in Table 12–1, occurs for $n = 2$, in perfect accord with the prediction obtained by theorem VIII.

Calcuation of items like $n!$ may sometimes pose a problem, although even the more inexpensive little hand calculators contain built-in routines for value of $n!$ up to about 69. Beyond that, the value of $n!$ exceeds the capacity of the usual eight-place registers of small calculators. Stirling developed a useful approximation for $n!$, which states that

$$n! = n^n e^{-n}\sqrt{2\pi n} \tag{12–7}$$

This equation gives values which are too low when $n < 15$, but the error is ordinarily tolerable.

DISTRIBUTION FUNCTIONS AND THE NORMAL ERROR CURVE

By extension of the ideas presented in example 7, above, one could calculate precise probability values over a large range of some variable. These probabilities constitute a description of the distribution of a random variable. A graph of the individual probabilities, plotted against n or some related characteristic, provides a picture of the distribution. Distribution curves, of course, may also be given in terms of some analytic function, just as a straight line has a characteristic function to describe it. By application of the binomial law to the general case, it is possible to formulate an expression for the distribution function of a theoretical random variable when n approaches infinity. Development of this expression will not be attempted here, but the function will simply be stated as

$$f(x) = \frac{1}{\sigma\sqrt{2\pi}} e^{-1/2\cdot(x - \mu)^2/\sigma^2} = \frac{1}{\sigma\sqrt{2\pi}} e^{-1/2(z^2)} \qquad (12\text{–}8)$$

where μ, σ, and z have a significance which will be discussed shortly. Equation 12–8 is often called the *normal error function* or the *Gaussian error function,* and one sometimes sees it abbreviated as *erf.* A plot of the error function has the appearance shown in Fig. 12–1. The power of the normal error function is that it describes quite well the distribution of a large number of variables. However, a word of caution is necessary. Many well-known variables do not obey the normal error law; instead they obey other distribution laws, including Poisson's law, or the log-normal law, etc. We shall confine ourselves here to cases that obey the Gaussian or normal error law.

The curve shown in Fig. 12–1 has several important properties. First, it is characterized completely by only two parameters, μ and σ. The parameter μ is a measure of the *central tendency* or mean of the values over which the variables may range. The parameter σ is a measure of the *deviation with respect to the mean* over which the variable may range.

STATISTICAL POPULATIONS AND STATISTICAL SAMPLES

The term *statistical population* refers to the totality of any set of quantities or objects that have some measurable characteristic in common. The size of a population may be fairly small or infinitely large. The key concept is that a population must contain all of the possible entities that compose the population. A *statistical sample* is any subset of a population. In day-to-day work in the laboratory, one deals with samples rather than with populations; the great advantage of statistical analysis is that it permits

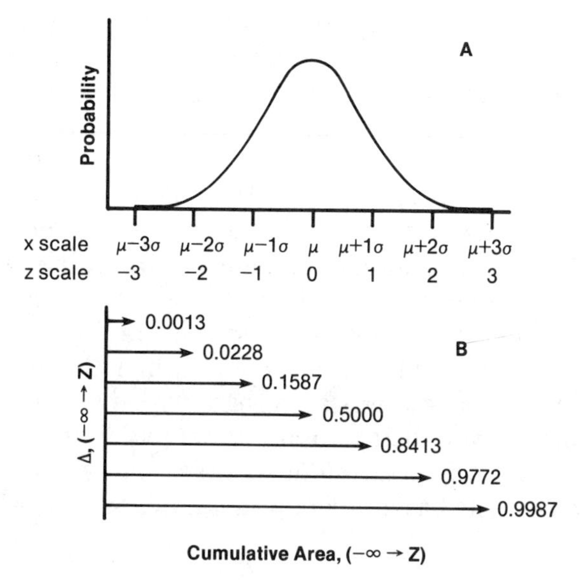

Figure 12–1. The Gaussian, or normal, error curve. (A) The curve describes the distribution of a theoretical random variable which ranges from zero to ∞ in either direction. The value of the random variable can be expressed on either of two scales. The x scale is a function of the population mean, μ, and the standard deviation, σ The z scale unit is $(x - \mu)/\sigma$. For any value of x, the probability of its occurrence is given by the corresponding value of the ordinate of the curve. Hence the total area under the curve is equal to unity, the sum of the probabilities of all possible values of x. (B) The distribution of the area under the normal curve is uniquely determined by μ and σ. The horizontal arrows, all beginning at $-\infty$ (on the left), show how the area under the curve is subtended by increasing values of x or z. Thus, all of the possible values of z ranging from $-\infty$ up to $z = -2$ together make up less than 3% of the area under the curve. For further details of the x and z scales, see the text.

one to infer, from data gathered on one or more small samples, that the data are in fact samples from the same population, or not. It is also possible to determine that the sample obeys the Gaussian error function, or not.

One can, of course, calculate a mean and a standard deviation for any set of numbers. If, when tested, the distribution of the numbers obeys the Gaussian error function, one can proceed according to the principles outlined here. If not, then a different body of statistical rules must be applied, or the data must be so transformed that they will obey the Gaussian law.

Is a Given Sample Distribution Normal?

As noted earlier, the graph in Fig. 12–1A shows the distribution of the Gaussian function. The value of the ordinate represents the probability of the abscissal coordinate. When the abscissa is expressed on the x scale, in terms of μ and σ, the most probable value is μ. When the abscissa is expressed in the z scale, the most probable value is 0. The choice between the x and z scales is largely a matter of convenience, since they inherently express the same thing. The probability of occurrence of values far removed from $x = \mu$ or $z = 0$ falls off rapidly in either direction. In simple terms, supposing that values of x different from μ may indicate some sort of error, small errors are more probable than very large errors.

Figure 12–1B shows the *cumulative probability* of z values ranging from $-\infty$ to $z = -3, -2, -1$, etc. Note that the right-hand ends of the arrows lie directly under the indicated z values of the z scale in Fig. 12–1A. Note also that if one lays a straight-edge along the arrow points, a single straight line would connect all of them. This property is characteristic of any normal distribution, and it is the basis of a special graph paper known as *normal probability paper*. The latter is paper on which the ordinate is a probability scale, with a linear abscissa. Probability paper is useful for quickly determining that a given distribution follows the normal error function.

Imagine a statistical population of six objects, lettered a through f, and that each of the objects has a different mass, in grams, as shown below.

a	b	c	d	e	f
13	18	23	15	19	17

For this population, the average or mean, μ, = 17.5, the standard deviation, σ, = 3.15 and the variance, σ_2, = 9.92.

One can also select samples of three out of the six objects of the population, and make similar determinations of the sample means, x, the sample standard deviations, s, and the sample variances, s^2. In choosing samples, one seeks only unique arrangements, so that none of the samples duplicates any other. According to theorem V, there are ${}_nC_r = {}_6C_3 = 6!/(3!)(3!) = 720/36 = 20$ unique samples.

Work Sheet No. 1 shows the data for this problem. The number of samples listed, as shown in the first column, satisfies theorem V. The sample codes, shown in the second column, demonstrate that each sample is unique. The calculated sample means, standard deviations, and variances are shown in the next successive columns. Based on these data, a z value for each sample can be calculated. The sample means are obviously not all the same, so they are next ranked in order of magnitude, the smallest being given a rank of 1, and so on, up through 20. The last column on the work sheet lists the cumulative probability for each sample. The reasoning goes as follows: There are only 20 possible values in the distribution of sample means. Each has an intrinsic probability of 0.05, but the cumulation of these would cover increasing areas under the normal curve, from $0.05 \rightarrow 1.00$. It is this value that locates each of the 20 collected means on the ordinate scale of the probability paper. The associated z value will serve as the coordinate on the z scale. The resulting probability plot is shown in Fig. 12–2.

Several comments can be made about the data in the work sheet.

1. Although the samples are unique, the observed sample means are not. Note that four values of the sample means occur twice.

Work Sheet No. 1. Analysis of Sample Means

Sample No.[a]	Sample Code[b]	Sample	Sample Std. Dev.[d]	Sample Variance[e]	z Value[f]	Rank[g]	Cum. p Value[h]
1	*abc*	18	5	25	0.16	12	.60
2	*abd*	15.3	2.5	6.35	−0.69	2	.10
3	*abe*	16.7	3.2	10.3	−0.26	7	.35
4	*abf*	16	2.7	7	−0.45	4	.20
5	*bcd*	18.7	4	16.32	0.37	16	.80
6	*bce*	20	2.7	7	0.79	20	1.00
7	*bcf*	19.3	3.2	10.3	0.57	18	.90
8	*cde*	19	4	16	0.48	17	.85
9	*cdf*	18.3	4.2	17.3	0.26	14	.70
10	*cda*	17	5.3	28	−0.16	8	.40
11	*def*	17	2	4	−0.16	9	.45
12	*dea*	15.7	3	9	−0.58	3	.15
13	*deb*	17.3	2	4	−0.05	10	.50
14	*efa*	16.3	3	9	−0.37	5	.25
15	*efb*	18	1	1	0.16	13	.65
16	*efc*	19.7	3	9	0.69	19	.95
17	*afc*	17.7	5	25	0.05	11	.55
18	*afd*	15	2	4	−0.68	1	.05
19	*afe*	16.3	3	9	−0.37	6	.30
20	*ace*	18.3	5	25	0.33	15	.75

[a]Numbers arbitrarily assigned to the 20 unique samples of three objects that can be made from the statistical population described in the text.

[b]Calling out the triplet sets of which each sample is composed.

[c]Calculated according to formula, with some rounding off.

[d]Calculated according to formula, with some rounding off.

[e]Calculated according to formula, with rounding off.

[f]Calculated as $(x - \mu)/\sigma$.

[g]Numbers assigned according to sample mean magnitudes, the small value of x being set as 1, up to 20.

[h]Since there are 20 possible values, each has a probability of 0.05. The cumulative probabilities, the numbers entered here, are obtained by multiplying the rank value by 0.05.

2. The mean of the sample means (i.e., the mean of the 20 values listed in the third column) is identical with the population mean.
3. The mean of the sample variances (or, for that matter, of the sample standard deviations) is not identical with the corresponding value(s) for the population.
4. When the cumulative probabilities of the sample means are plotted against z values on probability paper, the points fall reasonably well on a straight line, as shown in Fig. 12–2. This informs us that the distribution of sample means from a normal population is itself a normal distribution.
5. Although not shown here, the distribution of sample variances (or sample standard deviations) is not a normal distribution. Thus, when sample variances are treated as shown in the work sheet and plotted on probability paper, a straight line does not result.

This rather artificial example was set forth in considerable detail to explain the mechanics of probability plotting. Modern hand calculators do

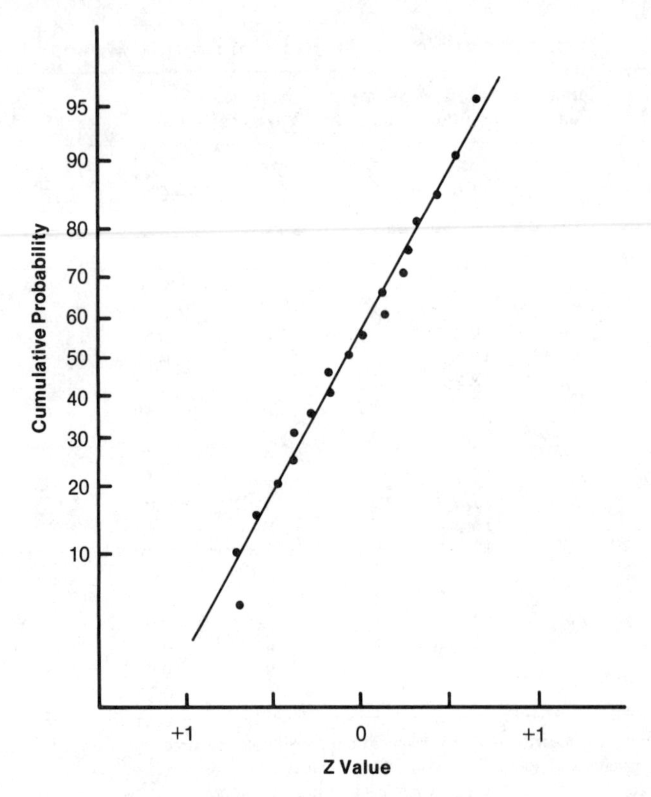

Figure 12–2. Graphic analysis of the distribution of a set of sample means. The ordinate is scaled according to cumulative probabilities, and the abscissa is in the *z* scale. The linearity of the indicated points shows that this distribution is a normal distribution. The probability scale, which on commercial paper runs from $p = 0.01$ to 99.99, has been cut off at either end for convenience, since no points fell beyond the range of values shown.

a great deal to simplify the calculations and allow a rapid appraisal of the normal behavior of a given data set. However, these plots are of little value unless five or six points are available.

CALCULATION OF THE MEAN, THE VARIANCE, AND THE STANDARD DEVIATION

Mean

The mean of a sample is defined as a weighted average, and the general formula for its calculation is given by:

$$x = \sum \frac{f_1 x_1 + f_2 x_2 + \cdots + f_n x_n}{f_1 + f_2 + \cdots + f_n} = \sum \frac{f_i x_i}{f_i} \qquad (12\text{–}9)$$

where f_i is the observed frequency of any value, x_i. If none of the individual values of x is repeated, Eq. 12–9 reduces to the usual calculation of an arithmetic mean. The symbol x_i ordinarily stands for a single value of the variable because most biochemical work involves small samples. In some instances, where large amounts of data are collected, they can be grouped into uniform intervals according to some consistent and appropriate rule. The midpoint of each uniform interval can then be represented by x_i, with a considerable saving in labor.

Variance

The most general formula for calculation of a sample variance is given by:

$$s^2 = \frac{\Sigma(x_i - x)^2}{n - 1} \qquad (12\text{–}10)$$

where the bracketed term is the deviation of each observation from the sample mean. The sign of the deviation is unimportant, since the squaring operation makes all signs positive.

If the available calculator does not have a preprogramed function for determination of the variance, the above formula becomes somewhat tedious to use. By expansion of the quadratic term of Eq. 12–10, followed by collection of like terms, it is easy to show an equivalent form, as:

$$s^2 = \frac{\Sigma(x_i^2) - (\Sigma x_i)^2/n}{n - 1} \qquad (12\text{–}11)$$

This equation can be used with even the simplest of solid-state calculators, or even with tables of squares.

A similar formula is available for use with paired observations, x_i and y_i, such as might result from "with and without" testing. If $(x_i - y_i)$ is defined as d, then:

$$s^2 = \frac{1}{2n} \Sigma d_i \qquad (12\text{–}12)$$

Note, by comparison of Eqs. 12–10 and 12–12, that from a statistical point of view, the same information is conveyed by n paired observations as from $2\ n + 1$ unpaired observations, a strong argument in favor of duplicate observations.

SOME IMPORTANT PROPERTIES OF THE VARIANCE AND STANDARD DEVIATION

The variance plays a role in statistics similar to the role played by the mole in chemistry; when errors combine, they do so in terms of the associated variances. However, since the formulae for the variance all contain a second power, units of variance are very awkward to manipulate in combination with units of the observed variable (i.e., original) data. For this reason, it is customary to deal not with the variance itself but with its square root, the *standard deviation.* As a single measure of dispersion, or error, the standard deviation is perfectly acceptable. When it is necessary to compare errors (except when dealing with paired observations) or when errors are to be considered in some combination, then only the variance is acceptable.

As noted earlier, the distribution of sample variances is not a normal distribution. It is also true that sample variances, unlike sample means, are sensitive to sample size. Equations 12–10 thrugh 12–12 therefore do not serve for calculation of population variances, represented by σ^2. To obtain the latter value, one must use the relation:

$$\sigma^2 = \frac{\Sigma(x_i - x)^2}{n} \tag{12–13}$$

from which it follows that:

$$\sigma^2 = ns^2 \tag{12–14}$$

DEGREES OF FREEDOM

The denominator of Eq. 12–10 was given as $n - 1$, whereas the denominator of Eqs. 12–13 and 12–14 (which relate to populations, not samples) was given as n. An explanation of the difference is essential. Whatever the value, the denominator of these equations can be identified as the *degrees of freedom* from the data. This term has essentially the same meaning here as in chemistry or thermodynamics. For every descriptive measure or constant that we fix by means of the data themselves, we lose a degree of freedom in further interpretation of the formation. Because the sample mean had to be determined (from the data) first in order to establish the variance, this cost a degree of freedom.

To illustrate the effect of degrees of freedom, we can refer back to the hypothetical population of six objects, identified as *a* through *f*, with the properties previously identified. Once again, the 20 unique samples of three objects each will be collected. Variances and standard deviations can again be calculated, but this time the exercise will involve two sets of estimates; in one set the denominator will be taken as n and in the other as $n - 1$. The essential details are set forth in Work Sheet No. 2. The figures entered in the lower right-hand corner of this work sheet are the means of the standard deviations obtained with n or $n - 1$ degrees of freedom. It is clear that "on the average" the first value is about 75% of the second. The value of σ is known to be 3.15 for this population; thus mistakenly using n degrees of freedom in place of the proper value of $n - 1$ results in gross underestimation of the sample standard deviations. When n approaches 50, the bias disappears, but for smaller samples one must use the correct value for degrees of freedom. Some calculators have programs built in for calcuations with n as well as $n - 1$. Care must be taken to use these calculators correctly.

Work Sheet No. 2. Two Different Estimates of Sample Variances and Standard Deviations

Sample No.[a]	Sample Code[b]	Sample Mean[c]	$\Sigma(x_i - x)^2$ [d]	$\Sigma(x_i - x)^2$ [e] n	$\Sigma(x_i - x)^2$ [f] $n - 1$	s_n [g]	s_{n-1} [h]
1	*abc*	18	50	16.7	25	4.08	5
2	*abd*	15.3	12.7	4.2	6.3	2.05	2.5
3	*abe*	16.7	20.7	6.9	10.3	2.63	3.21
4	*abf*	16	14	4.7	7	2.17	2.64
5	*bcd*	18.7	32.7	10.9	16.3	3.3	4.03
6	*bce*	20	14	4.7	7	2.17	2.64
7	*bcf*	19.3	20.7	6.9	10.3	2.63	3.21
8	*cde*	19	32	10.7	16	3.27	4
9	*cdf*	18.3	34.7	11.6	17.3	3.41	4.16
10	*cda*	17	56	18.7	28	4.32	5.29
11	*def*	17	16	5.3	8	2.3	2.83
12	*dea*	15.7	18.7	6.2	9.3	2.49	3.05
13	*deb*	17.3	8.7	2.9	4.3	1.7	2.07
14	*efa*	16.3	21.5	7.2	10.7	2.68	3.27
15	*efb*	18	2	0.6	1	0.77	1
16	*efc*	19.7	18.7	6.2	9.3	2.49	3.05
17	*afc*	17.7	50.7	16.9	25.3	4.11	5.03
18	*afd*	15	8	2.7	4	1.64	2
19	*afe*	16.3	18.7	6.2	9.3	2.49	3.05
20	*ace*	18.3	50.7	16.9	25.3	4.11	5.03
						2.52[i]	3.35[j]

[a]Numbers arbitrarily assigned to the 20 unique samples of three objects, which can be made from the statistical popualtion described in the text.
[b]Calling out the triplet sets of which each sample is composed.
[c]Calculated according to formula, with some rounding off.
[d]Sum of the squares of the deviation of each value from the sample mean.
[e]Sample variances calculated with a divisor of n.
[f]Sample variances calculated with a divisor of $n - 1$.
[g]Sample standard deviations, the square root of the variances, calculated from the variance estimates using a divisor of n.
[h]Sample standard deviations, the square root of the variances, calculated from the variance estimates using a divisor of $n - 1$.
[ij]The mean values of the respective columns.

SOME SPECIAL CASES

Thus far the discussion has assumed that the random variable is under no constraints. The assumption has been that a sample mean is an "accurate" reflection of a population mean, and that the observed variance is a proper estimate of sample dispersion or precision. In some instances an unrecognized constraint, in the form of a constant error, may exist. A balance pan may be dirty, or the balance may not be leveled. A vernier or other measuring scale may have slipped, or a pipet may have a broken tip. All of these would be constant errors in any series of analyses or complex measurements. In other cases, one may wish to deliberately alter the data by some coding process in an effort to simplify or speed up the required calculations. These possibilities give rise to the question: What is the effect

Table 12–2. Fast Calculation by Data Coding

x_i	$(x_i)^2$	Coded x_i	(Coded x_i)^2
147	21609	3	9
150	22500	6	36
149	22201	5	25
149	22201	5	25
149	22201	5	25
147	21609	4	9
891	132121	27	129

on sample means and variances of the presence of unforeseen or deliberate constraints of a constant nature? The answer to this question may be embodied in the following two theorems.

Theorem IX: Addition (or subtraction) of a constant, a, to (or from) each x_i will increase (or decrease) the mean to a new value, $x_i + a$ (or $x_i - a$), but will not change the variance.

Proof of this theorem can readily be demonstrated by inserting the values of $x_i \pm a$ into Eqs. 12–9 or 12–10, and 12–11.

Theorem X: Multiplication (or division) of the variable x_i by a constant will multiply (or divide) the mean to a new value, ax_i (or x_i/a), and will increase (or decrease) the variance to a new value, a^2s^2 (or s^2/a^2).

Proof of this theorem is similar to the argument used for theorem IX.

There are several important and useful consequences of theorems IX and X. First, they prove that constant errors cannot be detected directly by simple determinations of samples' means and variances. Tools for this purpose do exist, and they will be discussed later. A second consequence relates to deliberate alteration of data by coding. Coding of data may be very useful for simplifying calculations and manipulating unwieldy numbers, as the following example shows.

Example 8. Determine the mean and the variance of a set of six numbers, representing serum Na^+ analyses, by application of theorem IX and Eq. 12–11, with the results shown in Table 12–2. Note that the coding constant that was arbitrarily selected in a = 144 mmol/L.

STANDARD SCORES

Coding of sample data is sometimes done for reasons other than convenience of calculation. Standard scores, which are widely used in educational testing, represent a special kind of coding. The purpose of standard scoring is to adjust the properties of several distributions to have some desired means and standard deviations, *which are the same for each.* Such derived distributions are easily obtained by application of theorems IX and X sequentially. The need for standard scores can be put very simply.

To say that a certain performance earned a raw test score of 76 is to say very little. The performance may have been very poor, average, or very good, depending on the particular test, distracting factors such as proximity of vacations, and many others. The test itself may not be an effective instrument for measuring student accomplishment. Although no scoring method can eliminate all of these anxieties, standard scores do provide a uniform way of ranking scores in terms of probability of performance in a given class.

The theory of standard scores is as follows: If, from each item of a grade distribution, we subtract the mean of that distribution, we obtain a coded or derived distribution with $x = 0$ (theorem IX). If we now divide each item of the derived distribution by the standard deviation (or the variance), we obtain a third distribution which has $x = 0$, *and* s (or s^2) $= 1$ (theorem X). Lastly, we define the standard score, Z, as:

$$Z = A + B\frac{(x_i - \overline{x})}{s} \tag{12-15}$$

where A and B are constants and x_i, $\overline{x}$, and s have their usual significance with respect to the original data. Thus, if we choose $A = 50$, the mean of the standard scores will be 50. By choosing $B = 10$, we have automatically adjusted the standard deviation of the standard scores to be 10. In this way, test scores collected over a period of time can be compared on a common and uniform basis of probabilities.

The use of standard scores is not restricted to the classroom and examination grading. The technique is perfectly general; it may and should be used whenever it is required to compare several distributions with different sampling statistics. By proper selection of values for A and B, one can generate derived distributions with any desired properties consistent with the properties of the normal error function.

CONFIDENCE LIMITS AND CONFIDENCE RANGE: PREDICTIONS BASED ON SAMPLE STATISTICS

Figure 12–1 is the graph of Eq. 12–8; it gives some idea of how changes in the value of x, expressed in units of $\mu \pm \sigma$, are related to the area under the normal curve. It would be very inconvenient to rely on the data of Fig. 12–1 alone for computational purposes. Instead, one can rely on tables of data calculated by solution of Eq. 12–8. Table 12–I (all tables identified by Roman numerals are collected at the end of this chapter) consists of two parts. In the upper part, values of $\mu \pm \sigma$ are listed for constant increments of n. In the lower part, the values of n are selected to provide desired increments of the area under the curve. One must keep in mind that in all instances the areas are cumulative; that is, they are calculated from $-\infty$ to the indicated point on the abscissa.

Suppose now that one is interested in the part of a distribution that lies between $\mu + \sigma$ and $\mu - \sigma$. Reference to Table 12–I indicates that at $x =$

$\mu + \sigma$, the area under the curve $= 84.13\%$, whereas at $x = \mu - \sigma$, the corresponding area $= 15.87\%$. The difference between these values is 68.27%. One can interpret these results as follows: The chancсs are a little better than 2 to 1 ($68.27/31.73 = 2.15$) that a given value will lie in the interval, $\mu \pm \sigma$. By an inverse operation, if one is given a statement of probabilities, one can calculate the appropriate values of the variable, x. These ideas may be further clarified by some simple examples.

Example 9. A normal distribution is known to have the parameters $\mu = 60$ and $\sigma = 5$. What portion of the possible values fall in the range from 47 to 56? For $x = 47$, $z = (47 - 50)/5 = -0.6$; and for $x = 56$, $z = (56 - 50)/5 = 1.2$. By reference to Table 12–I, one can locate the areas under the curve associated with each of these z values. Subtraction of the first from the second gives the desired area, and hence the desired proportion of the cases. The answer is $88.49\% - 27.41\% = 61.08\%$. Thus, a little less than two-thirds of the total possible values fall in the desired range.

Example 10. Given the same parameters as in example 9, what is the range of values that just includes the middle 50% of all possible values? Reference to Table 12–I shows that $z = -0.674$ has just 25% of the area below it, and $z = 10.675$ has 25% of the area above it. Hence, the interval, $z = \pm 0.674$ includes exactly the central 50% of the possible cases. From the definition of z, one can write two expressions:

$$x = \mu - 0.674\sigma = 50 - (0.674)(5) = 46.63$$
$$x = \mu + 0.674\sigma = 50 + (0.674)(5) = 53.37$$

to obtain the values that include the central 50% of the cases.

Instead of dealing with the middle 50% of the area or the lowest 10%, it is sometimes useful to group the data in terms of *percentiles.* A percentile is simply a class marking of an arbitrary sort. When one speaks of the 10th percentile, one refers to a class mark chosen so that 10% of the values are less than this value; the 30th percentile is chosen so that 30% of the total values are less; and so forth. The lower portion of Table 12–I is particularly designed for use with percentile classification of data. For any normal distribution with known parameters, the percentile values of x can be found in the manner just outlined.

ESTIMATION OF POPULATION MEANS FROM SAMPLE MEANS: CONFIDENCE LIMITS EXTENDED

Although the population mean can be defined as the mean of all possible sample means, this is not very helpful when one has only a single sample with which to deal. Because the distribution of possible sample means is itself a normal distribution, with a standard deviation of its own, one can use that standard deviation as a measure of how the sample means would be dispersed. This can be formalized as the *confidence limits of the sample mean.* Between these limits, one may speak of the *confidence interval of the mean* obtained from any sample. Essentially, one regards the given sample mean as equivalent to x in a population of all possible sample means, and then treats that distribution as was done in the last few examples.

For example, the central 95% of the area under the normal curve is associated with a z value of ± 1.96; one can therefore write that:

$$z = \pm 1.96 = \frac{\bar{x} - \mu}{\sigma}$$

This can be restated in the form of an inequality:

$$-1.96\sigma < \bar{x} - \mu < +1.96\sigma$$

The inequality is not altered if one changes the algebraic signs and the direction of the carat signs:

$$1.96\sigma > \mu - \bar{x} > -1.96\sigma$$

Addition of $\bar{x}$ to each term of the inequality does not alter it, so that:

$$\bar{x} + 1.96\sigma < \mu < \bar{x} - 1.96\sigma$$

Since $z = \pm 1.96$, one would be confident that in 95 out of 100 repeated experiments the value of μ would fall within the specified limits. In other words, the prediction concerning the value of μ could be made with a 95% level of confidence. In the general case, any value of z (and thus any desired probability value) can be chosen. Accordingly, for any desired level of confidence, the definition becomes:

$$\bar{x} + z\sigma > \mu > \bar{x} - z\sigma \qquad (12\text{–}16)$$

The generalized confidence limits of μ can now be stated as $\bar{x}$ to $z\sigma$ and $\bar{x} - z\sigma$, and the confidence interval is the span of values between them.

The one awkward feature of Eq. 12–16 is that it involves the population parameter, σ, which in general is now known. To complete the formulation in terms of sample statistics only, recall that $\sigma = s\sqrt{n}$ (from Eq. 12–14) so that:

$$\bar{x} + zs\sqrt{n} > \mu > \bar{x} - zs\sqrt{n} \qquad (12\text{–}17)$$

It is important to understand what Eq. 12–17 implies and what is does not. By choice of some z value, Eq. 12–17 allows the prediction of some interval in which μ will be found. One cannot say exactly where, in the estimated interval, μ will be located. Neither can one say that μ will be found in the estimated interval in any given case. Indeed, the laws of probability require that in some number of cases, determined by the choice of z, μ will *not* lie in the estimated interval, but the chances of this occurrence can be made as unlikely as one wishes. If one wishes to make a statement with greater probability, one must increase the value of z, thereby lengthening the confidence interval. For the 95% level of confidence, z is 1.96; for the 99.5% level of confidence, z is 2.58; and for the 99.9% level of confidence, z is 3.09. Thus we do not get something for nothing, but Eq. 12–17 allows us to express what we have in a rather quantitative manner.

Table 12–3. Confidence Limits and Sample Sizes

Degrees of Freedom	95% Confidence Interval	99% Confidence Interval
2	12.71	63.66
5	2.70	4.60
10	2.26	3.25

As the sample size increases, the confidence limits converge. The more data one has, the more authoritatively one can speak. An idea of the effect of sample size can be gained from the following example in Table 12–3, in which two customary confidence intervals are set forth as a function of the degrees of freedom of a certain distribution. Note that although these samples are fairly small, the confidence intervals do not depend on sample size in a linear manner. Note also that even a relatively small increase in confidence level, from 95% to 99%, requires a considerable increase in the length of the confidence interval, or a spreading of the confidence limits.

THE *t*-TEST: STATISTICAL COMPARISON OF SAMPLE MEANS

In the situation where two sample means, $\bar{x}_1$ and $\bar{x}_2$, are known, how can it be determined if the samples are from the *same* population, that is, that the difference between $\bar{x}_1$ and $\bar{x}_2$, is *not* significant? Generally speaking, σ is not known in these instances, but may be estimated from Eq. 12–14 as $s\sqrt{n}$.

One reasons as follows: If $\bar{x}_1$ and $\bar{x}_2$ are each normally distributed, then the distribution of the difference, $\bar{x}_1 - \bar{x}_2$, ought also be normal. The z function, $(x - \mu)/\sigma$, is known to be normal with $\mu = 1$ and $\sigma = 1$. By analogy, one could write the so-called t function, where:

$$t = \frac{\bar{x}_1 - \bar{x}_2}{s\sqrt{n}} \qquad (12\text{–}18)$$

However, the distribution of s is not normal, and for a long time this fact posed a dilemma. In 1908, an English chemist-statistician named Gossett, who wrote under the pseudonym "Student," established that the distribution of t closely approximates the normal. From tables he constructed, it is now possible to evaluate the significance of observed differences in sample means, at desired levels of significance. This is an extremely valuable test, since experimental programs are frequently designed to include a control group of data and another in which some factor or procedural modification has been introduced. Experimental conclusions are then based on differences in the means of the two groups. One can, of course, proceed without the t-test. Each mean might be separately stated, and each might have an associated statement of a confidence interval. The confidence intervals might be widely separated, or they might overlap, or they might be identical. The last instance poses no problem; clearly it would signifiy that the means were from the same population. In the first two instances, the investigator would still face a burden of proof. The *t*-test increases one's powers of discrimination, as will be demonstrated by the next few examples.

Example 11. The addition of a surfactant is said to increase the yield of glucagon from a pancreatic extract. The original method gave a yield of 65 glucagon units/100 g pancreatic tissue. Nine pooled samples of tissue were then extracted by the modified, surfactant-containing process, with the following results: 55U, 62U, 67U, 63U, 70U, 59U, 70U, 63U, and 64U. These yields have a mean of 63.66 U and a standard deviation of 4.9. The hypothesis to be tested here is this: Is the population mean equal to a stated quantity (65U) or is it significantly different? The *t*-test is set up to give:

$$t = \frac{\bar{x} - \mu}{s\sqrt{n}} = \frac{(63.66 - 65)(3)}{4.90} = 0.82$$

The critical value in Table 12–II for $t_{0.95}$ and 8 degrees of freedom is − 1.86. Since the calculated value is less than the critical value, one concludes that the population mean is indeed equal to 65U; in other words, there is no significant effect of the surfactant, which could safely be omitted from the purification procedure.

The *t*-test can also be used with samples of different sizes from two different populations for which the variances are unknown. In such cases the estimate of variance must be pooled to reflect the contributions of samples from both populations and the difference in sample size. For any two samples, a pool variance estimate is defined by the equation.

$$s_p^2 = \frac{\Sigma x_1^2 - [(\Sigma x)^2/n_1] + \Sigma x_2^2 - [(\Sigma x_2)^2/n_2]}{n_1 + n_2 - 2} \qquad (12\text{–}19)$$

Accordingly, under these circumstances one computes the *t*-test as:

$$t = \frac{\bar{x}_1 - \bar{x}_2}{s_p\sqrt{(1/n_1 + 1/n_2)}} = \frac{\bar{x}_1 - \bar{x}_2}{s_p}\sqrt{\frac{n_1 n_2}{n_1 + n_2}} \qquad (12\text{–}20)$$

Example 12. Two groups of rats were fed diets identical except that one diet contained the L isomer of an amino acid while the other diet contained the D isomer of the same amino acid in the same concentration. At the beginning of the experiment, the same number of animals were randomly assigned to each group, but by the end of the feeding period two animals in the D-amino acid group had died. The weight gains of the surviving animals are recorded in Table 12-4. The numbers reflect weight gain in grams/100 g of initial body weight.

Table 12–4. Effect of Diet on Weight Gain

Diet	1	2	3	4	5	6	7	8	9	$\bar{x}$
L	31	34	29	26	32	35	38	34	30	32.11
D	26	24	28	29	30	29	32	—	—	28.28

The null hypothesis is that these populations have the same mean, and that there is no significance in the observed difference in sample means. The proper statement of the t-test is

$$t = \frac{32.11 - 28.28}{(3.21)(\sqrt{16/63})} = \frac{3.83}{(3.21)(0.504)} = 2.37$$

Table 12–II (see end of this chapter) must be entered at $t_{0.95}$ with 14 degrees of freedom, and the critical value is found to be 1.76, which is less than the calculated value. The hypothesis is therefore rejected, and the result is interpreted as a statement that the amino acid isomers have a significant effect on weight gain.

Example 13. Two detergents are to be tested in a mechanical dishwasher. The first, A, is cheaper than the second, B. Unless it can be shown that B is definitely better than A, the laboratory will naturally buy the less expensive detergent. Five typical loads of glassware are washed with each detergent, and the glassware is scored for appearance and for positive chemical tests for contamination by Cl^-, Na^+, and P_i. The scores are shown in Table 12–5.

Table 12–5. Dishwashing with Two Detergents

Detg.	1	2	3	4	5	$\bar{x}$
A	85	87	92	80	84	85.6
B	89	89	90	84	88	88.0

The hypothesis to be tested is that the mean of the A population is less than the mean of the B population. The t-test is given by

$$t = \frac{85.6 - 88.0}{(3.52)(0.4)} = \frac{-2.4}{1.408} = -1.70$$

The critical value of $t_{0.95}$ and 8 degrees of freedom is, from Table 12–II – 0.186. This result supports the hypothesis and furnishes good reason to buy detergent B.

A common feature of the examples given thus far is that they all involve measurements on separate sets of objects or individuals. In many situations it is necessary to make observations on the same set of individuals "before" and "after" some experimental procedure. Such paired observations tend to minimize extraneous effects that might otherwise obscure the factor to be studied. To handle paired observations, the form of the t-test can be modified as:

$$t = \frac{d \pm k}{s_d} \sqrt{n} \tag{12–21}$$

where n = the number of paired observations, d = the mean difference between members of the pairs, s_d = standard deviation of the differences, and k = a constant determined by the nature of the experiment. If one sets up a hypothesis that the "before" and "after" means do not differ significantly, then k has a positive sign. If the hypothesis states that the means are significantly different, then k has a negative sign.

Example 14. A certain compound reportedly lowered fasting blood sugar concentrations within 2 hours when administered at a dose of 50 mg/kg body weight. A group of ten volunteers served as subjects to verify this report. Their blood sugar concentrations were determined, they were given the drug, and then the analyses were repeated. The mean of the differences was found to be 27 mg/100 ml, and s_d was calculated to be 3.5 mg/100 ml. Do these observations support the hypothesis that this compound, at the indicated dosage, causes a decrease in blood sugar of at least 25 mg/100 ml? Under the circumstances of the hypothesis, the value of k = -25, and the t-test according to Eq. 12–21 would be set up as follows:

$$t = \frac{(27 - 25)}{(3.50)} (3.16) = 1.81$$

The critical value of $t_{0.95}$ and 10 degrees of freedom, found in Table 12–II, is 1.81. Since the critical value and the experimental value are identical, we cannot say whether the hypothesis is proved or disproved, but only that the data invite further experiment. If one wished to settle for k = $-$ 20, a conclusion could be reached.

THE *F*-TEST: COMPARISON OF SAMPLE VARIANCES

Just as it is possible to compare sample means by simple statistical tests, so it is possible to compare sample variances. This is done with a statistic called *F*, which is given by:

$$F = s_a^2/s_b^2 \tag{12–22}$$

The distribution of *F* has been studied and tables of critical values have been prepared for this statistic as for others. Since the *F* comparison involves two variances, entries in tables of the statistic must be made with three values in mind: (1) the level of confidence desired; (2) the degrees of freedom in sample a; and (3) the degrees of freedom in sample b. Any particular *F* distribution may be designated as $F(n_a - 1, n_b - 1)$, and in general this is *not* the same as $F(n_b - 1, n_a - 1)$. If there is no theoretical reason to place one or the other variance in the numerator, the rule is that the largest ratio is taken. If it is known that s_a^2 can be only $> s_b^2$, then s_a^2 must go in the numerator. The symbol for such a comparison would be $F(n-1,\infty)$, where the bracketed data describe the degrees of freedom of the numerator and the denominator.

Most tables of the F statistic contain data on one or more of the upper percentiles of the distribution, such as 95%, 97.5%, and 99%. But obviously the ratio of two variances may be very much larger or very much smaller than some critical value. The critical values for the lower percentiles, such as 5%, 2.5%, and 1%, are frequently omitted from reference tables because these values can readily be calculated by the following equation:

$$F_\alpha(n_1-1,\ n_2-1) = [F_{1-\alpha}(n_2-1,\ n_1-1)]^{-1} \qquad (12\text{–}23)$$

In other words, the critical value for 5% can be found by taking the reciprocal of the critical value for 95%, *provided* that the table is entered with inversion of the order of degrees of freedom of the numerator and the denominator. Tables 12–III and 12–IV (see end of this chapter) list the critical values of the F statistic for 95% and 97.5% probabilities, respectively. Both are presented because of the nature of the comparisons that can be made. Comparisons are sometimes termed one-sided or two-sided. One-sided comparisons are those in which one of the two variances is accepted as a reference mark for the particular purpose, usually because of extensive experience. Two-sided comparisons are those in which neither s_a^2 or s_b^2 can be given greater weight than the other. In two-sided comparisons, at 95% probability, one must therefore "split the difference" by using $F_{0.975}$ and $F_{0.025}$ values to embrace 95% of the possible cases. These ideas are illustrated by the next two examples.

Example 15. Based on a considerable body of results, the variance of a method for the determination of copper in proteins is claimed to be 5.0. A new, time-saving method is being considered. The variance of a series of five analyses by this new method was determined to be 7.6. Since the variance is a measure of precision, the question to be asked is whether the new method is significantly less precise than the old. However, it is simpler to set up the hypothesis that the variance of the new method is not greater than that of the old method; to do this, the proper F ratio is

$$F_{0.95}(4, \infty) = 7.6/5 = 1.52$$

From Table 12–III, the critical value is found to be 2.37. Since the experimental value does not exceed the critical value, the F-test supports the validity of the hypothesis. The newer, time-saving method may be safely adopted.

Example 16. Two individuals are asked to analyze samples of the same protein for its lysine content. Analyst A ran triplicate analyses with a variance of 0.053, whereas analyst B made five analyses with a variance of 0.032. Do these figures differ significantly? Here we have two small samples; we have no reason to assign different probabilities to either of them. The results obtained by analyst A show a poorer precision, but his series was the smaller of the two. Is the difference in number of analyses the only factor accounting for the variance difference, or does the difference exceed the bounds of normal probability? This problem involves a two-sided hypothesis, since the ratio of variances might be too large or too small to be probable at a 95% level of confidence. One therefore divides the 5% probability of making an erroneous conclusion into an upper 2.5% and a lower 2.5% zone.

The experimental F ratio may be written as

$$F_{0.975}(2,4) = 0.053/0.032 = 1.66$$

From Table 12–IV, the upper critical value is $F_{0.975}(2,4) = 39.2$. The lower critical value is also found from Table 12–IV by means of Eq. 12–23 to be $F_{0.025}(4,2) = 0.052$. Since the experimental F ratio falls within the limits set by the critical values, one concludes that the variance difference noted is not greater than the difference due to random chance. In other words, there is no significance to the variance differences noted by analysts A and B. Had the experimental value been greater than 39.2 (or less than 0.052), one would have been forced to draw the conclusion that analyst A (or B) had introduced some significant source of error.

ANALYSIS OF VARIANCE (ANOVA)

In the examples presented above, variance has been interpreted as a measure of error, but nothing was said about the individual factors that contribute to error. In example 11 (the effect of a surfactant on glucagon yields from pancreatic tissue), could more information have been gained by varying the surfactant concentration? In example 13 (a comparison of two detergents), were all of the loads uniformly soiled with material of equal difficulty to remove? Was there a bias in favor of detergent B due to differences in the kinds of glassware loaded? These questions obviously come to mind, and they point out that the techniques employed thus far fall short of yielding the maximum information contained in a *well-designed* experiment.

A well-designed experiment means a program laid out *in advance* which permits an investigator (1) to make and test statistical hypotheses about the factors that contribute to variance, as well as (2) to measure the total variance. This procedure is known as an analysis of variance. Experimental design, in preparation for analysis of variance, is a rather sophisticated subject; what is presented here must be regarded as a brief synopsis of its generalities.

Suppose that several methods, r in number, are used to test several materials, c in number. Applying each test to each material *in a random manner* would produce rc observations. These also are assumed to be random variables. These observations could be entered into a table in the form of a matrix, as is shown in Table 12–6. Each entry is identified by two subscripts, the first representing a matrix row, and the second a matrix column. To the right of the matrix, one enters totals for rows and means for rows, shown by the symbols, T_{1} and $\overline{x}_{1}$, respectively. Below the matrix one enters corresponding data for the columns, identified by $T_{.c}$ and $\overline{x}_{.c}$, respectively. Below the right-hand corner of the matrix, one also needs the

Table 12–6. *c* Materials by *r* Methods (Analysis of Variance)

x_{11}	x_{12}	x_{13}	x_{14}	→	x_{1c}	$T_{1.}$	$\overline{x}_{1.}$
x_{21}	x_{22}	x_{23}	x_{24}	→	x_{2c}	$T_{2.}$	$\overline{x}_{2.}$
x_{31}	x_{32}	x_{33}	x_{34}	→	x_{3c}	$T_{3.}$	$\overline{x}_{3.}$
x_{41}	x_{42}	x_{43}	x_{44}	→	$x_{4c.}$	$T_{4.}$	$\overline{x}_{4.}$
↓	↓	↓	↓		↓	↓	↓
x_{r1}	x_{r2}	x_{r3}	x_{r4}	→	x_{rc}	$T_{r.}$	$\overline{x}_{r.}$
$T_{.1}$	$T_{.2}$	$T_{.3}$	$T_{.4}$	→	$T_{.c}$	T	
$\overline{x}_{.1}$	$\overline{x}_{.2}$	$\overline{x}_{.3}$	$\overline{x}_{.4}$	→	$\overline{x}_{.c}$		$\overline{x}$

grand total, T, and the grand mean, $\overline{x}$; these are distinguished by a lack of subscripts.

The particular form of analysis of variance to be presented here is sometimes called the *linear hypothesis model.* It makes several assumptions about the data:

1. It is assumed that the random variables, x_{rc}, have a mean, μ_{rc}, which can be represented by the equation:

$$\mu_{rc} = a_r + b_c + c \tag{12–24}$$

   where a_r is some quantity related to the row arrangement, b_c is some quantity related to the column arrangement, and c is a quantity related to neither. Since a, b, and c are random variables, linear combinations of them like the one shown in Eq. 12–24 are also random variables. This is why the model is known as a linear model.
2. It is assumed that all of the random variables, x_{rc}, have the same variance. The initial hypothesis must be that all methods and all materials to be tested are free of built-in bias; otherwise, separate components of the total variance could not be fairly assessed.
3. It is not required that the matrix be complete; empty elements are permitted although they tend to lower the discriminating power of the analysis. Also, r and c need not be equal.

One can write an equation for the *sums of squares* (ss), equal to s^2 multiplied by the degrees of freedom, which is analogous to Eq. 12–24:

$$\mathrm{ss}_{rc} = \mathrm{ss}_r + \mathrm{ss}_c + \mathrm{ss}_R \tag{12–25}$$

where ss_R is a residual portion of the total ss; that is, a portion not due to row or column effects. One then proceeds to form F ratios between s_r^2 or s_c^2 and s_R^2 to determine if the row or column effects are greater than those due to chance, at some predetermined level of confidence. Calculation of a variance involves determination of the sums of squares of all deviations from a mean; this sum of squares is then divided by the degrees of freedom proper to the case. The result is a mean square of the deviations. For the pattern of analysis of variance discussed here, four mean squares are required; three must be calculated and the fourth determined by difference, as will be shown below. Also shown are the degrees of freedom that are to be employed with each sum of squares in order to determine the component of variance in question.

1a. Total sum of squares: This term is used to form a measure of the total variance.

$$\text{Total sum of squares} = \sum\sum^{rc} (x_{rc}^2) - T^2/N \qquad (12\text{–}26)$$

where $\sum\sum^{rc}$ signifies a double summation over all values of r and c, $N = rc$, and T is the grand total of all x_{rc}.

1b. Degrees of freedom for total sum of squares: $d.f. = N - 1$.

2a. Row effect sum of squares: This quantity is used to form a measure of variance due to arrangement of variables by rows. In some texts this is known as the *among groups* effect.

$$\text{Row effect sum of squares} = \sum^{r} (T_{r.}^2/n_{r.}) - T^2/N \qquad (12\text{–}27)$$

where Σ^r indicates summation over all values of r, $n_{r.}$ is the number of items in the rth row, and the other symbols have their usual significance.

2b. Degrees of freedom for row sum of squares: $d.f. = r - 1$.

3a. Column effect sum of squares: This quantity is used to form a measure of variance due to arrangement of variables by columns. In some texts, it is known as the *within groups* effect.

$$\text{Column effect sum of squares} = \sum\sum^{rc} x_{rc}^2 - \sum^{c} (T_{.c}^2/n_{.c}) \qquad (12\text{–}28)$$

3b. Degrees of freedom for column sum of squares: $d.f. = c - 1$.

4a. Residual sum of squares: This quantity is used to form a measure of variance not related to either the row or column effects. Residual effects are due only to the random errors of the processes or the materials. The residual sum of squares is usually taken as: Total sum of squares − (Row effect sum of squares + Column effect sum of squares).

4b. Degrees of freedom for residual sum of squares:

$$d.f. = N - 1 - (r - 1) - (c - 1) = N - r - c + 1.$$

The above display looks quite forbidding. In fact, the manipulation of the data is quite straightforward. Several examples and a brief discussion of the results follow.

Example 17. Three different compounds have been synthesized as possible antibiotics. A screening test is set up against four different organisms, the application of each compound to each organism being made in a random manner. The results were scored by the number of surviving colonies on each Petri dish. The matrix of colony counts is given below, along with the appropriate means and sums.

Table 12–7. Surviving Colony Counts

	Organism					
	A	*B*	*C*	*D*	$T_{r.}$	$\bar{x}_r$
Compound *a*	7	6	8	7	28	7
Compound *b*	2	4	4	4	14	3.5
Compound *c*	4	6	5	3	18	4.5
$T_{.c}$	13	16	17	14	60	
$\bar{x}_{.c}$	4.3	5.3	5.7	4.7		5

There is a certain random error in the above observations, as there is in any experiment. In addition, there may be a difference due to the compounds being tested. Superimposed on these is a possible variation due to the organisms being tested. The pertinent questions that the analysis of variance could answer are based on these issues. The computations are generally cast as shown in Table 12–7A, and are based on Eqs. 12–26 to 12–28. The italicized entries in Table 12–7A are determined by subtraction of the sum of row and column data from the appropriate total effects; all other entries are calculated from the matrix. The entry in square brackets, which is in fact the total variance, is *not* the sum of the identified variance factors. The value, $s^2 = 3.27$, is what one would determine if the 12 items in the body of the matrix had been considered as a data set in some random order, that is, without ordering into rows and columns. One gets this same value whether it is calculated from row data, column data, or from scrambled data. This is not so for the row or column mean sums of squares.

The calculated value for the *F* ratio, columns/residual, is much less than the critical value. This leads to the conclusion that variance due to the different organisms is not significant. The calculated *F* ratio, rows/residual, is distinctly greater than the critical value. This leads to the conclusion that the difference in compounds tested was significant at the 95% level of confidence.

Example 18. As part of a quality control program, 4 pooled serum samples were prepared. Each was analyzed for chloride content on 5 different days. Each day's work was performed by a different analyst, who was handed one of the samples at random. At the end of the program, the data were collected as shown in Table 12–8. All values are given in millequivalents per liter (meq/L).

Table 12–7A. Analysis of Variance

Effect	Sum of Squares	*d.f.*	Mean Sum of Squares	*F*-Ratio
Rows	26.00	2	13.00	11.70
Columns	3.33	3	1.11	1.00
Residual	*6.67*	*6*	1.11	
Total	36.00	11	[3.27]	

Critical values: $F_{0.95}(2,6) = 5.14$ $F_{0.95}(3,6) = 4.76$

Table 12–8. Serum Chloride Analyses

	Day 1	Day 2	Day 3	Day 4	Day 5	$T_{r.}$	$\bar{x}_{r.}$
Pool 1	103	118	122	99	122	564	112.8
Pool 2	95	98	112	88	105	498	99.6
Pool 3	102	102	113	103	126	546	109.2
Pool 4	89	103	104	89	113	498	99.6
$T_{.c}$	389	421	451	379	466	2106	
$\bar{x}_{.c}$	97.2	105.2	112.7	94.7	116.5		175.5

As in the previous problem, one starts with the assumption that items in the matrix are random variables. One wishes to determine if there is any significant difference in the values of the pooled samples themselves, and more importantly if there is a significant difference in the day-to-day results. As before, one computes the necessary sums of squares and sets up a table for the analysis of variance. In this instance both the row and column effects are significant at the 95% level of confidence. That the row effect exists is not too surprising; if the serum pools were prepared by mixing samples from sick individuals, the values may not have been normal to begin with, hence not random. The significance of the column effect clearly means that the day-to-day performance of the analysis (or of the analytical devices) is not well controlled.

In summary of our discussion on analysis of variance:

1. All variables must be regarded as random, since the underlying principles are those of probabilities of the normal error function.
2. ANOVA provides a technique for examination of two related factors as they vary jointly. It is neither necessary nor possible to "freeze" all factors but one.
3. If two methods are being tested against *identical* materials, simpler statistical techniques are adequate, but if the materials are *not identical* or if *three or more* methods are involved, then the analysis of variance becomes powerful and appropriate. The reader should keep in mind that the characterization by "methods" and "materials" is used only for purposes of discussion. Any similarly related factors can be subjected to variance analysis.
4. Values of r and c need not be identical, nor must the matrix be complete; however, incomplete matrices may lead to less incisive conclusions because the power of ANOVA lies in the ordering of the data.

Table 12–8A. Analysis of Variance

Effect	Sum of Squares	*d.f.*	Mean Sum of Squares	*F*-Ratio
Rows	682	3	227.3	10.1
Columns	1428	4	357	15.8
Residual	*270*	*12*	22.5	
Total	2380	19	[125.3]	
	Critical values: $F_{0.95}(3,12) = 3.49$		$F_{0.95}(4,12) = 3.26$	

CORRELATION OF VARIABLES

If a series of blood samples were analyzed for hemoglobin content, a certain distribution of values would be found. If the same samples were also subjected to a red blood cell count, a second distribution would be obtained. In general, it would be noticed that the hemoglobin content bears a definite relation to the red cell count, low hemoglobin concentrations being closely associated with low red cell counts and vice versa. The general impression could be strengthened by a graph of hemoglobin content versus red cell count. The points would be scattered over the two dimensions, but they would be distributed with maximum density along or about a straight line. Such a relation between variables is called *linear correlation.* The degree of correlation can be expressed in a quantitative way by means of the *correlation coefficient, r,* which ranges in magnitude from 0 to 1, with an algebraic sign dependent on the slope of the trend. Values of 1 indicate perfect correlation, or complete dependence of the variables, whereas values of 0 indicate no correlation, or complete independence of the variables. The value of r for any set of measurements can be calculated by the equation:

$$r = \frac{n\Sigma xy - \Sigma x \Sigma y}{\sqrt{\{[n\Sigma x^2 - (\Sigma x)^2]\,[n\Sigma y^2 - (\Sigma y)^2]\}}} \qquad (12\text{–}29)$$

where n is the number of paired observations and the other symbols are sums as indicated.

It must be clearly understood that the correlation coefficient *does not* imply any sense of cause and effect. The only relation implied is a relation

Figure 12–3. Correlation of red cell count and hemoglobin.

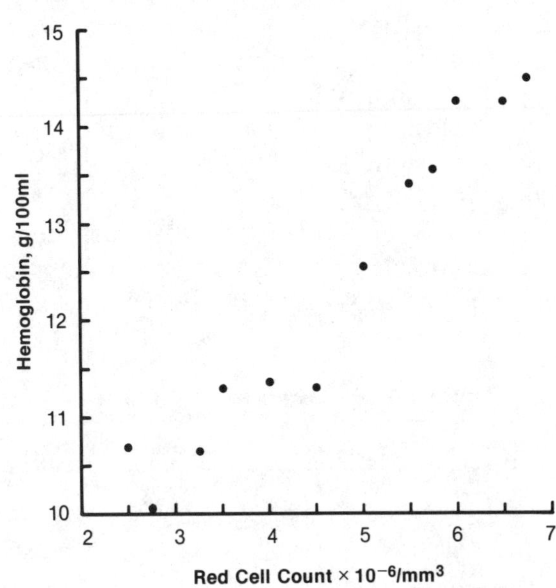

Work Sheet No. 3. Hemoglobin–Red Cell Count Correlation

Sample No.	Red Cell Count ($\times 10^{-6}/mm^3$), x	Hemoglobin (g/100 ml), y	xy
1	2.50	10.70	26.75
2	2.75	10.05	27.64
3	3.25	10.65	34.61
4	3.50	11.30	39.55
5	4.00	11.35	45.40
6	4.50	11.30	50.85
7	5.00	12.55	62.75
8	5.50	12.55	69.02
9	5.75	13.55	77.91
10	6.00	14.25	85.50
11	6.50	14.25	92.62
12	6.75	14.50	97.88

Note: $\Sigma x = 56$; $\Sigma y = 147$; $\Sigma xy = 710.49$. $\bar{x} = 4.67$; $\bar{y} = 12.25$. $\Sigma x^2 = 285$; $\Sigma y^2 = 1828.1$. $(\Sigma x)^2 = 3136$; $(\Sigma y)^2 = 21609$.

of magnitudes. It is true, for example, that weight gain and growth may be highly correlated variables, yet weight gain can and does occur in the absence of growth. In the data collected from some large city, a correlation might well exist between the number of reported murders per year and the number of school teachers on the payroll; this does not imply that any school teacher committed murder. More probably, both data reflect urban size.

The degree of correlation measured by r applies only to the *entire assembly* of paired values and not to relations between any *single* pair of values. For this reason, determination of correlation based on any 10 values does not permit prediction of a possible 11th value. This lack of power to make specific predictions indicates that correlation, like the analysis of variance, "freezes" no variables. Nevertheless, correlation studies are widely used to predict trends in the laboratory as well as in the political economy.

Example 19. A group of 12 blood samples was analyzed by determination of the individual hemoglobin concentrations and by counting the number of erythrocytes per cubic millimeter. When these paired observations were graphed, the results obtained appeared as in Fig. 12–3. The values appear to be highly correlated, but the graph, by itself, does not provide a value of r. Calculations that led to a value for r are shown in Work Sheet No. 3.

From Eq. 12–29,

$$r = \frac{12(710.49) - 56(147)}{\sqrt{\{[12(285)-3136][12(1828)-21609]\}}} = \frac{293.88}{304.7} = 0.964$$

REGRESSION AND THE STATISTICS OF A STRAIGHT LINE

As noted above, correlation does not imply a strict cause-and-effect relation between two variables. However, a very large number of some paired variables do have a causal connection. Among those that come quickly to mind are the concentration of a chromophore and spectrophotometric absorbance, or the relation between the boiling points of homologous alcohols and their carbon chain lengths. For dealing with causally related variables that have a linear relationship, one employs the methods of linear regression.

Regression is essentially an extension of correlation, with the singular difference that in regression one studies discrete and specific values of paired variables, one at a time, whereas in correlation one can only deal with the collection of pairs as a whole. It is this difference that gives to regression analysis a definite predictive capacity, lacking in correlation methods.

Regression lines are sometimes called *least square lines,* and regression methods are sometimes known as *methods of lease squares,* by virtue of an important property of these lines. For any chosen value of the abscissa, there will be a small subset of ordinate values that will have its own mean. That mean is the most probable value of the subset. Hence, the line that passes closest to the largest number of subset means is the most probable line relating all of the data. It minimizes the sum of squares of deviations from all of the plotted points to the line as drawn.

Returning to the cell count–hemoglobin problem of example 19, suppose one had a larger series of paired determinations, for example, 100 pairs of analyses. The mean of the hemoglobins for any single cell count would be known as the *regression of hemoglobin on the cell count.* In an experiment of this sort, the red cell count is known as the independent variable, and the hemoglobin concentration as the dependent variable.

Although it is easy to introduce regression as an extension of correlation of paired values, pairing as such is not required in least squares methods. Any time one analyzes a sample in triplicate, the three independent values for a given sample can be the basis of a least squares line. As usual, the more data one has, the more precisely can the line be located.

The analytical expression for a straight line is:

$$y = mx + b \tag{12–30}$$

where m is the slope of the line, and b is the ordinate intercept. The corresponding expression for a regression line is:

$$\mu_{y.x} = M(x - \mu_x) + B \tag{12–31}$$

where $\mu_{y.x}$ is the regression of y on x, x is any chosen value of the independent variable, and M and B are regression coefficients. Equation 12–31 is stated in terms of population parameters which are, as in previous instances, to be estimated from sample statistics. The following formulae for the sample regression coefficients will be stated without proof (which may be found in the sources cited in the bibliography).

$$m = \frac{n\Sigma xy - \Sigma x \Sigma y}{n\Sigma x^2 - (\Sigma x)^2} \tag{12–32}$$

$$b = \bar{y} - m\bar{x} \tag{12–33}$$

One can also determine variances for m and b in terms of a pooled variance defined by the symbol, $s^2_{y.x}$, which is read as the variance of the y values for a given value of x.

$$s^2_{y.x} = \frac{n-1}{n-2}(s^2_y - ms^2_x) \tag{12–34}$$

In some of the statistical literature $s_{y \cdot x}$ is known as the *standard error of estimate.* Using this pooled variance estimate, the variance of the sample regression coefficients can be stated as:

$$s^2_m = \frac{s^2_{x.y}}{\Sigma x^2 - nx^2} \tag{12–35}$$

$$s^2_b = \frac{\Sigma x^2}{n\Sigma x^2 - (\Sigma x)^2} \tag{12–36}$$

Having the variances, it is possible to calculate confidence limits for m and b, according to the following theorem:

Theorem XI: If the distribution of y for any given value of x is a normal distribution, and if the regression is linear, then the distributions of

$$t = \frac{(\bar{y} - B)\sqrt{n}}{s_{y.x}} \text{ or } \frac{(m - M)s_x(\sqrt{n-1})}{s_{y.x}}$$

are both t-distributions with $n - 2$ *degrees of freedom.* Accordingly, the confidence limits for the regression coefficients can be written as:

$$(M - z)\frac{x_{y.x}}{x_x\sqrt{n-1}} < M < (M + z)\frac{s_{y.x}}{s_x\sqrt{n-1}} \tag{12–37}$$

$$(b - z)\frac{s_{y.x}}{\sqrt{n}} < B < (b + z)\frac{s_{y.x}}{\sqrt{n}} \tag{12–38}$$

The form of Eqs. 12–37 and 12.38 should be compared with the form of Eq. 12–17.

These formulae allow a proper completion of the problem set in example 19, where a high correlation was found between the red cell count and the hemoglobin concentration of blood. A regression line can now be drawn for the data, as shown in Fig. 12–4. Furthermore, from the data contained in Work sheet No. 3 and the formulae of Eqs. 12–32 to 12–38, one can state that the equation of the regression line is:

$$Y = (1.034 \pm 0.022)X + 7.42 \pm 0.03 \qquad \text{(95\% confidence limits)}$$

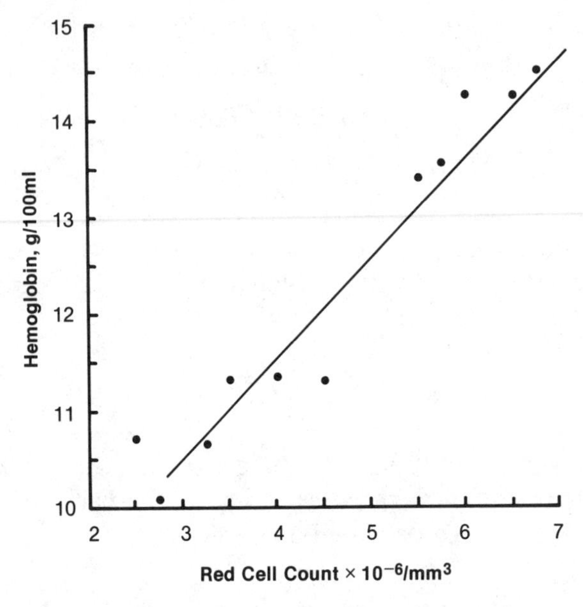

Figure 12–4. Regression of hemoglobin on cell count.

TRANSFORMATION OF DATA: A WARNING TO WOULD-BE ENZYME KINETICISTS

In setting up an enzyme assay, one typically measures reaction velocities as a function of substrate concentrations; all else is held constant. Reaction velocities are measured under *initial conditions,* and substrate concentrations are chosen to cover a fairly wide range of values, some of which are well below the K_m value and some of which are above it. From the collected data one seeks to determine K_m and V_{max}, the parameters that characterize any enzyme. A typical example of the data collected for an enzyme assay is shown in Table 12–9.

When the reaction rates are plotted as a function of the concentrations, the points fall along a hyperboloid curve, as predicted by the Michaelis–

Table 12–9. Data Collected for Enzyme Kinetic Analysis

Sample No.	Reaction Rate ($V_0 = \Delta A$/min/mg)	Substrate Conc. (μmol/ml)	$1/V_0$	$1/S$
1	1.19	0.20	0.840	5.00
2	1.67	0.30	0.598	3.33
3	1.82	0.35	0.549	2.86
4	2.30	0.50	0.435	2.00
5	2.53	0.60	0.395	1.67
6	3.25	1.00	0.307	1.00
7	4.01	2.00	0.249	0.50
8	4.41	4.00	0.226	0.22
9	5.00	5.00	0.200	0.20

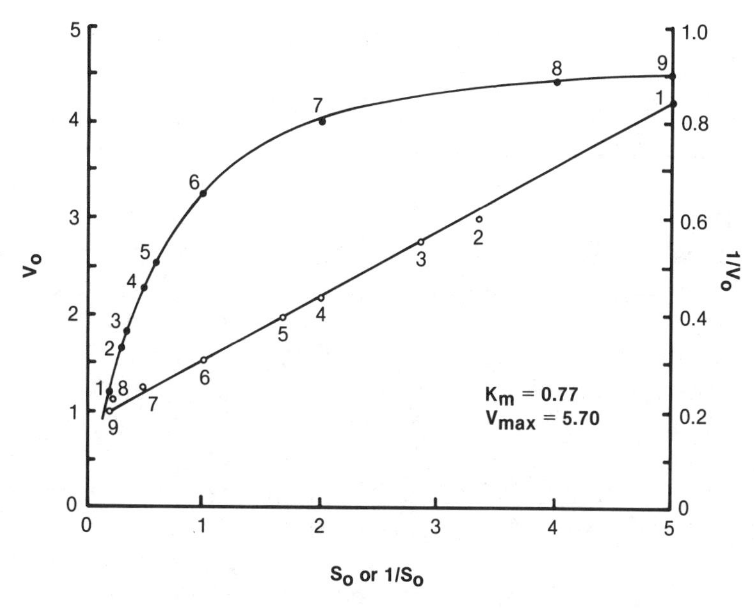

Figure 12–5. Two different plots of enzyme kinetic data. The upper curve is a Michaelis–Menten plot; the experimental points are identified by solid circles. These points were plotted using the ordinate scale to the left-hand side of the figure. The lower curve is a Lineweaver–Burk plot; the experimental points are indicated by the open circles. These points were plotted using the ordinate scale on the right-hand side of the figure. Note the difference in scale of the two ordinates. The abscissa represents either S_0 or $1/S_0$ since the scaling factors are identical. Note also that the order of the experimental points, shown by numbers relating to values in Table 12–9, run oppositely in the two curves; the reason is that scales of one curve are reciprocals of the scales of the other curve.

Menten equation. When reciprocal values are plotted, the points fall along a straight line, as predicted by the Lineweaver–Burk equation. Both of these curves are shown in Fig. 12–5. Such *transformation* of the data is perfectly legitimate and general (i.e., it can be employed for any set of data that follows a hyperbolic course).

There are several very good reasons for working with transformed data in the case of kinetic analysis of enzymes. First, the underlying theory states, for the Lineweaver–Burk plot, that the slope of the straight line is K_m/V_{max}, that the intercept on the reciprocal concentration axis is $-1/K_m$, and the intercept on the reaction rate axis is $1/V_{max}$. Second, the statistics of a straight line are much simpler to manipulate that the statistics of a hyperbola. As has already been demonstrated, it is possible to estimate quickly the precision of both the slope and the intercept of a straight line, with any desired level of confidence.

Unfortunately, there is a price to pay when one transforms the original data. On any scale, when one inverts a series of numbers, one also inverts their order. This is shown clearly by the small numbers in the body of Fig. 12–5, relating the locus of each point to its position in Table 12–9. It is clear that on the Michaelis–Menten plot, concentration values obtained where the initial rates were changing most rapidly are closely grouped together; concentrations at which the enzyme is nearly saturated are more widely spread apart. Just the reverse is true of the Lineweaver–Burk plot. Again, it must be emphasized that this has nothing to do with enzymology; it is entirely a property of numbers. Regardless, it is a warning to those interested in the subject of enzyme kinetics. Dependence on Lineweaver–Burk plots is quite adequate for approximate work, but it is *not* acceptable for results of publication quality in major biochemical journals.

Until about 1967, it was not practical to perform statistical anaysis of hyperboloid data; the methods available were too tedious and too costly to perform by hand. Until that time, one was constrained to rely on the linear transforms, and journals would accept such data. With the advent of digital computers, the labor of calculations directly from the hyperboloid data was very much reduced. Cleland demonstrated that the hyperbolic and linearized data sets could be brought into good statistical agreement by introduction of weighting factors to correct for the inversion of order and to give proper importance to values of reaction rates closest to K_m. He showed that the weighting factors were proportional to the fourth power of the initial velocities. Cleland's program was written in Fortran, and was designed for ready loading into main-frame computers by means of punched cards. The full details are given in Cleland's paper. Since that time, Barnes and Waring have published similar programs designed for use with programmable, hand-held calculators. There are no longer any good reasons to avoid the more accurate data-fitting process.

CONCLUSION

In this brief review of statistical fundamentals, we have deliberately omitted many simple and useful techniques. The calculations that have been covered are sufficient for the analysis of most general biochemical data. Little has been said regarding distributions other than the Gaussian distribution. No allusions to nonparametric statistics have been made. Ample information about these and other statistical data can be found in the appended references and other sources. From what has been said, the reader should be aware that even a modest acquaintance with statistical tools can result in better appreciation and analysis of experimental data.

REFERENCES

Barnes, J. E., and Waring, A. J. *Pocket Programmable Calculators in Biochemistry.* John Wiley and Sons, New York (1980).

Cleland, W. W. The statistical analysis of enzyme kinetic data, *Adv. Enzymol. 29:*1 (1967).

Cochran, W. G., and Cox, G. M. *Experimental Designs.* John Wiley and Sons, New York (1950).

Dixon, W. J., and Massey, F. J. *Introduction to Statistical Analysis,* 2nd ed. McGraw-Hill, New York (1957).

Rosander, A. C. *Elementary Principles of Statistics.* Van Nostrand, New York (1951).

"Student." The probable error of the mean. *Biometrika 6:*125 (1908).

Youden, W. J. *Statistical Methods for Chemists.* John Wiley and Sons, New York (1951).

Table 12–I. Ordinates and Cumulative Areas of the Normal Curve

$\mu + n\sigma$,			$\mu + n\Sigma$		
n	Ordinate	Area	n	Ordinate	Area
−3.0	.0044	.0013	3.0	.0044	.9987
−2.9	.0060	.0019	2.9	.0060	.9981
−2.8	.0079	.0026	2.8	.0079	.9974
−2.7	.0104	.0035	2.7	.0104	.9965
−2.6	.0119	.0047	2.6	.0119	.9953
−2.5	.0175	.0062	2.5	.0175	.9938
−2.4	.0224	.0082	2.4	.0224	.9918
−2.3	.0283	.0107	2.3	.0283	.9893
−2.2	.0355	.0139	2.2	.0355	.9861
−2.1	.0440	.0179	2.1	.0440	.9821
−2.0	.0540	.0228	2.0	.0540	.9772
−1.9	.0656	.0287	1.9	.0656	.9713
−1.8	.0790	.0359	1.8	.0790	.9641
−1.7	.0949	.0446	1.7	.0949	.9554
−1.6	.1109	.0548	1.6	.1109	.9452
−1.5	.1295	.0668	1.5	.1295	.9332
−1.4	.1500	.0808	1.4	.1500	.9192
−1.3	.1714	.0968	1.3	.1714	.9032
−1.2	.1942	.1151	1.2	.1942	.8849
−1.1	.2179	.1357	1.1	.2179	.8643
−1.0	.2420	.1587	1.0	.2420	.8413
−0.9	.2661	.1841	0.9	.2661	.8159
−0.8	.2897	.2118	0.8	.2897	.7881
−0.7	.3123	.2420	0.7	.3123	.7580
−0.6	.3332	.2741	0.6	.3332	.7257
−0.5	.3521	.3085	0.5	.3521	.6915
−0.4	.3683	.3446	0.4	.3683	.6554
−0.3	.3814	.3821	0.3	.3814	.6179
−0.2	.3910	.4207	0.2	.3910	.5793
−0.1	.3970	.5602	0.1	.3970	.5398
0	.3989	.5000			
−3.09	.0034	.001	3.09	.0034	.999
−2.576	.0145	.005	2.576	.0145	.995
−2.326	.0267	.010	2.326	.0267	.990
−1.96	.0584	.025	1.96	.0584	.975
−1.645	.1037	.050	1.645	.1037	.950
−1.282	.1731	.100	1.285	.1731	.900
−1.036	.2387	.150	1.036	.2387	.850
−0.842	.2799	.200	0.842	.2799	.800
−0.674	.3179	.250	0.674	.3179	.750
−0.524	.3478	.300	0.524	.3478	.700
−0.395	.3702	.350	0.395	.3702	.650
−0.260	.3739	.400	0.260	.3739	.600
−0.126	.3957	.450	0.126	.3957	.550

Table 12–II. Values of "Student's" t

d.f.	$t_{.95}$	$t_{.975}$	$t_{.9875}$	$t_{.995}$	$t_{.9975}$
1	6.31	12.7	25.5	63.7	127
2	2.92	4.30	6.21	9.92	14.1
3	2.35	3.18	4.18	5.84	7.45
4	2.13	2.78	3.5	4.6	5.6
5	2.01	2.57	3.16	4.03	4.77
6	1.94	2.45	2.97	3.71	4.32
7	1.89	2.36	2.84	3.5	4.03
8	1.86	2.31	2.75	3.36	3.83
9	1.83	2.26	2.69	3.25	3.69
10	1.81	2.23	2.63	3.17	3.58
11	1.80	2.20	2.59	3.11	3.5
12	1.78	2.18	2.56	3.05	3.43
13	1.77	2.16	2.53	3.01	3.37
14	1.76	2.14	2.51	2.98	3.33
15	1.75	2.13	2.49	2.95	3.29
16	1.75	2.12	2.47	2.92	3.25
17	1.74	2.11	2.46	2.90	3.22
18	1.73	2.1	2.45	2.88	3.2
19	1.73	2.09	2.43	2.86	3.17
20	1.72	2.09	2.42	2.85	3.15
21	1.72	2.08	2.41	2.83	3.14
22	1.72	2.07	2.41	2.82	3.12
23	1.71	2.07	2.4	2.81	3.1
24	1.71	2.06	2.39	2.8	3.09
25	1.71	2.06	2.38	2.79	3.08
26	1.71	2.04	2.36	2.75	3.03
27	1.7	2.05	2.37	2.77	3.06
28	1.7	2.05	2.37	2.76	3.05
29	1.7	2.05	2.36	2.76	3.05
30	1.7	2.04	2.36	2.75	3.03
d.f.	$t_{.05}$	$t_{.025}$	$t_{.0125}$	$t_{.005}$	$t_{.0025}$

Note: When this table is entered from its foot, the tabulated values of t are to be changed in sign. For example, the value of $t_{.025}$ for 20 degrees of freedom should be read as -2.09.

Table 12–III. *F* Distribution, Upper 5% Points ($F_{0.95}$)

	Degrees of Freedom for the Numerator												
	1	2	3	4	5	6	7	8	9	10	12[a]	15[a]	∞
1	161	200	216	225	230	234	237	239	241	242	246	254	254
2	18.5	19.0	19.2	19.2	19.3	19.4	19.4	19.4	19.4	19.4	19.4	19.4	19.5
3	10.1	9.55	9.28	9.12	9.01	8.94	8.89	8.84	8.81	8.79	8.74	8.70	8.53
4	7.71	6.94	6.59	6.39	6.26	6.16	6.09	6.04	6.00	5.96	5.91	5.86	5.63
5	6.61	5.79	5.41	5.19	5.05	4.95	4.88	4.82	4.77	4.74	4.68	4.62	4.37
6	5.59	5.14	4.76	4.53	4.39	4.28	4.21	4.15	4.10	4.06	4.00	3.94	3.67
7	5.59	4.74	4.35	4.12	3.97	3.87	3.79	3.73	3.68	3.64	3.57	3.51	3.23
8	5.32	4.46	4.07	3.84	3.69	3.58	3.50	3.44	3.39	3.35	3.28	3.22	2.93
9	5.12	4.26	3.86	3.63	3.48	3.37	3.29	3.23	3.18	3.14	3.07	3.01	2.71
10	4.96	4.10	3.71	3.48	3.33	3.22	3.14	3.07	3.02	2.98	2.91	2.85	2.54
11	4.84	3.98	3.59	3.36	3.20	3.09	3.01	2.95	2.90	2.85	2.79	2.72	2.40
12	4.75	3.89	3.49	3.26	3.11	3.00	2.91	2.86	2.80	2.75	2.69	2.62	2.30
13	4.67	3.81	3.41	3.18	3.03	2.92	2.83	2.77	2.71	2.67	2.60	2.53	2.21
14	4.60	3.74	3.34	3.11	2.96	2.85	2.76	2.70	2.65	2.60	2.53	2.46	2.13
15	4.54	3.68	3.29	3.06	2.90	2.79	2.71	2.64	2.59	2.54	2.48	2.40	2.07
∞	3.84	3.00	2.60	2.37	2.21	2.10	2.01	1.94	1.88	1.83	1.75	1.67	1.00

[a]To interpolate between the values, use *reciprocals* of the degrees of freedom.

Table 12–IV. *F* Distribution, Upper 2.5% Points ($F_{0.975}$)

	Degrees of Freedom for the Numerator												
	1	2	3	4	5	6	7	8	9	10	12[a]	15[a]	∞
1	648	800	864	900	922	937	948	957	963	969	977	985	1018
2	39.5	39.0	39.2	39.2	39.3	39.3	39.4	39.4	39.4	39.4	39.4	39.4	39.5
3	17.4	16.0	15.4	15.1	14.9	14.7	14.6	14.5	14.5	14.4	14.3	14.3	13.9
4	12.2	10.6	9.98	9.60	9.36	9.20	9.70	8.98	8.90	8.84	8.75	8.66	8.26
5	10.0	8.43	7.76	7.39	7.15	6.98	6.85	6.76	6.68	6.62	6.52	6.43	6.02
6	8.81	7.26	6.60	6.23	5.99	5.82	5.70	5.60	5.52	5.46	5.37	5.27	4.85
7	8.07	6.54	5.89	5.52	5.29	5.12	4.99	4.90	4.82	4.76	4.67	4.57	4.14
8	7.67	6.06	5.42	5.05	4.82	4.65	4.53	4.43	4.36	4.30	4.20	4.10	3.67
9	7.21	5.71	5.08	4.72	4.48	4.32	4.20	4.10	4.03	3.96	3.87	3.77	3.33
10	6.94	5.46	4.83	4.47	4.24	4.07	3.95	3.85	3.78	3.72	3.62	3.52	3.08
11	6.72	5.26	4.63	4.28	4.04	3.88	3.76	3.66	3.59	3.53	3.43	3.33	2.88
12	6.55	5.10	4.47	4.12	3.89	3.73	3.61	3.51	3.44	3.37	3.28	3.18	2.72
13	6.41	4.97	4.35	4.00	3.77	3.60	3.48	3.39	3.31	3.25	3.15	3.05	2.60
14	6.30	4.85	4.24	3.89	3.66	3.50	3.38	3.28	3.21	3.15	3.05	2.95	2.49
14	6.20	4.77	4.15	3.80	3.58	3.41	3.29	3.20	3.12	3.06	2.96	2.86	2.40
∞	5.02	3.69	3.12	2.79	2.57	2.41	2.29	2.19	2.11	2.0 5	1.94	1.83	1.00

[a]To interpolate between these values, use *reciprocals* of the degrees of freedom.

SECTION II

Experiments

Guide to the Experiments

Section II of this book contains a selection of unit operations of laboratory research in biochemistry, virtually all of which are equally applicable to other areas of biological science. By a unit operation, we mean a single technique with a significant theoretical basis of its own. Thus, spectrophotometry, dialysis, ion-exchange chromatography, and the binding of a ligand to some macromolecule may each be regarded as a typical unit operation. When combined in various ways, these unit operations permit attack and solution of many complicated experimental problems.

We shall apply these operations to a variety of biochemical substrates, including amino acids and small peptides, carbohydrates, fatty acids, enzymic and nonenzymic proteins, and nucleic acids. You will have an opportunity to purify and partially characterize the protein, hemoglobin, which does not have enzymic activity, and you will explore an interesting chemical modification of hemoglobin. You will also isolate, purify, and partially characterize an enzyme known as lactate dehydrogenase. Through this series of experiments, you will learn additional unit operations and something of the propeties of lactate dehydrogenase as well. In further experiments, you will examine the effects of a certain inhibitor on the catalytic activity of the enzyme citrate synthase, using commercial preparations of the enzyme. The preparations of the synthase are highly purified. In these experiments, the emphasis will be on the kinetics of the system. In another experiment, you will explore the effect of an inhibitor on an enzyme(s) in intact cells. The final experiment concerned with enzyme activity will examine how enzyme induction in intact bacterial cells is affected by the addition of certain chemical substances related to metabolites. You will perform two experiments properly assigned to the field of molecular biology: (1) in vitro protein synthesis by a cell free system and (2) purification and endonuclease mapping of a plasmid, as well as of a plasmid with an inserted DNA sequence.

The program outlined is a busy one. You will benefit greatly by careful review of the written protocols and relevant readings in Section I *before* coming to the laboratory. Experience indicates that you will have time to complete the experiments as scheduled provided that you are adequately organized in advance.

A number of the experiments were designed to be as quantitative as possible; in some cases you will be provided with unknown(s), and you will be expected to determine or calculate a particular result.

Some of the experiments will be performed on an individual basis; others will be performed in small groups of two or three. Irrespective of group participation, all students are *individually* responsible for collecting or exchanging the total data required. Even when working in groups, each student is required to submit his/her own laboratory report in complete and final form.

ORGANIZATION OF LABORATORY REPORTS

A good laboratory report will demonstrate (1) your understanding of the basic principles involved; (2) your ability to follow directions; (3) your ability to perform basic manipulations in a manner that gives results of estimated reliability; and (4) your ability to reason from raw data to some desired conclusion in an acceptable manner.

You will be provided with outlines of the form to be followed in composing your reports; these will vary from experiment to experiment. You will find well-defined places for entry of data and results, as well as a provision for such extended verbal comment as you may wish to submit. Neatness, clarity, and good English will be favorably prejudicial to your cause.

When a report requires a graph or graphs, assume that all axes are to be labeled, with units clearly defined. Each graph should be submitted on a separate sheet (8.5 × 11 in.) unless you are otherwise instructed.

PERSONAL PERFORMANCE IN THE LABORATORY

1. Treat all reagents and materials as if they were caustic or toxic. If you spill anything, clean it up at once. If you spill it on yourself, wash it off at once. Familiarize yourself with the location of the fire extinguishers and with the eye-washing stations, in case of an emergency. Special precautions are noted in the detailed protocols you will be furnished. *Heed these carefully!*
2. Federal regulations require, when radioactive materials are handled, that disposable gloves be worn, that protective laboratory coats or aprons be worn, and that bench tops be covered with nonabsorbent material. Radioactive materials may not be transferred by mouth pipetting. These rules also prohibit eating, drinking, or smoking in the laboratory area.
3. Students are expected to provide themselves with laboratory coats or aprons to protect themselves and their clothing. (Safety goggles are supplied and must be worn when required by the protocol.) Disposable items such as gloves will also be provided for you.
4. Any accident resulting in personal injury, no matter how slight, must be promptly reported to the staff. This is for your own protection.
5. Certain instruments will be preset; others not. Do not be a "fiddler." If you have difficulty with any of the equipment, please report the difficulty to the staff.
6. Material stored in refrigerator/freezer space, not properly identified as to owner or content, stands in danger of being discarded by the staff.
7. Part of your laboratory responsibility is to clean up your working area when you are finished. This includes proper disposition of waste materials,

dirty glassware, etc. Specific instructions will be given to you from time to time. Your rule of thumb should be to *leave the site cleaner than you found it.*

8. Although not specifically stated each time, *all water used to prepare reagents has been purified* by distillation, followed by adsorption and ion-exchange column treatment with a final membrane filtration.

Methods for Quantitative Protein Determination*

USE OF AUTOMATIC PIPETS

Competence in quantitative transfer of fluids is essential for success in these laboratory exercises.

The pipets we shall employ are based on the principle of positive piston displacement. Depressing a plunger displaces a certain volume of air; releasing the plunger sucks up that volume of liquid into a replaceable plastic tip. It is important that liquid never be drawn up into the body of the instrument, for this could lead to corrosion of the piston and loss of volume calibration. Once liquid has been drawn up into the plastic tip, depressing the button to a second stop point will eject the liquid into a tube or other container (Fig. E1–1).

These instruments fall into two classes. The first class is designed to deliver a fixed volume, clearly marked on the body of the instrument. The second class includes several models, each clearly identified across the top of the plunger button, which can be adjusted over some volume range. Calibration of the adjustable volume is made by a knurled wheel which adjusts a micrometer. As noted above, all of these instruments must be fitted with either a yellow or a blue plastic disposable tip. Yellow tips are used for smaller volumes, and blue tips for larger volumes; the size of the openings of the tips makes their fit to a given pipet self-evident. Note also that some of the instruments carry a second button; depressing this button automatically disengages the disposable tip without need for touching it by hand. This is useful when one is dealing with radioactive materials, or toxic materials.

Calibration of the micrometer that adjusts delivery volumes depends on the model of the instrument, as follows:

MODEL P-20D:

Volume adjustable from 0 to 20 μl in 0.1-μl increments.

*Preparatory reading: Section I, Chapters 1 and 2; Appendix I.

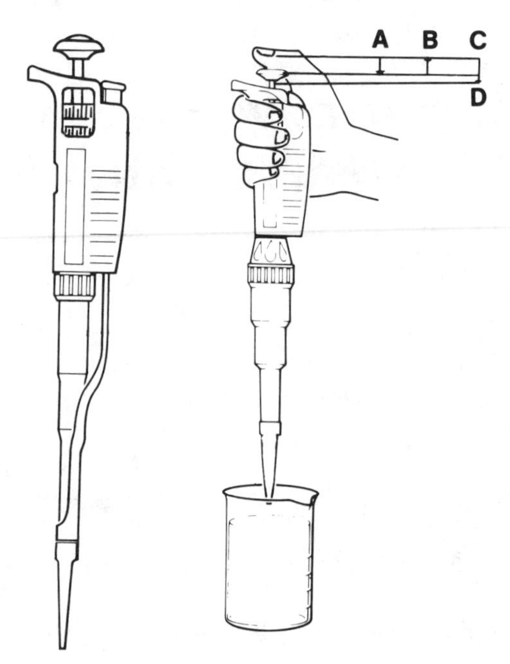

Figure E1–1. Example of adjustable piston displacement pipet. Button is first depressed to position A, the tip is immersed in the fluid to be transferred, then the button is smoothly released (position B). To effect transfer, the button is depressed again (positions C to D). (Reproduced by courtesy of Gilson Medical Electronics.)

MODEL P-200D:

Volume adjustable from 0 to 200 μl in 1.0-μl increments. The scale on the micrometer is calibrated in 0.2-μl increments. This is the model to be used for greatest precision in delivery of 20–200 μl.

MODEL P-1000D:

Volume adjustable from 0 to 1000 μl in 10-μl increments. The scale of the micrometer is calibrated in 2-μl increments. This is the model to be used for the range 200–1000 μl. Although the instrument can be set to deliver smaller volumes, the error of delivery increases significantly below 200 μl.

To adjust an instrument, fit a proper tip to the delivery tube, using a slight twisting motion to seat the tip firmly. Set the micrometer by holding the body of the pipet in one hand while rotating the knurled knob to the desired micrometer reading. Depress the plunger button to the *first* positive stop. This part of the stroke has the calibrated volume displayed on the micrometer scale. Immerse the end of the tip into the liquid to be dispensed, then *slowly* release the plunger while holding the pipet in a vertical position. This draws liquid into the tip. Never allow the plunger to snap back, since this could suck liquid into the calibrated mechanism, as mentioned above. It is well to wait a few seconds to ensure complete filling of the tip, especially with viscous liquids, to prevent introduction of air bubbles.

To dispense the measured sample, depress the plunger button to the *second* positive stop, withdrawing the tip from the container by drawing the free tip end along the side of the vessel. Discard the tip by depressing the tip ejection button or by a slight twisting motion. Tips need not be changed if repeated sampling from a single solution is required. Tips should be changed when going from one liquid sample to another.

These units are warranted to be within 1% accurate when employed at the recommended delivery volumes. The reproducibility is within the limit of 3% for the P-20D and 1.5% for the P-1000D. These performance estimates apply in the temperature range of 22 ± 5°C, provided that the other precautions regarding the vertical position of the pipet and smoothness of plunger release are observed. Use of the instruments at much higher or much lower temperatures requires that the tips be prerinsed at least twice with the liquid to be dispensed.

Air bubbles may occasionally be noted within the tip during liquid intake. If this occurs, return the liquid to its container, then refill the tip. If air bubbles can still be seen in the tip, it should be discarded as faulty, and a new tip should be fitted.

PROTEIN DETERMINATIONS: MISCELLANEOUS METHODS

A variety of methods, none of which is entirely specific for proteins, are available for determination of the protein content of a given sample. The methods employ different principles, and may be sensitive to interferences by certain salts, by buffer components, and by some solvents. Each method therefore has certain unique and useful characteristics as well as certain limitations.

Good general references detailing the most common methods are the papers by Layne and by Peterson in *Methods in Enzymology*. All of the methods cited there have one feature in common; they must be standardized against some pure, known protein, usually crystalline bovine serum albumin. Two additional methods are available which do not require secondary standardization. The first of these is the Kjeldahl method, in which the nitrogen content of a protein is converted to NH_3. The NH_3 can be collected and titrated directly with gravimetrically standardized HCl. The second direct procedure is that of Dumas. Here the protein is thermally oxidized in the presence of CuO which serves as an oxygen source. The gaseous oxidation products include N_2 gas, resulting from the protein breakdown. It is the volume of the bubble of N_2 that is measured, under standard conditions of temperature and pressure.

Fluorescence methods of high sensitivity are available in which a reagent—for example, fluorescamine or *o*-phthalaldehyde—becomes intensely fluorescent only after reacting with a functional group in the protein. For the two reagents just mentioned, free amino groups are so derivatized.

Many dyes are noncovalently bound to proteins. Methyl orange, phenolphthalin (*not* the indicator, phenolphthal*ein*), 4′-hydroxyazobenzene-2-carboxylic acid (HABA), and Coomassie Blue (anazolene sodium) are examples. Various analytical methods based on dye binding have been proposed, but few have stood the test of time, for various reasons. A revival of interest in these procedures is exemplified by the method of

Bradford, based on use of Coomassie Blue. In this procedure the absorption maximum of Coomassie Blue shifts from 465 nm to 595 nm when the dye binds to a protein. You will have the opportunity to test this method.

What must be clearly understood is that each method capitalizes on the presence in the protein of a particular structure—for example, the aromatic amino acid content (ultraviolet method) or lysine content (*o*-phthalaldehyde method)—and this structure may vary in occurrence among various proteins.

PRECAUTIONS REGARDING PROTEIN STANDARDS

The bovine serum albumin, which is generally employed as a standard, must also be selected and handled with care. The preparation, preferably fat-free, should be carefully dialyzed to rid it of salts and nonprotein nitrogenous compounds. It should, of course, be stored in the cold. It should be thoroughly dried, preferably over a dish of P_2O_5 in a tightly sealed dessicator. The dry weight may be obtained as follows: The protein to be used as a standard is removed from its storage over P_2O_5; the time of exposure of protein to room air is "zero time" or t_0. The sample is transferred to a tared weighing dish, the dish placed on the pan of a balance, and the weight noted at timed intervals. Usually five or six measurements of increasing weights are made. The weights are plotted against time, and the initial linear portion of the plot regressed to t_0. The intercept on the ordinate is defined as the dry weight. Depending upon the atmospheric humidity, the difference in weight obtained by this method as compared to a weight obtained after the protein has equilibrated in room air may vary from 1% to 5% or more.

An alternate procedure, which minimizes the need for exact dry weight determinations, depends on spectrophotometric assay of the standard solution. Foster et al. (1965) have shown that an albumin solution containing 100 mg per 100 ml has a value for $A_{279nm} = 0.667$ in a 1-cm light path. This nongravimetric standardization also provides a useful check on weighings. It obviously is not an acceptable standardization procedure for ultraviolet absorption-based methods.

Lowry Protein Determination*

MATERIALS AND REAGENTS

1. 0.1% (w/v) $CuSO_4 \cdot 5H_2O$, 0.2% (w/v) sodium and/or potassium tartrate, 10% (w/v) Na_2CO_3. Prepare by dissolving Na_2CO_3 in about one half the final desired volume (for this experiment, in about 1000 ml ÷ 2 = 500

*This procedure is essentially that of Peterson (1977).

Table E1–1. Comparison of Some Methods for Protein Determination

	Phenol	Biuret	UV Absorption at: λ = 260/280	UV Absorption at: λ = 230/210	Dye Binding	Ninhydrin
Basis of method	Biuret rxn. *plus* Tyr, Trp reduction of Folin–Ciocalteau reagent	Cu^{2+} complex with peptide bonds	Tyr, Trp	peptide bonds, aromatics	color change of dye upon binding	amino groups released by hydrolysis in base (NaOH)
Useful range[a] Sensitivity[b] ($A_\lambda^{1\%}$)	10–200 μg $A_{750} = 3.3$ $A_{650} = 2.5$	1–10 mg $A_{545} = 0.04$	0.2–2 mg $A_{280} = 1.0$	10–100 μg $A_{210} = 20$ $A_{225} = 14$	0.3–2.5 μg $A_{395} = 9.0$	10–100 μg $A_{570} = 20$
Advantages	quick, simple, widely used, sensitive	A_{545} rather insensitive to reaction time and protein structure	fastest method; needs UV spectrophotometer; sample can be recovered; best for pure proteins		sensitive; one reagent needed; rapid reading time	sensitive; A_{570} independent of protein
Disadvantages	nonlinear standard curve; color depends on protein	insensitive; reagents not too stable	absorbance depends on protein structure		corrosive reagent required; responses of different proteins often vary	slow; needs oven, water bath, etc.
Interfering substances	phenol, uric acid, xanthine, EDTA, guanine, glycine, histidine, tris, mercaptoethanol, P_i, $(NH_4)_2SO_4$	lipids, $(NH_4)_2SO_4$		nucleic acids, aromatic compounds, mercaptoethanol, many buffer ions at λ < 230 nm	alkali; detergents	amino compounds
Noninterfering substances	0.5% urea or guanidine, or most salts		most salts at λ = 280 nm			

[a]Most linear portion of the calibration curve obtained on common spectrophotometers.
[b]Based on 1-cm light path, calculated for a 1% (w/v) crystalline serum albumin preparation.

ml). Add this slowly and with stirring to the copper sulfate–tartrate solution (also in about 500 ml). This is *solution A;* it is stable at about 10°C; at lower temperatures, carbonates may crystallize.

2. 5% (w/v) sodium dodecyl sulfate (SDS) *(solution B).*
3. 0.8 M NaOH *(solution C).*
4. BSA (bovine serum albumin), 0.3 mg/ml water = 300 μg/ml water.
5. Lysozyme, 0.3 mg/ml water.
6. Unknown protein solution.
7. 3 M KCl.
8. 1 M Tris-HCl (pH 7.5).
9. 5% (w/v) Triton X-100.
10. 0.25 M dithiothreitol.
11. 2 N Folin–Ciocalteau phenol reagent.

PROCEDURE

1. Turn on spectrophotometer and allow it to warm up.
2. Number ten 10 × 150 mm test tubes and place them in a test-tube rack. In each tube, carefully pipet one of the following volumes of a 0.3 mg/ml BSA solution: 0, 0.05, 0.1, 0.2, 0.3, 0.4, 0.5, 0.6, 0.8, and 1.0 ml, and bring all tubes to a final volume of 1 ml by adding the appropriate amount of water. If a double-beam spectrophotometer is used, prepare *double the volume* of the "0" ml or blank tube.
3. Set up two tubes containing 0.14 ml of lysozyme solution in a final volume of 1.0 ml.
4. The protein concentration of your unknown solution will be between 0.1 mg/ml and 1.0 mg/ml. Set up three tubes with different amounts of your unknown in a final volume of 1.0 ml. You are using different amounts of protein in hopes of finding at least one that falls within the range of standards.
5. Set up and number four tubes, each containing 0.14 ml BSA and 0.66 ml H_2O. Then:
   a. To tube 1 only, add 0.1 ml of 3 M KCl.
   b. To tube 2 only, add 0.1 ml of 5% Triton X-100.
   c. To tube 3 only, add 0.1 ml of 0.25 M dithiothreitol.
   d. To tube 4 only, add 0.2 ml of 1 M Tris·HCl (pH 7.5).
   Bring all tubes to a final volume of 1.0 ml with purified water.

6. In a 125-ml Ehrlenmeyer flask mix 10 ml of solution A with 20 ml of solution B and 10 ml of solution C. Mix carefully.

7. Add 1.0 ml of the solution from step 6 to each of your sample tubes. (For double-beam instruments add 2.0 ml to the blank tube.) Vortex them to mix thoroughly after each addition. Incubate the tubes at room temperature for 10 min.

8. While the tubes are being incubated, add 10 ml of 2 N Folin–Ciocalteau phenol reagent to 50 ml distilled water in a 125-ml Ehrlenmeyer flask and mix thoroughly.

9. At the conclusion of the 10-min incubation period, forcibly pipet 0.5 ml of the yellow solution from step 8 into each sample tube and vortex *immediately* and *thoroughly.* (For double-beam instruments, add 1.0 ml into the blank tube.)

10. Incubate the samples at room temperature for 30 min.

11. Determine the absorbance of each sample at a wavelength of 740 nm. Since all absorbance readings will be made at wavelengths in the visible range, disposable plastic cuvets may be used. Polystyrene cuvets will transmit about 80% of incident light for wavelengths of $\geq$330 nm; acrylic has the same transmittance for wavelengths $\geq$300 nm. Scan the visible spectrum (400–650nm) of the 0.2-, 0.4-, and 0.6-ml BSA standard tubes. Record each scan, superimposing each curve on the same portion of the chart paper.

REPORT

1. On a sheet of graph paper, construct a graph of A_{540} (y axis) vs. [BSA] (x axis). Label the x axis in micrograms (μg) of BSA. Calculate the micromolar extinction coefficient for the protein:Cu:phosphotungstate–phosphomolybdate complex.

2. Determine the protein concentration of the standard lysozyme solution on the basis of the BSA standard curve. Explain any difference from the predicted value.

3. Determine the concentration of your unknown protein solution.

4. Report the results from the four tubes in step 5:
   a. KCl
   b. Triton X-100
   c. Dithiothreitol
   d. Tris-HCl (pH 7.5)

   Explain briefly the differences noted relative to the material added.

Bradford Protein Determination

Note: *This formulation is based on the use of a Coomassie Blue sample that contains 65–75% dye. Commercially available Coomassie Blue G250 dye products vary in dye content.*

MATERIALS AND REAGENTS

Dissolve 100 mg Coomassie Blue G250 in 50 ml 95% ethanol. To this solution, add 100 ml of 85% phosphoric acid. Dilute the concentrated solution to 0.8 L with distilled water and then filter through Whatman No. 2 paper.

PROCEDURE

1. In seven 16 × 150 mm tubes, place the following amounts of a 0.3 mg/ml BSA solution: 0.0, 0.05, 0.1, 0.2, 0.3, 0.4, and 0.5 ml.
2. In two tubes, dispense 0.25 ml of a 0.3 mg/ml lysozyme solution.
3. Add water to each tube from steps 1 and 2 to bring the final volume to 1 ml. (If a double-beam spectrophotometer is used, *double* the volumes in the blank or "0" ml tube.)
4. Add 4 ml of reagent rapidly to each tube and vortex; avoid excessive foaming.
5. After 5 min to 1 hour (but no longer), read the absorbance at 595 nm.

REPORT

1. On a sheet of graph paper, plot a standard curve of A_{595} (y axis) vs. μg BSA (x axis).
2. What is the experimentally determined concentration of your 0.3 mg/ml lysozyme solution based on a BSA standard?
3. Compare the results with those from the Lowry procedure. Explain any difference.

REFERENCES

General Reviews

Layne, E. *Methods Enzymol. 3:*447 (1957).
Peterson, G. L. *Methods Enzymol. 91:*95 (1983).

Kjeldahl Methods: Determination as NH_3

Bock, J. C., and Benedict, S. R. *J. Biol. Chem. 20:*47 (1915).

Conway, E. J. *Microdiffusion Analysis and Volumetric Error,* 3rd ed. Crosby Lockwood & Sons, London (1950).
Kjeldahl, J. *Anal. Chem. 22:*366 (1883).
Koch, F. C., and McMeekin, T. L. *J. Am. Chem. Soc. 42:*2066 (1924).

Dumas Methods: Determination as N_2
Dumas, J.B.A. *Anal. Chim. Phys. 47:*198 (1831).
Kirsten, W. In *Comprehensive Analytical Chemistry* (Wilson, C. L., and Wilson, D. W., eds.), Vol. 1b, p. 494. Elsevier, Amsterdam (1960).

Biuret Methods
Beisenherz, G., Boltze, H. J., Bücher, T., Czok, R., Garbade, K. H., Meyer-Arendt, E., and Pfleiderer, G. *Naturforschung 8b*(10):555 (1953).
Gornall, A. G., Bardawill, C. J., and David, M. M. *J. Biol. Chem. 177:*751 (1953).
Robinson, H. W., and Hogden, C. G. *J. Biol. Chem. 135:*707 (1940).
Weichselbaum, T. E. *Am. J. Clin. Pathol.* (Tech. Sec. 10) *16:*40 (1946).

Phenol–Alkaline Copper Methods
Chou, S-C., and Goldstein, A. *Biochem. J. 75:*109 (1960).
Lowry, O. H., Rosebrough, N. J., and Randall, R. *J. Biol. Chem. 193:*265 (1951).
Peterson, G. L. *Anal. Biochem. 83:*346 (1977).

Bicinchoninic Acid–Alkaline Copper
Smith, P. K., Krohn, R. I., Hermanson, G. T., Mallia, A. K., Gartner, F. H., Provenzano, M. D., Fujimoto, E. K., Goeke, N. M., Olson, B. J., and Klenk, D. C. *Anal. Biochem. 150:*75 (1985).

Ultraviolet Absorption Methods
Foster, J. F., Sogami, M., Petersen, H. A., and Leonard, W. J. *J. Biol. Chem. 240:*2495 (1965).
Kalckar, H. *J. Biol. Chem. 167:*461 (1947).
Waddell, W. J. *J. Lab. Clin. Med. 48:*311 (1956).
Warburg, O., and Christian, W. *Biochem. Z. 310:*384 (1941).

Turbidometric Methods
Bücher, T. *Biochim. Biophys. Acta 1:*292 (1947).
Heppe, F., Karte, H., and Lambrecht, E. *Hoppe-Seyler's Z. Physiol. Chem. 286:*207 (1951).
Sieber, A., and Gross, J. *Laboratoriumsblätter (Behring) 26:*117 (1976). [Laser method]

Ninhydrin Methods
Fruchter, S., and Crestfield, A. M. *J. Biol. Chem.* 240:3868 (1965).
Moore, S. *J. Biol. Chem. 243:*6281 (1968).
Moore, S., and Stein, W. *J. Biol. Chem. 211:*907 (1954).

Fluorescence-Linked Methods
Böhlen, P., Stein, S., Dairman, W., and Udenfriend, S. *Arch. Biochem. Biophys. 155:*213 (1973). [Fluorescamine]
Lee, H.-M. Forde, M. D., Lee, M. C., and Bucher, D. J. *Anal. Biochem. 96:*298 (1979). [*o*-Phthalaldehyde]

Dye Binding Methods
Bradford, M. *Anal. Biochem. 72:*248 (1976).
Read, S. M., and Northcote, D. H. *Anal. Biochem. 116:*53 (1981).

pH Measurements*

A pH meter with hydronium ion (H_3O^+)-sensitive electrode(s) is used for direct potentiometric measurement of equilibrium hydronium ion activity and for potentiometric titration of sample solutions. The electrode potentials developed in standard and unknown solutions are proportional to the pH. In this laboratory exercise, direct reading pH meters are used together with combination electrodes (i.e., electrodes in which both the glass and reference electrodes are combined).

pH meters require a warm-up time of 15–30 min, and in most biochemical laboratories they are left plugged in ("On") with the function switch on "Standby." Both manual and automatic temperature control is possible, but for most applications temperature control is manually adjusted by turning the control knob to the temperature of the standard and unknown solutions (which should be equal).

To measure pH, the well-rinsed (H_2O) and dried (absorbant tissue) electrodes are placed in the sample and the function control switch is placed in the "pH" position. The "calibration" knob is used (with standard pH solutions) to set the meter to the pH of the standard solution. To avoid parallax errors in the readings, the needle of the meter should be aligned with its image in the mirror behind the scale. There is no such problem, of course, with digital readout instruments.

PROCEDURE

Care of the Combined Glass Electrode

Note: *Be careful! The electrodes are fragile and expensive.*

1. Inspect the electrodes; the level of liquid in the outer jacket should be at the "fill hole." If it is not, add some 4 M KCl saturated with AgCl or add saturated KCl, depending upon the electrode used. (This solution is provided.)
2. Readings are made with the "fill hole" unstoppered.

*Preparatory reading: Section I, Chapters 1 and 4.

3. When the pH meter is not in use, stopper the "fill hole" and immerse the lower one-third of the electrode in buffer solution (pH 7).

4. When transferring the electrode from one solution to another, always rinse the electrode with water several times (discard the washings) and dry it carefully with tissue.

5. When using a magnetic stir bar, *do not* allow the stir bar to hit the electrodes, and always turn the stirrer off when taking the pH reading.

6. When taking pH readings, do not allow the electrodes to touch the side of the sample vessel.

Calibration of the pH Meter

1. Set the temperature control to room temperature (a thermometer will be available).

2. Fill a small beaker or test tube with the standard pH 7 buffer.

3. Raise the electrode from the storage solution (have beaker ready for waste), and wash and dry the electrode.

4. Unstopper the "fill hole," and raise the beaker or tube containing the pH 7.0 standard up under the electrode until the bulb portion and the reference contact port are submerged. **Care!**

5. Place the meter function switch to "pH"; wait a few seconds, and then set the meter to read 7.0, using the calibration knob.

6. Turn the switch to "Standby," remove the sample, and wash and dry the electrode.

7. Without touching the calibration knob, read the pH of the standard solutions provided (pH 2.0, 4.0, 8.0, 10, 12). Record the meter reading for each standard solution.

8. Construct a correction curve by plotting the difference in pH (ΔpH = meter reading − standard pH) on the y axis against the standard (x axis). In subsequent titrations add or subtract these differences to give the correct pH; for example, if the reading is 4.02 and the standard pH is 4.01, substract 0.01 from readings at pH 4 during titrations to give the correct pH. Use this curve to correct all pH readings. (If a different meter is used on different days, a new correction curve will be needed for each pH meter used.)

Determination of the pH of an Unknown Solution

Proceed as in (7) above.

Potentiometric Titrations Using the pH Meter

Solvent Blank

Measure 15.0 ml of 0.2 M KCl and place it in a small (20 ml) beaker. Add a magnetic stir bar, place the beaker atop a magnetic stirrer, and carefully insert the electrode into the liquid. Read the initial pH and adjust it to pH 12.0, if necessary, with a KOH solution as required. Now add 1 M HCl to decrease the pH by about 0.5 pH units. Record the pH and the volume of acid added. Continue down to about pH 1.5.

Titration of Unknown

1. Discard the acidified 0.2 M KCl solution. *Do not discard the stir bar!* Wash it with distilled water and dry with tissue.
2. Rinse both the beaker and electrode with distilled water and dry.
3. Place 15.0 ml of your unknown solution in the beaker. Note the concentration of the unknown solution (the solvent is a 0.2 M KCl solution).
4. Adjust the pH of the unknown solution to pH 12.0 and then perform the incremental titration as above to a pH of about 1.5. (**Hint:** Do the titration rapidly the first time, then repeat slowly with another sample to collect more data in the pH region bracketing the pK values of your unknown.)

Determination of the pH of Two Buffer Solutions

1. Measure the pH of the 1 M phosphate buffer solution. Record the answer. Make a 1:100 dilution of this buffer and remeasure the pH. Record your answer.
2. Repeat the above procedure with the 1 M sodium acetate buffer.

Henderson–Hasselbach Equation

This equation is:

$$\mathrm{pH} = \mathrm{p}K + \log \frac{[\text{base}]}{[\text{acid}]}$$

The pK_{a2} for phosphate is 7.20; the pK_a for acetate is 4.75; and the pK_a for Tris is 8.30.

1. Calculate the relative amounts of 0.1 M base and 0.1 M acid for each buffer that have to be mixed to give buffers of the following pH values:

Sodium phosphate:	7.68
Sodium acetate:	5.45
Tris-HCl	8.00

2. Make up the buffers according to your calculations. Record the experimental pH values obtained.

REPORT

1. Prepare a pH meter correction curve by plotting ΔpH (meter reading − standard pH) on the *y* axis against the standard pH on the *x* axis.
2. Report the number and pH of your unknown solution.
3. Plot the volume of acid added to the solvent (curve 1) and the volume of acid added to your unknown (curve 2) to achieve a given pH (as the *y* axis) vs. the correct pH (as the *x* axis).
4. Subtract the volume of acid added to the solvent from the volume of acid added to your unknown to achieve the same pH. Calculate this value (Δ-value) for at least each 0.5 pH unit. Convert the Δ volume of acid to millimoles of acid added per millimole of unknown titrated. Plot this value (*y* axis) against pH (*x* axis).
5. Report p*K* (or p*K*s) of your unknown.
6. Report the pH values of dilutions from step 5.
7. Calculate the amounts of acid and base needed in step 6, using the Henderson–Hasselbach equation and the experimental pH values obtained with these ratios.
8. Explain any differences between the predicted and experimental results in steps 5 and 6.

Liquid Scintillation Counting of Radioactive Decompositions*

This experiment will demonstrate the use of a liquid scintillation spectrometer or "counter" (LSC). The particular instrument used in this exercise has a capacity of 300 samples. Each position on a continuous conveyer train is numbered; corresponding numbers will appear on the printout tape. In more recent instruments, racks serve as holders for the vials containing the scintillation "cocktail" and the sample to be counted. Such counters may have a capacity of over 300 vials of 20-ml volume, or over 600 of the so-called minivials (about 7-ml volume). The vials are lifted into the instrument for counting rather than being lowered from positions as in the older conveyer train LSC models.

Each series of samples must be preceded by a *channel tower* residing in the position immediately ahead of the first sample. The tower has five openings through which a red light may shine. For the lower four openings, a filter may be rotated into position by means of a knurled wheel, thereby blocking the light from reaching a photocell. The system will now count all following samples in channel(s) corresponding to the blocked opening(s). The blocking filters are color-coded to correspond to the colored channel selector buttons. In sequence these are usually external standard (yellow); ^{3}H (red), channel A; ^{14}C or ^{35}S (green), Channel B; and ^{32}P (blue), channel C. If no light is intercepted and all photocells are illuminated, as is the case for an empty conveyer position, the conveyer belt will continue to advance. A channel tower with all openings clear is placed at the end of the sample group to signal termination of counting in the channel just used.

Windows for the three counting channels have been preset. *No adjustments need be or should be made.* The Gain and Preset Error controls should also be left undisturbed.

Each sample may be counted for the time selected on the Preset Minutes dial. The color-coded pushbuttons are pressed to select the channels to be counted. If one of these buttons is pressed while counting is proceeding according to channel tower programming, counting will be interrupted and will recommence in the channel corresponding to the button

*Preparatory Reading: Section I, Chapter 3.

depressed. The Log Ratemeter displays the counting rate in counts per minute (upper scale). When the Preset Error button is pushed, the meter needle displays the 2-sigma (2σ; 95% confidence) statistical counting error (bottom scale).

To begin the counting procedure, load up the samples in the conveyer chain or racks, *making note of the numbered positions they occupy.* Insert appropriately adjusted channel towers immediately before and after the sample group. Usually the mode-selector switch is set at the *single cycle* automatic position. Depress the channel selector pushbutton(s) as desired for the isotope(s) being counted. Depressing the Count pushbutton initiates counting in all modes. When it is depressed again at any time during operation of the spectrometer, the counting will cease and the count data will be printed out.

The printout format is as follows (this may vary with different instruments):

Column I	Col. II	Col. III	Col. IV	Col. V
Sample No.	External standard ratio	Channel A cpm (% error)	Channel B cpm (% error)	Channel C cpm (% error)

For this experiment, common and individual samples will be located at several points in the laboratory. Please return all common samples to the appropriate area when you are finished with them. All waste materials must be disposed of in the proper containers marked for radioactive wastes. Gloves must be worn for all operations.

PROCEDURE

Part I

Take five vials from the box marked "quenching experiment," each of which contains a known amount of ^{3}H-labeled compound and 10 ml of scintillation cocktail. Label the caps 1–5, load the vials into the counter, wait 5 min, then count each vial for 2 min. Remove the vials from the LSC and take them to your bench. Open the vials, but make certain that each cap remains clearly associated with its own vial. Add 0.2, 0.6, 1.0, 1.4, or 1.8 ml of acetone to each vial. Make certain that during the addition you do not contact the contents of the vial. Mark the caps as to the volume of acetone added, close the vials, and shake to mix. Load the vials again into the LSC, wait 5 min, then count each sample for 2 min as above. These samples are to be discarded when you are done. Report any spillage to the staff.

Part II

Locate a vial labeled "STAT." Count the same vial three times for each of the following times: 0.1, 0.4, 1.0, 4.0, and 10 min. After loading the vials, wait 5 min before counting.

Part III

Collect a set of three standard vials containing known amounts of ^{3}H, ^{14}C, and ^{32}P; also collect your own set of three unknown vials. Count each vial for 5 min on each of the three channels. The counting will not be completed before you leave the laboratory and will continue overnight. Retrieve your data from the teaching staff the next day.

Part IV (Optional): Generating an External Standard Channels Ratio (ESCR) Quench Correction Curve

Obtain a set of sealed vials, each containing a known amount of either ^{14}C or ^{3}H in 10 ml of scintillation cocktail. To each of these vials has also been added different amounts of a quencher, usually nitromethane. Such a set of quenched standards is commercially available. Alternatively, at the direction of the course instructor, a student may use the quenched samples from Part I of this experiment.

Count each vial in two channels—for example, ^{3}H and ^{14}C—with the instrumental external standard (gamma source) positioned next to the sample for 1 min. Count them as well *without* the gamma source, for about 12 sec (the exact time will be given to you). This counting program can easily be set up in modern LSCs, but the particular settings to be made will vary with the model of instrument used.

CALCULATIONS

Subtract the counts (properly normalized for the same time interval) obtained with the sample alone (A), from the counts obtained with the sample plus external standard (B). Repeat for each sample and for each channel. Calculate the ratio of the two net counts to give the external standard channels ratio (ESCR)

$$\text{ESCR} = \frac{B_{\text{channel}_1} - A_{\text{channel}_1}}{B_{\text{channel}_2} - A_{\text{channel}_2}}$$

Next calculate the extent of quenching of each sample in the appropriate channel, that is, the ^{3}H channel for ^{3}H quenched series, as follows:

$$\%\ \text{Counting efficiency} = \frac{\text{Sample cpm} - \text{Background cpm}}{\text{dpm}}$$

The percent counting efficiency (ordinate) is plotted against the ESCR (abscissa) to obtain the correction curve.

REPORT

The report should contain:

1. For the quenching experiment, a plot of dpm (ordinate) vs. volume of acetone (abscissa).
2. Statistical data: a table to include counting time, average counts, standard deviations.
3. Standards and unknowns:
   a. From the known specific activities of the standard samples (values to be supplied to you), calculation of the efficiency of counting in each channel for each sample.
   b. From the counting patterns of your unknowns in all channels, identity of the isotopic species in each sample.
4. A plot of the ESCR values (abscissa) vs. the counting efficiency for ^{3}H (ordinate). [Optional]

Avidin–Biotin Complexation: Titration of Ligand Binding*

Biotin is an essential cofactor in the conversion of carbon dioxide from bicarbonate to carboxylate groups. For example, the carboxylation of pyruvate to oxaloacetate is catalyzed by the enzyme, pyruvate carboxylase, to which biotin is covalently bound through the amino-terminal of a lysine residue in the apoenzyme. Conversion of biotin to N^1-carboxybiotin occurs at the site indicated by the arrow on the nitrogen in the structural formula shown below. The energy to drive the conversion of bicarbonate plus biotin into N^1-carboxybiotin is obtained from ATP. Details of this two-step carboxylation of pyruvate may be found in any standard biochemistry text.

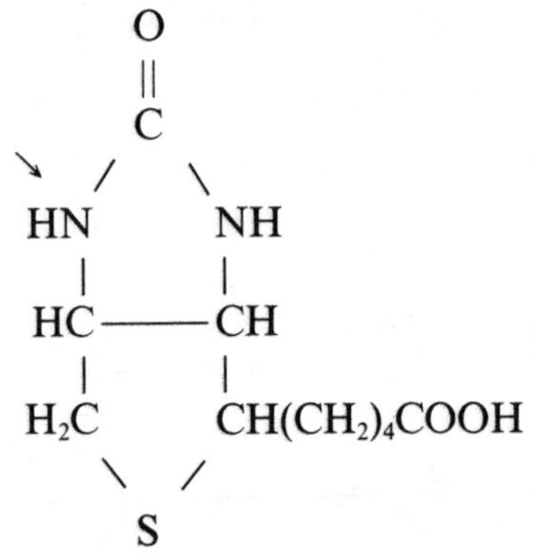

```
              O
              ||
              C
      ↘   /       \
      HN           NH
       |            |
      HC ————— CH
       |            |
      H2C          CH(CH2)4COOH
           \      /
              S
            Biotin
```

Avidin is a glycoprotein (MW = 66–70 × 10^3; there are several different forms present in egg whites) possessing an extremely high affinity for biotin (K_d for biotin = 10^{-15}). There are four binding sites per mole of avidin. Recent evidence suggests that the biotin binding sites are in the form of clefts in the hydrophobic areas of the avidin. The biological function of avidin is unknown, and the only cases of mammalian biotin deficiency have occurred in humans on diets rich in raw egg whites. Cooking, apparently, causes heat denaturation of avidin.

Avidin is also known to bind certain anionic dyes, such as 4′-hydroxy-

*Preparatory reading: Section I, Chapters 1, 2, 5, and 7.

azobenzene-2-carboxylic acid (HABA) at the same sites that bind biotin, although the binding of these dyes is much weaker (K_d for HABA = 5.8 $\times 10^{-6}$)

4′-Hydroxyazobenzene-2-carboxylic Acid

HABA itself absorbs ultraviolet (UV) light with a λ_{max} = 348 nm. When HABA is complexed with avidin, the λ_{max} is shifted to 500 nm, and this change affords the basis of a spectrophotometric measurement. When λ_{max} is measured at 500 nm, the molar extinction coefficients for the free and bound forms of HABA are 600 and 34,500, respectively.

A convenient and accurate assay for avidin involves the displacement of HABA from avidin by biotin, while monitoring A_{500} nm. In this experiment you will first titrate pure avidin, then a sample of avidin diluted with other proteins. These data, together with measurement of the total protein in the second sample, will provide an assay of the concentration of avidin in that sample.

MATERIALS AND REAGENTS

1. *Pure avidin reagent:* Dissolve the compound in 0.1 M Tris-HCl buffer (pH 7.0), to give a solution containing 0.2 mg/ml.
2. *HABA reagent:* Dissolve the pure reagent in 0.1 M Tris-HCl buffer (pH 7.0), to make a 0.56 mM solution of the dye. The molecular weight of HABA = 242.2.
3. *Biotin:* Make a 0.4 mM solution in Tris-HCl buffer (pH 7.0). The molecular weight of biotin = 244.3.
4. *Unknown avidin solutions:* A variety of these will be prepared by the staff. Some will contain pure avidin at concentrations other than the concentration of the reagent mentioned above. Others may contain deliberate additions of proteins other than avidin. Each student group will receive one such solution to serve as a sample unknown.

PROCEDURE A

Mix 1.0 ml of HABA reagent with 1.0 ml of water and measure the value of A_{500} against a reference cuvet containing water only. Record this value; then carefully clean the sample cuvet.

Mix 1.0 ml of HABA reagent with 1.0 ml of pure avidin reagent, mix well, then measure A_{500} against a water reference cuvet. Record the value.

Without removing the sample from the spectrophotometer, add 5 μl of biotin reagent, then mix by means of a Pasteur pipet fitted with a rubber bulb. Aspirate and eject the solution several times to mix the contents of the cuvet. Record the value of A_{500} again.

Add another 5-μl aliquot of biotin reagent, mix the cuvet contents as described above, and again measure A_{500}. Repeat this sequence of biotin addition, mixing, and measurement of A_{500} until the absorbance change becomes substantially constant, reflecting only slight dilution of the cuvet contents.

PROCEDURE B

Mix 1.0 ml of HABA reagent with 1.0 ml of the unknown avidin solution. Measure the value of A_{500} against a reagent blank as before. Record the value of A_{500}.

Repeat the sequence of steps outlined in (A); that is, add 5-μl aliquots of biotin with mixing and measurement of A_{500} after each addition. Thus you will generate a second set of data relative to the interaction of the impure avidin with biotin in the presence of HABA.

Using the Bradford method, determine the concentration of total protein in the impure avidin solution, expressing the result as milligrams of protein per milliliter of solution.

REPORT

1. Submit your titration graphs, where A_{500} (y axis) is plotted against quantity (μg) of biotin added (x axis).
2. An avidin unit is defined as the quantity that binds 1.0 μg of biotin. What is the activity (U/mg) of the pure and the impure avidin preparations?
3. What is the significance of the intersection points on the graphs you have made?
4. What factors might interfere with HABA binding?
5. Calculate the expected absorbance (A_{500}) of a 0.28 mM solution of HABA, measured in a cuvet with a 1-cm light path.
6. Using the ΔA_{500} on addition of avidin to the HABA solution, what is the concentration of HABA binding sites occupied by HABA? How many micrograms of biotin are needed to replace HABA at all these binding sites?

REFERENCES

Baxter, J. H. *Arch. Biochem. Biophys. 108:*376 (1964).
Chingnell, C. F. *J. Biol. Chem. 250:*5622 (1975).
Green, N. M. *Biochem. J. 89:*599 (1963).
Green, N. M. *Biochem. J. 94:*23c (1965).
Ness, A. T., Dickerson, H. C., and Pastewka, J. V. *Clin. Chem. Acta 12:*532 (1965).
Rutstein, D. D., Ingenito, E. F., and Reynolds, W. E. *J. Clin. Invest. 33:*211 (1954).

Separation of Fatty Acids by Argentation Chromatography*

In this form of thin-layer chromatography (TLC), the adsorbent layer is made of silica gel H (finely powdered SiO_2 without any binder) instead of cellulose (cf. Expt. 14). It demonstrates that the adsorbent properties of the thin layer can be modified by addition of a reagent (Ag^+) so as to distinguish selectively between differing degrees of unsaturation in long-chain fatty acids.

The surface of silica gel contains many hydroxylic groups; these play a significant role in adsorption of solutes. Carboxylic acids, such as the long-chain fatty acids, may bind to silica so tightly that development of the plate (separation of solutes) becomes inconveniently slow. To overcome this difficulty, fatty acids are usually converted to their methyl esters before their application to the plate. Most interest in fatty acids relates to their occurrence in lipid esters, so it is customary to *transesterify* lipids, freeing the fatty acids and generating their esters simultaneously. A very similar methodology is employed for GLC or GLC–MS studies. One of the most reliable methods for transesterification is credited to Metcalfe and Schmitz. It employs a solution of BF_3 in a mixture of ether and methanol. The reaction is typified by the following equation:

$$\begin{array}{l} H_2C{-}O{-}\overset{O}{\overset{\|}{C}}{-}R \\ \quad | \\ HC{-}O{-}\overset{O}{\overset{\|}{C}}{-}R' \\ \quad | \\ H_2C{-}O{-}\overset{O}{\overset{\|}{C}}{-}R'' \end{array} + CH_3OH \xrightarrow{BF_3} \begin{array}{l} H_2COH \\ \quad | \\ HCOH \\ \quad | \\ H_2COH \end{array} + \begin{array}{c} CH_3{-}O\overset{O}{\overset{\|}{C}}R \\ + \\ CH_3O\overset{O}{\overset{\|}{C}}R' \\ + \\ CH_3O\overset{O}{\overset{\|}{C}}R'' \end{array}$$

One simply mixes an aliquot of lipid (or fatty acid) with an excess of the reagent and allows the mixture to stand for 1 hour at room temperature (or at 40°C for complex lipids), destroys the excess reagent with water, then extracts the residue with $CHCl_3$ to obtain a clean solution of the fatty acid esters.

*Preparatory Reading: Section I, Chapter 7.

Unmodified silica gel layers will separate fatty acid esters according to chain length, using solvent systems of $CHCl_3/CH_3OH$ in various proportions. Similar results can be obtained with systems composed of benzene, ether, and petroleum ether. The effect of unsaturation is less important than the effect of chain length.

Silica gel layers can be modified in various ways—for example, as in *reversed-phase* chromatography. It should be evident that there the function of the silica is only to hold the silicone oil on the backing of the plate. Similar technology is now quite popular in preparation of HPLC columns.

Another simple and effective layer modification is to incorporate certain ions into the silica. The plates we shall use were made by addition of $AgNO_3$ into the silica gel slurry from which the plates were cast. The presence of Ag^+ makes a dramatic change in the separation scheme. Now, the degree of unsaturation plays a pronounced part in the separation as a result of π-bond formation between the region(s) of unsaturation and the Ag^+. One of the foremost advocates of this method is Lindsey Morris, who coined the term *argentation chromatography* for these procedures. Morris has reported separations by position and number of double bonds in an extensive series of fatty acids with 16 or 18 carbon atoms.

Silica gel layers can also be modified by incorporation of H_3BO_3, which permits the formation of borate-ester-like adducts with dihydroxy fatty acids. Dihydroxy fatty acids can be separated into threo or erythro forms on layers containing both Ag^+ and H_3BO_3. In the present experiment, only Ag-impregnated plates will be used.

MATERIALS AND REAGENTS

1. *Commercially prepared Ag–silica gel TLC plates, on glass backing* (**Note:** These plates have a somewhat darkened surface, due to the photosensitivity of silver salts. This will not impair their use.)
2. *Fatty acid methyl esters:*
   a. Methyl palmitate
   b. Methyl stearate
   c. Methyl oleate
   d. Methyl linoleate

   } all made up in $CHCl_3$ to 1 mg/100 μl (w/v)
3. *Solvent system:* diethyl ether:hexane (5:95, v/v) (**CAUTION: Flammable!**)
4. *Visualizing spray:* 0.025% (w/v) dichlorofluorescein in 95% alcohol

PROCEDURE

Obtain a single TLC plate and prepare it as depicted below; that is, mark an origin line and three running lanes.

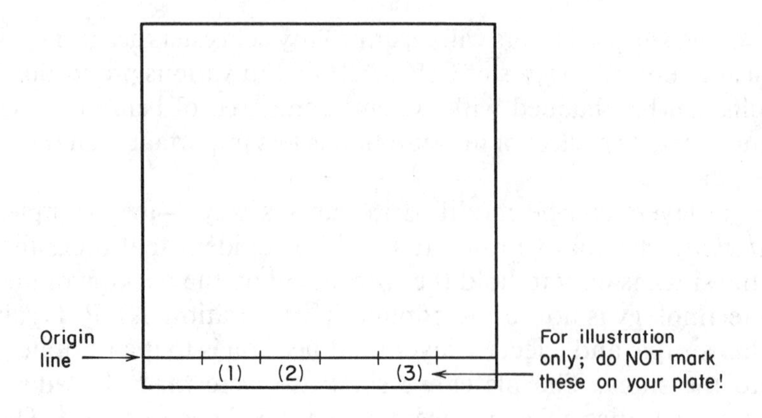

In handling these plates, keep in mind that the silica layers are considerably more fragile than the cellulose layers. If you are not careful, you may chip off large chunks of the layer, which will ruin the plate. These plates are quite expensive, so try to avoid waste.

Place the plate, properly marked, into the 110°C oven for 30 min to activate it (i.e., drive off excess water vapor). Remove the plate from the oven, lay it on the flat surface, and allow it to cool until it can be handled comfortably.

At the origin in lane 1, apply 5 μl of methyl palmitate. When this is dry, apply a similar amount of methyl stearate directly over the first application.

At the origin in lane 2, apply 5μl of methyl oleate solution. When it is dry, overlay with 5μl of methyl linoleate solution.

At the origin in lane 3, apply 2 $\times$ 5 μl of the unknown mixture you will be given. Write the number of the unknown mixture in the upper right-hand corner of the plate, and place your initials in the upper left-hand corner of the plate.

Place the spotted and identified plate into one of the developing tanks, which will be located in one of the fume hoods. Allow the solvent system to travel approximately half the distance between the origin and the top of the plate. When migration is complete, remove the plate and quickly make a small mark with a pencil to indicate the height to which the solvent front rose. Allow the plate to air dry.

Lightly spray the developed and dried plate with the dichlorofluorescein reagent provided. Keep the spray stream moving over the plate, first in one direction, then in the other, but avoid soaking the surface.

Examine the plate under UV light, using the lamp provided. **(CAUTION: Wear protective goggles!)** While observing it in the ultraviolet, lightly sketch around the observed spots with the sharp tip of a pencil. This allows you to make further observations later and calculations under the comfort of room light.

REPORT

1. Indicate the R_f's of all known compounds, as you measured them.
2. Indicate the R_f's of all separated material(s) in your unknown sample.
3. Compare the relative R_f values obtained for C16:0 and C18:0 with the relative values for C18:1 and C18:2. Does the result of the comparison give you a basis for determining the predominant separation factor in this system?
4. Is there another way in which the separated spots might have been visualized without having to resort to the expense of radioactively tagged materials?
5. Answer the following question as part of your report: Could this system distinguish between cis and trans unsaturated fatty acid methyl esters?

REFERENCE

Morris, L. *J. Lipid Res.* 7:717–732 (1966). [See also the handbooks of chromatography. Those by Stahl and Randerath are especially good.]

Molecular Weight Estimation by Exclusion Chromatography*

As presented in Section I, Chapter 7, gel or exclusion chromatography utilizes a particulate phase of porous beads immersed, usually in a column, in a solute solution. The solutes may differ in size and shape, and, as the solvent phase moves relative to the porous beads, some of the solutes may be retarded by entering the pores of the beads before passing on. The degree of retardation, for a given bead porosity, is related to the molecular weight (M_r) and/or shape of the solute. By use of solutes of known molecular weight, a given column (or other mass) of porous beads may be used to evaluate the molecular weight of an unknown. Equation 7–17 may be rearranged to the form:

$$K_d = \frac{(V_e - V_o)}{V_i}$$

where:

V_e = elution volume for the solute of interest

V_o = outer volume (i.e., elution volume of a completely excluded solute such as "Blue Dextran"

V_i = inner volume = $V_t - V_o$

V_t = total volume [i.e., elution volume of a completely included solute such as dinitrophenyl (DNP)-aspartate]; volume of gel matrix is ignored.

By means of this equation, one can recognize that a given solute is distributed in its concentrations between the space *within* the porous beads and the space between the porous beads. The value, K_d, will be calculated in this experiment for standard proteins and for an unknown protein. A high-molecular-weight, blue-colored substance, Blue Dextran, is completely excluded from the beads; therefore, its elution time or volume from a column gives a measure of V_o. In contrast, a small molecule, DNP-aspartate, enters all solvent spaces, and thus its elution volume gives a measure of V_i. Molecules of intermediate size, or which are long and narrow, are able to penetrate the gel particles to some extent and so are eluted from the gel at intermediate posi-

*Preparatory reading: Section I, Chapters 2 and 7.

tions. These concepts may be expressed quantitatively by the distribution coefficient, K_d.

MATERIALS AND REAGENTS

Note: *All buffers must be degassed before use.*

1. *Sephadex G-75,* hydrated in 0.05 M Tris-HCl (pH 7.5).
2. *Elution buffer:* 0.1 M NaCl in 0.05 M Tris-HCl (pH 7.5).
3. *Separation mixture:* 3 mg "Blue Dextran," 8 mg myoglobin, and 0.3 mg DNP-aspartate per ml. Dissolve the above solutes in elution buffer which also contains 15% (v/v) of glycerol.
4. *Chromatographic column:* 2.5 × 30 cm, with tubing at tip controlled by a screw clamp.
5. *Unknown protein solution.*

PROCEDURE

The swirled or gently stirred suspension of Sephadex should be poured into the columns provided, to form a gel bed of dimensions 2.5 × 20–25 cm., with tubing at tip controlled by a screw clamp. Take care to avoid trapping air bubbles. When the gel has settled, add or remove buffer so the liquid level in the column is about 5 cm above the top of the gel bed. Be sure, at this point, that the outlet of the column is securely closed.

You are now ready to apply the separation mixture to the top of the column, using either method A or method B, below. It is often helpful to cut a filter paper wafer just a bit smaller than the inside diameter of the column (e.g., 2.4 mm), and insert this into the liquid above the column bed. Let the wafer settle evenly on top of the Sephadex. Now when the sample is applied, it will not disturb the column.

METHOD A:

Fill a 1-ml pipet with 1.0 of the mixture. Carefully place the tip of the pipet about 1 cm beneath the surface of the liquid in the column and allow the sample to layer slowly onto the top of the gel bed. It should take about 2 min to accomplish this application.

METHOD B:

Let buffer level drop so that the top of the gel bed is just exposed. Attach a Pasteur pipet (5- to 6-in. size) to a screw-type propipet. Fill the pipet with the separation mixture and place the pipet tip approximately 1 cm above the gel bed. Holding it firmly in place, operate the screw so as to extrude the sample as a layer over the gel bed. Turn the screw mechanism slowly but continuously.

After removing the sample-loading pipet, place a graduated 10-ml container (cylinder or centrifuge tube) in position to collect liquid emerging from the bottom of the column. *Cautiously* open the screw clamp so that the liquid flow forms a drop every 4 sec. (This corresponds to a flow rate of about 60 ml/hour.)

Do not add any more buffer to the top of the column until the layer of sample has just passed completely into the gel bed. After the sample has done so, carefully pipet additional buffer into the top of the column as needed to maintain a fairly constant flow rate. (**Note:** Do not disturb the top of the gel bed, and *do not allow the liquid level to drop below the top of the gel bed.*)

When 28 ml has been collected, start to collect individual 4-ml fractions. Transfer these, as they are collected, into a series of tubes. Number the first of these "tube 8" and thereafter serially. When all of the color has been eluted from the column, shut off the flow by tightening the screw clamp. Read the absorbances of fractions 8 → X at the appropriate wavelengths:

1. "Blue Dextran" (blue): 650 nm
2. Myoglobin (amber): 500 nm
3. DNP-aspartate (yellow): 440 nm

After elution of the last band (yellow), apply 1 ml of your unknown solution to the column, collect 28 ml, and then start to collect 4-ml fractions as before. *Read absorbance at 280 nm* if protein is colorless (or at another wavelength to be given to you by the course instructor).

CALCULATIONS

Determination of K_d

On a single graph, plot the absorbance of each fraction as a function of the fraction number (remember that the first seven fractions were collected as a single, 28-ml volume). Indicate the wavelength at which each peak was measured. Determine the midpoint of each peak, either by inspection or by the equation for estimation of the arithmetic mean,

$$\bar{n}_i = \frac{\Sigma n_i A_i}{\Sigma A_i}$$

where:

$\bar{n}_i$ = peak midpoint

n_i = number of each fraction in the peak

A_i = absorbance of each successive fraction in a given peak

Repeat this procedure for the 1 ml of your unknown solution. Calculate the K_d for myoglobin from your data and the information given above.

Estimation of the Molecular Weight of Myoglobin and Your Unknown

The following data have been collected from the results available in the literature.

Protein	Molecular Weight	K_d^a
Trypsin inhibitor (pancreas)	6,500	0.70
Trypsin inhibitor (lima bean)	9,000	0.60
Cytochrome c	12,400	0.50
α-Lactalbumin	15,500	0.43
α-Chymotrypsin	22,500	0.32
Carbonic anhydrase	30,000	0.23
Ovalbumin	45,000	0.12

[a] K_d values were determined with Sephadex G-75 in 0.05 M Tris-HCl (pH 7.5) plus 0.1 M KCl.
Source: Andrews (1970).

Using the above information, prepare a graph of K_d (ordinate) as a function of $\log_{10}$ MW for the known proteins. Estimate the molecular weight of the myoglobin and your unknown in the preparations you examined by interpolation of your estimated K_d value on this semilog plot.

REPORT

Include the specified graphs. Make sure your estimates of K_d and MW for myoglobin and the unknown are clearly marked in the appropriate places.

REFERENCES

Andrews, P. Estimation of molecular size and molecular weights of biological compounds by gel filtration. In *Methods Biochem. Anal.* (D. Glick, ed.). Interscience, New York (1970).

Fischer, L. An introduction to gel chromatography. In *Laboratory Techniques in Biochemistry and Molecular Biology* (T. S. Work and E. Work, eds.). American Elsevier, New York (1969).

Giddings, J. C., and Mallik, K. L. *Anal. Chem. 38*:997 (1966).

Polyacrylamide Gel Electrophoresis (PAGE)*

Charged biopolymers can migrate through the pores of a gel formed by polymerizing and cross-linking acrylamide in the presence of bisacrylamide. Gels in this experiment will be formed in lengths of glass tubing 4 mm in diameter, each of which can be loaded with a single sample. (For slab gels, see Expt. 10.) The apparatus to be used was sketched in Chapter 6 and is now shown in Fig. E7–1. Migration of the biopolymers is induced by application of a voltage (150–300 volts) across the ends of the gel. Since the flow of current (I) through the resistance of the gel (R) produces heat, it is imperative that the current be limited to avoid thermal denaturation of the sample. (Remember, $W = I^2R$, so small changes in the current can produce significant changes in the heat which must be dissipated.)

Other things being equal, movement of a macromolecular species through an acrylamide gel depends on the total acrylamide concentration and on the degree of cross-linking, since these are the factors that determine gel pore size. Provided one has an idea of the type of macromolecules in a sample mixture, it is possible to optimize gel formulation to yield the best possible separation. In the present experiment, we shall keep the acrylamide/bisacrylamide mass ratio constant at 20:1, but the total concentration will be varied.

In neutral or native gels, the movement of charged macromolecules is a function of their net charge at the given pH, of the strength of the applied field, and of the sieving effect of the gel pore size. The term "neutral" implies only the lack of added detergent or other denaturing agent.

In SDS gels, the detergent sodium dodecyl sulfate (SDS) is added in sufficient quantity to coat the macromolecules, giving all of them essentially the same net charge. Studies have shown that the number of SDS molecules bound to a polypeptide chain is roughly half the number of amino acid residues in the polypeptide, giving an approximately constant negative charge per unit of polypeptide mass. In addition, SDS causes disaggregation of polymeric proteins in which the subunits are not covalently linked. Consequently, movement of the charged species is a function solely of molecular or subunit size, of the strength of the applied field, and of the sieving effect of the gel pore size.

*Preparatory Reading: Section I, Chapter 6.

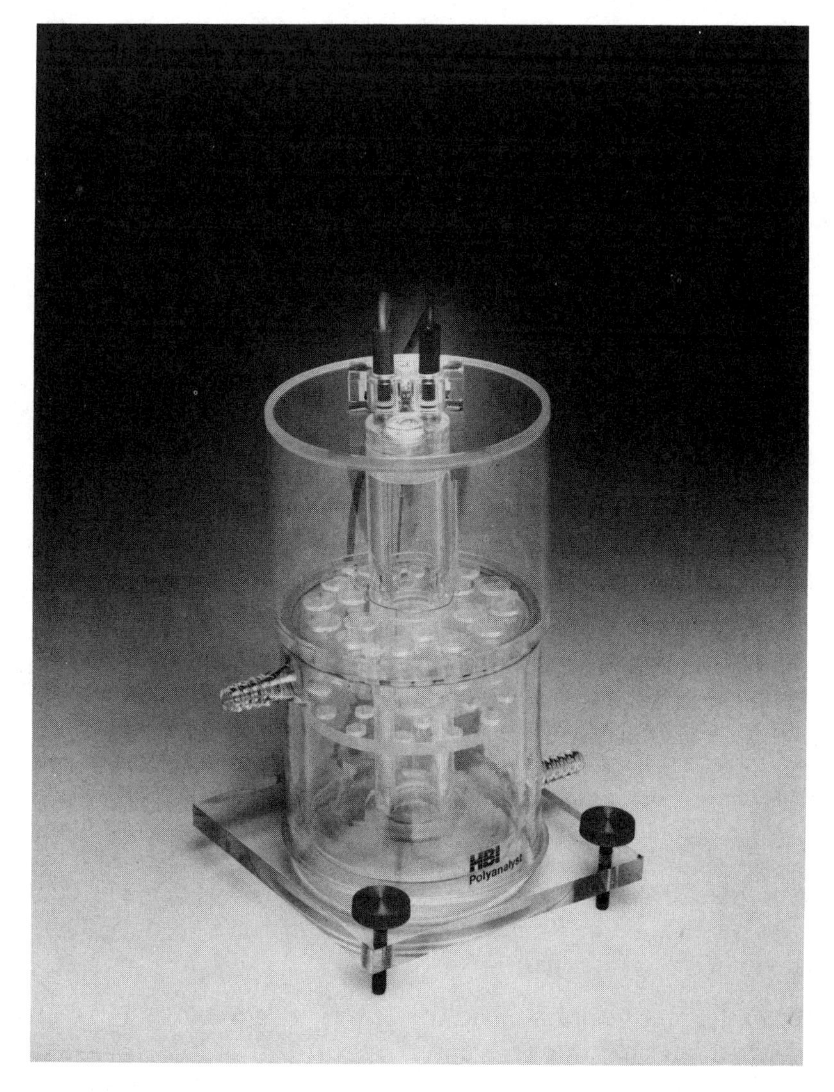

Figure E7–1. Apparatus for tube gel electrophoresis. Neither tubes nor electrode solutions are in place. (Reproduced by courtesy of Haake Buchler Instruments, Inc.)

Measurements have shown that if the acrylamide/bisacrylamide ratio is held constant at 20:1, then the pore size is relatively constant over a total acrylamide concentration from approximately 5% to 10%. Increasing the ratio gives larger pores, and vice versa.

MATERIALS AND REAGENTS

Gel Reagents

1. *Solution A:* 3 M Tris-HCl + 0.3% (v/v) *N,N*-tetramethylethylenediamine (TEMED), adjusted to pH 8.3.

2. *Solution B:* 40% (w/v) acrylamide + 2% (w/v) bisacrylamide in water. (**Note:** See Chapter 6, Gelling Agents, and this chapter, Procedure, below, for precautions in handling acrylamide.)

3. *Solution C:* 10% (w/v) ammonium persulfate; this should be made fresh each day; it does not keep well. (Some sources claim that it can be kept longer, but experience indicates that much difficulty in casting good gels can be avoided by use of fresh solutions.)

4. *Solution D:* 10% (w/v) sodium dodecyl sulfate (SDS) in water.

Electrode Buffers

5. *Solution E (10× neutral gel buffer)*: Tris-HCl (0.5 M) + glycine (0.02M) in water, adjusted to pH 8.3. **Note:** *This is ten times the working concentration. Before use in the electrode chambers of the electrophoresis apparatus, it must be diluted accordingly.*

6. *Solution F (5× SDS gel buffer):* Mix 500 ml of solution E (i.e., the 10 × neutral gel buffer) with 50 ml of solution D and bring to a final volume of 1 L. **Note:** *This is five times the working concentration; before use, it must be diluted accordingly.*

Protein Staining Solutions

7. *Coomassie Blue stain:* Mix 450 ml of CH_3OH with 90 ml CH_3COOH; add 2.5 g Coomassie Blue R and stir. Dilute the mixture to 1 L. Filter through Whatman No. 2 filter paper.

8. *Destaining solution:* Mix 50 ml of CH_3OH and 75 ml of CH_3COOH; add sufficient water to bring the volume up to 1 L.

Protein Standards

Protein	Mol. wt.	Isoelectric point
9. Phosphorylase B	97,400	5.9–6.0
10. Bovine serum albumin	66,200	4.5–4.6
11. Carbonic anhydrase	31,000	6.0–6.5
12. Egg white lysozyme	17,200	10.9–11.0

Native Sample Buffer

13. 0.005% of bromphenol blue (BPB) in Tris-HCl, 0.2 M (pH 8.3). Mix this buffer 1:1 (v/v) with the aqueous protein samples.

Sample Denaturing Buffer

14. 2.5 SDS, 5% β-mercaptoethanol (v/v), 10% glycerol (v/v), and 0.005% BPB in Tris-HCl, 0.2 M (pH 8.3). Aqueous solutions of protein samples are mixed 1:1 with this denaturing buffer and heated in a boiling water bath for 2 min **(CAUTION: In the hood!)** to reduce disulfide bonds.

Protein Mixtures

15. *Protein mixture A:* Mix phosphorylase *b*, bovine serum albumin, bovine carbonic anhydrase, and egg white lysozyme, each at a final concentration of 2 mg/ml, with 10% glycerol (v/v) and 0.005% BPB in Tris-HCl, 0.2 M (pH 8.3).
16. *Protein mixture B:* Mix phosphorylase *b* and carbonic anhydrase, each at a final concentration of 2 mg/ml, with glycerol and BPB, as above.
17. *Protein mixture C:* Mix aliquots of your unknown protein at 2 mg/ml, with glycerol and BPB, as above.

PROCEDURE

Collect 12 glass tubes, 12 × 75 mm, which have been prepared for electrophoresis. Preparation includes scrupulous cleaning by immersion for several hours either in chromic acid cleaning solution or in ethanolic KOH (0.5 M KOH in 95% alcohol). This is followed by thorough rinsing, after which the tubes are dried in an oven. While still warm, the tubes are immersed briefly in a 1:200 dilution of a proprietary wetting agent. The prepared tubes are then allowed to dry.

Note that these tubes have carefully squared ends; treat them in a manner that will avoid chipping of the ends. Have ready a support board for these tubes, consisting of inverted serum stoppers glued to a flat board. Mount the tubes vertically by plugging each into a stopper. Using a twister motion will facilitate insertion (and later removal) of the tubes.

Prepare an 8% acrylamide solution for neutral gels by mixing 2 ml of solution B **(CAUTION: Unpolymerized acrylamide is reported to be a skin irritant and a neurotoxin. When handling solution B, wear gloves; do not mouth-pipet this solution!)**, 2 ml of solution E, and 5 ml of water. Mix thoroughly, degas for 3 min, then add 1 ml of solution A, and 0.05 ml of solution C. Mix again. As soon as this mixture is prepared, fill *six* of the mounted electrophoresis tubes to within 0.8–1.0 cm of the top. A Pasteur pipet or a 10-ml disposable syringe may be used for filling the tubes. Tap each tube to dislodge air bubbles. Discard the remainder of the gel solution, promptly rinsing out the container. With a fine-tipped Pasteur pipet or 1-ml syringe, carefully layer a little water over the top of the acrylamide solution in each of the six tubes. The water layer should be about 5 mm high. Leave the tubes undisturbed until polymerization is complete. This requires about 45 min.

Prepare a 12.5% acrylamide solution for SDS gels by mixing 2.5 ml of solution B, 1 ml of solution A, 2 ml of solution D, and 2.5 ml of water. Mix well, then add 0.55 ml of solution C and mix again. *Use the precautions noted above concerning the toxicity of acrylamide.* Fill the remaining six tubes with this solution, overlaying each gel column with about 5 mm of water. Discard the remainder of the gel solution, promptly rinsing out the container. Let the tubes stand undisturbed until polymerization is complete. This will require 30–45 min.

While the gels are polymerizing, prepare properly diluted electrode buffers for both the neutral gel and SDS gel apparatus. Pour buffer into the lower electrode chambers to a level that will allow the lower ends of the gel tubes to be immersed in the buffer as much as possible, when they are loaded into the instrument. The electrode buffer level must still be low enough to prevent contact with the upper compartment, which would short-circuit the system. Reserve the remainder of the diluted buffers for use in the upper electrode chambers.

When the gels in the tubes have properly hardened, remove the tubes from the casting stand. With a sharp flick of the wrist, discard the water layer that overlaid the gel columns. Insert *four* of the six neutral gel tubes into the apparatus designated for neutral gels. (The remaining two tubes will not be loaded, but used for practice in removing the gel columns from their glass tubes.) Orient the tubes in the gel holder so that the incompletely filled ends point upward. Any unused ports in the gel holder must be plugged with rubber stoppers. Check to be sure that the lower ends of the gel tubes are positioned beneath the buffer surface in the lower chamber. If not, adjust the tubes in the holder or add more buffer to the lower chamber. With the gel holder and upper electrode chamber in place, fill the upper electrode chamber with diluted buffer until its level is above the tops of the gel tubes. This will allow completion of an electrical circuit when power is applied. Dislodge any air bubbles adhering to either end of the gel tubes, either by gentle jarring of the apparatus or by means of a Pasteur pipet (whose tip has been curved to a U-shape) filled with buffer.

Connect the electrodes so that the cathode *(negative electrode, black wire)* is in the *upper* chamber.

Rotate the high-voltage control of the power supply to its extreme counterclockwise position ("Off"). Make certain that any "DC/Standby" switch is in the standby position and that the "Voltmeter" switch is off. Turn the main power switch on; the red pilot light should be illuminated. Within a few moments, move the proper switch to the position marked "DC On." Finally, adjust the instrument until the milliammeter indicates a current flow of about 1 mA per gel tube. Preelectrophorese the tubes for about 20 min to remove any undecomposed ammonium persulfate.

While preelectrophoresis of the neutral gels is proceeding, mount the six SDS gel tubes in the separate apparatus designated for these gels. Fill the upper and lower electrode chambers with suitable buffer solution as described above. Be sure any unused ports in the instrument are stoppered. The SDS gels do not require preelectrophoresis.

Denature small samples (20–30 μl) by mixing aliquots of *protein mixtures A–C* with equal volumes of *sample denaturing buffer.* Immerse these mixtures in a boiling water bath for 2 min, then cool the solutions to room temperature in a beaker of cold water. Mix an additional 20–30 μl of sample with the *native sample buffer,* but do *not* heat these samples.

When all is ready, the gel tubes must be loaded with samples. Remove the power from the neutral gel apparatus by turning the DC On/Standby switch to the standby position, then remove the top from the apparatus. With a 10-μl pipet, load one of the tubes with *protein mixture A,* another with *protein mixture B,* and a third with *protein mixture C.* The samples will, if application is properly made, form a thin layer that sinks just to the top of the gel column. The applications are preferably made by lowering the tip of the pipet through the overlaying electrode buffer. Replace the top of the apparatus, apply power, and allow electrophoresis to continue until the BPB band has moved to within 0.5 cm of the lower end of the gel column.

Denatured samples are similarly applied to SDS gels in the designated instrument. When samples have been applied, turn the power on and allow electrophoresis to continue until the BPB is near the bottom end of the gels. If the dye leaves the tube as a result of overly prolonged migration, it will be impossible to estimate relative movement of the other components in the loaded samples.

While electrophoresis proceeds, practice removing the gels from their glass tubes. Use the two remaining gel tubes for this purpose. Fill a syringe, fitted with a 4-in. 28-gauge needle, with water. Holding the gel tube in a nearly horizontal position over a water-filled glass dish, insert the needle between the glass wall and the acrylamide gel, forcing a film of water into the interface. Rotate the tube around the needle to introduce water around the circumference of the interface. This will loosen the gel, and continued application of pressure will cause the gel to fall into the water-filled dish. The gels may then be scooped out of the water with a small test tube. Be careful in handling the gels; they tend to be brittle and will break if roughly handled.

It is now necessary to make a permanent record of the position of the BPB front. This can readily be done by taking a small needle, dipped into a drop of India ink, or a 1-ml syringe (with a 25-gauge needle) filled with the ink and plunging the needle into and through the gel at the point where you see the BPB. "Tatooing" the gel in this manner provides a permanent reference which will not be lost during the staining–destaining process.

To stain for protein in the gels, immerse each in Coomassie Blue staining solution for about 30 min, but not longer than 1 hour. Then decant the staining solution and wash each gel several times with destaining solution. Place the racks bearing the gel tubes on a rotary shaker and set for gentle shaking. During the next 24 hours, repeatedly replace the destaining solution with fresh reagent until the gel backgrounds are reasonably clear and the colored bands are well defined. Store the finished gels in destaining solution; close each tube with a square of Parafilm.®

REPORT

1. Plot the mobility of each of the known protein standards (relative to the mobility of the BPB tracking dye) as ordinate versus the log of the molecular weights as abscissa.
2. From this plot, estimate the molecular weight of your unknown protein and of any other bands you detected by the Coomassie Blue stain.
3. Submit a sketch of a typical neutral gel separation and of a typical SDS gel separation. Identify as many bands as you can.

REFERENCE

Weber, K., Pringle, J. R., and Osborn, M. *Methods Enzymol. 26* (Part C):3–27 (1972).

Ion-Exchange Chromatographic Purification of Native and Chemically Modified Hemoglobin: Evaluation of Purity by Isoelectric Focusing*

Hemoglobin, found in erythrocytes of many species, is essential for the proper transport of oxygen from the lungs to other tissues. It is also important in the transport of carbon dioxide as carbaminohemoglobin and so to maintenance of acid–base balance. These subjects are discussed in many biochemistry texts. Stryer's *Biochemistry* is especially recommended for its discussion of the structure of hemoglobin.

Hemoglobin consists of four polypeptide chains, noncovalently associated. Each bears a ferroporphyrin, or heme, group to which a molecule of oxygen can attach, also by noncovalent forces. The oxygen-bearing molecule is known as *oxyhemoglobin* (sometimes abbreviated as HbO_2). The nonoxygenated form is known simply as *hemoglobin,* although a better designation is *deoxyhemoglobin* (sometimes abbreviated as deoxyHb, or Hb, or HbA, etc.). Note that none of these designations properly describes the tetrameric structure. (It is easy to become confused by the various ways of describing these compounds. See the last section of this experiment, Notes on the Nomenclature of Hemoglobin and Related Compounds.)

It has been shown that various types of hemoglobin differ in the subunit chains they contain. The major adult hemoglobin of humans is hemoglobin A (HbA), which consists of two α and two β chains; HbA can therefore be described as $\alpha_2\beta_2$. A variant form makes up about 2% of the total in normal individuals; it is known as HbA_2, and it has the subunit structure $\alpha_2\delta_2$ since it contains the subunit identified as δ. The normal human fetus contains HbF, or fetal hemoglobin, which has the subunit structure $\alpha_2\gamma_2$. The subunits α, β, γ, and δ differ in amino acid composition; these differences are reflected in their isoelectric pH values and in the net charge on each subunit at a given pH. The dissimilarity in net charge makes possible

*Preparatory Reading: Section I, Chapters 6 and 7.

the separation of the hemoglobins by ion-exchange chromatography or by gel electrophoresis.

In addition to these normally occurring hemoglobins, there are abnormal species which differ as a result of the genetically determined primary structures of their component subunits. Among the more prominent of those that have been studied is hemoglobin S (HbS), present in the condition known as sickle-cell anemia. In the β subunits of HbS, amino acid residue No. 6 is valine instead of glutamic acid. This substitution is accompanied by an increased isoelectric pH of oxyHbS (pI = 7.09) and deoxyHbS (pI = 6.91). The pI values for oxyHbA and deoxyHbA are 6.87 and 6.68, respectively. Consequently HbS can be separated from HbA either by ion-exchange chromatography or by electrophoresis.

Another consequence of the Glu → Val substitution in HbS is the decreased solubility of deoxyHbS due to increased interaction between the Val-6 region of one β subunit with the corresponding region of the other. The amino acid substitution cited above makes these regions more hydrophobic than the coresponding regions in HbA. As a result, deoxyHbS polymerizes with formation of long fibers that distort the red cells into the characteristic sickle-cell shape, part of the pathologic process.

The properties of proteins can be altered covalently by modifying one or more of the functional groups, such as the amino, carboxylate, or sulfhydryl groups. For example, acylation of a protein amino group would eliminate a positive charge, and the pI of the acylated protein would be lowered. In this experiment, we shall examine the effect of a reagent developed with the special intent of altering the properties of HbS so as to inhibit polymerization which leads to sickling, as noted above (see Walder et al., 1980).

The reagent, bis(3,5-dibromosalicyl)fumarate, reacts with amino groups as shown below:

$$\text{HOOC(Br)}_2\text{C}_6\text{H}_2\text{–O–C(=O)–CH=CH–C(=O)–O–C}_6\text{H}_2\text{(Br)}_2\text{COOH} + 2\,\text{RNH}_2 \rightarrow 2\,\text{HO(Br)}_2\text{C}_6\text{H}_2\text{COOH} + \text{RNH–C(=O)–CH=CH–NR}$$

Since it is bifunctional, the reagent can form a bridge between two amino groups. In hemoglobin, the amino groups specifically attacked are Lys-82β_1 and Lys-82β_2. This results in cross-linking of the two β subunits. Cross-linking has been found to interfere with HbS polymerization and may have clinical potential in treatment of sickle-cell anemia. Because HbS is not readily available, we will demonstrate this reaction with HbA.

The lysine residues at which the reagent cross-links the β subunits are also the sites of binding of 2,3-diphosphoglycerate (2,3-DPG). Since binding of 2,3-DPG is a normal means of decreasing the affinity of hemoglobin for oxygen, it follows that the derivatized hemoglobin is no longer subject to modulation of its oxygen affinity by 2,3-DPG.

MATERIALS AND REAGENTS

1. *For washing red blood cells:* 0.15 M NaCl, cooled to 4°C.
2. *For estimation of Hb concentration:* Modified Drabkin's reagent, made by dissolving 220 mg of $K_3Fe(CN)_6$ and 170 mg KCN in about 900 ml of 0.015 M potassium phosphate buffer (pH 7.2). Add 60 mg of saponin, stir until it is dissolved, then adjust the pH to 7.2 ± 0.1. Finally, dilute the solution to 1 L with phosphate buffer. (A commercial version of this reagent is available.) **CAUTION: This reagent is very toxic; do not use with a mouth pipet!**
3. 5% $K_3Fe(CN)_6$: Dissolve 5 g of the solid in 100 ml of 0.015 M potassium phosphate buffer (pH 7.2).
4. Neutralized KCN: Mix equal volumes of a freshly prepared 10% KCN solution with an equal volume of a 12% (v/v) acetic acid solution. **(CAUTION: In the hood!)** Add the acetic acid to the cyanide solution, *not* the reverse. This solution is suitable for use for only an hour or so. Do not prepare large quantities. **CAUTION: This reagent is very toxic; do not use with a mouth pipet! Avoid inhalation of the vapors from this solution. Dispense it only in the hood!**
5. Phosphate buffer (0.015 M, pH 7.2): Prepare this from the potassium salts, or a mixture of Na or K salts, whichever is most convenient.
6. *For derivatizing hemoglobin:* bis(3,5-dibromosalicyl)fumarate (MW = 640), prepared as described by Walder et al. (1980).
7. 1 N NaOH: used for solubilizing the above reagent.
8. 1-cm diameter dialysis tubing; this is boiled in ~0.1 M trisodium phosphate–0.1 M EDTA mixture for one-half hour. Rinse the tubing extensively with hot purified water. Store under water until used.
9. 0.2 M glycine buffer, pH 7.8: used for dialysis.
10. *For ion-exchange purification of hemoglobin:* The exchanger used is DEAE–cellulose, which will be equilibrated with 0.2 M glycine buffer (pH 7.8). For the elution gradient one also needs two additional solutions made from the above glycine buffer, one containing 0.03 M NaCl and the

other 0.06 M NaCl. The total volume of the linear gradient should be approximately 100 ml.

11. *For isoelectric focusing gels:* Dissolve 19.23 g of acrylamide plus 0.77 g of bisacrylamide in water to make 200 ml. **(CAUTION!)**

12. 30% (v/v) glycerol; dilute pure glycerol with water.

13. 20% (w/v) ammonium persulfate; dissolve the solid in water, but make this fresh on the day of use since it does not keep well.

14. 40% Ampholine®, pH 6–8. (The term *Ampholine* is a proprietary designation for a series of synthetic polyelectrolytes produced by condensation of polyamines with polycarboxylic acids.)

15. *Reservoir solutions:* Bottom (anodic) reservoir; dilute 1 ml of 85% phosphoric acid (H_3PO_4) to 500 ml with water. Upper (cathodic) reservoir; add 5 ml of 1 N NaOH to 500 ml of water. (These solutions take the place of the usual electrode buffers in electrophoresis.)

16. *Sample diluting reagent:* 0.2 M KCN in 0.01 M bis-Tris buffer, pH 7.2. **CAUTION: This reagent is very toxic; do not use with a mouth pipet!**

PROCEDURE

CAUTION: Before proceeding, dispose of all waste cyanide-containing liquids in the sodium hypochlorite solution in the carboy located in the hood.

Separation of Erythrocytes from Whole Blood

Outdated whole blood, obtained from a blood bank, is centrifuged at 4°C at 1000 × *g* until the red blood cells are packed (20–30 min). The supernatant is discarded, and the cells are carefully resuspended in an equal volume of cold 0.15 M NaCl and centrifuged again. The NaCl washing is repeated four times.

Extraction of Hemoglobin from the Erythrocytes

The packed, washed cells are lysed by addition of an equal volume of cold purified water. The mixture is stirred for 30 min at 4°C, then sufficient solid NaCl is added to bring the salt concentration to 1%. The lysed cells are then centrifuged for 1 hour at 4°C and 9000 × *g* (use heavy glass or metal tubes).

The hemoglobin solution is decanted into a flask; this leaves behind a viscous phase and some cell debris amounting to about 10% of the original volume. The quality of the resulting hemoglobin solution is roughly checked by making a 1:10 dilution of a small aliquot in water; if the diluted solution is not clear, the remainder must be centrifuged again.

Analysis of Hemoglobin and Its Storage

The greatest part of the Fe contained in hemoglobin is in the form of a fer*ro*heme (Fe^{2+}), which can transport oxygen. However, in vitro as in vivo, a small fraction of the Fe is present as a non-oxygen-transporting species, which contains a ferr*i*heme (Fe^{3+}). This oxidized hemoglobin is known as *methemoglobin* (MetHb or Hi). It is useful to examine hemoglobin solutions to determine the extent of oxidation that may have occurred during storage of the whole blood or during hemoglobin isolation. A second problem, of importance here, is the determination of total hemoglobin concentration, since one will be treating the protein with a stoichiometric quantity of the derivatizing reagent.

Methods for analysis of hemoglobins fall into three major categories. The first is based on the oxygen-binding capacity of Hb, but this clearly is not satisfactory for estimation of MetHb. One can also estimate the Hb content by quantitative Fe analysis, but these methods are tedious and subject to error if reagents and water are not scrupulously Fe free. Further, such methods are not suited for distinction between Hb and MetHb. The third category of methods depends on the spectral properties of Hb, HbO_2, MetHb, or derivatives of these molecules. It is the latter that we will use in this experiment. (Spectral properties of the important species are discussed below in the section, Notes on the Nomenclature of Hemoglobin and Related Compounds.)

Total Hemoglobin

Drabkin's reagent contains ferr*i*cyanide ion, which oxidizes the ferr*o*heme of HbA to the ferr*i*heme of MetHbA. The reagent also contains CN^-, which converts MetHbA to cyanMetHbA. CyanMetHbA has a strong absorbance at 540 nm, with a millimolar extinction coefficient (ϵ^{mM}) of 44. This is the principle of the method we shall use to determine total Hb.

To 5.0 ml of the modified Drabkin's reagent, add 20 μl of the hemoglobin solution. Mix well, then transfer to a cuvet and measure A_{540} against a blank cuvet containing the Drabkin's reagent plus 20 μl of water. Allow the solutions to stand for a few minutes before recording the absorbance, to ensure complete reaction. The millimolar concentration of total hemoglobin is based on the equation:

$$[\text{cyanMetHb}] = \frac{A_{540}}{44}$$

Oxidized Hemoglobin (MetHb)

MetHb normally comprises about 5% of the total Hb content and has a moderate absorbance at 630 nm, which is substantially abolished by conversion to cyanMetHb. This is the principle employed to determine MetHb in mixtures with Hb.

Place 5.0 ml of 0.015 M phosphate buffer in a tube, and add 100 μl of the hemoglobin solution. Mix well, let stand for 2–3 min, then divide the solution into approximately equal volumes. Using one of the aliquots, measure A_{630} against a water blank. Record the value as A_1.

Add 50 μl of a neutralized KCN solution to the second aliquot; mix well and let stand for 2–3 min. Read the A_{630} against a water blank. Record this value as A_2. The concentration of MetHb is given by the equation

$$[\text{metHb}] = \frac{A_1 - A_2}{3.29}$$

Storage of Hemoglobin

The hemoglobin solution is concentrated by ultrafiltration in an Amicon® pressure cell, using a PM-10 membrane, until the Hb concentration is 2.5 mM. The solution is then added dropwise to a suitable container of liquid nitrogen, in which the drops freeze virtually immediately. The liquid nitrogen is drained off, and the frozen Hb pellets are stored at −70°C. They remain stable for many months under these conditions. When needed, a few pellets can be quickly dissolved by thawing in a warm water bath.

Derivatization of Hemoglobin

Determine the total hemoglobin concentration of the solution you prepared or were supplied. Adjust the concentration to 2 mM with bis-Tris buffer. You will be adding to this adjusted solution a stoichiometric quantity of derivatizing reagent. **Note:** *Reserve a 200-μl portion of this hemoglobin solution for later use in isoelectric focusing.*

Weigh out (or it will be supplied to you) 1.28 mg of the derivatizing reagent into a capped plastic 2- to 3-ml tube. Add sufficient 1 N NaOH so that the solution will be 4 mM in NaOH when diluted to 1 ml with water. Vortex-mix vigorously to dissolve the fumarate ester; it may be necessary to pulverize the compound by grinding it in the tip of the tube with the rounded end of a small glass rod. *Work rapidly since the compound is easily hydrolyzed!*

Add 1 ml of the adjusted fumarate ester solution to 1 ml of the 2 mM hemoglobin solution, in bis-Tris buffer (or, in any case, equal volumes of each) and mix well.

Allow reaction to proceed for 1 hour at 37°C. Transfer the reaction mixture to a 1-cm diameter dialysis tube that is long enough to contain the 2 ml of reaction mixture and also allow for about 1 ml more space for possible osmotic expansion. Dialyze exhaustively against glycine buffer in the cold, e.g., two changes of 2 L each, over 24 hours.

Reserve 100 μl of the dialyzed mixture for isoelectric focusing.

Ion-Exchange Purification and Separation of Hemoglobin and Its Derivatives

Pour a 6-ml column with the DEAE-cellulose provided. Check the quality of the packing by monitoring the pH of its effluent. If the pH is not close to 7.8, wash the column with buffer until that value is reached. This may require continuous washing for several hours, using a system such as that shown in Fig. E8–1A.

Add about 6 mg of "total hemoglobin"—that is, Hb, plus Hi, plus derivatized Hb—to the column (see Expt. 6, Procedure, for technique of addition), then wash with glycine buffer until the A_{280} of the effluent is negligible (<0.005). About 6 ml of glycine buffer should remain on top of the column bed; if necessary, add buffer to provide that volume.

Set up a salt-gradient elution system (see Fig. E8–1B) and switch elution to a gradient of 0.03–0.06 M NaCl in glycine buffer. The volume of the total gradient should be about 100 ml. Collect small fractions (about 2 ml), and monitor the position of the eluting bands by measuring the A_{577} of each, although the general position of the hemoglobins can be observed from their colors. Save those fractions corresponding to the centers of the eluted peaks.

Isoelectric Focusing

Mix 36 ml of acrylamide solution, 3 ml of 40% (stock concentration) of pH 6–8 Ampholines, 10 ml of 30% glycerol, and 11 ml of water; mix well, then degas this solution with the vacuum of a water aspirator. Add 150 μl of ammonium persulfate solution and 20 μl of TEMED. Mix well, then pour into prepared gel tubes, as described previously (Expt. 7). The volume of final solution is sufficient to make 36 gel tubes at an acrylamide concentration of 6%.

Overlay each filled gel tube with a little water to produce meniscus-free tops to the gel columns. After polymerization is complete, close both ends of the gel tubes with some Parafilm® wrap, and store under buffer at 4°C for a week before use. For reasons not altogether clear, aged gels give better results than fresh gels.

Mount the aged gel tubes in the electrophoresis apparatus used in Experiment 7, and fill the upper and lower electrode reservoirs with appropriate solutions.

Prepare the previously reserved samples of isolated hemoglobin and hemoglobin fumarate, separately, by diluting each to a protein concentration of about 40 μg/μl, using bis-Tris buffer. Add 10 μl of diluted sample to an equal volume of 0.2 M KCN in bis-Tris buffer **(CAUTION: Poison!).** Complete preparation of the samples by addition of 20 μl of an 8% Ampholines® solution (pH 6–8) made by a 1:5 dilution of the stock reagent. Finally, layer 20 μl of the finished sample solutions onto the tops of separate gel tubes.

Start electrophoresis by applying a field of 200 volts (but do not exceed 2 mA/tube during the entire electrofocusing period). After 15 min, increase the voltage to 300 volts, and after an additional 15 min, increase it to 500 volts. After 1 hour, increase the voltage to 700 volts, and run the separation at that voltage for an additional 30 min. Turn off the power supply, disconnect the plugs from the electrophoresis stand, and momentarily short-circuit the leads to ensure discharge of the capacitors of the power supply.

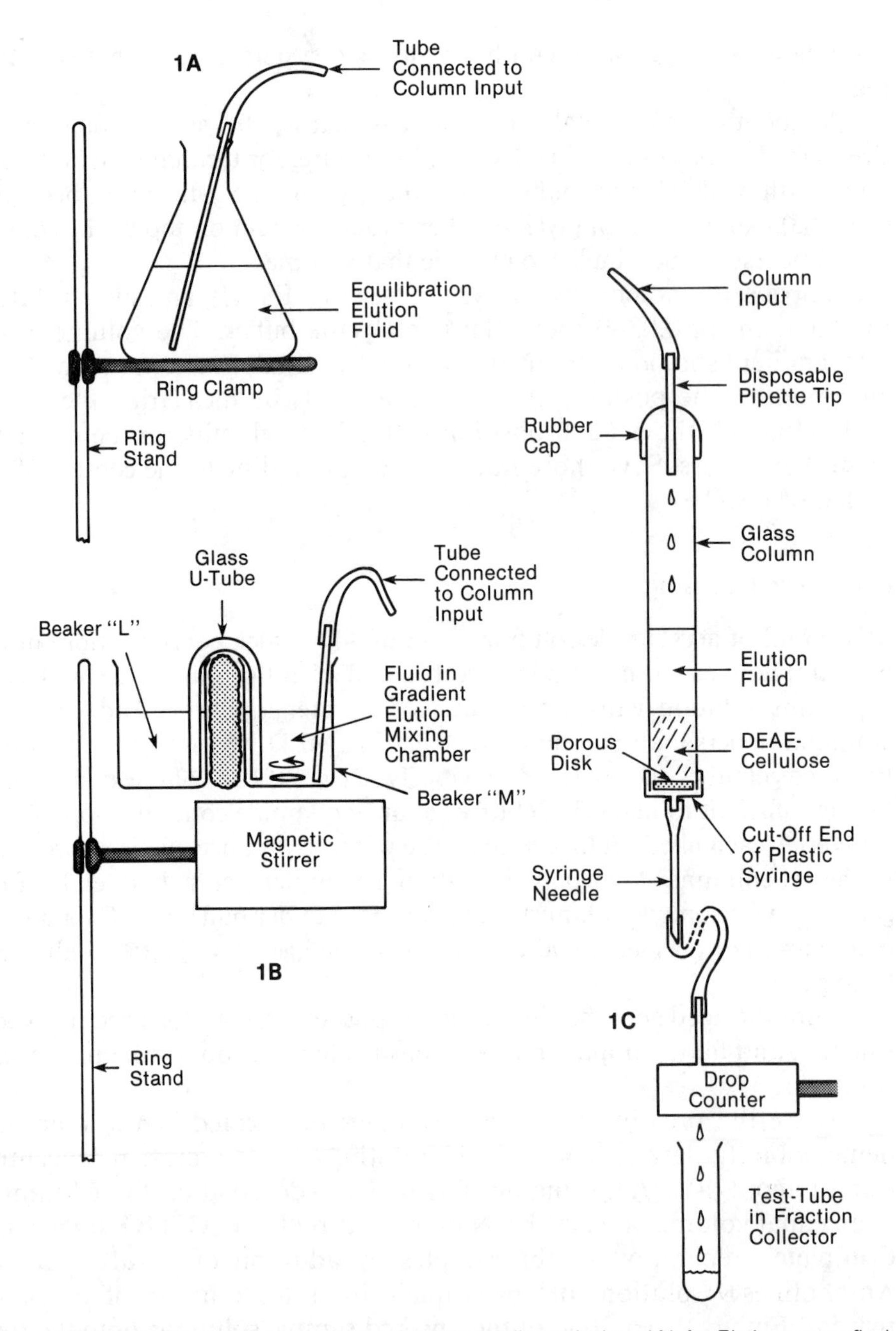

Figure E8–1. Chromatographic system for salt gradient elution. (A) An Ehrlenmeyer flask contains the buffer used to equilibrate the DEAE-cellulose column. In this mode of operation, tubing leads from the flask to the top of the column, as shown. The head pressure is controlled by increasing or decreasing the relative heights of the flask and the column. The tubing is initially filled by suction with a rubber bulb.

(B) The setup for the simplified gradient elution device. Two 100-ml beakers, separated slightly by a piece of foam plastic, are taped together. A small stirring bar is placed in the beaker marked "M" (mixing chamber). This beaker also contains 50 ml 0.03 M NaCl in glycine buffer. The other beaker, marked "L" (limit buffer chamber) contains 50 ml of 0.06 M NaCl in glycine buffer. At the start, the level of fluid in both beakers should be the same. A glass U-tube connects the two beakers.

Remove the gels, still in their tubes, from the apparatus. The red protein bands should be clearly visible without further processing. (You will be provided with data concerning the pH gradient of the gel.)

REPORT

Submit data and calculations for the following:

1. Concentration of hemoglobin and methemoglobin in your sample
2. Plot of elution data for salt-gradient ion-exchange purification of hemoglobin reaction mixture
3. A copy of the pH gradient graph
4. Sketches of your isoelectric focusing gels, showing the anodic and cathodic ends, the positions of native and derivatized hemoglobin bands, and the pI values calculated for each.

REFERENCES

Bunn, F. H., Forget, B. G., and Ranney, H. M. *Human Hemoglobins.* W. B. Saunders, Philadelphia (1977).
Haglund, H. *Methods Biochem. Anal. 19:*1 (1971).
Walder, J. A., Walder, R. Y., and Arnone, A. *J. Mol. Biol. 141:*195–216 (1980).

NOTES ON THE NOMENCLATURE OF HEMOGLOBIN AND RELATED COMPOUNDS

The preceding experiment alluded to the difficulties in describing hemoglobin and related compounds so that the names and abbreviations employed would give the greatest amount of information regarding the oxidation state of the heme iron and the presence of ligands other than oxygen. As the molecular architecture of the hemoglobin molecule was elucidated, it became increasingly difficult to be precise using terminology that originated in an era when the major concern was quantitative measurements.

Specialists in hemoglobin chemistry have therefore adopted a new system

---

A second tube leads from the mixing chamber to the top of the column. When the gradient is to be started, exert gentle suction with a rubber bulb on the tube leading to the column top, then quickly connect that tube to the column.

(C) The column should be in such a position, relative to the gradient device, that the surface of the fluid in the mixing chamber is slightly above the end of the tube which leads to the drop counter of the fraction collector. The drop rate should be low and adjusted so that the fluid flowing from beaker L to beaker M maintains equal fluid heights in the beakers.

Table E8–1. Systems of Nomenclature for Erythrocyte Heme Proteins

Old System		New System	
Oxyhemoglobin	HbO_2	Oxyhemoglobin	HbO_2
Deoxyhemoglobin	Hb	Hem*o*globin[a]	Hb
Methemoglobin	MetHb	Hem*i*globin[a]	Hi
Cyanmethemoglobin	CNMetHb	Hem*i*globin cyanide[a]	HiCN
Carbon monoxide hemoglobin	COHb	Carbon monoxide hemoglobin	HbCO[b]

[a]Many, but not all, sources proceed as we have here and use an italicized letter to emphasize the valence state of the heme iron.
[b]Not to be confused with HbCOOH, representing carbaminohemoglobin.

of terminology which has a more rational chemical basis. This is now, slowly, supplanting the older designations. Table E8-1 may make it easier for students to correlate readings in the literature, where both systems may be encountered.

A second, recently adopted convention relates to expression of spectrophotometric extinction coefficients. None of the symbols shown above properly emphasize the tetrameric nature of the molecular species. In rec-

Figure E8–2. Absorbance spectra of various hemoglobin derivatives. Absorbance values are expressed as quarter-millimolar extinction coefficients. Panel A shows most of the visible spectral range. Panel B shows the Soret bands, regions of very strong absorbance, characteristic of porphyrin pigments generally. Note the change in the extinction scale relative to panel A. Individual compounds are represented as follows: (–· · –· ·) Hb; (–) HbO_2; (––––) HbCO; (––·––·) Hi; (–· –·) HiCN; (––· ·––· ·) coproporphyrin III (pH 7.7); (–· · ·–· · ·) coproporphyrin III (pH 7.0). (Reproduced by permission from E. J. van Kampen and W. G. Zijlstra, Determination of hemoglobin and its derivatives, in *Advances in Clinical Chemistry*, H. Sobotka and C. P. Stewart, eds., *8*:142–187, 1965.) (*See also* A. Zwart, A. Buursma, E. J. van Kampen, and W. G. Zijlstra, Multicomponent analysis of hemoglobin derivatives with a reversed-optics spectrophotometer, *Clin. Chem. 30*:373–379, 1984.)

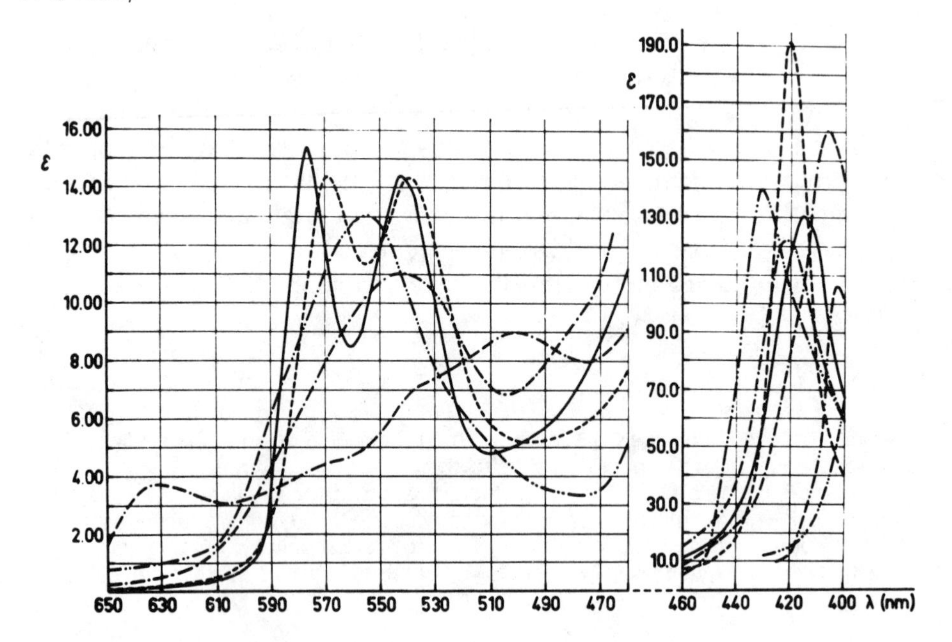

ognition of the fact, the new convention expresses extinction as *quarter-millimolar values,* even though the spectral values were probably obtained from tetramers. This has the merit of rationally relating abbreviations with numerical data. (In the preceding experiment, the cited extinction values were already multiplied by 4.)

Visible spectra of the major hemoglobin derivatives are shown in Fig. E8–2. These should assist your understanding of the analytical procedures employed in this experiment.

Affinity Chromatography and Isolation of Lactate Dehydrogenase*

Lactate dehydrogenase (LDH) is a major dehydrogenase which catalyzes the reversible interconversion of pyruvate to lactate. LDH is a tetrameric enzyme; each tetramer is composed of different combinations of two types of subunits, M and H, giving rise to five isozymes which may be distinguished by electrophoresis. Assay of LDH activity may be performed by measuring the disappearance of pyruvate (or by measuring lactate production), but the chemical measurement of these materials at substrate concentrations is tedious and difficult. It is far more expedient to measure changes in the cosubstrate pair, NAD^+/NADH, since there is a pronounced change in the molar absorbance as the nucleotide is converted from one of its forms to the other. Thus, NAD(P)H has a molar absorptivity $= 6.22 \times 10^3$ in a 1-cm cell at 340 nm, whereas the molar absorptivity of $NAD(P)^+$ under the same conditions is negligible. These facts are the basis of rapid and sensitive spectrophotometric assays of many dehydrogenases which can be linked to either of the pyridine nucleotides. It is the basis of the assays we will do here.

The general plan of the experiment is as follows: An extract of chicken breast muscle, containing significant amounts of LDH, is applied to an affinity chromatography column. The column is packed with agarose, to which is covalently bound Cibacron Blue, FeGA, sometimes known as Reactive Blue. This ligand has shape and charge properties that mimic those of the pyridine nucleotides, so it serves as an affixed moiety, complementary to the nucleotides, around which may wrap that portion of the enzyme that has been termed the "dinucleotide fold." In this way, the enzyme is immobilized on the column, whereas proteins lacking such conformational features are washed through. Subsequently, the enzyme may be liberated either by a salt gradient or by elution with NADH in moderately high concentration.

Very rapid isolation of LDH, with a fairly high degree of purification, can be accomplished by this or similar affinity procedures. Progress of the scheme is monitored by measuring the enzymic activity in all of the collected fractions. Determination of the protein content of fractions allows determination of the specific activity of the product. Finally, as a check,

*Preparatory reading: Section I, Chapters 2 and 7; and Appendix 2.

the purity (homogeneity) of the product is examined by gel electrophoresis, by immunodiffusion, and by an ELISA procedure. To do this, the enzyme product must be saved for later experiments. The enzyme should be carefully stored in accordance with directions appearing below.

MATERIALS AND REAGENTS

For Affinity Chromatography

1. 20 mm Tris-HCl buffer (pH 8.6)
2. Reactive Blue–agarose
3. 20 mM Tris-HCl (pH 8.6), with 1 mM β-mercaptoethanol
4. 20 mM Tris-HCl (pH 8.6), with 1 mM β-mercaptoethanol and 1 mM phenylmethylsulfonyl fluoride (PMSF)
5. 20 mM Tris-HCl (pH 8.6), with 0.5 mM β-mercaptoethanol and 1 mM lithium lactate
6. 10 mM Tris-HCl (pH 8.6), with 0.5 mM β-mercaptoethanol
7. 10 mM Tris-HCl (pH 8.6), with 0.5 mM β-mercaptoethanol and 1 mM NADH

For Protein Determination

8. Bradford reagent (see Expt. 1), Bradford Protein Determination)

For Assay of LDH Activity

9. 99 mM lithium lactate in 10 mM Tris-HCl (pH 8.6)
10. 0.7 mM NAD^+ in 10 mM Tris-HCl (pH 8.6)
11. 500 mM NaCl containing 18 mM $NaHCO_3$

PROCEDURE

Preparation of Chicken Breast Muscle Extract

Remove skin from a chicken breast and pass the breast through a meat grinder (twice) in the cold room; suspend ground meat in 1.5 volumes of cold 20 mM Tris-HCl (pH 8.6) buffer containing 1 mM β-mercaptoethanol and 1 mM phenylmethylsulfonyl fluoride (PMSF). The latter is an inhibitor of serine proteases. It has been reported, however, that PMSF may lose effectiveness within an hour under certain conditions (James, 1978); therefore, do not delay these initial steps. Extract by stirring for 1 hour at 4°C.

Filter extract through a double layer of cheese cloth and centrifuge filtrate at 10,000 × *g* for 45 min. Repeat centrifugation of decanted supernatant fluid. The final supernatant is used in the experiment.

Pouring the Affinity Column

Note: *Be sure to degas all buffers to be used for chromatography.* Obtain a 10/12-ml plastic syringe; remove and discard the plunger, then fit the tip with a plastic stopcock. Place a small plug of glass wool into the barrel and push it tightly against the lower end. Alternatively one may use a cork borer to cut a disc from a porous polyethylene sheet with a 70-μm porosity, ¹⁄₁₆-inch thick. The disc should fit snugly into the end of the syringe.

Mount the barrel in a vertical position, then wet the inside with about 5 ml of Tris-HCl buffer (20 mM, pH 8.6), Next, check carefully to see that no air bubbles are trapped in the glass wool or beneath the plastic disc. If you see bubbles, dislodge them. Then open the stopcock and let the fluid level drop to just above the top of the glass wool or disc.

Take about 4 ml of the stock suspension of Reactive Blue–agarose in Tris buffer (approximately two Pasteur "pipetsful"); pour this into the column and allow to settle before again draining the liquid to just above the gel surface. Check again for the absence of air bubbles in the glass wool, below the disc or in the gel. If none are evident, equilibrate the column by passing through it 10 ml of Tris-HCl (20 mM, pH 8.6), containing 1 mM β-mercaptoethanol. Do not allow the liquid to drop below the top surface of the gel bed. The flow rate during equilibration should be approximately 1 ml/min.

Loading and Eluting the Column

Obtain a sample of 6–7 ml of the chicken breast muscle extract, then carefully add 5 ml to the top of the column in such a way that the top of the bed is not disturbed. The device of adding a wafer of filter paper to the top of the column may be used as described in Experiment 6. Collect a single 5-ml fraction, labeling the eluate "flow-through." Set this tube aside.

FIRST ELUTION:

Wash the column with 5 × 5 ml portions of the Tris–mercaptoethanol buffer used in preparing the tissue extract. Collect these fractions separately and label the tubes "wash 1, tube 1," "wash 1, tube 2," etc. Set these tubes aside.

SECOND ELUTION:

Next, wash the column with 1 × 5 ml of a mixture of 1 mM lithium lactate and 1 mM NAD^+ in Tris-HCl containing 0.5 mM mercaptoethanol. This elutes weakly bound dehydrogenases. Collect the fraction, label it "lithium lactate wash" and set it aside.

THIRD ELUTION:

Wash the column further with 2 × 5 ml portions of Tris-HCl buffer (10 mM, pH 8.6) containing 0.5 mM mercaptoethanol. Collect the two fractions and label them as "wash 2, tube 1," and "wash 2, tube 2." Set these aside.

FOURTH ELUTION:

Elute the LDH by washing with 1 × 5 ml of Tris-HCl (10 mM, pH 8.6 + 0.5 mM mercaptoethanol), to which has been added 1 mM NADH. Additional enzyme can be recovered by washing with 2 × 5 ml of Tris-HCl with mercaptoethanol. These three last-mentioned fractions should contain the bulk of the LDH activity.

Determination of Protein Concentrations

For Solutions Containing Protein and Nucleotides

Protein determinations in solutions containing nucleotides as well as protein cannot be performed simply by measuring the absorbancy at 280 nm. As was discussed earlier (Expt. 1), the Warburg method, using the A_{280}/A_{260} ratio, gives a close first approximation. This may be applied to the fractions eluted from the column, even though NAD^+ and NADH are present but nucleic acids are not.

Read A_{280} for each of the reserved column fractions and record the data. *Be sure to use as blanks the appropriate elution buffers.* Rezero the instruments and read A_{260} for each fraction. Calculate A_{280}/A_{260} and, using either the Warburg tables or a special formula to be given to you, calculate the protein concentrations.

For Solutions Not Containing Nucleotides

For solutions not containing added nucleotides, two extinction values may be used, derived from past experience:

$\epsilon_{280}^{1\%}$ (whole homogenate) = 10.0
$\epsilon_{280}^{1\%}$ (rabbit muscle LDH) = 14.9

Use the first value for the crude extract and for the "first wash" fractions, and use the second for the eluted LDH fractions.

Use of Lowry or Bradford Procedure

If time permits, protein concentrations will be more exactly determined using either the Lowry or Bradford procedure (see Expt. 1). This is the preferred protocol.

Assay of Lactate Dehydrogenase Activity

Note: *Either quartz or acrylic plastic disposable cuvets are acceptable for use in these assays.*

Method A

The assay may be carried out by an elapsed-time procedure as described, using:

LITHIUM LACTATE:

0.099 M in 10 mM Tris-HCl (pH 8.6)

NAD^+:

7×10^{-4} M in 10 mM Tris-HCl (pH 8.6)

DILUTION FLUID:

0.5 M NaCl, containing 0.018 M $NaHCO_3$

There should be 11 reserved fractions as detailed above. Set up a numbered series of 13 tubes (the extra tubes are for spectrophotometer blanks). To each tube add 1.4 ml lithium lactate, 0.70 ml NAD^+, and 0.4 ml of diluent. Mix the contents of the tubes. To tubes 1 and 2, add 10 μl of buffer; use them to zero the spectrophotometer at 340 nm.

Using separate disposable micropipets or pipet tips for each fraction, add 10 μl to the corresponding assay tube, one at a time, mix, and measure A_{340} at (as nearly as possible) 1 min after addition of the enzyme. If the absorbance is too high, dilute an aliquot of the fraction and repeat the assay. Record the absorbance values.

Method B

For the kinetic assay, the course of the reaction is continuously monitored by recording the increase in absorbancy at 340 nm due to the increase in [NADH], one of the reaction products.

A recording spectrophotometer is set to read at 340 nm, using either the UV or the visible lamp. As an initial setting, the recorder speed should be set at approximately 1 in./20 sec or 60 mm/min, although the setting may have to be increased or decreased depending upon the speed of the reaction. A single-beam spectrophotometer is zeroed with a cuvet containing all of the enzyme system components (see Method A) except for the sample to be assayed. Double-beam instruments are zeroed with two cuvets both containing the same components as above. At zero time, 10 μl of the enzyme sample is added to the sample cuvet, the cuvet is inverted (use a small square of Parafilm® as cover) to mix the system and is inserted into the instrument. The addition and mixing should take no more than 10 sec. Record the A_{340} until the trace is no longer linear. This should take about 0.5 to 1.0 min. If the trace exhibits no linear phase, either the substrate concentration is too low or the enzyme preparation is too concentrated. To remedy this, first dilute a portion of the enzyme preparation with 20 mM Tris-HCl, pH 8.5 (consult with an instructor), and try again. Once a linear trace has been obtained, measure the slope of the line ($\Delta A/\Delta$ time) and calculate the micromoles NADH formed per minute per milligram (see Calculation, below).

Activity determinations should be repeated *(only with the most active fraction)* using a range of substrate concentrations. Consult with instructor concerning choice of lactate concentrations to use.

Use these data to calculate K_m (lactate) and V_{max} (lactate) by an appropriate linear plot (e.g., Eadie–Hofstee).

Preservation of the Product Remaining

Carefully label the remainder of your fractions as to their identity (and your own). Close each tube tightly with a small square of Parafilm®. Place the assembled tubes in a rack and give them to a teaching assistant for storage in a freezer. In a later experiment, these samples will be analyzed by electrophoresis, immunodiffusion, and ELISA.

CALCULATION OF SPECIFIC ACTIVITY

We shall adopt the convention of the *international unit;* thus the activity is expressed as μmoles product formed/min/mg protein. To estimate specific activity, use the protein values obtained, the ΔA_{340}, the micromolar extinction coefficient, and the elapsed time.

Disposal of the Affinity Gel

Carefully expel the contents of your column into the container provided for this purpose. The affinity gels are quite expensive. They can be washed and stored for future use.

REPORT

The report for this experiment will include (1) the completed table (see Table E9–1), (2) the recordings of ΔA_{340} for each of the active fractions, (3) data for

Table E9–1. LDH Purification Data

Fraction A_{280}	Protein (mg/ml)	Activity (μmol/min)	Sp. Activity (μmol/min/mg)	Yield (%)
Whole tissue extract				
Column flowthrough				
First elution				
1				
2				
3				
4				
5				
Second elution (lactate wash)				
Third elution				
1				
2				
Fourth elution (LDH)				
1				
2				
3				

protein determinations, (4) calculations for percent yield and for specific activity, (5) data and plots used for graphic determination of K_m and V_{max}.

REFERENCES

Lactate Dehydrogenase

Everse, J., and Kaplan, N. V. *Adv. Enzymol. 37:*61–133 (1973).

James, G. T. *Anal. Biochem. 86:*574–579 (1978).

Cibacron Blue Binding

Stellwagen, E. *Accounts Chem. Res. 10:*92–98 (1977).

Robinson, J. B., Strottman, J. M., and Stellwagen, E. *Proc. Natl. Acad. Sci. U.S.A. 78:*2287–2291 (1981).

Ryan, L. D., and Vestling, C. S. *Arch. Biochem. Biophys. 160:*279(1974).

Slab Gel Electrophoresis of Lactate Dehydrogenase*

In this experiment, the LDH preparation reserved in Experiment 9 will be subjected to slab acrylamide gel electrophoresis under both native and denaturing conditions. The protein bands will be stained with Coomassie Blue (as in Expt. 7); a portion of the native gel will be cut out and subjected to a special staining mixture which will detect LDH activity. By comparing the displays of stained protein bands and of the sites of enzyme activity, the position and number of LDH bands can be observed. It should then be possible to identify the LDH enzymic forms as well as obtain an indication of the heterogeneity or purity of the preparation.

MATERIALS AND REAGENTS

Reagents are different from those used in Experiment 7.

1. *Buffer for native gel:* 1.5 M Tris-HCl (pH 8.5).
2. *Buffer for denaturing gel:* 1.5 M Tris-HCl (pH 8.5); 0.4% (w/v) SDS.
3. 24% (w/v) acrylamide, 1% (w/v) bisacrylamide.
4. 10% (w/v) ammonium persulfate.
5. TEMED.
6. Water.
7. LDH preparation (from Expt. 9): Adjust the concentration of total protein to 1–2 mg/ml; concentrate by lyophilization or dilute with 0.2 M Tris-HCl, pH 8.5, as needed.
8. Protein standards (as in Expt. 7, Materials and Reagents, #9–11).
9. LDH activity stain reagents
   a. *Solution A:* Combine 0.15 M lithium lactate, 1 mM NAD^+, 15 mM NaCl, and 8.5 mM $MgCl_2$, all in 80 mM potassium phosphate buffer (pH 7.4).

*Preparatory reading: Section I, Chapter 6.

b. *Solution B:* Nitroblue tetrazolium (MW = 817.7), 5 mM in 80 mM potassium phosphate buffer (pH 7.4).
c. *Solution C:* Phenazine methosulfate (MW = 306.3), 1 mM in 80 mM potassium phosphate buffer (pH 7.4).

Note: *Solutions B and C are light sensitive. They should be stored in foil-wrapped or dark-brown bottles at 4°C. Prepare shortly before use. The mixed activity stain should be similarly handled with respect to light.*

PROCEDURE

Preparation of the Slab Gel

The glass plates used to form the gel measure 6½ in. × 6½ in.; the slot in the top of one of the plates is ¾ in. × 5¼ in. The plates must first be thoroughly cleaned to remove all traces of oil or grease. This may be done in several ways: (1) Use a mixture of methanol, ammonia, and ~5% (w/v) Na_3PO_4 [30:30:40 (v/v/v)] and scrub with a soft pad; or (2) scrub with so-called "soft" cleaner containing calcium carbonate plus detergents, or (3) scrub with more recently available strong laboratory alkaline detergents of unspecified composition. After such cleaning, the plates must be thoroughly rinsed with purified water and permitted to air-dry in a vertical position. Handle the plates only at the edges with the tips of your fingers.

The two plastic spacers separating the plates, one on each of the two opposite sides of the plates, will determine the thickness of the gel slab. A thinner slab will have less capacity, but the proteins will migrate faster; a 0.6-mm-thick spacer is recommended so that the electrophoresis may be completed within the time limits of a laboratory session. The silicone rubber tubing, of diameter ~1.5 mm, is positioned between the plates on the bottom and two sides as close to the edge as possible. This sandwich of plates, spacers and tubing is sketched in Fig. E10–1. As mentioned in Chapter 6, it is necessary only to use water to position the spacers until they are clamped. Alternatively, grease or the special tape may be used. Large spring clamps are applied to the two sides and the bottom. The assembly may be kept in an upright position by resting on the bottom two extended clamp grips. The tightness of the seal may be tested by adding water to the space between the plates; if there is a leak, no matter how small, the plate setup must be reassembled. Filter paper may be slipped between the plates to dry the chamber after this water test.

Mix the gel reagents as follows: 4.5 ml buffer (native or denaturing), 6.0 ml acrylamide/bisacrylamide, 7.6 ml water. Mix thoroughly. Degas. Then add 22.5 μl of ammonium persulfate and 22.5 μl of TEMED. Swirl to mix. This preparation will produce an 8% gel. Pour the gel mixture between the plates so that the level is 2–3 mm below the bottom of the slot. Be sure that no air bubbles are trapped; if so, remove them with a thin wire. Insert a plastic comb between the plates, dipping to a depth of about 1.5 cm below the surface of the gel solution (see Fig. E10–2). The comb must be

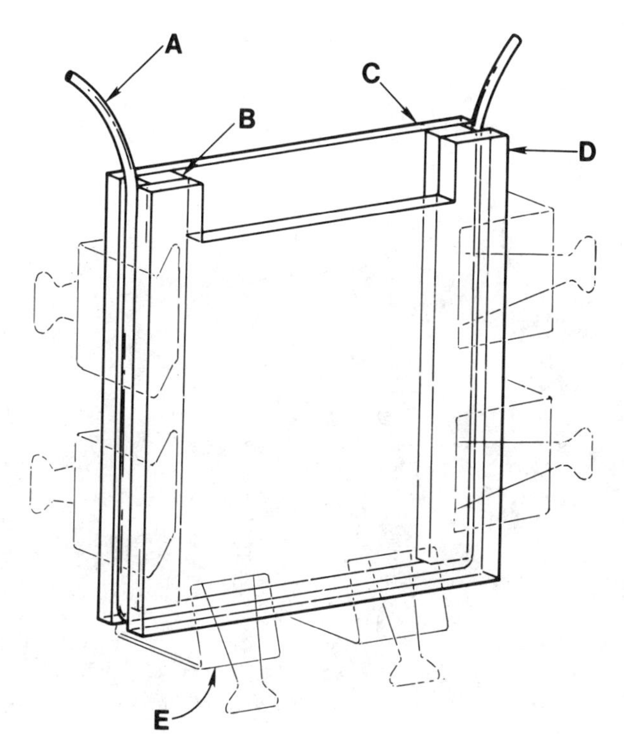

Figure E10–1. Arrangement of glass plates, spacers, plastic tubing, and metal clips in preparation for pouring acrylamide mixture for slab gel electrophoresis: A, silicone rubber tubing; B, spacer; C, rear rectangular glass plate; D, front-notched glass plate; E, large spring clip. (Thicknesses of A and B are exaggerated for clarity.)

level, as must the entire plate assembly. Be certain no air is trapped by the comb. The number of teeth in the comb will, of course, determine the number of sample slots in the completed gel. The comb should be left in place while the gel sets (about 40 min) and then be removed carefully.

Rinse with water the space formerly occupied by the comb, shake out (or blot out) any excess drops, and, while maintaining pressure on the plate "sandwich" with your fingers, remove the spring clamps. Pull out the plastic tube, rinse the bottom edge of the gel with water.

Mount the plate–gel unit on the electrophoresis apparatus with the notched plate facing the upper electrode chamber and flat against the foam rubber facing of the unit (Fig. E10–3). Clamp the plate–gel unit in place, using two large spring clamps for each side. Pour the electrode solutions into the upper (cathode) and lower (anode) chambers. There should be enough solution in the upper chamber so that the level is about 1 cm or more above the crenellated edge of the gel; remove any air trapped in the sample slots by squirting buffer in and out with a Pasteur pipet. The level of thc solution in the lower chamber should be about 2 cm above the bottom of the gel; remove any trapped air by use of a Pasteur pipet with tip bent to a U-shape.

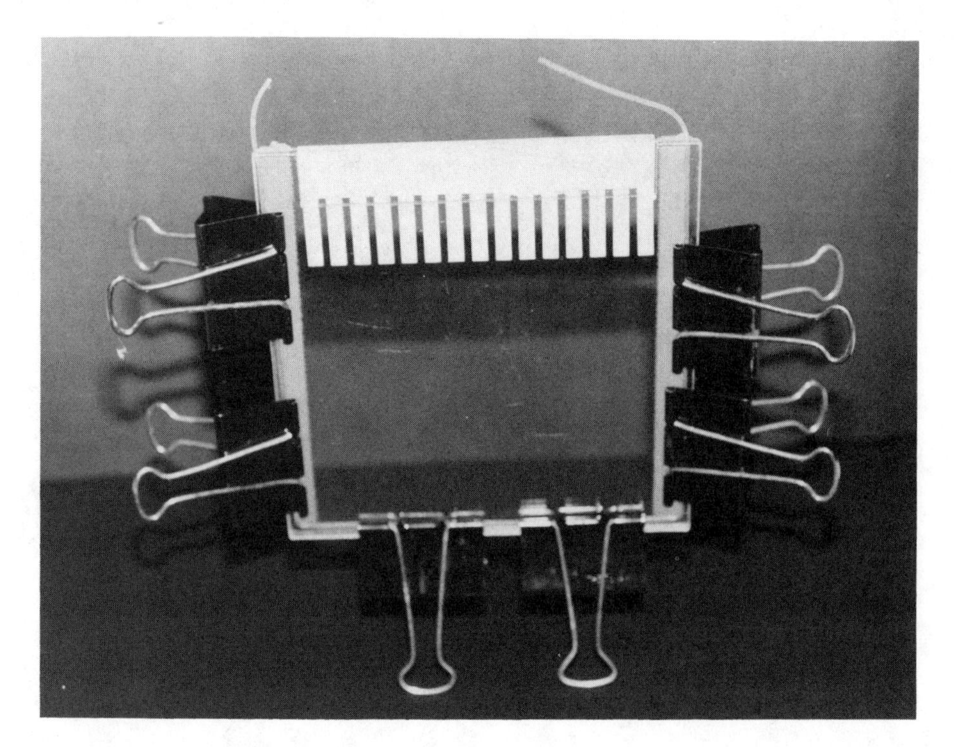

Figure E10–2. Apparatus as in Figure E10–1 with added nylon comb as mold for forming sample slots.

Application of the Sample to the Gel

Dilute your LDH sample 1:1 with the native and with the denaturing sample buffer mixture; prepare sufficient amounts of these two solutions so that 10–20 μl can be used for each sample application.

Using either a Hamilton syringe, an adjustable-volume pipet with plastic tip, or other microsyringe or micropipet, introduce the sample into one of the slots in the top of the gel, keeping the tip of the needle or pipet beneath the electrode buffer surface. The blue-dyed sample should settle to the bottom of the slot. Apply samples of 5, 10, and 20 μl in separate slots. The mixture of protein standards should be applied to one of the central positions on the gel.

Electrophoresis and Staining

The power supply should be connected to the electrophoresis unit and about 20 milliamps (mA) applied; the voltage will be about 75–100 volts. Discontinue electrophoresis when the tracking dye approaches the bottom of the gel ($\sim$2–3 cm from edge); this should take about 2 hours. Dismount the gel assembly and place it on a flat surface. Separate the glass plates by gently inserting a thin metal spatula between the plates at the bottom and pry them apart. **(Take care! Do not crack the glass!)** Remove one glass plate; with the gel resting on the other plate, mark with India ink or snip one corner of the

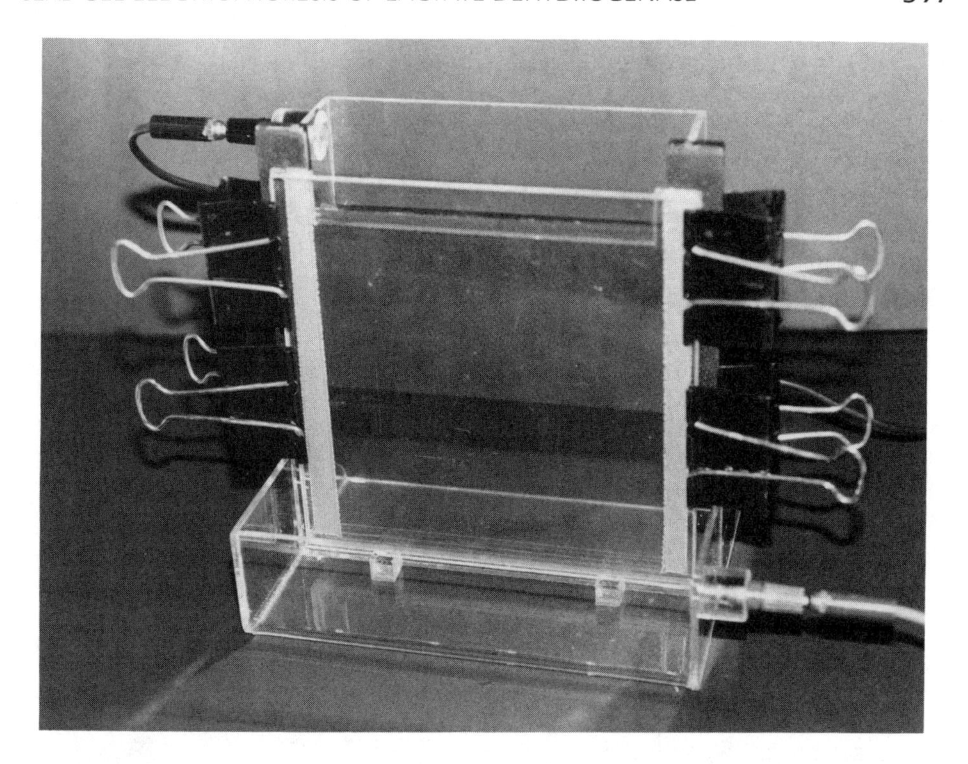

Figure E10–3. Electrophoresis apparatus with slab gel plates in position. The electrode buffers have not been poured; the electrode wires lead off to a power supply. (Apparatus was constructed in the Medical Instrument Shop of the University of Iowa College of Medicine, Iowa City.)

gel to permit identification of the sequence of the lanes in which the proteins traveled.

With a blade, slice off two lanes from the native gel and set aside for activity staining. Remove the gels from the plates by gently separating the gels with a thin spatula. Immerse the remaining portion of the native gel as well as the denaturing gel in the Coomassie Blue staining solution in a flat glass dish. Stain for about ½ hour, but no longer than 1 hour. Destain with the CH_3OH/HOAc mixture in flat glass dishes placed on a gently moving rotary shaker.

Activity Stain

A brief discussion of the theory of this activity stain is appended at the end of this experiment. The procedure is as follows: Place the reserved strip of the native gel on a sheet of aluminum foil folded to a small boat shape. Add in order, the following: 6.5 ml of solution A, 2.5 ml of solution B, and 0.25 ml of solution C. Tip the boat back and forth to mix the contents and bathe the gel. Fold the aluminum foil to shield the contents completely from light. After 15–20 min, examine the gel. The position of LDH will be indicated by a blue color.

REPORT

1. Plot the mobility of the known protein standards relative to the mobility of the tracking dye for the native and denaturing gels.
2. For the denaturing gel, evaluate the molecular weights of the major protein bands. How do they compare to the reported molecular weights of LDH subunits?
3. For the native gel, to what molecular weights do the major protein bands seem to correspond?
4. Compare the location of the activity-stained band(s) with the protein-stained bands in the native gel.
5. From these data, can you determine which LDH isozymes predominate?
6. Submit sketches of all gels with measured mobilities indicated.

NOTES ON ACTIVITY STAIN FOR LACTATE DEHYDROGENASE

The reaction catalyzed is:

$$\begin{matrix} CO_2^- & & & CO_2^- & & \\ | & & & | & & \\ CHOH & + \; NAD^+ & \rightleftharpoons & C{=}O & + \; NADH & + \; H^+ \\ | & & & | & & \\ CH_3 & & & CH_3 & & \end{matrix}$$

Although this reaction was monitored in a soluble system in Experiment 9 by measuring the change in absorbance at 340 nm with time, such a procedure would not be practical with a gel since readings in the near ultraviolet (340 nm) would be obviated by the absorbance of the gel material. The technique used is to couple the $NAD^+/NADH$ system with another oxidation–reduction system or systems. The NADH formed would be used to reduce another molecule, which in turn would reduce a third molecule. The third molecule is a dye which is visibly blue in the reduced state.

In order for this oxidation–reduction chain of electron transfers to occur, there should be a favorable relationship between the oxidation–reduction potentials of the unit cells. The chain of reactions in this system is:

lactate	NAD^+	Phen. Metho.	Tetr. Blue (oxidized)
$\mathscr{E}_0' = -0.19V$	$\mathscr{E}_0' = -0.32V$	$\mathscr{E}_0' = +0.08V$	$\mathscr{E}_0' = -0.08V$
pyruvate	NADH	Phen. Metho. (oxidized)	Tetr. Blue (reduced)

where $\mathscr{E}_0'$ = standard oxidation–reduction potential in volts, Phen. metho. = phenazine methosulfate, and Tetr. Blue = nitroblue tetrazolium.

Under *standard conditions* this reaction sequence would not occur spontaneously since electrons tend to flow from the more electronegative to the more electropositive systems,

$$\Delta G^0 = -n\mathscr{F}\,\Delta\mathscr{E}_0'$$

where:

ΔG^0 = standard free energy change

n = number of electrons transfered

$\mathscr{F}$ = Faraday's constant = 96,500 coulombs (per equivalent)

$\Delta\mathscr{E}_0'$ = difference in standard oxidation–reduction potentials of two coupled systems

If $\Delta\mathscr{E}_0'$ is negative, then ΔG^0 is positive and the reaction cannot proceed spontaneously. However, the Nernst equation

$$\mathscr{E}' = \mathscr{E}_0' + \frac{RT}{n}\ln\frac{[\text{oxidant}]}{[\text{reductant}]}$$

tells us that the *effective* or "real" electromotive force depends upon both the $\mathscr{E}_0'$, or standard potential, and the logarithm of the ratio of oxidants to reductants. Therefore, in the activity stain for the LDH system, the concentrations of lactate and NAD^+ are such as to yield a positive $\mathscr{E}'$. The same is true for the phenazine methosulfate + tetrazolium blue-coupled reactions; thus the net electron flow favors the final reduction of tetrazolium blue which is blue in color.

REFERENCES

Allen, R. C., Saravis, C. A., and Maurer, H. R. *Gel Electrophoresis and Isolectric Focusing of Proteins.* Walter de Gruyter, Berlin, New York (1984).

Latner, A. L., and Skillen, A. W. *Proc. Assoc. Clin. Chem. 2*:3 (1962).

Latner, A. L., and Skillen, A. W. *Isoenzymes in Biology and Medicine.* Academic Press, New York (1968).

Characterization of LDH by Immunodiffusion and by an Enzyme-Linked Immunosorbent Assay (ELISA)*

The homogeneity of the lactate dehydrogenase (LDH) preparation from Experiment 9, which was examined by slab gel electrophoresis (Expt. 10), will be further tested by two immunochemical procedures: immunodiffusion and ELISA.

Immunodiffusion

This process, a variation of the Ouchterlony procedure, is termed double diffusion because both the antibody (with which the student is usually provided) and the antigen (the LDH prepared previously) each diffuses some distance through an agarose gel from a point of application to the regions in which they interact. Precipitation will occur where optimal antibody–antigen ratios have been achieved. The patterns of the precipitin lines observed give a clue to the homogeneity and concentration of antigenically active LDH in your preparation. You may observe a multiplicity of bands, particularly at higher antigen concentrations; or a single band which becomes visible above a certain antigen concentration; or no band at all.

The gels will be cast on a standard microscope glass slide, 25 × 75 mm. Each gel has cut into it a series of wells, made by aspirating from the surface small, circular segments of the agarose. This can be done with an aspirator and a small metal or glass tube. (Be sure there is a trap in the vacuum line!) These wells will be filled with either the antibody or the antigen solution, then set aside for 24 hours in a humidified chamber until diffusion is complete. The chamber consists of a plastic Petri dish, 90 × 90 × 10 mm. In the bottom of this dish is placed a piece of plastic or paper sponge thoroughly moistened with phosphate-buffered saline. The gel slides are placed on *top* of this moistened material.

*Preparatory Reading: Section I, Chapters 9 and 10.

When diffusion is complete, the plates are removed from the moist chamber and examined over a piece of black paper to locate the precipitin band(s), which should be located in the space between the parallel rows of wells. If it is necessary to enhance detection, the plates may also be stained with dilute Coomassie Blue in the usual way to detect proteins.

PROCEDURE

Purification of Chicken Breast Muscle LDH Antigen for Antibody Production

This is a scaled up protocol similar in its first few steps to that in Experiment 9.

1. Equilibrate a 20-ml Reactive Blue column with 20 mM Tris (pH 8.6), 1 mM β-mercaptoethanol.

2. Remove the skin from the boned chicken breast, and pass the breast through a meat grinder twice in the cold room. Place meat in 1.5 volumes of cold 20 mM Tris, 1 mM β-mercaptoethanol, 1 mM PMSF (pH 8.6). Extract with stirring for 1 hour in the cold room.

3. Filter extract through a double layer of cheese cloth and centrifuge the filtrate at 10,000 × *g* for 45 min.

4. Apply 40 ml of the resulting supernatant to the equilibrated Reactive Blue column. Wash the column with extraction buffer until A_{280} is less than 0.05. Elute weakly bound dehydrogenases with 20 ml of 10 mM Tris (pH 8.6), 1 mM lithium lactate, 1 mM NAD^+, 0.5 mM β-mercaptoethanol. Wash the column with 40 ml of 10 mM Tris (pH 8.6), 0.5 mM β-mercaptoethanol.

5. Elute the LDH by washing with 20 ml of Tris-HCl (10 mM, pH 8.6 plus 0.5 mM mercaptoethanol), to which has been added 1 mM NADH. Additional enzyme can be recovered by washing with 2 × 20 ml of Tris-HCl. Check the LDH activity of all samples. Combine active fractions and concentrate to about 20 ml using an Amicon® or similar membrane concentrator (PM-10 membrane).

6. Dialyze the concentrated sample against 2 L of 10 mM Tris-HCl (pH 7.0), 15 mM β-mercaptoethanol overnight in the cold room.

7. Equilibrate a 15-ml DEAE-cellulose column with 10 mM Tris-HCl (pH 7.0), 0.5 mM β-mercaptoethanol. [It may help to pass ~100 ml 0.5 M Tris-Hu (pH 7.0) through the column first, and then wash with 10 mM Tris-HCl (pH 7.0).] The column is equilibrated when the pH of the effluent is at pH 7.0.

8. Determine the activity and protein concentration of the dialyzed LDH preparation.

9. Apply the concentrated, dialyzed LDH to the column and wash with 10 mM Tris-HCl (pH 7.0), 0.5 mM β-mercaptoethanol. Collect 5-ml fractions.

10. Determine protein concentrations and assay each of the fractions. The bulk of the activity should elute in the second through fifth fractions. These fractions contain fairly pure LDH. If necessary, concentrate the LDH to about 2 mg/ml, using the membrane concentrator.

Production and Collection of Rabbit Antibody Directed Against Chicken LDH

Mix 1 ml of the LDH preparation with 1 ml of *Freund's complete adjuvant* (a mixture of mannide oleate, 1.5 ml; dried *Mycobacterium butyricum,* 5 mg; and paraffin oil, 8.5 ml). Homogenize completely. Load the resulting mixture into a 3-ml syringe fitted with a 22-gauge needle. Inject subcutaneously into the back of a rabbit at five or six locations. In two weeks, repeat with 0.5 ml (1 mg) LDH and 0.5 ml Freund's adjuvant.

After an additional week, obtain a 1- to 2-ml blood sample from the ear vein of the rabbit using a 23-gauge needle. Permit the blood to clot, centrifuge at ~7000 × *g*, and use the serum to test for the presence of LDH antibody (see procedures below). The antibody reaction should be clear and strong, that is, show sharp precipitin bands or a +5 ELISA value. If not, inject another LDH booster and test the serum again after one week. Once the antibody reaction is acceptable, the rabbit should be bled from an auricular vein; about 60 ml of blood may be collected at one time (see Kaplan and Timmons, 1979). Let blood clot, and centrifuge to obtain serum.

Collect the serum and recentrifuge at ~1000 x *g* for 10 min to remove any contaminating red cells. To the cleared serum add an equal volume of saturated (R.T.) ammonium sulfate (pH 7.0) in a slow steady stream with stirring. Continue stirring 30 min at room temperature, then centrifuge at ~10,000 × *g* for 20 min to pellet the precipitate. (This pellet can be frozen or refrigerated for several months before further purification without appreciable loss of activity.) Redissolve the pellet in Tris-buffered saline (TBS) [10 mM Tris-HCl (pH 7.0), 150 mM NaCl] to roughly the same volume as that of the original serum volume, and dialyze against at least 200 volumes of the same buffer overnight at 4°C.

Casting the Gel Slides

Materials and Reagents

1. *Phosphate-buffered saline (PBS):* sodium phosphate buffer, 0.01 M (pH 7.5), which contains 0.14 NaCl

2. *1% Stock agarose gel:* Warm 40 ml of PBS to approximately 50°C, then sift over the top of the warm solution 400 mg of powdered agarose. The agarose should have a gel point of 36–40°C. Mix well, and continue to warm

the suspension until the agarose is completely dissolved and the solution is clear. When cool, it sets to a stiff gel.

3. *Working agarose gel:* Warm the stock gel and an equal volume of PBS in separate vessels. When the stock gel has completely liquified, add an aliquot to an equal volume of warm PBS to make the needed quantity of working gel. Mix well before sampling.

4. *Glass microscope slides:* Clean the requisite number of glass microscope slides, which should be free of deep scratches or other imperfections. Dip the slides individually in a dish of 95% ethanol, then wipe each one dry with a fresh tissue sheet, or lint-free cloth square. Lay the slides on a clean, level, flat surface. Once the cleaned slides have been set down, do not touch the top surfaces with your bare fingers.

Method

Warm a 5-ml serologic pipet to about 55°, either by aspiration of heated PBS or by quickly passing the empty pipet through a small burner flame. Draw up about 3 ml of the liquified working gel, then quickly transfer some of the liquid to the outer edges of a cleaned glass slide. As liquified gel hits the cold glass, it will begin to harden, forming a moat into which the remainder of the liquid gel can be delivered. With a little care (and a little practice) the gel fluid will spread to form a substantially uniform layer over the surface of the slide. Repeat this process until all of the slides have been coated.

Allow the coated slides to cool thoroughly and the gel to reach maximum hardness before attempting to move or disturb them. This will take 30–45 min. Discard any slides that do not have a smooth and bubble-free coating. If necessary, the gel may be carefully scraped off a slide, put back into the container, and remelted.

As an alternative to the above procedure, a commercial gel casting apparatus can be used, but it is quite costly and, in fact, does very little better than the manual method described above.

Cutting the Wells

In order to obtain reproducible results, it is important that the wells be cut in a uniform line down the long axes of the glass slide, and as parallel as possible. The cutting of the wells is made much easier by a template, which is reproduced below. By laying the slide within the rectangle, it is possible to apply the well punch over each of the heavy black circles that mark the proper locations. Note the odd circle at the right-hand end. This is not a well to be filled; it serves merely to distinguish one end of the plate from the other.

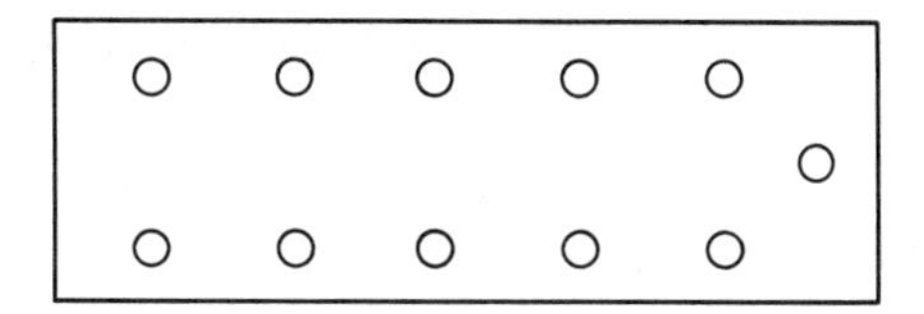

The well punch itself is merely a short length of metal tubing, sharpened at one end (as in a cork borer). The other is connected to a water aspirator. It is important that the punch be pressed through the template and be lowered onto the surface of the gel as nearly vertically as possible in order to make clean cuts through the gel.

When all the wells are punched, the diffusion plate should be a replica of the template. We shall adopt the convention that the end with the center-cut well will be the top of the plate.

Immunodiffusion Process and Staining

1. *Antibody:* The antibody (Ab) is prepared as outlined above; the Ab concentration is adjusted to 1 mg/ml.

2. *Antigen:* In this case the antigen is your own LDH preparation (Expt. 9). If you are in doubt concerning its concentration, get a fresh estimate by spectrophotometry, using the value of $A_{280} = 1$ as approximately equivalent to a concentration of 1 mg/ml.

If the protein concentration in your LDH preparation is less than required, it may be increased by lyophilization; proceed as follows: Transfer the LDH solution to a dialyzing tube of a size appropriate to the sample volume. Dialyze against 1 mM Tris-HCl (pH 8.6), 0.5 mM β-mercaptoethanol at 4°C overnight, then pour the dialyzed LDH solution into a 13 mm × 100 mm test tube. Freeze the sample in a shell in the bottom of the tube using an acetone–dry ice mixture (see Chapter 9). Place the tube in a flask that can be attached to a lyophilizer, and turn on the apparatus; freeze-drying should be complete within 2 hours if the original sample volume was 2 ml or less. Redissolve the dried residue in water to a volume sufficient to attain a protein concentration of 1 mg/ml.

Setting Up the Plates

Two plates are required, hereafter identified as plates I and II. On plate I, the antibody is employed in stock concentration. On plate II, the antibody is serially diluted with PBS.

On plate I, the antibody can be applied to all of the wells with the same 10-μl pipet, since the concentration used is the same. When setting up plate II, use a fresh 10-μl pipet for each addition, since the concentration is being serially reduced by a factor of 2 in each of the successive wells. *Do not cross-contaminate the antibody dilutions.*

When you have applied 10 μl of antibody to all of the wells, the plates should appear as shown below. The right-hand rows of wells are omitted, since they contain nothing as yet.

Be sure to make the antibody additions *directly* into the wells; do not slop the solution over the remainder of the gel surface.

Plate I		Plate II	
	O	Indexing well	O
Stock	O	Stock	O
Stock	O	1→2 diln.	O
Stock	O	1→4 diln.	O
Stock	O	1→8 diln.	O
Stock	O	1→16 diln.	O
Varying Ag		Varying Ab	

Adding Antigen to Plate I

Using PBS, make serial dilutions of a small amount of your antigen and apply them to the right-hand row of wells. In descending order, the antigen concentrations you should apply to plate I should correspond to the values shown above for plate II. When antigen application is complete, carefully place plate I into the moist chamber for diffusion.

Adding Antigen to Plate II

Make a 1 → 3 dilution of an aliquot of stock antigen. Put a 10-μl aliquot of this dilution into each of the right-hand row of wells on plate II. This completes the preparation of the plate; it can now join plate I in the moist chamber for diffusion.

Make sure the chamber is still moist; cover it, and leave undisturbed for the remainder of the diffusion period.

Check the plates at the end of 24 hours; look for the presence of precipitin bands in the space between the rows of wells on each. The bands may be faint; they are best seen over a dark background when the plates are examined at a grazing angle. If you cannot see the bands, it may be necessary to stain the plates.

Staining

Remove the plates from the moist chamber, and *carefully* rinse them off with *cold* water. Discard the sponge in the moist chamber, put the plates back into the Petri dish, then add sufficient Coomassie Blue staining reagent to cover the plates. Leave the plates to stain for one-half hour, then decant the stain and destain the plates at *room temperature.* Change the destaining solution (discard) until the backgrounds of the plates are clear.

REPORT

1. Submit sketches of each of your plates.
2. Calculate the initial antibody/antigen ratios for each of the well combinations on plates I and II.

3. Interpret your results in terms of any valid conclusions you may reach concerning the homogeneity of your LDH preparation.

REFERENCES

Crowle, A. J., *Immunodiffusion,* 2nd ed. Academic Press, New York (1973).

Clausen, J. Immunochemical techniques for identification and estimation of macromolecules. In *Laboratory Techniques in Biochemistry and Molecular Biology* (T. S. Work and E. Work, eds.), *1*:397–505. American Elsevier, New York (1970).

Kaplan, H. M., and Timmons, E. H. *The Rabbit, a Model for the Principles of Mammalian Physiology and Surgery.* Academic Press, New York (1979).

Enzyme-Linked Immunosorbent Assay (ELISA)

The ELISA used in this experiment is a BAELISA (biotin–avidin ELISA) in which the enzyme conjugated to avidin is alkaline phosphatase. The substrate is 4-nitrophenyl phosphate; upon hydrolysis of this ester, 4-nitrophenol is released according to the equation:

$$O_2N-C_6H_4-O-PO_3^{2-} + H_2O \xrightarrow{\text{alkaline phosphatase}} O_2N-C_6H_4-O^- + H^+ + HPO_4^{2-}$$

The nitrophenolate ion has a λ_{max} = 400 nm and ϵ_{400}^{M} = 18,600, whereas the phosphate ester substrate absorbs only in the ultraviolet.

MATERIALS AND REAGENTS

1. 50 mM Tris-HCl (pH 9.5)
2. 0.02% bovine serum ablumin (BSA) in 50 mM Tris-HCl (pH 9.5)
3. 0.05% Tween-20 in PBS
4. LDH preparation from Experiment 9
5. Rabbit antibody to chicken LDH
6. Goat anti-rabbit antibody, biotinylated (BGARA); commercially available
7. Avidin–alkaline phosphatase conjugate (AAPC); commercially available

8. 4-Nitrophenyl phosphate (4-NPP), 1 mg/ml, in 0.05 M K_2CO_3, 0.001 M $MgCl_2$ (pH 9.5)
9. 96-well Immulon® I ELISA plates
10. 37°C humidity chamber
11. Pipette tips treated with BSA solution
12. Micro ELISA autoreader

PROCEDURE

1. Prepare three dilutions of your LDH preparation, 0.5, 1.0, and 1.5 μg protein per ml, using the 50 mM Tris-HCl buffer as the diluent.
2. Add 200 μl of each of the three LDH solutions to three separate wells in the Immulon® plate; each well now contains 200 μl of a different LDH solution. Run in triplicate.
3. Cover and allow to "sensitize" (i.e., remain undisturbed) overnight at room temperature.
4. Wash the plate thoroughly with PBS/Tween, using a wash bottle to direct the stream of buffer. Wash three times and, between each wash, sharply flick out the remaining buffer. Ignore any slight residue in the form of bubbles.
5. Add 100 μl of diluted rabbit anti-LDH antibody to each well. Commonly used dilutions are 1:1000 to 1:5000 (this must be determined empirically by using known LDH samples).
6. Incubate in the humid chamber at 37°C for 2 hours.*
7. Wash the plate with PBS/Tween as before.
8. Add 100 μl of the BGARA to each well.
9. Incubate for 2 hours at 37°C.*
10. Wash the plate with PBS/Tween as before.
11. Add 100 μl of AAPC to each well.
12. Incubate at 37°C for 2 hours.*
13. Wash the plate 3 times with PBS/Tween.
14. Add 100 μl of 4-NPP to each well.
15. Incubate at 37°C until distinct reaction is seen (~30 min).
16. Stop the reaction by adding 50 μl of 0.3 N NaOH to each well.

*The incubation times may be reduced to 1 hour each, but there is some loss in sensitivity.

17. Estimate the intensity of the yellow color in each well with a +1 to +5 rating.

18. Read plates on an automatic plate reading photometer; or have them read for you. A number of such instruments are available.

REPORT

1. Include an evaluation of the intensity of the yellow color in each reaction well, as +1 to +5.
2. If the ELISA plate autoreader is used, submit the computer tape readout.
3. How well do the assay results from (1) and (2) above correlate?

Inhibition of Citrate Synthase*

Our general understanding of enzyme-catalyzed reactions includes the following points.

1. Any enzyme has one or more sites to which substrate(s) bind and one or more sites to which product(s) bind; these comprise at least a portion of the active site(s) at which catalysis occurs.
2. Molecules (analogs) that have a structure similar to that of a substrate may also bind to an active site, but are not able to participate in the chemical reaction catalyzed. They have the apparent effect of inhibiting the reaction.
3. Enzymes show a variable degree of specificity with respect to their substrates; that is, some enzymes accept only a single molecular species as substrate, whereas others accept one of several closely related species, and still others act on a fairly wide range of substrates, provided they all contain some particular bond type or particular structural feature.

Citrate synthase catalyzes the condensation of acetyl-CoA with oxaloacetate to form citrate.

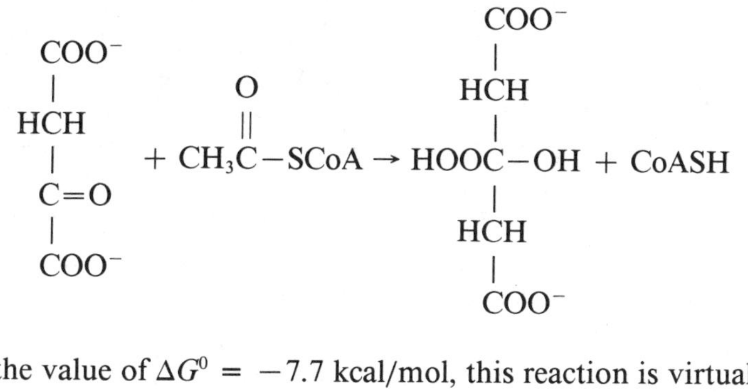

$$
\begin{array}{c} COO^- \\ | \\ HCH \\ | \\ C{=}O \\ | \\ COO^- \end{array}
+ CH_3\overset{\overset{\displaystyle O}{\|}}{C}{-}SCoA \rightarrow
\begin{array}{c} COO^- \\ | \\ HCH \\ | \\ HOOC{-}OH \\ | \\ HCH \\ | \\ COO^- \end{array}
+ CoASH
$$

Because the value of $\Delta G^0 = -7.7$ kcal/mol, this reaction is virtually a "one-way" system. Besides acetyl-CoA, only fluoroacetyl-CoA is known as an enzymatically acceptable partner for condensation with oxaloacetate, giving rise to fluorocitrate. Fluorocitrate cannot be metabolized further; thus cells

*Preparatory reading: Section I, Chapter 2; Experiment 9; and Appendix 2.

exposed to it quickly die from inhibition of the Krebs cycle, a major source of metabolic energy and essential intermediates.

Although they cannot participate in the catalyzed reaction, certain other derivatives of CoASH can also bind to the synthase. We shall examine the effect of one of these on the kinetics of the synthase reaction. This particular derivative, known as acetonyl-CoA, is prepared by alkylation of CoASH with $BrCH_2COCH_3$.

The citrate synthase assay will be performed by measuring appearance of the product, CoASH, by means of an accessory reagent, 5,5′-dithiobis(2-nitrobenzoic acid) (DTNB; *Ellman's Reagent*). This disulfide reagent reacts with free thiols to form mixed disulfides and free 2-nitro-5-thiobenzoic acid, which has a yellow color with a λ_{max} at 412 nm. The absorbance of the solution is, therefore, proportional to the quantity of free CoASH formed per unit time. Note carefully that this form of assay would not be acceptable if the enzyme contained an essential -SH group with which the reagent could react.

MATERIALS AND REAGENTS

1. Preparation of *S*-Acetonyl-CoA* (MW = 818.55)
   a. The starting material is a commercial Na or Li salt of CoASH. Either salt is acceptable, provided the molecular weights are correctly estimated. Since, on the average, the commercial Na salt actually contains 2.5 g-atoms of Na per mole, the presumed MW of the starting material is estimated as 822.6. Similar estimates can quickly be made for the Li salt.
   b. Place about 15 ml water in a 50-ml beaker containing a small magnetic stirring bar. Add 20 μmol dithiothreitol to the water, and stir while passing a gentle stream of nitrogen gas over the surface of the solution. The purpose of the dithiothreitol is to reduce any CoASH that might have oxidized. Add 150 mg $CoASH(Na)_{2.5}$ and allow it to dissolve. Cautiously add 1 N NaOH to bring the pH to 8.0–8.2.
   c. Alkylation is done by addition of bromoacetone (MW = 137, sp. gr. = 1.634), which has a molar volume = 83.84 ml/mol. Since any excess reagent will be removed in subsequent processing, take 20 μl in approximately 5 ml of 95% ethanol. **CAUTION: Great excess of the reagent should be avoided, since bromoacetone is a powerful lachrymator, and the excess will be vented to the atmosphere during subsequent lyophilization.**
   d. Add the ethanolic solution of bromacetone to the reduced CoASH solution. The pH will show a marked fall. Add 1 N NaOH dropwise to bring the pH back to 8.0–8.2 while stirring for a period of at least 3

*Procedure from Rubenstein and Dryer (1980).

min. Finally, add 6 N HCl, dropwise, to bring the pH to approximately 3.

e. The acidified solution is lyophilized to remove solvent and excess reagent. The dried product is then dissolved in about 2 ml of water for desalting.

f. Desalting is accomplished with a long, thin column of G-15 Sephadex, which should be operated at 4°C. The product is eluted with water. Collect those fractions which show the earliest A_{257} absorbance; pool those with significant absorbance at this wavelength. The collected fractions are then lyophilized for final storage. A column with dimensions of 1.5 × 2.5 cm is sufficient to desalt about one-half of a preparation on the scale given above.

2. *Citrate synthase:* A commercial preparation is usually dispensed at a concentration of 80 U/mg. It is sold as a crystalline suspension in 2.2 M $(NH_4)_2SO_4$ at pH 7. Make a stock solution of 40 μg/ml in water.

3. *DTNB:* 1 mM solution in Tris-HCl buffer, 0.05 M (pH 8.0). $\epsilon_{412}^{mM} = 14.4$. This should be made shortly before use, since the reagent hydrolyzes slowly at pH 8, taking on the typical yellow color. To avoid high blanks, keep the reagent cold.

4. *Oxaloacetate:* 2 mM stock solution in Tris-HCl buffer, 0.05 M (pH 8.0). It is important that this be freshly made and that it be kept in an ice bath when not in use. This is *not* a stable reagent.

5. *Tris-HCl buffer:* 0.05 M (pH 8.0).

6. *Acetyl-CoA:* Approximately 0.4 mM; make this by dissolving 0.95 mg in 2.5 ml of water. This should also be made fresh and kept in an ice bath.

7. *Acetonyl-CoA:* Made by dilution of the stock preparation to a concentration of 2 mM in water, based on the millimolar extinction coefficient $\epsilon_{260} = 15.4$.

PROCEDURE

These assays will be done with the recording spectrophotometers, and several special adjustments of them must be made. The instruments should be turned on at least 30 min before use, to allow the cuvet compartments to come to thermal equilibrium, when they will be at approximately 30°C. The recorder sensitivity should be set so that 0.02 *A* (absorbance) causes full-scale deflection, and the chart speed should be set at 5 in./min or as near that value as is practical. A water bath at 30 ± 2°C, utilizing a large beaker and a hot plate or other arrangement, should be set up nearby.

Each assay system requires four cuvets *plus* two blanks: One of the blanks is required only to zero the spectrophotometer (two are required for double-beam instruments); the second (or third) is in use while reading absorbance values of the remaining four cuvets. Since all readings will be made at λ = 412 nm, plastic disposable cuvets are used.

Four spectrophotometric systems are described below. System I will provide a set of curves dealing with the uninhibited reaction. Systems II–IV will provide data on the effect of increasing inhibitor concentrations. The detailed outlines of the systems should be followed closely.

System I

Reagent addition	Volume to be added (μl)				
	Blank	S_1	S_2	S_3	S_4
1. Tris buffer	600	600	600	600	600
2. OAA	100	100	100	100	100
3. DTNB	100	100	100	100	100
4. AcCoA	0	30	50	100	150
5. Inhibitor	0	0	0	0	0
6. H_2O	200	170	150	100	50

Mix the above by cuvet inversion, then place all cuvets in the water bath for at least 5 min. Remove the cuvets marked *Blank,* add 5 μl of enzyme to each, mix, then use the cuvets to zero the instrument and recorder. Remove the "blank" cuvet in the sample channel of the recorder, then proceed as indicated below.

Reagent addition	Blank	S_1	S_2	S_3	S_4
7. Enzyme	0	5	0	0	0

As soon as the enzyme is added, mix by inversion, avoiding bubbles. Place cuvet S_1 immediately in the spectrophotometer and record the abosrbance for 1–5 min. Turn off the recorder, roll the chart backwards until the pen is at the starting point of the first recording. Offset the pen by about 0.005 *A* (absorbance) *above* the previous starting point.

Reagent addition	Blank	S_1	S_2	S_3	S_4
8. Enzyme	0	0	5	0	0

Proceed with cuvet S_2 as described in step 7 for S_1; record for 1–5 min, then roll back the chart and offset the pen for the next recording to follow.

Reagent addition	Blank	S_1	S_2	S_3	S_4
9. Enzyme	0	0	0	5	0

Record the absorbance, roll back the chart, and offset the pen.

Reagent addition	Blank	S_1	S_2	S_3	S_4
10. Enzyme	0	0	0	0	5

Record the absorbance. You will now have a chart strip containing four tracings. This completes examination of System I.

System II

In a new set of cuvets, make the same additions 1 through 4 as listed for System I; then add inhibitor acetonyl-CoA as indicated:

11. Inhibitor	0	10	10	10	10
12. H_2O	200	160	140	90	40

Repeat steps 7 through 10 as outlined for System I. You will recover a chart with four tracings showing the effect of a low inhibitor concentration.

System III

In a new set of cuvets, make the same additions 1 through 4 as listed for System I, then add the acetonyl-CoA:

13. Inhibitor	0	20	20	20	20
14. H_2O	200	150	130	80	30

Repeat steps 7 through 10 as outlined for System I. You will recover a chart with four tracings showing the effect of a larger inhibitor concentration.

System IV

In a new set of cuvets, make the same additions 1 through 4 as listed for System I, then add the inhibitor acetonyl-CoA;

15. Inhibitor	0	40	40	40	40
16. H_2O	200	130	110	60	10

Repeat steps 7 through 10 as outlined for System I. This chart will show the effects of a higher inhibitor concentration. This chart completes the experimental protocol.

DATA ANALYSIS

The reaction between CoASH and DTNB is stoichiometric and rapid. It follows that the increase in absorbance should bear a linear relation to micromoles of CoASH formed, and that this, in turn, should be equivalent to micromoles of citrate formed. The absorbance tracings should be linear over a significant portion of their lengths, so it should be easy to determine ΔA/min. Given that the millimolar extinction coefficient of the chromophoric product—2-nitro-5-thiobenzoic acid—is 17.5, you should be able to calculate micromoles citrate formed per minute for each of the lines in each of the data sets comprising Systems I–IV. You should also be able to relate these values to acetyl-CoA and acetonyl-CoA concentrations.

REPORT

1. Prepare and submit a plot of reciprocal velocities (ordinate) versus reciprocal substrate (acetyl-CoA) for the four systems.
2. Calculate and clearly indicate in your report the best values for V_{max} and for K_m, obtained from your data.
3. From your data, make a best estimate of the value of K_I for acetonyl-CoA.
4. Indicate whether the pattern of inhibition you observed is competitive or noncompetitive.
5. In a sentence or two (no dissertations, please), state how your answer to (4) might be different if oxaloacetate concentration had been varied while the concentration of acetyl-CoA had been held fixed.

REFERENCE

Rubenstein, P., and Dryer, R. L. *J. Biol. Chem.* *255*:7858 (1980).

Substrate Specificity of a Snake Venom Phosphatase*

The diverse group of enzymes known as phosphatases, or *phosphohydrolases,* can be broadly subdivided into three classes depending on the pH optima of the particular enzymes: (1) acid phosphatases (optimum pH range = 4.8–6.5), (2) neutral phosphatases (optimum pH range = 6–8), and (3) alkaline phosphatases (optimum pH range = 8.6–10.3). Obviously, this kind of classification is not rigorous. A second classification scheme divides these enzymes into *phosphomonoesterases* and *phosphodiesterases,* depending on the chemical nature of the phosphorylated substrates. Five such structures are shown in Fig. E13–1; the two sugar phosphates shown on the left are phosphomonoesters. Cyclic AMP and the typical lecithin, both shown on the right, are phosphodiesters.

Fructose 1,6-bisphosphate [D-fructose-1,6-bis(dihydrogenphosphate)] is hydrolyzed by a specific alkaline phosphomonoesterase to yield fructose-6-phosphate. No common sugar monophosphate is a substrate for the fructose 1,6-bisphosphatase. Glucose-1-phosphate is hydrolyzed by a number of nonspecific phosphatases.

The specificity of phosphodiesterases is, as a rule, quite high. In the case of phosphatidylcholines, two specific types of diesterases are known. Phospholipases D (there are several) yield a phosphatidic acid and free choline, whereas phospholipases C (there are several) yield a diglyceride and phosphocholine. The phosphodiesterase that converts cyclic AMP to 5′-AMP is also quite specific. Methylxanthines, such a caffeine, are good inhibitors of this phosphodiesterase; according to some, the "lift" obtained from drinking coffee may be due to diminished hydrolysis of cyclic AMP. Note that the phosphodiesterases break only one bond and that they do not liberate inorganic phosphate, as do the phosphomonoesterases.

As suggested by the above discussion, phosphorylated compounds are widely distributed in living systems. They serve as storage forms for energy (e.g., ATP, phosphocreatine, etc.), as components of informational macromolecules (i.e., dNTP), as allosteric effectors of certain enzymes (e.g., fructose-1,6-bisphosphate, phosphoenolypyruvate, etc.), and as so-

*Preparatory reading: Section I, Chapters 2 and 4; and Appendix 2.

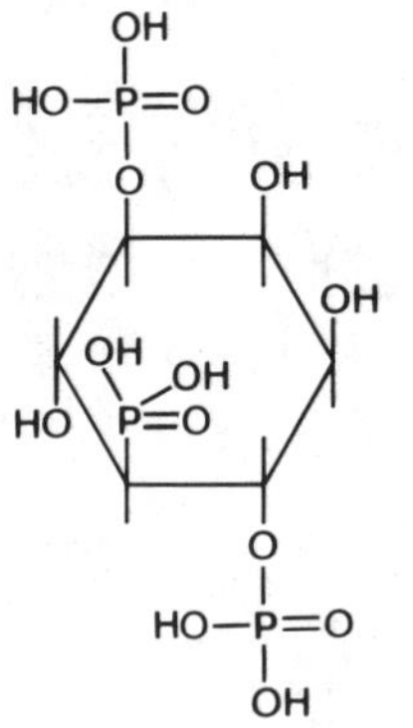

MYOINOSITOL-1,4,5-TRIPHOSPHATE

PHOSPHOMONOESTERS

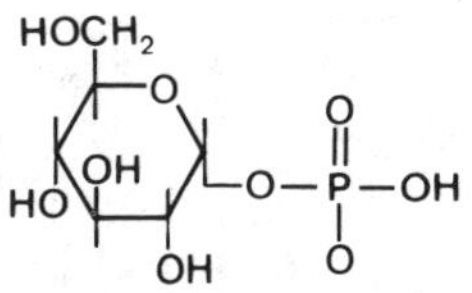

GLUCOSE-1-PHOSPHATE

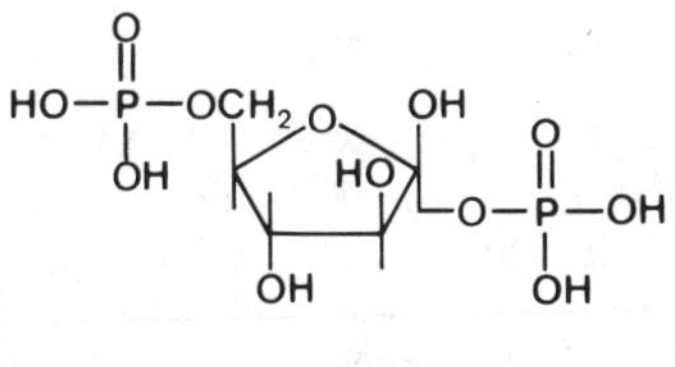

FRUCTOSE 1,6-BISPHOSPHATE

PHOSPHODIESTERS

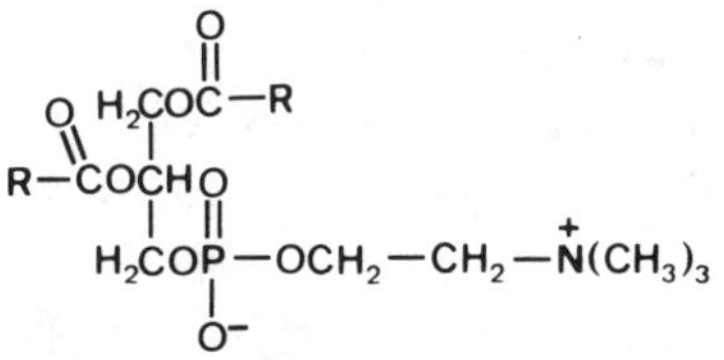

PHOSPHATIDYL CHOLINE (LECITHIN)

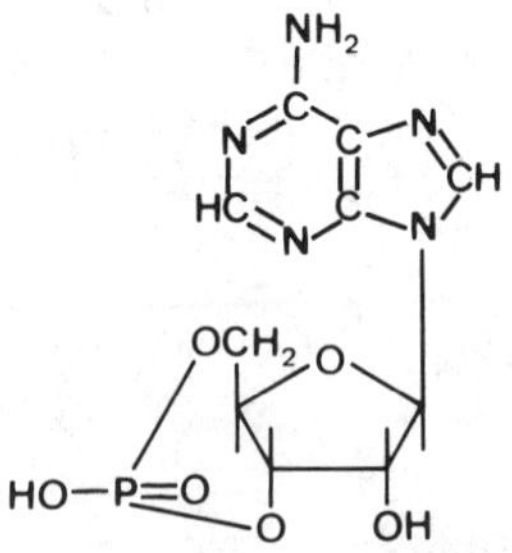

3′,5′-ADENYLIC ACID (CYCLIC AMP)

Figure E13–1. Some phosphate ester structures.

called "second messengers" (i.e., cAMP, cGMP, inositol phosphates). Phosphorylation–dephosphorylation reactions of proteins, mediated by protein kinases and protein phosphatases, modulate some enzyme activities (phosphorylase, pyruvate dehydrogenase, etc.). It is therefore not surprising that phosphatases of many kinds can be extracted from many tissues. Various phosphatases can also be found in extracellular or bodily fluids such as milk, blood plasma, and serum, and also in urine and animal venoms. These phosphatases are frequently quite *nonspecific* with respect to their substrates. Since determination of phosphatase activity in blood is a valuable diagnostic tool, this lack of specificity is fortunate. It allows enzymologists to construct a variety of synthetic substrates that permit accurate and rapid spectrophotometric assays. Phenolphthalein phosphate, a colorless compound, can be hydrolyzed by the blood enzyme(s) to free phenolphthalein. Addition of alkali stops the reaction and produces a red color whose intensity is proportional to the concentration of free indicator.

The present experiment deals with a phosphomonoesterase derived from rattlesnake venom, and with a series of potential substrates including some nucleotides, a nucleotide polyphosphate, and some sugar phosphates. You will examine first the effect of the enzyme on these substances, and then the effect of strong acid and elevated temperature on the nonenzymic hydrolysis of some of the same compounds.

REAGENTS

1. *Buffer + Mg^{2+}:* 20 mM $MgCl_2$ in 0.2 M glycine–NaOH buffer (pH 9.0).
2. *Potential substrates:* All potential substrates will be supplied as aqueous solutions (at 0.25 mM) which are to be adjusted to pH 6–8 *promptly* after solution.
3. *Diluted snake venom:* Lyophilized venom collected from *Crotalus adamanteus* (commercially available) is dissolved in water at a concentration of 10 mg/ml. This is clarified by centrifugation for 15 min at 10,000 × *g*, at 4°C, then diluted 200 fold with ice-cold water. Because the venom contains powerful proteases which can degrade the phosphatases, it is important to prepare this solution just before use, and to keep it as cold as possible on an ice bath.
4. *Phosphate assay reagent:* 3.5 N H_2SO_4. Make a 1:10 dilution of concentrated sulfuric acid with water.
5. *3.5% ammonium molybdate:* Dissolve the solid salt in water made slightly acidic by addition of a few drops of concentrated H_2SO_4. This is sometimes required to prevent hydrolysis of the salt. The final solution should be perfectly clear.
6. *Monoethyl-*p*-aminophenol* (also known as Elon®, a photographic developer or reducing agent): 0.7% in 2.1% $NaHSO_3$.
7. *Inorganic orthophosphate (P_i) standard:* This is prepared by dissolving KH_2PO_4 in water to make a 0.250 mM solution.

PROCEDURE

Standard Curve for Determination of Inorganic Phosphate

To a series of tubes, add 0, 1.0, 2.0, 3.0, and 4.0 ml of inorganic phosphate standard solution, then adjust the volume in all of the tubes to 4.0 ml with distilled water. To each tube, add 1.0 ml of 3.5 N H_2SO_4, 1.0 ml of ammonium molybdate solution, and 1.0 ml of reducing reagent. Vortex-mix thoroughly between the additions. Allow the tubes to stand at room temperature for 20 min, then read the A_{660} against the reagent blank. (**Note:** For double-beam spectrophotometers prepare *two* blank tubes.)

Enzymatic Hydrolysis of Phosphomonoesters

Number a series of tubes, 1 through 18, and add 1.0 ml of *buffer* + Mg^{2+} to each tube. Place 2.0 ml of potential substrate into the tubes, according to the following table:

Tube No.	Solution to be added
1 + 2	Water (these serve as "no substrate" controls)
3 + 4	2′-AMP
5 + 6	3′-AMP
7 + 8	5′-AMP
9 + 10	5′-UMP
11 + 12	6′-dTMP
13 + 14	Glucose-6-phosphate
15 + 16	Glucose-1-phosphate
17 + 18	5′-ATP

To each of the *odd*-numbered tubes add 1.0 ml of water. These tubes will serve as "no-enzyme" controls. To each of the *even*-numbered tubes add 1.0 ml of the cold, diluted venom solution. **CAUTION: It is not likely that accidental ingestion of this diluted venom could cause any problems (Why?), but you should not mouth-pipet any hazardous reagent; therefore, handle this reagent with automatic pipets!**

Let all of the tubes in this series of 18 stand at room temperature for 30 min, then add 1.0 ml of 3.5 N H_2SO_4 to stop the enzyme reaction and to prepare the tubes for phosphate analysis. Promptly add 1.0 ml of the ammonium molybdate reagent and 1.0 ml of reducing reagent (Elon®) to all tubes. Between additions, vortex-mix thoroughly. Allow the tubes to stand for an additional 20 min, then measure the A_{660} against the reagent blank.

Acid-Catalyzed Hydrolysis of Nucleotides

Place 3.0 ml of water in each of two labeled tubes. To the first, add 1.0 ml of 5′-AMP solution; to the second, add 1.0 ml of 5′-ATP. Add 1.0 ml of 3.5 N H_2SO_4, mix well, and place both tubes in a gently boiling water bath for 10 min. Remove the tubes and cool them to room temperature in a beaker of cold water. Add 1.0 ml of ammonium molybdate and 1.0 ml of reducing reagent, as before. Let stand for 20 min, then measure A_{660} against the reagent blank.

CALCULATIONS

Construct a standard curve by plotting A_{660} as a function of micromoles of P_i per milliliter of colored solution. Correct the A_{660} values measured in the Enzymatic Hydrolysis of Phosphomonoesters section, for all necessary blanks, including any P_i present in the venom solution.

REPORT

1. Draw the structures of all compounds that were extensively hydrolyzed by the enzyme in the second part of the Procedure.
2. Attach your standard curve for inorganic phosphate to your report.
3. What can you conclude from your results about the specificity of this phosphatase?
4. Calculate and report the number of acid-labile phosphate groups per molecule of ATP, after correcting for hydrolysis of 5′-AMP.
5. What can you conclude about the relative stabilities of phosphoric anhydride bonds relative to phosphomonoester bonds?
6. Assume that the compound showing the most extensive hydrolysis by the venom phosphatase is an ideal substrate; what is the specific activity of the diluted venom preparation (μmoles/min/mg protein)?

REFERENCES

Drummond, G. I., and Yamamoto, M. Nucleotide phosphomonoesterases. In *The Enzymes,* 3rd ed. (P. Boyer, ed.), *4*:355–371. Academic Press, New York (1971).

Laskowski, M., Sr. Venom exonucleases. In *The Enzymes,* 3rd ed. (P. Boyer, ed.), *4*:313–328. Academic Press, New York (1971).

Stereospecificity of Acylase I*

The enzyme, acylase I, specifically hydrolyzes *N*-acetyl-L-amino acids to yield acetate + L-amino acids. *N*-Acetyl-D-amino acids are not readily attacked by this enzyme. Acylase I is thus a useful tool in resolving racemic mixtures of D,L-amino acids or in determining the configuration of an amino acid when the quantities of material are not sufficient to use other means. The capacity of this enzyme to distinguish between optical isomers is also a good illustration of a particular aspect of enzyme specificity in general.

Two other concepts are demonstrated by this experiment. First, it provides an example of the use of thin-layer chromatography (TLC), a powerful and relatively inexpensive means of separating certain mixtures into their individual components. Lastly, it allows a demonstration of autoradiography, whereby localization of components tagged with a radioisotope can be visualized using x-ray film.

In this experiment you will be given two vials. Each will contain either *N*-acetyl-L-[U-^{14}C]alanine or *N*-acetyl-D-[U-^{14}C] alanine. You are to determine which is which by using acylase I. **CAUTION: Since these samples are radioactive, take all needed precautions. Report any spills to the attending staff member at once. Be sure, also, to wear disposable gloves, to dispose of all contaminated materials in the proper containers, and to protect your desk top with proper protective paper when using these materials in solution or on the TLC plates.**

Because a finite time is required for proper exposure of the x-ray film, this experiment is most conveniently performed over two laboratory periods. The protocol has accordingly been divided into two sections.

MATERIALS AND REAGENTS

1. *Acylase I:* a commercial preparation from hog kidney, provided as a lyophilized powder. Weigh out 50 μg and dissolve it in 50 μl of Na phosphate buffer, 0.1 M (pH 7). An enzyme unit of this preparation is defined as the activity that hydrolyzes 1 μmol of substrate per hour at 25°C and at pH 7. The preparation should have an activity of approximately 250 U/μl.

*Preparatory reading: Section I, Chapters 2, 3, 7, and 9.

2. *Preparation of* N*-acetyl-*D/L*-[U-*^{14}*C]alanine*
   a. The starting material is L- or D-[U-^{14}C]alanine, which should have a specific activity of at least 10 mCi/mmol. The labeled material is usually sold as an aqueous ethanol solution, which should be lyophilized. The best container for lyophilization is a small (12 × 75 mm, or similar) test tube.
   b. Add to each tube, 500 μl of *saturated* $NaOOCCH_3$(NaOAc). To be sure that this solution is saturated, warm it to some temperature above the ambient, then allow it to cool slowly. Crystals should form as the solution comes to room temperature, and should remain at room temperature.
   c. To each tube, add 2.5 μl of acetic anhydride, mixing well and allowing each tube to remain for at least 15 min in an ice bath. Add a second 2.5 μl of acetic anhydride, mix well, and again allow to stand for 15 min in an ice bath. Do not use the acetic anhydride unless it is from a fairly fresh bottle.
   d. Prepare a column (10 ml) of Bio-Rad AG50W-×2 resin, properly freed of fines (fine particles). If new resin is used, be sure it is in the H^+ form. If the resin is not new, cycle it from the Na^+ to the H^+ form several times, batchwise, before it is packed into the column. Wash the column well with at least 200 ml of water. The final eluate should be colorless; the A_{260} should be less than 0.005.
   e. Transfer the reaction mixture to the top of the column, washing the sample into the column with water. Collect the eluate in small fractions (0.25–0.50 ml), and monitor each fraction for radioactivity. Pool and lyophilize those fractions that contain significant radioactivity.
   f. Dissolve the product in 0.1 M Na phosphate buffer (pH 7.0), adjusting the radioactivity so that each microliter of the final solution contains approximately 30,000 cpm.
   g. Add sufficient unlabeled acetyl L- or D-alanine to give a final concentration of approximately 1 μmol/μl.

3. *Unknowns: N*-acetyl-D- or -L-[U-^{14}C]alanine; each tube should contain 15 μl at a specific activity of 30 μCi/mmol.

4. *Standard markers:* D- or -L-[U-^{14}C]alanine, as above, which should contain in addition 0.1 mM L-alanine.

5. *TLC plates:* Commercially prepared cellulose thin layers coated on plastic backing. These may be cut with scissors.

6. *TLC tanks:* These contain a solvent system composed of *n*-butanol:acetic acid:water (4:1:1, v/v/v).

7. *Ninhydrin spray reagent:* Pure ninhydrin (0.3 g) is dissolved in 100 ml *n*-butanol + 3 ml of glacial acetic acid. Store in a brown bottle or in the dark until needed.

8. *X-ray film:* Kodak X-omat AR film is recommended for this purpose. It may be processed in the darkroom manually or by use of an automatic x-

ray film processor. Developed TLC plates may be lightly taped to the x-ray film under a yellow safe light.

9. *Radioactive ink:* This is best made as needed by addition of a drop of ^{35}S- or ^{32}P-containing solution to a small amount of India ink. It is used solely for visualing initials on the x-ray films.

10. *"Microcap" micropipets:* These are disposable pipets that require a rubber bulb fitted with a special carrier for use. This device will be demonstrated.

PROCEDURE

Day 1

Obtain two unknown sample vials and carefully note their numbers. To each vial add 2 μl of the enzyme solution, using a fresh microcap each time so as not to cross-contaminate the contents of the unknown tubes. Carefully vortex both tubes and incubate them for 1 hour at room temperature.

While the samples are incubating, obtain one TLC plate from the stock box, and lay it on a clean, dry surface, cellulose layer uppermost. Lightly draw a pencil line across the bottom of the plate, 2 cm from the edge closest to you. This line will become the starting point of your separation, and all mobilities will be calculated from this line, or origin. Take care when drawing this line to use a gentle touch; do not use the plate if a great deal of the layer chips off the backing.

Further prepare your plate by making a series of light pencil marks at right angles to the origin line, and intersecting with it. These lines need not be more than 1–2 cm long, and should not be so heavy as to go through the cellulose layer. Make these marks at distances measuring 1.5, 3.5, 4, 6, 6.5, and 8.5 cm from the left-hand edge of the plate. You now have marked out three lanes, each 2 cm wide, separated by 0.5 cm spaces which will remain empty. The finished plate should appear roughly as follows:

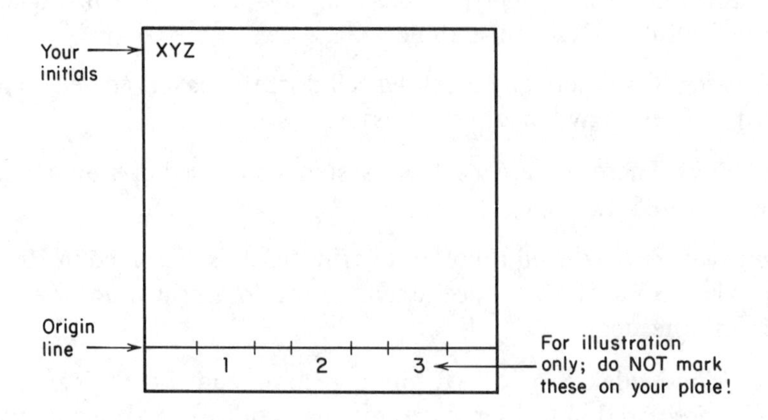

The quality of a TLC separation, other things being equal, is a direct function of the quality of sample application. One must aim to keep the spot or band of applied material as narrow as possible with respect to the vertical, or running, dimension of the plate. Capillarity effects cause rapid diffusion in all directions of applied liquids, so it is most important to limit the rate of application. Furthermore, application of additional solution to an area of the plate already wet can make matters worse. To avoid these problems, you are urged to be sure that the flow of solution from the tip of the pipet is carefully controlled. The layer can be kept fairly dry by blowing over the plate surface a gentle stream of air from a hair dryer or similar device. Direct the air stream in such a way that free drops of liquid are not thrown over the general surface of the plate. While applying the sample, hold the pipet in a position normal to the plane of the TLC plate. It will be worth some time to practice on a small section of a TLC plate. Several scraps will be provided for this purpose.

You are now ready to spot your plate. In lane 1, apply $3 \times 2\ \mu l$ of the standard marker mixture, heeding the precautions given above. In lane 2 apply $4 \times 2\ \mu l$ of your first reaction mixture and in lane 3 apply $4 \times 2\ \mu l$ of your second reaction mixture. Try to apply the materials so that the bands are as narrow as possible, but so they extend from edge to edge of the marked lanes.

After the spots have dried, carefully place the TLC plate into one of the prepared tanks, origin line lowermost. *Before* you immerse your plate, check that the solvent level in the tank is lower than your origin line, but deep enough to contact the lowermost edge of your plate. The solvent will rise up the vertical dimension of the plate, causing separation to occur. Allow the solvent to rise approximately half the distance between the origin line and the top of the plate, then remove the plate, mark the line of solvent advance at the edges with a pin or pencil, and allow it to air-dry in the hood. This process may be expedited by a gentle stream of warm air from the hair dryer.

Place a small piece of adhesive label in the upper right-hand corner of your plate; have the staff member mark your initials on the tape with radioactive ink. Lay your finished plate on a clean piece of protective paper and cover it with a similar piece. Be sure that the plastic side of the protective paper contacts the TLC layer. Weigh down the entire assembly with an empty beaker or similar mass. This completes the first portion of the exercise.

Interim Procedure

In the interim between day 1 and day 2, the staff will collect the TLC layers and clamp each to a sheet of x-ray film. After proper exposure, these films will be processed. Radioactivity adherent to the TLC layer will cause blackening of the x-ray film, constituting an autoradiogram, or replica of the loci of radioactivity on your layer. You will receive both the TLC layer and the developed film on day 2.

Day 2

Take your TLC plate to the hood and cover it with a piece of clean, dry glass so that *only lane 1* is exposed; then spray the exposed portion of the plate with ninhydrin reagent. **NOTE:** *Be careful to avoid getting the ninhydrin on your skin (wear disposable gloves); otherwise you will turn a ghastly purple where the ninhydrin has reacted with skin surface proteins or amino acids.* Allow the sprayed plate to air-dry for 45 min. Examine the plate for any colored spots; make a sketch or otherwise carefully note the lcoation of spots with respect to the origin and the lane in which they are.

Using the autoradiogram as a guide, locate the presence of all radioactive spots on the surface of your TLC plate. These may or may not be visible. With a pencil, lightly sketch in the perimeter of the spots. With a scissors or surgical blade, cut out the individual spots, allowing some clear margin around your marked perimeter. Put each cut-out section into its own scintillation vial, add 5–10 ml of scintillation fluid, shake to dislodge and disperse the TLC material, and count the vials for 1 or 2 min. If no radioactive band was present on the autoradiogram where you might have expected it (i.e., no conversion of substrate to product), cut out the region where the product should have been and count it anyway.

Properly dispose of all radioactive and contaminated materials in the containers provided. Remove any possibly contaminated desktop covering and dispose of it also. This ends the exercise for day 2.

REPORT

1. Calculate the R_f's of alanine and of *N*-acetylalanine in this system.
2. What is your conclusion concerning the configuration of the *N*-acetylalanine in the sample you obtained? (Include unknown number.)
3. Calculate the minimum degree of conversion of substrate to product for each of your unknown samples, expressed as a percentage of the sample applied.
4. With a marker pen, indicate on your autoradiogram the location of the alanine spot observed on the original TLC plate. Submit the autoradiogram as part of the report.

Note: In addition to the reference cited below, see also handbooks on TLC. Those by Egon Stahl are the most comprehensive and thorough. The first edition contains an extended discussion of theory, but less applied information. The second edition does not repeat the theoretical discussion in full form, but does contain a larger amount of applied information.

REFERENCE

Levy, M. *Methods Enzymol.* *4*:251. Academic Press, New York (1957).

Partial Purification of Corticosteroid-Binding Globulin (CBG) from Human Blood*

The adrenal cortex is the site of synthesis of two major groups of steroid hormones. These include the *glucocorticoids* such as cortisol (hydrocortisone), and *mineralocorticoids* such as aldosterone. Lesser amounts of some *androgens,* such as dehydroepiandrosterone and androstenedione, are also produced by both male and female adrenal cortices. Structures of these compounds are shown in Fig. E15–1. After secretion into the blood, these hormones are transported to their respective hormone-sensitive tissues where they exert controls on specific physiologic processes. A primary effect of glucocorticoids is to facilitate gluconeogenesis, increasing the blood glucose concentration and the deposition of glycogen in the liver. Glucocorticoids promote protein catabolism in most tissues. In the liver they stimulate the de novo synthesis of tyrosine aminotransferase and tryptophan oxygenase, enzymes that degrade their respective amino acids. Mineralocorticoids act primarily on the excretion–resorption exchange of Na^+ in the kidney, although they are not entirely free of other physiologic effects. For further details of the functions and effects of the adrenal cortical steroids, consult any major biochemistry or endocrinology textbook.

The major glucocorticoid produced by adrenal tissue varies with the species; in the rat it is corticosterone, whereas in the human it is cortisol. In the human, under basal conditions, about 20 mg of cortisol are produced per day. Blood cortisol concentrations vary with the time of day, from about 5 μg/100 ml in early evening and night to about 15 μg/100 ml in the morning. About 90% of the blood cortisol is noncovalently bound to protein, and about 10% is unbound, or free in solution. Of the bound cortisol, some 20% forms a comparatively loose and nonspecific complex with serum albumin; some 80% apparently binds quite specifically and with high affinity to a glycoprotein called *corticosteroid-binding globulin (CBG)* or *transcortin.* The present experiment is concerned with the partial purification of this protein from human blood plasma or serum.

*Preparatory reading: Section I, Chapter 4, 8, 9; and Appendix 3.

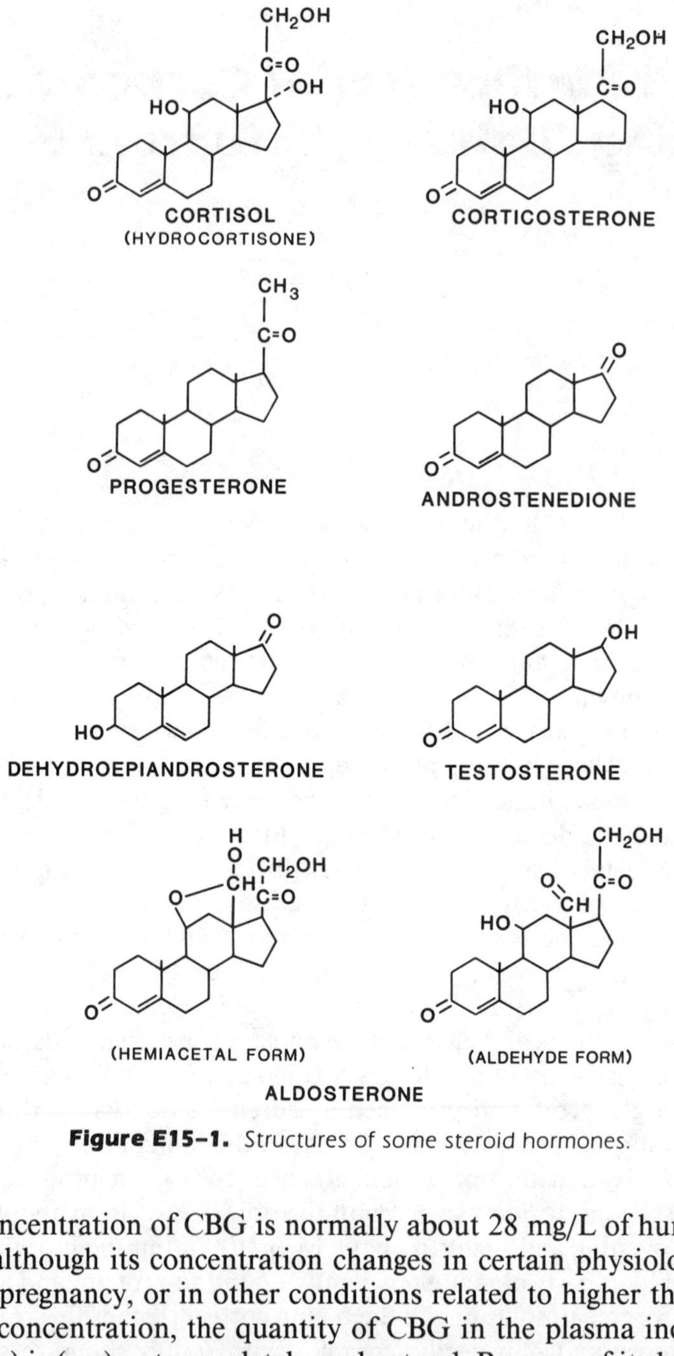

CORTISOL
(HYDROCORTISONE)

CORTICOSTERONE

PROGESTERONE

ANDROSTENEDIONE

DEHYDROEPIANDROSTERONE

TESTOSTERONE

(HEMIACETAL FORM)

(ALDEHYDE FORM)

ALDOSTERONE

Figure E15–1. Structures of some steroid hormones.

The concentration of CBG is normally about 28 mg/L of human blood plasma, although its concentration changes in certain physiologic conditions. In pregnancy, or in other conditions related to higher than normal estrogen concentration, the quantity of CBG in the plasma increases. Its function(s) is (are) not completely understood. Because of its high affinity ([K_a (cortisol) = 3.5×10^9 M^{-1} at 4°C] and specificity (only cortisol and progesterone are tightly bound), CBG may have the effect of modulating the concentration of free cortisol, thereby controlling its availability to the tissues. This may result in minimizing catabolism of cortisol in the body. However, at 37°C the affinity constant is significantly less than at 4°C, and recent work suggests a possible role of CBG in a specific delivery system for transfer of steroids to "target" tissues (Hyrb et al., 1986).

Human CBG, putatively formed in the liver, consists of a single polypeptide chain which contains 30% carbohydrate and has a molecular weight of 51,700. It binds cortisol or progesterone, and, to a lesser extent, testosterone, with a stoichiometry of 1 mol of steroid per mole of protein. The isoelectric pH of CBG has been reported to lie between 4.1 and 4.5, a low figure due, in part, to sialic acid residues in the carbohydrate appendage. Corticosteroid-binding globulins are found in virtually all species examined and are among the most ubiquitous of plasma steroid-binding proteins.

CBG from human plasma has been purified to homogeneity by a number of methods. These include a variety of more or less classic ion-exchange and adsorption chromatographic steps. Recently, several affinity chromatographic agents have been developed and employed. The agents are tedious and expensive to prepare and are somewhat unstable. Some, at least, are subject to degradation by plasma or serum enzymes. However, because CBG has a high specificity and a high affinity for cortisol, it is not necessary to purify the protein completely in order to determine an affinity constant for the CBG–cortisol interaction, or to measure the amount of CBG in a blood plasma or serum sample. Indeed, the commercial availability of [^{3}H]cortisol with reasonably high specific radioactivity makes it possible to perform these measurements on unfractionated or partially fractionated samples. In this experiment we shall partially purify the CBG by ammonium sulfate fractionation. The product will be dialyzed to free it of the salt; then it will be dried by exposing a frozen solution to a high vacuum *(freeze-drying)* and held for further study at a later date.

MATERIALS AND REAGENTS

Human Blood Plasma or Serum

Plasma or serum is obtained as outdated samples from a blood bank, although better results are obtained with freshly drawn blood. One unit is sufficient.

Calcium Chloride (Solid)

Calcium chloride is used to convert plasma to serum; if the starting material is obtained as serum, no $CaCl_2$ is needed.

Ammonium Sulfate (Saturated Solution)

Ammonium sulfate is to be used in fractionating the serum proteins to concentrate the CBG. Prepare this solution by adding 770 g of solid $(NH_4)_2SO_4$, weighed carefully, to 1 L of water at a temperature of 40–50°C. Stir the water while the salt is being added. When the solute has dissolved, remove the stirring magnet and allow the solution to cool. Saturation is indicated by the mass of crystals that form in the bottom of the container. The pH of a 1:50 dilution of this solution ($\sim$0.1 M) is about 5.5.

Amberlite XAD-2 Resin

This is a preparation of polystyrene beads which carry no charged groups. It is used as a neutral adsorbent to bind relatively nonpolar materials from aqueous solutions. The resin is furnished as an aqueous suspension, which should be well shaken before an aliquot is taken.

XAD-2 preparations must be thoroughly washed before use in this experiment. Note the volume of a suitable portion of the preparation, then remove the water by suction through a filter funnel. Wash the residue by suspension, with vigorous shaking, in a volume of CH_3OH equal to that of the water removed. Repeat the CH_3OH washing at least three times; after the last CH_3OH wash is removed, resuspend the resin in water and repeat the water washing at least three times.

Washed Dialysis Tubing

Dialysis tubing is a modified cellulose acetate film formed into a continuous tube. In the dry state, it is rather brittle and inflexible, but when wet, it becomes pliable and mechanically strong. The membrane is permeable to water, to small molecules, and to ions, but it retains most proteins.

As it is manufactured, dialysis tubing is coated with lubricants and other materials which may inactivate or denature sensitive proteins. The tubing is best prepared for biochemical use by boiling lengths (20–25 cm) in a solution containing about 5 mM EDTA plus 10 mM Na_3PO_4 for 20–30 min. Discard this solution, then boil the tubing in water for 5 min. Repeat the boiling water wash at least three times to be sure the EDTA and Na_3PO_3 have been completely removed. The washed tubing is best stored under water in the cold.

Dry Ice–Acetone Bath

Place a quantity of dry ice in the bottom of an ice bucket, then *carefully* pour over the crushed ice a quantity of acetone sufficient to form a 1- to 2-in. layer above the ice. As the acetone hits the ice, there will be a vigorous evolution of CO_2, which subsides as the acetone is cooled. **CAUTION: The temperature of this mixture is about −70°C. It is cold enough to inflict severe burns if it contacts one's skin.** Use of this bath will be demonstrated.

Sodium Azide (NaN_3, Solid)

Sodium azide is used to inhibit bacterial growth in plasma or serum samples during processing. **CAUTION: Sodium azide is quite toxic, and HN_3 is even more so. Do not dispose of excess azide where acidification is possible. Heavy metal azides are explosive; avoid letting NaN_3 solutions come in contact with easily soluble metals (Fe, Cu, Ag, and the like).**

Dialysis Solution

This is a 0.15 M KCl solution containing 5 mM benzamidine, 1 mM dithiothreitol, and 50 mM Tris, adjusted to pH 9.

PROCEDURE

When starting with serum, add sufficient solid NaN_3 to give a concentration of 0.02%; then proceed directly to removal of endogenous steroids.

When starting with plasma, add solid $CaCl_2$ to give a concentration of 30 mM. This activates the clotting mechanism of plasma and leads to formation of a mass of fibrin. Complete clot formation requires 1 hour at 37°C or 6 hours at 4°C; retraction may require several hours. The serum is then separated by centrifugation at 4°C for 15 min at 15,000 $\times$ *g*. The collected serum should be a limpid solution; if not, it must be centrifuged a second time. Add NaN_3 to give a concentration of 0.02%. The serum will usually be prepared by the teaching staff and supplied to the class.

Removal of Endogenous Steroids

To each 100 ml of serum, add 20 ml of washed XAD-2 resin. Stir the mixture gently for 7–8 hours (or overnight) at 25°C. Filter off the resin through a coarse (40–60 μM)-fritted glass disc Buchner funnel *using minimal vacuum to avoid frothing* and collect the serum. This treatment removes endogenous steroids by adsorption, making binding sites on the CBG available for interaction with cortisol or other steroids to be added later on during the experiment. Acceptable results are obtained, however, even if this step is omitted.

First Dialysis

After XAD-2 resin treatment, the serum is dialyzed overnight against 20 volumes of Tris-HCl solution to further ensure removal of endogenous steroids and of other materials that might inhibit cortisol binding by CBG. A double knot is tied at one end of a length of washed dialysis tubing, the serum is added, and the free end of the dialysis tubing is also closed with a double knot or a special clamp. An extra length of tubing, beyond that required to actually contain the sample, is left between the ends. (This process will be demonstrated.) At the end of the dialysis period, the tubing is held over a funnel inserted into a beaker or a flask. One end of the tubing is cut or punctured, and the serum is collected. (A suggested technique for this operation will also be demonstrated.)

Ammonium Sulfate Fractionation

Measure, to the nearest milliliter, the recovered volume of dialyzed serum. Transfer 50 ml to a beaker containing a magnetic stirring bar. All of the following steps are to be performed at 4°C. Add, in a dropwise manner, sufficient saturated ammonium sulfate solution (previously brought to 4°C) to bring the mixture to 40% of saturation (consult Appendix 3 for further details). Stir the mixture for an additional 30 min after the calculated salt volume is added, then centrifuge the mixture at 4°C for 20 min at 10,000 $\times$ *g*. Carefully collect the supernatant, and discard the precipitate.

Adjust the pH of the supernatant to 6.4 with dilute (~0.1 N) H_2SO_4, then measure the volume to the nearest milliliter. Add saturated ammonium sulfate solution, as before, to bring the mixture to 63% of saturation. Stir for at least an additional 30 min, then collect the precipitate by centrifugation. Dissolve the precipitate in a minimal volume of water; then transfer the solution to a length of prepared dialysis tubing. Wash the centrifuge tube with a little additional water, add the washings to the dialysis tubing, then close it off as described above in preparation for the second dialysis step.

Second Dialysis

The solution is dialyzed against water to remove the ammonium sulfate. The ratio of dialysis fluid volume to sample volume should be at least 100:1. Frequent changes of the water and good stirring will speed the removal of dissolved salt. A quick and simple test for completeness of salt removal may be performed by addition of a small sample of the external fluid to a few drops of $CaCl_2$ solution (1% in water). If removal of the ammonium sulfate is substantially complete, there will be no precipitate or only a faint haze formed on mixing the dialysis fluid with this reagent.

Lyophilization (Freeze-Drying)

Collect the second dialysis product as described earlier. If the solution is turbid, or if it contains an obvious precipitate, remove it by centrifugation at 4°C for 15 min at 15,000 × *g*. The material to be lyophilized should be clear.

Add the solution to be lyophilized to a proper flask. Make the addition in such a way that, insofar as is possible, no liquid is allowed to cling to the neck of the flask. Carefully immerse the lower portion of the flask into the dry ice–acetone bath, with the flask neck at approximately a 45° angle to the vertical. Rotate the flask rapidly to form a thin layer of frozen solution that coats the inside of the bulb of the flask. Remember, the aim is to maximize the surface of the frozen solution with respect to its volume. The thinner the layer of frozen liquid, the sooner lyophilization will be completed.

When the solution is completely frozen, remove the flask from the dry ice–acetone bath; connect the flask to the lyophilizer. Open the connection between the flask and the instrument. Check that the vacuum is maintained—that is, that your connection does not leak room air into the instrument. In a little while, you will notice formation of a film of ice on the outside of the flask. This ice coat will persist until the sample is completely dry.

When your sample is dry, close the stopcock, disconnecting your flask from the vacuum of the instrument. *Carefully* allow air to enter the flask. **Note:** *If the inrush of air is too sudden, your product is likely to be lost in a cloud of dust!* With due heed to the precautions cited earlier, collect and store the lyophilized product. It should contain about 85% of the total CBG activity in about 17% of the total serum protein.

REPORT

No report will be required until the CBG studies are complete (see Expts. 16 and 17).

REFERENCES

Hyrb, D. J., Khan, M. S., Romas, N. A., and Rosner, W. *Proc. Natl. Acad. Sci. (U.S.A. 83:*3255–3256 (1986).

Lata, G. F., Hu, H-F., Bagshaw, G., and Tucker, R. F. *Arch. Biochem. Biophys. 199:*220 (1980).

Mickelson, K. E., Teller, D. C., and Petra, P. H. *Biochemistry 17:*1409 (1978).

Muldoon, T.S., and Westphal, U. *J. Biol. Chem 242:*5635 (1967).

Rosner, W., and Bradlow, H. L. *J. Clin. Endocrinol. 33:*1193 (1971).

Schiller, H. S., and Petra, P. H. *J. Steroid Biochem. 7:*55 (1976).

Westphal, U., Burton, R. M., and Harding, B. G. *Methods Enzymol. 36:*91. Academic Press, New York (1975).

DEAE-Cellulose Disc Binding Assay for CBG*

This assay can be used to determine the concentration of CBG in a sample of serum or of a serum fraction. It can also be used to determine the affinity of a given steroid for CBG. In the pH range 7.0–8.9, the net charge on the CBG molecule is negative, whereas the net charge on DEAE-cellulose is positive. In this procedure, filter paper discs made of DEAE-cellulose fibers are used to tightly bind the CBG contained in a sample filtered through the disc. The CBG has previously been saturated with [^{3}H]cortisol (1 mol per mole); since the affinity of CBG for this ligand is high ($K_a = 3.5 \times 10^8$ per mole at 4°C), the bound cortisol is retained on the filter disc along with the protein(s); the unbound steroid is washed away under carefully prescribed conditions. After washing, the filter paper disc is subjected to liquid scintillation counting in the usual way.

Unlabeled steroids added to an incubation mixture together with the [^{3}H]cortisol compete (to varying degrees, depending on their affinity) for binding sites on the CBG. This competition permits use of the assay for evaluation of the specificity of CBG binding of other steroid ligands; it is an example of a competitive binding assay. In this experiment we will use only unlabeled cortisol as competing molecule to determine the binding parameters for CBG.

MATERIALS AND REAGENTS

1. *Stock [1,2,6,7-^{3}H]Cortisol:* This is usually provided in $CH_3OH–C_6H_6$ at a specific activity of about 80–100 Ci/mmol, or 248 μCi/μg and a concentration of 1 mCi/ml. (The exact values will be furnished at the time of the experiment.)

2. *Working solution of [^{3}H]Cortisol:* Place a 500-μl aliquot of the stock labeled-cortisol solution in a 25-ml volumetric flask, and add methanol to the mark. **Note:** On the basis of the actual specific activity of the prepara-

*Preparatory reading: Section I, Chapter 5; Section II, Expt. 1.

tion delivered, you will be advised as to the exact recommended aliquot to be taken, at the time of the experiment.)

3. *Cortisol methanol diluent (CMD):* This is employed as a diluent for the labeled material. Place 5.5 mg of unlabeled cortisol in a 100-ml volumetric flask. Dissolve the material and bring to the mark with CH_3OH. This gives a 15 μM solution. Prepare secondary dilutions with MeOH of 1:10, 1:100, 1:1000, and 1:10000, which are then labeled CMD-A, CMD-B, CMD-C, and CMD-D, respectively.

4. *10 mM Tris-HCl buffer (pH 7.2):* This should be stored in the refrigerator or kept in an ice bath when not in actual use. It is important that it be cold when used to wash the DEAE-discs.

5. *Protein assay solutions:* The Bradford reagent is suitable for use here.

6. *Whatman DE-81 filter paper Discs:* These are 2.4 cm in diameter and are made of DEAE-cellulose fibers. They should *not* be touched with bare hands; forceps should be used in picking them up and putting them into the Gooch crucibles.

7. *Amberlite XAD-2 resin:* Prepared as described earlier (see Expt. 15, Reagents and Materials). This resin is required only if native serum samples are to be analyzed. Since partially purified CBG preparations (serum fractions) will have already been stripped of endogenous steroids, resin treatment will not be needed in these cases.

8. *Gooch crucibles on holders:* These are small crucibles, usually made of procelain (fragile!), with perforated bottoms. Their dimensions are 4 cm, top diameter; 2.5 cm, bottom diameter; and 4.8 cm deep. Filter paper discs placed in these crucibles are supported on the bottoms, and the perforations allow filtration to occur. The holders permit a vacuum-tight seal between the crucible and a suction filter flask. To minimize possible spillage due to top-heaviness, the filter flask employed should be at least 250 ml (or, preferably, 500 ml) in volume. The apparatus is assembled as shown in (Fig. E16–1).

9. *Manifold for use with nitrogen gas:* To evaporate the solvent from the aliquote of steroid solutions, some form of gas conduit system must be used. A commercially available setup may be obtained from Organomation Associates, Inc., called the Meyer N-Evap®. We have constructed an effective alternate unit consisting of a plastic block drilled to accept tubing from a nitrogen tank and lead the gas to a number of outlet holes. Each of these holes bears a small plastic tube adapter over the tip of which fits a 16–80-gauge, 1½-inch hypodermic needle. The outlet holes were drilled into the plastic block in a configuration to match the positions of the 13 × 100 mm glass tubes residing in a test-tube rack. The hypodermic needles are inserted into the necks of the corresponding tubes, and the nitrogen is turned on with sufficient flow to pass a *gentle* stream of gas into each tube.

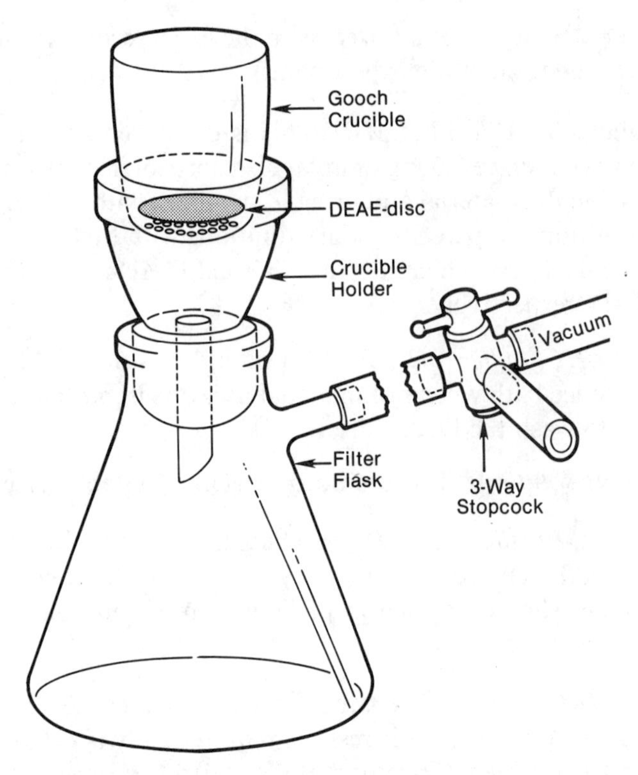

Figure E16–1. Apparatus for the disc binding assay. The crucible is made of porcelain; the holder is rubber with a glass tube protruding into the flask. The suction flask is connected to a water aspirator with a glass or plastic three-way stopcock control.

PROCEDURE

For CBG-Enriched Serum Fractions (or Pure CBG)

Samples prepared by the ammonium sulfate fractionation procedure described in Experiment 15 are not pure, but only CBG enriched. The freeze-dried fraction contains about 20% of the protein content of native serum but retains about 85% of the total serum CBG. This means that the CBG has been concentrated a little more than four fold [(100/20)(0.85) = 4.25]. However, the degree of enrichment may vary, depending on the original serum. In any event, these data do not provide sufficiently specific information for one to proceed with the preparation in hand. The following is a procedure for reconstitution of a CBG-containing solution at a concentration best suited for the assay.

The optimum protein concentration for the stock protein solution to be used in this disc binding assay is about 1 mg/ml. The suitability of this concentration results from several factors, including (1) the finite binding capacity of the DEAE-cellulose disc, (2) the need for a proper range of molar ratios between the steroid ligand(s) and the protein content of the samples, and (3) the solubility characteristics of the steroid ligands in aqueous media.

It has been shown that $A_{280}^{0.1\%} = 0.74$ for purified human CBG. Most serum proteins have higher values of A_{280}. The compromise suggested here ensures that the protein concentration will not exceed 1 mg/ml. You may therefore proceed by dissolving about 1 g of the freeze-dried powder in 5 ml of Tris buffer. Centrifuge at 10,000 × *g* for 15 min if solution is not clear. Carry out a Bradford protein determination on this preparation. From this primary stock solution, prepare dilutions of 1:50, 1:100, 1:200, and 1:400, using Tris buffer as diluent. Check each of the serial dilutions for its protein content by measuring A_{280} in quartz cuvets. Select the dilution that is closest to $A_{280} = 0.74$; if the actual value is more than 10% too high, adjust the concentration by appropriate dilution. Prepare about 12 ml of the working CBG-enriched solution.

Keep all of the solutions cold by storing in a refrigerator or in an ice bath.

Preparation of the Incubation Systems

Saturating amounts of cortisol must be present in each system. Prepare the following systems by placing the indicated volumes of the various reagents into dry, 13 × 100 mm glass test tubes. (Alternatively one may use 3-ml plastic conical capped tubes of the type used in high-speed table-top centrifuges.) Note that *all volumes are given in microliters (μl).*

Tube No.	Methanol	[^{3}H]Cortisol	Unlabeled steroid
1	50	50	0
2	—	50	50 CMD-D
3	—	50	50 CMD-C
4	—	50	50 CMD-B
5	—	50	50 CMD-A

Evaporate the solvent from the tubes, using a gentle stream of nitrogen directed into the tubes from the manifold provided. The tubes may be placed into a small rack immersed in a 30–35°C water bath while the solvent is being removed. After evaporation (about 15–30 min), add 500 μl of Tris-HCl (pH 7.2) buffer to each tube and cover each with a small square of Parafilm®. Puncture the Parafilm coverings to avoid pressure buildup, then incubate the tubes for 45 min in a water bath set at 40°C, agitating the tubes every 5–10 min. (This will dissolve the steroids.) Cool the tubes to room temperature (water bath), and add 500 μl of the 1 mg/ml solution of CBG-enriched protein to each tube. Incubate all of the tubes for 10 min at room temperature, then transfer the tubes to an ice bath for 30 min.

Measurement of Binding

Connect the suction filter flask to a water aspirator, with a Gooch crucible in its holder and the holder in the neck of the flask. With the water running, place your gloved hand over the top of the crucible to test that a reasonable vacuum has been established. If not, carefully check for leaks between the various parts of the apparatus. *Turn off the in-line three-way stopcock to release the vacuum.*

With the aid of forceps, pick up a single DE-81 filter paper disc and place it squarely in the bottom of the crucible, so as to cover all of the holes in the bottom of the crucible.

Apply 1 50-μl aliquot of the incubation mixture as nearly as possible in the center of the disc. Wait 30 sec, then apply the vacuum for 10 sec. With the vacuum still applied, rapidly add 10 ml of *ice-cold* Tris buffer. Wait an additional 10 sec to allow excess buffer to drain from the disc. Turn the stopcock to break the vacuum. Finally, using forceps again, remove the filter disc from the crucible, and place the disc into a properly labeled scintillation vial. Proceed in a similar fashion for each of the incubation tubes in the series. Lastly, when all of the discs have been collected, add 10 ml of scintillation cocktail to each of the scintillation vials and count them in the usual way.

Note: *It is important that all samples be handled as nearly as possible in an identical manner with respect to time as well as temperature; the temperature of the 10-ml wash must not be higher than* 4°C.

CALCULATIONS

The binding capacity exhibited by the CBG contained in a given sample is the resultant of several factors. These may be defined as:

A = the total bound radioactivity as measured by liquid scintillation spectrometry.

B = the radioactivity due to nospecific binding, also as measured by liquid scintillation spectrometry. This is the value obtained from tube 5.

C = the efficiency of the scintillation counter determined in Experiment 3, or given to you.

D = the number of decompositions per minute per microcurie, or dpm/μCi; this is a constant with the value 2.2×10^6.

E = the specific activity of the [^{3}H]cortisol preparation, expressed as μCi/μmol.

F = the efficiency of a DE-81 disc in retaining the CBG–steroid complex passed through it. Experimentally, this value is known to be 0.82.

G = the stoichiometry of the CBG–steroid complex. This has been shown to have a value of 1:1.

Given these definitions and using the data from tube 1, it should be clear that:

$$\mu\text{moles radioactive steroid bound} = \frac{A - B}{C} \cdot \frac{1}{DEF}$$

Lumping the constants together (assuming $C = 0.37$), this reduces to:

$$\mu\text{moles radioactive steroid bound} = (A - B)(E^{-1})(1.50 \times 10^{-6})$$

Since the molecular weight of CBG is known to be 51,800, and since the stoichiometry of the complex is known to be 1:1, one can calculate the quantity of CBG as:

$$\mu g \text{ CBG} = (\mu\text{moles radioactive steroid bound})(51{,}800)$$

One can then express the concentration of CBG as μg/100 ml of sample or as μg CBG/mg protein. We shall adopt the latter convention here.

The radioactivity on the disc from tube 1 represents the total bound [^{3}H]cortisol. It includes cortisol bound to CBG, to other proteins, and to the filter itself. Of all the tubes in the series, it will have the largest value for the factor, A, as defined above. Radioactivity on the disc from tube 5, a system to which a large excess of unlabeled cortisol was added, represents *nonspecifically bound cortisol.* It therefore provides a value for the factor, B, defined above. Note that the value of $A - B$ ($\text{cmp}_{\text{tube1}} - \text{cpm}_{\text{tube2}}$) is the net cpm in the absence of competing unlabeled cortisol.

As unlabeled steroid is added to the system, it will bind to the disc in competition with the radioactive steroid. The extent of "displacement" of the [^{3}H]cortisol is a measure of the binding of unlabeled cortisol; as the concentration of unlabeled compound increases, therefore, the radioactivity measured in each disc decreases. The binding of "cold" cortisol is obtained by subtracting the radioactivity values for each *competed* system from that for the *noncompeted* system; that is, tube 1 − tube 2, tube 1 − tube 3, etc. Or, if the counting rates of tubes 2, 3, and 4 are identified as A_2, A_3, and A_4, respectively, one can determine net counting rates for those tubes by calculating the quantities $(A_1 - B) - (A_2 - B) = A_1 - A_2$; $(A_1 - B) - (A_3 - B) = A_1 - A_3$, etc. in order to get measures of cortisol bound with increasing concentrations of competitive unlabeled cortisol.

Labeled and unlabeled cortisol have indistinguishable K_a values, so both free steroid forms compete on equal terms for CBG binding sites. Under the experimental conditions, both free forms are in equilibrium with the CBG–steroid complex. Again, displacement of [^{3}H]cortisol by unlabeled cortisol is a measure of cortisol affinity for CBG.

The concentration of *total steroid* added is known or can be calculated; this includes both labeled and unlabeled cortisol. The concentration of bound steroid can be measured as outlined above, so the concentration of free steroid can be determined by the difference. From the collected data, you will be able to generate a plot similar to that shown below.

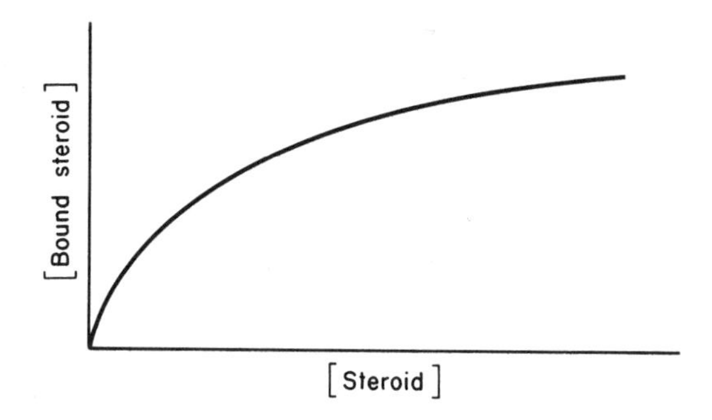

REPORT

No separate report is required for this section of the CBG studies.

For references, consult those appended to the notes on CBG fractionation (Expt. 15).

Determination of Affinity Constants for CBG–Cortisol Binding*

The DEAE-cellulose disc assay procedure (Expt. 16) is repeated for a series of systems in which an increasing concentration of unlabeled cortisol is used in the presence of a constant, saturating concentration of [^{3}H]cortisol and a constant concentration of CBG. Each incubation system is at equilibrium when sampled for assay. The unlabeled cortisol will compete with the CBG-bound [^{3}H]cortisol, so that the decreases in bound radioactivity observed with increasing concentrations of unlabeled cortisol is a measure of the CBG–steroid interaction. You will also determine the extent to which steroids other than cortisol can compete with cortisol for binding to CBG.

REAGENTS AND MATERIALS

These are identical to those set forth in the text of the DEAE-cellulose disc binding assay (Expt. 16) with the addition of the following diluents:

1. *Progesterone–methanol diluent (PMD):* Place 45 mg of progesterone in a 100-ml volumetric flask; dissolve and dilute to the mark with CH_3OH. This makes a 1.43 mM solution. Prepare secondary dilutions with methanol of 1:10, 1:100, 1:1000, and 1:10000 with methanol; mark these as PMD-A, PMD-B, PMD-C, and PMD-D, respectively.

2. *Testosterone–methanol diluent (TMD):* Place 450 mg testosterone in a 100-ml volumetric flask; dissolve and bring to the mark with CH_3OH. This makes a 15.6 mM solution. Prepare secondary dilutions with methanol of 1:10, 1:100, 1:1000, and 1:10000 with CH_3OH; mark these as TMD-A, TMD-B, TMD-C and TMD-D, respectively.

Steroids other than progesterone and testosterone may also be included. However, for this experiment to be completed in a reasonable amount of time, *only cortisol plus one other steroid* should be tested by each laboratory group.

*Preparatory reading: Section I, Chapter 5.

PROCEDURE

The systems are set up as described in the disc binding assay, but according to the following protocol. In addition to the systems containing labeled and unlabeled cortisol, an additional series containing labeled cortisol plus unlabeled progesterone or testosterone (or other assigned steroid) is included. Note that, as previously, *all volumes are cited in microliters* (μl).

Tube No.	[^{3}H]Cortisol	Unlabeled steroid
1	50	0
2	50	16 CMD-D
3	50	33 CMD-D
4	50	50 CMD-D
5	50	16 CMD-C
6	50	33 CMD-C
7	50	50 CMD-C
8	50	16 CMD-B
9	50	33 CMD-B
10	50	50 CMD-B
11	50	50 CMD-A
12	50	0 (either PMD or TMD or other)
13	50	16-D
14	50	33-D
15	50	50-D
16	50	16-C
17	50	33-C
18	50	50-C
19	50	16-B
20	50	33-B
21	50	50-B
22	50	50-A

Proceed with the assays as previously described (see Experiment 16).

CALCULATIONS

The data from the progesterone or testosterone or other systems are processed exactly as are the cortisol data. The results will give affinity constants and binding capacities for these steroids.

Binding Parameters (Affinity Constants, Binding Capacities)

From the collected data, calculate for each incubation system the concentration of bound steroid, $\bar{\nu}$, and of free steroid; [S].

To generate a *Scatchard* plot, plot $\bar{\nu}$/[S] versus $\bar{\nu}$; a straight line should result, the slope of which is equal to $-K_a$. The intercept on the abscissa gives the binding capacity. The plot you obtain should resemble the one shown below.

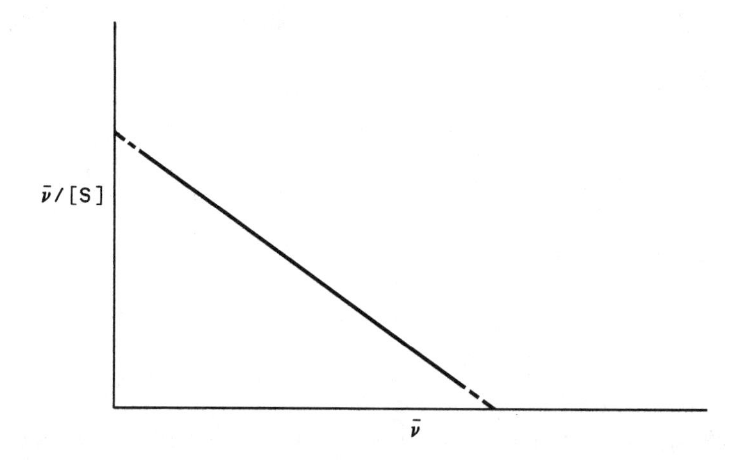

REPORT

For simplicity, a single report will be used to summarize all of the data relating to CBG fractionation and ligand binding to CBG—that is, from Experiments 15, 16, and 17. Your report should include the following:

1. All data relating to the serum fractionation procedure
2. The specific binding activity of your CBG preparation
3. The concentration of CBG in your preparation
4. The calculated affinity constant of your preparation for cortisol, as well as the Scatchard plot from which it was derived
5. Comparative binding data for steroid ligands other than cortisol with respect to CBG, including affinity constants.

Metabolic Production of Carbon Dioxide by Yeast and Its Inhibition by Ethanol*

Ordinary bakers' yeast *(Saccharomyces cerevisiae)* ferments certain mono- and disaccharides to produce CO_2 and ethanol. The overall conversion can be summarized by the equation:

$$C_6H_{12}O_6 + 2P_i + 2ADP \rightleftharpoons 2C_2H_5OH + 2CO_2 + 2ATP + H_2O$$

This process has been important to humankind since time immemorial; it has leavened not only our bread but also our spirits; it may also be called upon to operate our internal combustion engines. Sociology aside, the study of fermentation has been most rewarding from a purely biochemical view, as the work of Pasteur, Embden, Meyerhof, Buchner, Harden, and many others will attest. Our basic understanding of the process of glycolysis has greatly evolved from many studies of yeast metabolism attendant on fermentation.

The above equation, which summarizes the overall process, does not indicate the numerous intermediate reactions that function together as a coupled system, from which the yeast cells derive useful energy in the form of ATP. Consequently, even if it is possible to demonstrate some inhibitory effect of ethanol on CO_2 production, the exact locus of the inhibition cannot be determined by this single experiment. Further, it does not address the question of the effect of ethanol on yeast grown under different conditions—for example, under vigorous oxygenation. Any conclusions you may draw apply *only* to the experimental conditions elaborated below.

This experiment makes several important points: First, CO_2 can be a normal metabolic product without intervention of molecular oxygen. Second, production of CO_2 is characteristic of a great many metabolic systems; therefore, any method for its measurement may be of general applicability. Third, in this system, CO_2 does not accumulate whereas ethanol does. Thus any effects noted *may* be attributed to the ethanol but *cannot* be due to the CO_2.

*Preparatory reading: Section I, Chapters 3 and 4.

MATERIALS AND REAGENTS

1. *0.5 M KOH solution:* Dissolve the proper quantity of solid pellets in water that has been freshly boiled and then cooled *in vacuo* to minimize uptake of CO_2 from air. Store this reagent in a flask with a tight-fitting rubber stopper.

2. *Sulfuric acid, 6 M:* This is made by dilution of concentrated (18 M) sulfuric acid.

3. *Yeast growth medium:* Two *separate* media are required. These must be sterilized, either by heat sterilization or by passage through a sterile, 0.22-μm bacteriologic filter.
   a. *Basic medium:* contains 1% Bacto Yeast Extract® + 2% Bacto Peptone® in water.
   b. *Glucose solution:* a 50% solution of glucose in water.
   c. *Glucose-enriched medium:* made by mixing 25 volumes of basic medium with 1 volume of glucose. Take care to maintain sterility during mixing.

4. *Substrate solution:* Prepare a 0.2% glucose solution. Add sufficient [U-^{14}C] glucose to give approximately 10,000 cpm when 25 μl is added to 5 ml of scintillation cocktail.

5. *95% ethanol*

Initial Yeast Culture

It is best to start with an agar slant of a pure yeast culture; alternatively, one can make do with a cake of dry, commercial bakers' yeast, rubbed up in a mortar with a little of the growth medium. Such cultures will almost certainly be contaminated with other organisms; thus it is essential to streak out this solution on an agar plate and to use an isolated colony from that plate.

The diploid strain BJ926 has been shown to work well for this experiment. A 50-ml starter culture in glucose-enriched medium is grown for 12–24 hours at 25–30°C. The cells should be used while they are in the log phase of growth—the phase during which their metabolic activity is highest.

Fresh medium is inoculated from the starter culture. Cells are allowed to grow to a concentration of not more than 7×10^7 cells/ml ($A_{660} = 7.0$). Collect the cells by centrifugation at about $2000 \times g$ for 5 min. Resuspend the sedimented cells in *basic medium (no added glucose)* to a concentration of about 2×10^7 cells/ml ($A_{660} = 2$). This suspension will be used in the incubations described below.

Incubation Materials

1. *Scintillation cocktail:* Bray's solution (or the equivalent).

2. *Fluted filter paper wicks for CO_2 absorption:* Cut 1-in. wide filter paper strips into lengths of approximately 1.5 in., then fold each piece at approx-

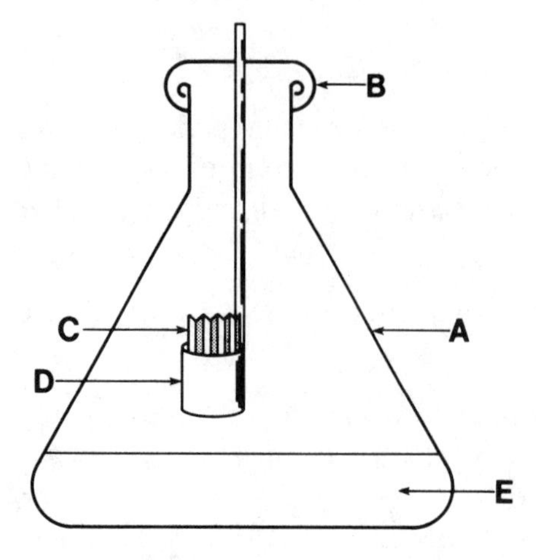

Figure E18–1. Flask arrangement for trapping of $^{14}CO_2$ during incubation of yeast with [U-^{14}C] glucose. Flasks are secured with clips (not shown) to the platform of a water bath shaker. A, 25-ml Erlenmeyer flask; B, rubber serum stopper; C, fluted filter paper; D, plastic cup suspended by a plastic rod and containing the NaOH solution; E, yeast in the incubation system.

imately ¼-in. intervals, accordion-fashion. Use of these will be demonstrated.

3. *Incubation flasks with CO_2 traps* (see Fig. E18–1): Prepare a series of 25-ml Erlenmeyer flasks, numbered 1 through 8. Each flask will require a rubber septum punched to support a Kontes center well attached to a small plastic rod. There should be a snug fit between the plastic rod and the hole in the septum. Each of the plastic wells should contain an accordion-folded strip of filter paper.

4. *Metabolic shaker bath:* This will be equipped with clips to contain a number of metabolic flasks so they may be gently shaken simultaneously. The bath temperature should be set at 25–30°C.

PROCEDURE

Make additions to the numbered flasks according to the following table. *All indicated volumes are expressed in milliliters.*

Flask No.	Basic medium (no glucose)	95% ethanol	0.2% glucose + [^{14}C]glucose
1	2.00	0	1.0
2	1.90	0.10	1.0
3	1.80	0.20	1.0
4	1.70	0.30	1.0
5	1.50	0.50	1.0
6	1.35	0.65	1.0
7	1.25	0.75	1.0
8	1.00	1.00	1.0

Gently swirl the flasks to mix the contents; then set them aside while proceeding to the next step.

Take up one of the assembled CO_2 traps, and make certain that the filter paper wick is well seated in the plastic cup. Holding the assembled trap in a vertical position, wet the filter paper with 0.25 ml of the KOH solution. Do this carefully; none of the alkali should wet the outside of the cup. Any KOH not immediately soaked up by the filter paper should fall into the cup. **CAUTION: Do not get any of the KOH on your skin; it is quite caustic!**

Swirl the stock yeast cell suspension to ensure that the cells are uniformly dispersed, then add 1.0 ml of the suspension to the metabolic flask. Immediately insert the rubber septum, folding the lip of the septum down over the top of the flask to make a tight seal. In making the seal, be careful not to splash any of the liquid contained in the bottom of the flask into the suspended CO_2 trap.

Proceed in like manner with the remainder of the flasks in the series, then place all of them into the metabolic shaker clips. When doing this, it is important to turn the shaker motor off to avoid the hazard of splashing. Check to be sure, when power is restored, that the shaking is gentle and not so violent as to risk splashing within the flasks.

When all of the flasks have been incubated for 1 hour, remove them from the shaker. With a syringe and needle, add 0.5 ml of 6 M H_2SO_4 to each flask by penetrating the rubber septum **(Not by opening the flask!).** Make sure that the acid is directed into the flask and *not* into the suspended well. In adding the acid, you will find that the outer edge of the septum is quite thick and difficult to penetrate with the needle; more central portions of the septum are thinner and designed to be penetrated easily.

Addition of the sulfuric acid kills the cells, stopping the reaction, and liberates any dissolved CO_2. After acidification, return the flasks to the metabolic shaker for another 10 min to be sure that all volatile radioactivity has been trapped.

Prepare a series of scintillation vials by numbering them to correspond with the metabolic flasks. Remove the septa from the flasks, one by one, then cut the trap from its suspending rod so that the well plus its wick falls directly into the scintillation vial. Add 15 ml of scintillation cocktail to all of the flasks and count the series for 2 min. Dispose of the wastes and contaminated materials according to directions.

REPORT

Submit a graph of CO_2 production as a function of added ethanol concentration.

REFERENCES

Gadden, E. L. Production methods in industrial microbiology. *Sci. Am. 245:*181–196 (1981).

Peppler, H. J., and Perlman, D., eds. *Microbial Technology,* 2nd ed. Academic Press, New York (1979).

Radioimmunoassay of Dehydroepiandrosterone (Prasterone)*

All of the mammalian steroid hormones are derived from cholesterol. Depending on the hormone-producing tissue and its circumstances at a given moment, the cholesterol may be produced de novo from precursors such as acetate, or it may be obtained preformed from cellular sources or the blood.

The initial reaction (and the rate-limiting step) in steroid hormone biosynthesis involves oxidative cleavage of cholesterol at C-20 by the enzyme *desmolase,* which removes a large part of the side chain on the D ring of cholesterol, with introduction of a keto group. The product formed is pregnenolene (see Fig. E19–1). Desmolase action requires NADPH and molecular oxygen, among other factors. In the adrenals, desmolase activity is stimulated by ACTH; in other tissues it presumably responds to similar and appropriate trophic hormones.

Pregnenolone is readily converted to progesterone, which itself is a major hormone of the corpus luteum of the ovaries. In most other hormone-producing tissues, progesterone is not an end product but an intermediate in the biosynthesis of other steroid hormones. Thus, progesterone is readily converted to 17α-hydroxyprogesterone and thence to androstenedione. Androstenedione can be converted to the triad of estrogenic hormones, estrone, estradiol, and estriol. Alternatively, androstenedione may be converted to the androgenic hormones, testosterone and 5α-dihydrotestosterone. Although we shall not consider the matter further here, it should be noted that progesterone is also the major precursor of the adrenal cortical hormones responsible for regulation of salt and carbohydrate metabolism (mineralo- and glucocorticoids). Details of all these processes can be found in most biochemistry textbooks.

Figure E19–1 shows that pregnenolone can be converted to androstenedione by more than one pathway. In the first pathway, it can end up as androgens or estrogens, as described in the paragraph above. In the second pathway, pregnenolone is initially converted to 17α-hydroxypregnenolone, thence to dehydroepiandrosterone (DHEA), which serves as an alter-

*Preparatory reading: Section I, Chapters 3 and 10; and Experiments 11 and 16.

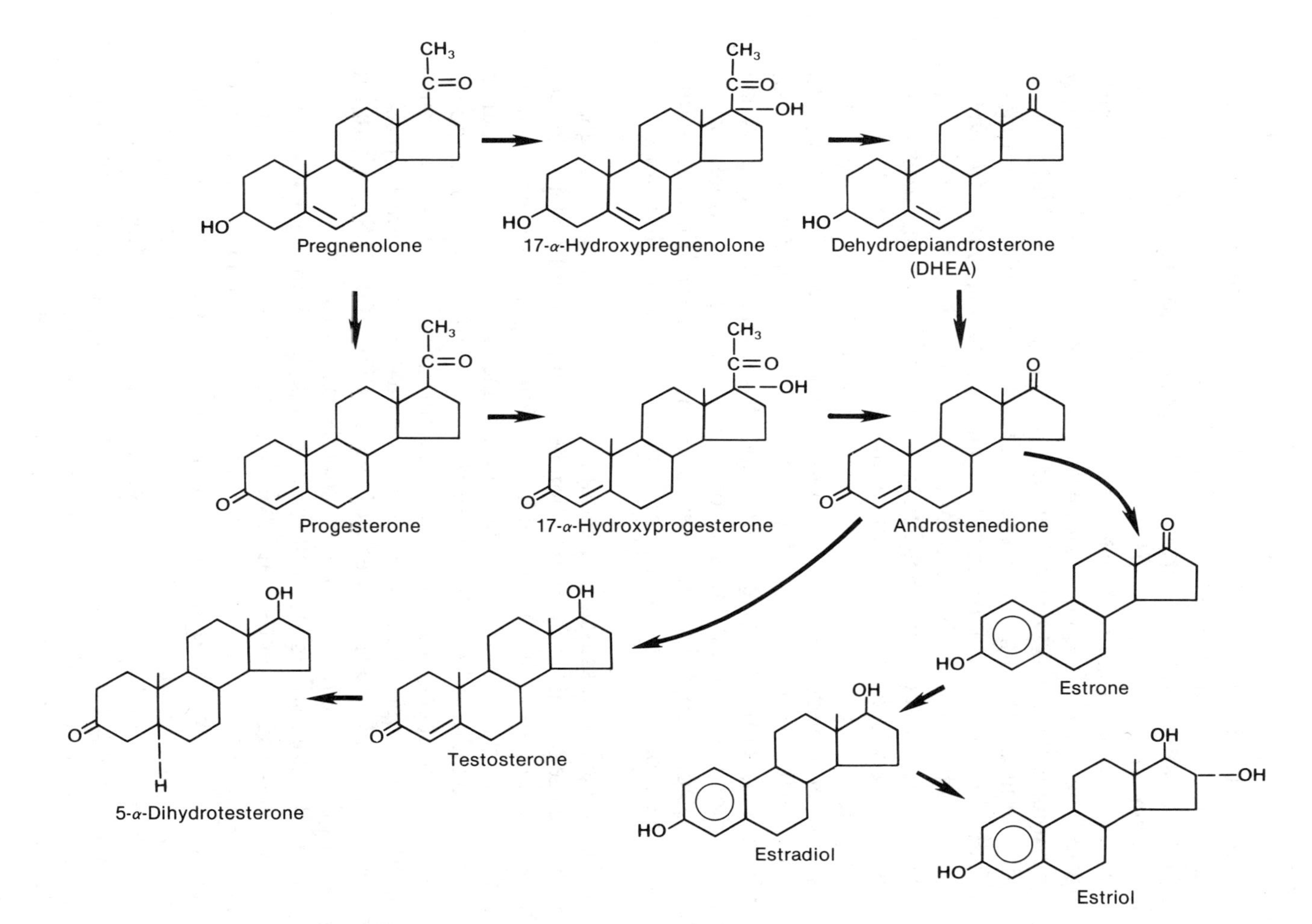

Figure E19–1. Metabolic pathways for conversion of pregnenolone to dehydroepiandrosterone (DHEA) and to testosterone and 17β-estradiol.

nate precursor of androstenedione. This second pathway is particularly important during pregnancy, when tissues of the fetus and of the mother are coordinately involved in the production of estrogens (of which, during pregnancy, estriol constitutes about 90%). The high concentration of estrogenic activity required to maintain a proper environment for growth and development of the fetus is thus ensured, provided that the fetal and placental tissues are both capable of normal function.

The fetus obtains at least part of its cholesterol and pregnenolone from the mother via the placenta. However, the fetal adrenals lack the 3β-hydroxysteroid-Δ^5-isomerase system necessary to convert pregnenolone to progesterone. Instead, pregnenolone is first converted to 17α-hydroxypregnenolone, thence to DHEA (see top line of Fig. E19–1). Fetal liver can rapidly conjugate the 3-hydroxy group of DHEA with sulfate, forming the 3-sulfate ester. A major purpose of this conjugation is to make the DHEA molecule sufficiently soluble so it may pass to the placenta and from there to the maternal circulation. Fetal tissues lack *sulfatases* and *aromatases,* so, by themselves, could not complete the biosynthesis of any estrogens.

The placenta, curiously, lacks *desmolase* and *17-hydroxylase* activities; thus it depends very largely on the fetus for a supply of DHEA and 16-hydroxy-DHEA (not shown in Fig. E19–1). On the other hand, the placenta does contain *sulfatase* activity (which hydrolyzes the conjugated steroid hormone precursors), and it also contains *aromatases* (which convert the A rings of estrogenic precursors to aromatic rings). It can, therefore, complete the biosynthesis of the entire triad of estrogenic hormones.

From the above discussion, it should be clear that a normal pregnancy involves close biochemical cooperation between fetal and material tissues; the fetus is not an innocent bystander. Equally important is proper functioning of the placenta. If either the fetal or placental tissues lack a particular enzyme involved in estrogen formation, then the threat of a spontaneous abortion becomes very large. For this reason, it is of particular importance that one be able to measure not only the estrogens themselves but also such key intermediates as DHEA sulfate (DHEAS), in order to pinpoint the problem and to adopt suitable measures.

Given the close similarity of the compounds shown in Fig. E19–1, there are few analytical methods that have the requisite sensitivity, specificity, and speed to be useful. Fortunately, many of these compounds can be chemically conjugated to simple proteins such as bovine serum albumin (BSA), and the conjugates can be employed as antigens. Following immunization of a small animal with such an antigen, antibodies can be recovered from the serum of the host animal in fairly purified form. These antibodies then form the basis of immunoassays that allow the detection and quantitation of the free haptenic group in a biological matrix such as serum.

In comparing the present experiment with the earlier LDH immunoassay and ELISA, we find several unique differences and similarities. The DHEAS assay is a *double-antibody, competitive, ligand-binding assay* and therefore resembles the LDH ELISA. The first antibody, against DHEAS, is a globulin raised by immunization of a rabbit with a synthetic complex prepared by chemical coupling of DHEAS to serum albumin. It recognizes and *binds tightly free DHEAS.* This complex remains soluble.

The first stage of the assay involves competitive ligand binding, since we shall add a small amount of pure, radiolabeled DHEAS to the sample. Both the radioactively labeled, and the unlabeled or "cold" DHEAS will compete for binding sites available on the antibody, in proportion to their relative concentrations in the solution. Since the quantity of antibody added has a constant binding capacity, the lower the amount of intrinsic DHEAS the more radioactive DHEAS will be bound, and vice versa.

To separate the bound DHEAs from that still free in solution, there are two main approaches that may be used.

The first is to use a *second antibody* raised against rabbit serum, usually in goats; interaction of these two antibodies results in precipitin formation. This precipitation is promoted by polyethylene glycol. The precipitate is carefully collected by centrifugation, and the pellet is washed and then counted for radioactivity.

The second method is to use *dextran-coated charcoal.* This material (commercially available) will sequester the free, unbound DHEAS, leaving the antibody-bound steroid in solution. Centrifugation will remove the charcoal, and the supernatant may be sampled for determination of radioactivity.

Both methods are described below: The results from these assays are quite sensitive to variations in temperature, to the volume of solution at each stage of the assay, and to contamination of the final pellet or fluid. You must pay special attention to these sources of error as you proceed. Constancy of technique, particularly in pipetting the various reagents, is essential for good results.

Note: *Gloves should be worn for all steps in this experiment.*

Double-Antibody Assay

MATERIALS AND REAGENTS

Tris-Buffered Saline (TBS)

Dissolve 6.055 g of Tris base and 8.8 g NaCl in 750 ml of water. Adjust the pH to 7.2 with 1 N HCl, then dilute to 1 L.

Physiologic Saline (0.15 M)

Dissolve 9.0 g NaCl in water to make 1 L.

Bovine γ-Globulin (BGG)

Suitable starting material to make this solution is obtained commercially. It is fraction II of bovine plasma proteins. Weigh out exactly 0.300 g and gently sift the powder over the surface of 50 ml of TBS in a 100-ml beaker. Do not stir this solution vigorously, since the protein is readily denatured by foaming. The solution must be limpid and clear. If not, remove any particulate matter by filtration through a Swinney 0.2-μm filter attached to a syringe. Do this gently. Do not filter by suction for the reason given above. Keep this solution at room temperature.

Radiolabeled DHEAS in BGG

[^{3}H]DHEAS is usually available at a higher specific activity than the ^{14}C-labeled product. Both are provided in an ethanol solution. Depending on the exact specific activity, transfer a sufficient quantity to a 100-ml beaker so that when the solvent is removed by gentle warming and the residue is dissolved in 10 ml of BGG reagent, a 50-μl aliquot will give a counting rate of 40–50 $\times$ 10^3 cpm. The exact amount of solute in the final solution is less important than its counting rate, since the absolute quantity of DHEAS is so very small in any case. Keep this solution cold until it is to be used. It should be made up just prior to use, and any excess should be discarded with other radioactive wastes.

Polyethylene Glycol (PEG) "4000"

Dissolve 250 g in water and dilute to 1 L. (The destination "4000" roughly describes the degree of polymerization in terms of the mean molecular weight.) Do not use other forms of polyethylene glycol for this experiment. Keep the final solution in a refrigerator at 4°C or an ice bath.

DHEAS Standards in TBS

In order to prepare a suitable standard curve, one needs an extended series of standard solutions. These are made from either of two primary solutions, which may be stored frozen. The thawed solutions should be well mixed before sampling to ensure uniformity of the aliquots taken.

STOCK A:

Weigh out exactly 5.00 mg of DHEAS; dissolve it in 50.0 ml of 5% (w/v) Triton X-100 in methanol.

STOCK B:

Dilute 100 μl of stock A to 10 ml with TBS. This has a concentration of 1 μg/ml.

From stock solution B, prepare two working standards, which should be made fresh on the date of intended use.

WORKING STANDARD #1:

Dilute 500 μl of working standard #2 to 2 ml with TBS; mix well. Working standard #1 has a final concentration of 0.005 ng/μl.

WORKING STANDARD #2:

Dilute 400 μl of stock B to 2 ml with TBS; mix well. Working standard #2 has a final concentration of 0.2 ng/μl.

First Antibody (Ab-1)

This antibody could be prepared by condensing DHEA hemisuccinate and bovine serum albumin by means of a carbodiimide condensing reagent. The condensation product is dissolved in 1 ml of saline and emulsified with 2 ml of Freund's complete adjuvant. (See Chard, 1982, for components of Freund's adjuvants.) The emulsion is injected in small aliquots into 16 separate sites on the back of a rabbit. The rabbit is subsequently given two intravenous injections of 1 mg of condensate in saline, at monthly intervals, to boost antibody production.

Second Antibody (Ab-2)

This antibody may be prepared by dissolving 50 mg of commercial rabbit IgG in saline followed by emulsification with 2 ml of Freund's complete adjuvant. The emulsion is injected into multiple sites on the back of a sheep. Two successive intravenous doses of 25 mg IgG in saline are given at monthly intervals.

(**Note:** Both Ab-1 and Ab-2 are available from several commercial sources).

Unknowns

These are serum samples from the Department of Obstetrics and Gynecology; the DHEAS concentrations have already been determined.

PROCEDURE

Setting Up the Standard Curve (Day 1)

Because of the high sensitivity of immunoassays, one is constrained to work with very small aliquots of the analyte in samples as well as in the standard solutions. This is fortunate, since host animals do not always produce antibodies of acceptable titer; even when they do so, the production of antibodies is tedious and expensive. Finally, the product must always be checked for sensitivity and cross-reactivity.

The table that follows gives, in the first column, the actual quantity of DHEAS in the 200-μl aliquot of standard. The second column indicates from which working standard a given solution is to be made. The third column indicates the volume of working standard to be taken, and the last column indicates the volume of TBS to be added in order to keep all volumes constant. It is most convenient to use ~2 ml plastic tubes with attached caps.

DHEAS (ng/200μl)	Working standard	Volume of standard (μl)	TBS to be added (μl)
0	—	0	200
0.1	1	5	395
0.25	1	5	195
0.50	1	10	190
1.00	1	20	180
2.00	2	10	190
4.00	2	20	180
8.00	2	40	160

For each of the above, mix well, transfer 200 μl to a clean tube and discard the remainder. **Note carefully:** As mentioned earlier, and as demonstrated by the above table, the volumes to be transfered are small. You must use great care in setting the adjustable pipets and in delivering the specified volumes. Solutions should be mixed by gentle inversion; avoid violent vortex mixing and splashing of the reagents.

Make sure that each of the above tubes is carefully labeled as S_x, where the subscript indicates which of the standards a given tube contains.

Preparation of Sample Dilutions (Day 1)

Thaw the sample(s) if frozen, and bring to room temperature. An initial dilution of the sample(s) is made by adding 20 μl to 4.00 ml of TBS. Mix by inversion. This initial dilution may be regarded as essentially a 1:200 dilution.

Because a given sample may have a high, low, or normal content of DHEAS, several secondary dilutions are made to optimize the chance of getting a final dilution that falls on the linear portion of the standard curve. Set up a series of three tubes, labeled U_1, U_2, and U_3. To the first, add 50 μl of initial sample dilution, to the second add 100 μl, and to the third add 150 μl. Now bring the volume in each of the "U" tubes to 200 μl with TBS. Mix carefully the contents of all of the tubes.

Setting Up the Assay: First Stage (Day 1)

In addition to the series of standard tubes, S_x, and the unknown sample tubes, U_x, one needs a measure of the nonspecific binding of the radiolabeled ligand, DHEA. Therefore, label one more tube, containing 200 μl of TBS, as NSB (for nonspecific binding).

For an exact measure of the total radioactivity in each of the assay tubes, label one more tube as T. For the moment it will remain empty.

Place all of the tubes in series in a rack, with the tube marked NSB at the left-hand end of the series. Next add 50 μl of BGG reagent to the NSB tube *only*.

Warm a 1:10 dilution of Ab-1 to room temperature, and make a 1:1000 dilution of Ab-1 with BGG reagent. Example: Take 100 μl of the 1:10 dilution, and bring it to a volume of 10 ml with BGG reagent. The final dilution of Ab-1 is thus 1:1000.

Add 50 μl of the 1:1000 dilution of Ab-1 to each assay tube *except* the tubes marked NSB and T.

Add 50 μl of radiolabeled DHEAS to *all* tubes. Close and cap each tube, mix by gentle inversion, and place the covered tubes in a refrigerator overnight (or over the weekend). This ends the operations for day 1.

Completing the Assay: Second Stage (Day 2)

Remove the tubes from the refrigerator and allow them to come to room temperature. To all tubes *except* T, add 200 μl of a 1:40 dilution of Ab-2 in BGG. Vortex-mix the tubes, then allow them to stand at room temperature for 2 hours.

Cool the tubes in an ice bath, then add 0.5 ml of polyethylene glycol to all tubes *except* T. Vortex-mix well, then return the tubes to the ice bath for an additional 30 min.

Except for T, centrifuge all tubes at 4°C for 10–15 min at 12,000 rpm in a table-top microcentrifuge.

Carefully decant the supernatants from the centrifuged tubes; put the discarded liquid into waste containers marked for radioactive wastes. Allow the tubes to drain the last drops that fall freely; then wipe away any remaining liquid at the tops of the tubes with a cotton-wrapped applicator stick. Dispose of the used applicator sticks in the container marked for dry radioactive wastes.

Add 200 μl of physiologic saline to all tubes that contain an immunoprecipitate, cap the tubes, and vortex-mix the contents until the packed precipitate is dispersed or dissolved.

Remove the caps, transfer each tube to a separate and properly labeled scintillation vial, add 10 ml of scintillation cocktail, and count each vial for 1 min to determine the radioactivity of all the tubes in the series. This ends the manipulations of the first immunoassay procedure.

CALCULATIONS

One must first generate a standard curve for the assay, then determine the concentration of the samples from the standard curve. If the data are plotted on log–log graph paper (1 × 2 cycles), a straight line is obtained, and this is by far the easiest way of dealing with the results. Because the handling of the data is somewhat different from calculations used in earlier experiments, a typical example is provided at the end of this section, together with a log–log plot drawn from the example. Examination of these examples will assist you in manipulating your own data, as described below.

1. Determine the *net cpm* for all standards and unknowns by subtracting the count of the *NSB* tube from each of the others *except* the tube marked T.

2. Determine the *net % bound* for the zero standard. This is designated as B_0. It is obtained by dividing the net count for the zero standard by the count obtained in T. (The example provided shows that B_0 is approximately half the counting rate of T. This is in a good working range, and it is also the highest value of binding to be observed in the entire assay system.)

3. Similarly, determine the net % bound for each of the remaining tubes. (Note that as the quantity of added DHEAS increases, so the counting rate of radioactive DHEAS decreases. This results from the competitive nature of the assay system.)

4. Determine the *relative % bound* for each of the standards, by the equation:

   $$\text{Relative \% bound} = ([\text{Net \% bound}]/[B_0]) \times 100$$

   Plot the relative % bound vs. the quantity of added DHEAS on log–log graph paper. As shown in the example, this should yield a straight line.

5. Determine the relative % bound for the unknown sample(s); read the quantity of DHEAS in the unknowns from the standard curve. The information contained in the following graph and in Table E19-1 typify the results that can be obtained from this radioimmunoassay. The examples are given to assist you in analysis of your own data.

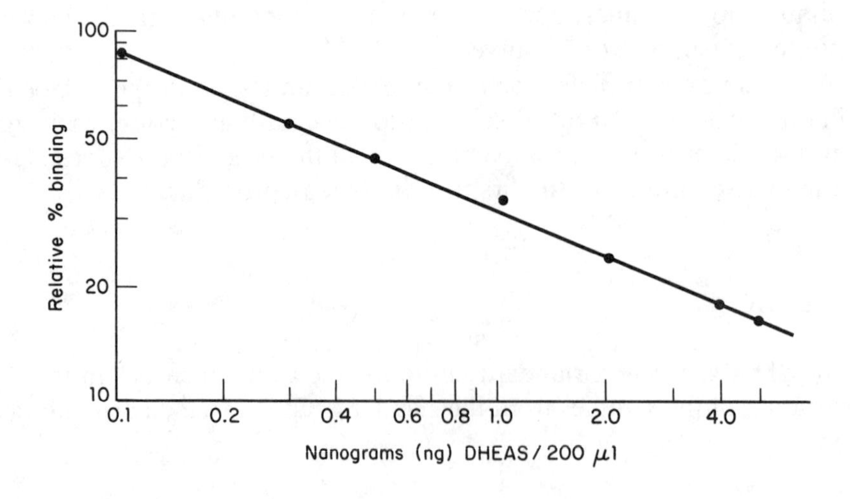

Table E19-1. Typical Calculations for the DHEAS Assay

Tube Identification	Gross cpm[a]	Net cpm[b]	% Bound (net cpm/total)[c]	Rel. % bound[d]
Total	33,172	31,832		
NSB ng	1,340			
Standards				
0	18,317	16,977	.5333	100
0.1	17,020	15,680	.4925	92.5
0.25	12,215	10,875	.3416	64.1
0.50	9,674	8,334	.2618	49.1
1.00	8,263	6,923	.2174	40.8
2.00	6,243	4,903	.1540	28.9
4.00	5,058	3,718	.1168	21.9
8.00	4,182	2,842	.0892	16.8
Unknowns, ad hoc				

[a]Values as read from the LSC tape or other printout. Except for the tubes identified as Total and NSB, all values are the means of replicates.
[b]Values obtained by subtraction of the cpm for the NSB tube.
[c]Values in this column are expressed with respect to the Total.
[d]Values here are expressed with respect to the zero standard.

Single-Antibody Assay with Dextran-Coated Charcoal

This procedure differs from the first in that it utilizes activated charcoal to remove the *free (unbound)* steroid from the reaction mixture. The steroid–antibody complex remains in solution in the supernatant after centrifugation to remove the charcoal. (In the previous procedure, a double-antibody method was used to precipitate the radioactive steroid–antibody complex, and this precipitate was counted in the LSC to give values for the *bound* steroid).

MATERIALS AND REAGENTS

1. Sheep anti-DHEAS antibody.
2. [^{3}H]DHEAS in ethanol (specific activity = 10–25 Ci/mmol). This is evaporated to dryness under nitrogen and then dissolved in buffer (solution 6) by equilibrating for 15 min to give solution 3.
3. Working solution of [^{3}H]DHEAS: ~10,000 cpm/0.1 ml.
4. DHEAS standards. These contain the following nanogram quantities of DHEAS per 0.5 ml: 0.1, 0.025, 0.05, 0.1, 0.25, 0.5.
5. Charcoal–dextran suspension in PBS, 0.2% (w/v) gelatin (pH 7.0).
6. Buffer for dilution; PBS, 0.2% (w/v) gelatin (pH 7.0).
7. Unknown (serum samples).

PROCEDURE

1. Dilute serum sample 1:201 with diluting buffer—for example, 10 μl of serum and 2000 μl of buffer. Further dilute the 1:201 solution to 1:21—for example, 100 μl plus 2.0 ml of diluent. The final dilution is 1:4221.
2. Bring reagents to room temperature prior to use.
3. Using 3-ml microfuge plastic tubes, add 0.6 ml of solution 6 to tube Nos. 1 and 2 and 0.5 ml to tube Nos. 3 and 4.
4. Add 0.5 ml (in duplicate) of each *DHEAS standard* (0.01–0.5 ng/0.5 ml) to tube Nos. 5–16.
5. Add 0.5 ml (in duplicate) of the 1:4221 diluted *sample* or *control* to tube No. 17, etc., to end of assay.
6. *With the exception of tube Nos. 1 and 2,* add 0.1 ml of anti-DHEAS antibody to the rest of the assay tubes.
7. Add 0.1 ml of "working" (i.e., diluted) [^{3}H]DHEAS containing ~10,000 cpm/0.1 ml to *all* of the assay tubes. Mix and incubate for a minimum of 1 hour at 4°C.
8. After incubation, add 0.2 ml of *cold (4°C) charcoal–dextran* suspension. (*Shake well* initially then mix continuously.) Mix for 20 sec by shaking the test-tube rack *vigorously* and incubate for 20 min at 4°C.
9. Centrifuge in a table-top microcentrifuge for 15 min, then decant the entire supernatant into 10 ml of scintillation cocktail in a counting vial. Count for a minimum of 1 min in the LSC.

CALCULATIONS

Take the average of all duplicate tubes. Subtract the averaged NSB (blank) counts from the averages obtained. This yields the corrected values. Divide the corrected values (B) by the corrected zero standard value (B_0) to obtain the percent bound. The formula is:

$$\%B/B_0 = \frac{\overline{\text{cpm}}_{\text{sample}} - \overline{\text{cpm}}_{\text{NSB}}}{\overline{\text{cpm}}_{0\ \text{stnd.}} - \overline{\text{cpm}}_{\text{NSB}}} \times 100$$

where:

$\overline{\text{cpm}}$ = average counts of duplicates

sample = particular serum or standard being calculated

NSB = nonspecific binding tube

0 stnd. = 0 tube (also known as the 100% binding tube)

Final Results

Plot the percent bound, using 100% as the starting point, against the DHEAS standards ranging from 0.01 to 0.5 ng/0.5 ml. Then, nanograms (ng) as read off of the standard curve × 8000 = ng DHEAS/ml.

REFERENCES

Buster, J. E., and Abraham, C. E. Radioimmunoassay of plasma dehydroepiandrosterone sulfate. *Anal. Lett. 5*:543 (1972).

Chard, T. An Introduction to radioimmunoassay and related techniques. *Anal. Lett. 6*:Part 2 (1972).

Clausen, J. Immunochemical techniques for identification and estimation of macromolecules. In *Laboratory Techniques in Biochemistry and Molecular Biology* (T. S. Work and E. Work, eds.), Vol. 1, pp 397–572. American Elsevier, New York (1970).

Kricka, L. *Ligand-Binding Assays.* Marcel Dekker, New York (1985).

Norman, A. W., and Litwack, G. *Hormones.* Academic Press, Orlando (1987).

Inhibition of Protein Biosynthesis in *Escherichia coli**

β-Galactosidase catalyzes hydrolysis of glycosidic bonds involving the β-anomer of galactose. In *E. coli,* β-galactosidase is an inducible enzyme; that is, the activity of the enzyme is increased in the presence of high concentrations of lactose or of isopropyl β-D-thiogalactoside (IPTG). ITPG is an analog of allolactose, a metabolite formed during lactose metabolism.

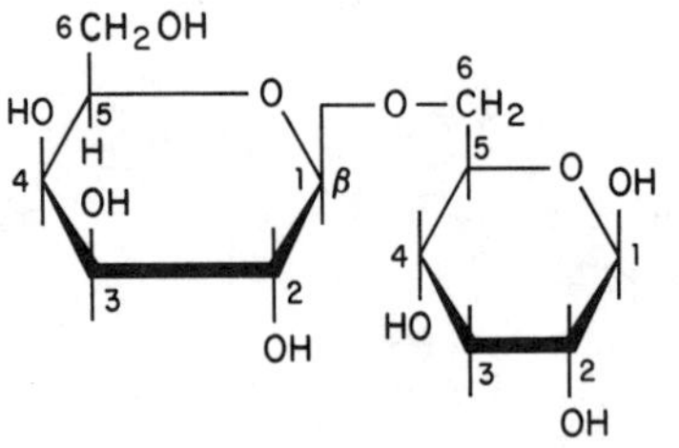

Allolactose

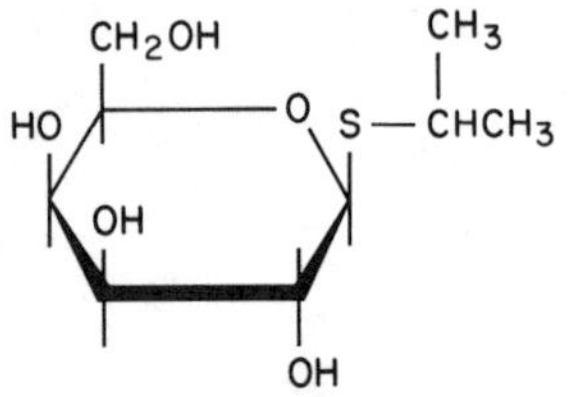

Isopropyl β-D-thiogalactoside (IPTG)

The increased β-galactosidase activity could come about through increased catalytic effectiveness of preexisting protein, through increased protein biosynthesis, or through decreased inactivation of existing protein. Increased catalytic effectiveness could be the result of allosteric effects, of activation by phosphorylation, etc. Increased protein biosynthesis could result from increased translation of existing mRNA that codes for the enzyme, or from increased transcription of the galactosidase gene with increased subsequent translation.

A number of agents are known which act at one or another of the several steps in protein biosynthesis. Chloramphenicol (CAP), obtained from *Streptomyces venezuelae,* inhibits peptide chain elongation and therefore translation. Rifampicin (RIF), obtained from the fermentation broth of *Streptomyces mediterranei,* inhibits RNA synthesis and therefore transcription. We shall use CAP and RIF to learn something of the mechanism

*This experiment is based on the classic work of Jacob and Monod. Preparatory reading: Section I, Chapters 2 and 11.

that may account for increased β-galactosidase activity in IPTG-treated *E. coli* cultures.

A simple spectrophotometric assay may be used to determine the galactosidase activity. Incubation of a galactosidase-containing bacterial lysate with *o*-nitrophenyl-β-galactoside (*o*-NPG) results in hydrolysis of the glycosidic bond, liberating galactose and *o*-nitrophenol, whereas *o*-NPG is colorless, free *o*-nitrophenol absorbs strongly with a λ_{max} at 420 nm. However, RIF is also a colored material; thus special care is required, by means of proper blanks, to correct for that part of ΔA_{420} due to the presence of the antimetabolite.

MATERIALS AND REAGENTS

1. E. coli *cells:* Many strains show galactosidase induction, but some strains work better than others. WR3102 and WR1485, which we have recently obtained, appear to work well. They give reasonable growth on minimal medium.
2. *Minimal medium:* To prepare 1 L of medium, weigh out 10.5 g K_2HPO_4, 4.5 g KH_2PO_4, 1.0 g $(NH_4)_2SO_4$, and 0.5 g trisodium citrate·$2H_2O$. Dissolve the salts in about 900 ml of water and autoclave. When the solution is cool, add 5 ml of glycerol, 1 ml of 1 M $MgSO_4 \cdot 7H_2O$ and 1 mg of thiamine hydrochloride. Bring the volume to 1 L with sterile distilled water.
3. *Sodium dodecyl sulfate:* Make a 0.1% (w/v) solution in water.
4. *Chloroform:* Used together with the SDS solution, to rupture cells.
5. *Isopropyl β-D-thiogalactoside* (IPTG): Make a 0.1 M solution in minimal medium. Keep frozen until just before use.
6. *Chloroamphenicol (CAP):* Make a solution containing 1 mg/ml in minimal medium by first dissolving the reagent in the smallest possible volume of methanol, then diluting with minimal medium to the required volume.
7. *Rifampicin (RIF):* Make a solution containing 1 mg/ml in minimal medium. Proceed as described for the chloroamphenicol solution.
8. *Glucose:* Make a 20% (w/v) solution in minimal medium.
9. o-*Nitrophenyl β-galactoside:* Make a solution containing 4 mg/ml in minimal medium.
10. *Sodium carbonate:* Make a 1 M solution in water. This is used to terminate the galactosidase reaction by a rapid change in pH. It also produces a color by ionization of the free *o*-nitrophenol.
11. *Buffer Z:* To make 1 L of this buffer, weigh out 16.1 g $Na_2HPO_4 \cdot 7H_20$, 5.5 g $NaH_2PO_4 \cdot H_2O$, 0.75 g KCl, 0.246 g $MgSO_4 \cdot 7H_2O$, and dissolve the

mixed salts in about 750 ml of water. Add 2.7 ml of β-mercaptoethanol, mix, and adjust the pH to 7.0. Finally, bring to 1L with water.

PROCEDURE

It is important to employ cells growing exponentially. An overnight culture of a suitable strain should be subcultured on fresh minimal medium and incubated at 37°C to a population of 2–5 × 10^8 cells/ml, which will give an A_{600} of 0.25–0.70. The cultures can then be cooled in an ice bath to prevent overgrowth.

Each student group should set up four series, each consisting of 13 disposable glass test tubes, 12 × 75 mm in size. The first series, *CON,* will be used as controls; a second series, *CAP,* will be used for the chloroamphenicol incubations; a third, *RIF,* for the rifampicin incubations; and the fourth, Glc for glucose incubations, Two additional (53rd and 54th) tubes will be used only to obtain a correction for background absorption due to RIF alone. It is useful to arrange these tubes in four rows in a rack, so that they will correspond to the separate incubation conditions, with a fifth row containing the last tubes used only to measure the ΔA_{420}. To each tube add 0.5 ml of buffer Z and 0.05 ml of SDS solution.

All student groups should also obtain four large test tubes, each containing 10 ml of the *E. coli* culture. These are to be labeled CON, CAP, RIF, and Glc. These tubes are placed in a 37°C water bath.

At t_0, begin β-galactosidase induction by adding 0.05 ml of IPTG solution to each of the four tubes. Mix thoroughly.

At 0, 3, 6, 9, and 12 min after IPTG addition, transfer 0.5 ml from each of the four bacterial culture tubes into the appropriate 12 × 75 mm tubes. *Immediately* add 2 drops of chloroform and vortex-mix each tube for 10–15 sec.

Immediately after sampling at 12 min, add (a) 0.75 ml of minimal medium to the CON culture; (b) 0.75 ml of CAP to the CAP culture; (c) 0.75 ml of RIF to the RIF culture; and (d) 0.75 ml of Glc to that culture. *Immediately* after RIF addition take two 0.5-ml samples into the extra control tubes, (Nos. 53 and 54) set aside for the ΔA measurement mentioned above. Add 2 drops of chloroform to these tubes and vortex-mix, like the others.

Resume sampling of the four culture systems at 15, 18, 22, 26, 30, 34, and 42 min, again adding 2 drops of chloroform to each 12 × 75 mm tube and vortex-mixing.

At the end of the incubation period, enzyme activity may be measured by addition of 0.2 ml of the substrate, *o*-NPG, to each of the four sets of tubes corresponding to the 13 time points, and 0.2 ml water into blank tubes 53 and 54. Following substrate addition, vortex-mix all tubes briefly, then let them stand at room temperature for 20 min. Stop the enzyme reaction by addition of 0.5 ml of sodium carbonate solution to each of the tubes.

Using a water blank and disposable 1-ml cuvets, measure A_{420} for all of the samples contained in the 12 × 75 mm tubes. Be careful not to transfer any of the chloroform to the cuvets since, if disturbed, it can introduce a cloudiness. Periodically clean (or replace) the cuvet containing the samples, since insoluble, gummy debris may accumulate on its walls. Check the sample cuvet frequently to be sure that the observed absorbance values represent that of the sample and not that of deposited matter. Use A_{420} values from the 53rd and 54th tubes to correct the A_{420} values for the RIF series of tubes, or use the RIF blanks to zero the instrument.

REPORT

Submit a single graph on which is displayed the A_{420} (ordinate) vs. IPTG induction time (abscissa) for each of the four growth media (CON, CAP, RIF, and Glc) additions. Make the proper correction for RIF absorbance.

REFERENCES

Lewin, G. *Genes.* John Wiley and Son, New York (1983).

Miller, J. H. *Experiments in Molecular Genetics.* Cold Spring Harbor Laboratory, Cold Spring Harbor, NY (1972).

Pardee, A. B., Jacob, F., and Monod, J. *J. Mol. Biol. 1*:165 (1959).

Watson, J. D. *Molecular Biology of the Gene,* 3rd ed. W. A. Benjamin, Menlo Park, CA (1976).

In Vitro Protein Biosynthesis by a Cell-Free System*

Protein biosynthesis in vitro, by cell-free systems, requires three distinct components. The first component includes competent ribosomes, a mixture of the twenty common amino acids and the corresponding tRNAs, aminoacyl-tRNA synthetases, and appropriate initiation and elongation factors. The second component is an intact mRNA to direct the synthesis of a specified protein. The third component is an energy source suitable to drive the system. The energy source must include ATP and GTP, and it is usual to add creatine phosphate plus creatine kinase (CK) to maintain the nucleotide concentrations over a period of time.

By radiolabeling one of the amino acids at a high specific activity, it is possible to measure trichloroacetic acid (TCA)-precipitable material even in the minute amounts produced by a reaction in a volume of 20–30 μl. Other aliquots of the product can be examined by electrophoretic or immunochemical methods. Commonly, the amino acid selected for labeling is L-[^{35}S]methionine, chiefly because of the good counting characteristics of ^{35}S, and also because methionine is available with a specific activity of >800 Ci/mmol and because methionine is more directly used for protein synthesis and not dispersed among various metabolites. Although this highly radioactive amino acid gives great sensitivity, there are some drawbacks to its use; unless handled with care, the methionine is easily oxidized to methionine sulfoxide, which cannot participate in protein biosynthesis. For this reason, some workers prefer to employ ^{14}C-labeled amino acids. They accept the lower sensitivity in favor of greater substrate stability.

MATERIALS AND REAGENTS

Reagents used in translation systems do not have a long shelf life even when stored at −70°C. As described in Chapter 11, their preparation requires a significant investment of time, effort, and materials. Considerations of cost and

*Preparatory reading: Section I, Chapters 3 and 11; also Ancillary Information, at the end of this experiment.

convenience make it appropriate to use commercially produced reagent kits. Although there are minor variations in kits purchased from different suppliers, the main features of all the translation kits are much the same (see list at the end of this experiment). You will be advised, at the time of the experiment, of any departures from the specifications below, and more exact details will be provided about concentrations, etc., which may vary from kit to kit.

1. *Reticulocyte lysate (nuclease-treated):* Prepared from rabbit blood, this preparation contains the ribosomes and soluble factors essential for protein biosynthesis. It was treated with micrococcal nuclease (plus Ca^{2+}), which was subsequently inactivated by chelation of the Ca^{2+} with EGTA. One of the special reasons for careful glassware preparation is to avoid reintroduction of Ca^{2+} that would reactivate the nuclease which is still present in this reagent.

2. *Energy source:* This consists of a mixture of ATP, GTP, creatine phosphate, and CK.

3. *Magnesium acetate, potassium acetate:* Both Mg^{2+} and K^+ are essential ions in protein biosynthesis, but the optimal concentration of each varies with the mRNA being translated. In order to adjust these ion concentrations to their optimal values, vials of each in sterile aqueous solution are provided as part of the kit, usually at a concentration of 800 mM to 1 M. You will be informed about the actual values at the time of the experiment.

4. *Sodium acetate, 800 mM:* This solution is not ordinarily provided with the kit, but will be used here to examine the effect of added Na^+ on protein biosynthesis. Because it will be added to the incubation mixtures during biosynthesis, it must be made in sterile purified water.

5. *Control mRNA:* This is a purified preparation, usually a sample of the message used by the manufacturer to calibrate the kit. Different manufacturers use different mRNA species for this purpose. In the user's hands, the control can provide evidence that the kit was not damaged during transit or storage.

6. *Experimental mRNA:* Prepared by standard methods from some tissue to provide sufficient material for laboratory study, this is an enriched but not purified preparation. It will be employed to study the effect of deliberate experimental variations on protein biosynthesis by the reticulocyte lysate preparation. We will usually use globin mRNA.

7. L-[^{35}*S*]*Methionine:* A specially purified, translation-grade preparation in aqueous or ethanol solution with added dithiothreitol as a preservative, it is labeled to a high specific activity, the exact value of which will be provided at the time of the experiment. **Note:** *Ethanol is a strong inhibitor of in vitro protein biosynthesis. If the methionine provided is in ethanol-containing solution, the alcohol must be removed by lyophilization, then the original volume restored with sterile water.*

8. *Sterile highly purified water* (see Chapter 4): Carefully freed of metal ions and RNase, only this water should be used in diluting reagents or in setting up the incubation mixtures described in the procedure section.

All of the above components (except for the sodium acetate solution, which must be made locally) are shipped in the frozen state in a container of dry ice. The kit should be inspected, then immediately stored at −70°C. The biological components should not be subjected to repeated freeze–thaw cycles. Instead, the thawed components should be subdivided into appropriate aliquots, and these should be refrozen promptly at −70°C. Even at this low temperature, the shelf life is only a few months. When thawing the components, use either the heat of your hand or a water bath set at no higher than 37°C; work quickly to avoid thermal damage. The thawed reagents should be kept in an ice bath at all times. Use great care to avoid contamination by unexpected sources of Ca^{2+} or RNase.

All of the above reagents are used in the biosynthetic system, and they must be made with sterile water. Care must be taken to avoid contamination of them. Several additional reagents, used only in processing the products of the biosynthetic systems, are also required. The items listed below need *not* be made with sterile water.

9. *Trichloroacetic acid (TCA), 100% (w/v):* **CAUTION: Avoid getting trichloroacetic acid (solid or liquid) on your skin. It is very corrosive and can cause severe burns.** For this reason, it is preferable to purchase this solution rather than to prepare it.

10. L-*Methionine, 0.1 M:* Dissolve unlabeled material in water to make 10 ml of solution. This will be used to massively dilute the radioactive pool, as explained in the procedure section.

11. *Sodium hydroxide–hydrogen peroxide solution:* Place 9.0 ml of 30% hydrogen peroxide in a 100-ml volumetric flask, then make to volume with 0.1 M sodium hydroxide solution. This will give a peroxide concentration of approximately 0.25 M. **CAUTION: Do not get any of the 30% hydrogen peroxide on your skin; it is a powerful oxidant and can cause severe burns. This reagent should be mixed from its components just before use and should be discarded at the end of the day.**

12. *Sodium hydroxide, 0.1 M:* Dissolve a sufficient quantity of the solid NaOH in water. This will be used in the final washings of the precipitated protein mixture generated as detailed in the procedure section (Part I).

13. *Carrier protein solution:* A 5% (w/v) solution of bovine serum albumin will suffice for this purpose, which is merely to increase the bulk of the protein precipitate formed on addition of trichloroacetic acid.

14. *Tissue solubilizer:* Any of the proprietary preparations, Soluene®, Protosol®, or the like, is acceptable. Although the exact compositions of the commercial solubilizers are not revealed, they generally contain alkyl quaternary ammonium compounds, the purpose of which is to dissolve

precipitated proteins in preparation for liquid scintillation counting. They may also contain detergents, such as Triton X-100, so that even on addition of a toluene-based cocktail phase separation does not occur.

15. *Scintillation cocktail:* Any toluene-based preparation will suffice.
16. *Presterilized polyethylene microfuge tubes, 1.4 ml:* If commercially presterilized tubes are not available, tubes bought in bulk (and their caps) can be thoroughly washed (see below) and finally sterilized by autoclaving at 15 psi for at least 15 min. Do not bake these for sterilization, since they will melt.
17. *Presterilized, disposable pipet tips:* It is best to purchase these in presterilized packages. They can be prepared as noted above for the bullet tubes, but that is a great nuisance and the quality of sterilization achieved is somewhat uncertain. Tips should be kept in their protective wrapper until needed, and should not be allowed to touch the bench top (or anything else) between uses. Keep in mind the absolute necessity of avoiding contamination by RNase all through this experiment.
18. *0.2 N KOH*
19. *95% ethanol*
20. *Glass filters*
21. *Glass filter holders* (Millipore or Gelman). Alternatively, one may use the Gooch crucible setup employed in Experiment 17.

PROCEDURE

Part I: Setting Up the "Premix"

The reagents used in this experiment are expensive, and a number of them are provided in quite limited supply. There will be virtually no excess for class use, so plan your operations with care. The individual assays of protein biosynthesis will be performed in final volumes of 50 μl. Preparation of a "premix" allows for greater consistency in the composition of the basic solution by minimizing volume transfer errors. It is still important that the volume of each reagent be measured very carefully, not only to avoid waste but also because the synthetic systems are quite sensitive to concentration changes.

Each student group should collect 10 presterilized, 1.4-ml polyethylene microfuge tubes, complete with caps. The tubes should be serially numbered, 0 through 9, and prechilled in an ice bath. Arrange the tubes so that the caps are not wet by the ice or water in the bath.

The tube numbered 0 will be used only to prepare the "premix," made by combining the [^{35}S]methionine, the synthesis reaction mixture, and the essential mineral ions Mg^{2+} and K^{+}. Aliquots of the "premix" will then be distributed among the remaining tubes of the series. In working with the "premix," keep in mind the high specific activity of the methionine! Be careful not to contaminate yourself or your surroundings with ^{35}S.

To the tube marked 0 add, in the order cited, the following volumes of the indicated reagents:

[^{35}S]Methionine (12.5 mCi/ml)	50 μl
Synthesis reaction mixture*	55 μl
Potassium acetate (1 M)	20 μl
Magnesium acetate (32.5 mM)	5 μl

Cap the tube, vortex-mix, then centrifuge for a few seconds in a table-top microcentrifuge to collect the contents as a single drop in the bottom of the tube. Then transfer 13 μl of the "premix" to each of the remaining tubes. It is prudent to keep all of the tubes capped, except when making an addition, in order to avoid contamination, and it is important to keep all of the tubes well chilled in an ice bath.

Tube 0 has now served its purpose. Any remaining contents can be discarded in the proper container for liquid radioactive wastes, and the empty tube can be placed in the proper container for dry radioactive waste.

Part II: Demonstration of mRNA Concentration Dependence

Using the sterile solution of mRNA provided, the sterile water, and the reticulocyte lysate preparation, make the indicated additions to the indicated tubes. (All volumes shown are given in microliters.) The additions should be made in the order shown from left to right.

Tube No.	mRNA	Sodium acetate	Sterile water	Lysate
1	0	0	27	10
2	1	0	26	10
3	2	0	25	10
4	3	0	24	10
5	3	2	22	10

Cap the tubes, vortex-mix, and briefly centrifuge as before, then incubate tubes 1 through 5 in a water bath at 37°C for 1 hour.

At the end of the incubation period, add 5 μl of *unlabeled,* 0.1 M methionine, vortex-mix the contents, and place the five tubes in an ice bath. The addition of the very large excess of unlabeled methionine, plus cooling the tubes, effectively stops incorporation of label into protein. These tubes will now be held in the ice bath until completion of Part III.

*Manufacturers of the kits frequently regard the exact composition of their synthesis reaction mixture as proprietary information and are reluctant to state its precise composition. At the end of this experiment, you will find some representative, nonproprietary data taken from the open research literature or from manufacturers who do reveal the composition of this reagent.

Part III: Time Dependence of Protein Biosynthesis

The remaining tubes containing "premix" will now be used to examine the time dependence of protein biosynthesis. All of the tubes will contain identical mixtures but will be incubated for different lengths of time. Make the additions shown below *in the order given from left to right.*

Tube No.	mRNA	Sterile water	Lysate	Incubation time (min)
6	3	24	10	10
7	3	24	10	20
8	3	24	10	30
9	3	24	10	45

Proceed as in Part II to ensure uniformity of contents, then incubate each tube for the time shown. At the end of the incubation period, add 5 μl of unlabeled methionine, vortex-mix, and chill the tube in an ice bath.

Part IV: Collection and Washing of the Synthesized Protein

The newly synthesized protein remains in solution and is grossly contaminated by the excess, unincorporated [^{35}S]methionine. The mixture is also contaminated by hemin, which gives it an obvious color. Unless that color is removed, it will produce significant quenching during liquid scintillation spectrometry. The procedures that follow address these matters; use Method A, B, or C.

Method A

Collect a second set of 3-ml plastic microfuge tubes, and number them *1′* through *9′* to correspond with the first set. Into each of the empty tubes, place 1 ml of the sodium hydroxide-hydrogen peroxide reagent. Next, transfer 10 μl* of solution from tube 1 to tube *1′*, from tube 2 to tube *2′*, and so forth. The second set of tubes is then incubated at 37°C for 30 min (or less, if the color disappears sooner). The alkaline peroxide destroys the hemin and discharges the color.

To each of the digestion tubes, add 25 μl of bovine serum albumin (or other carrier protein, as provided) and vortex-mix to ensure uniformity of the contents. Then, add 100 μl of the 100% trichloroacetic acid reagent to each tube. This should result in the formation of a distinct precipitate. The precipitate is pelleted in a table-top microcentrifuge. *Take care that the precipitate is firmly packed.*

*The volume of the aliquot to be taken depends on the efficiency with which the experimental mRNA is translated. The volume specified here in the text is hypothetical; you will be given more precise information at the time of the experiments.

The supernatants are carefully decanted into a container provided for liquid radioactive waste. Using small pieces of lint-free tissue paper, remove the last, clinging drop of liquid from all tubes, taking pains not to disturb the packed precipitates. The paper should then be deposited in a container for dry radioactive waste.

To each tube add 1 ml of 0.1 M sodium hydroxide (**Note:** *This should not contain hydrogen peroxide*); vortex-mix to dissolve the pellets. When solution is complete, again add 100 μl of trichloroacetic acid to reprecipitate the protein, and centrifuge as before. The new supernatant solutions are discarded with the precautions already mentioned. This accomplishes the first washing of the original pellet. The operations necessary to dissolve and reprecipitate the proteins are repeated *twice more,* so that the original pellet will have been washed three times in all. This procedure is essential to lower the nonspecific contamination by unbound radiolabeled methionine to acceptable values.

The final, thrice-washed pellet is dissolved by addition of 1 ml of tissue solubilizer, with vortex-mixing as needed. Each of the bullet tubes is then transferred to a scintillation minivial, along with 3 ml of a suitable scintillation cocktail. The vials are capped and placed into *empty* standard and scintillation vials, in preparation for counting. (You will be advised as to which channel of the LSC instrument has been set for proper recording of ^{35}S, and as to counting times.)

Method B

Obtain from the instructor as many 1.5 cm × 1.5 cm squares of Whatman 3 MM filter paper as there are reaction systems in Parts II and III above. Using a graphite pencil, mark each square with a code corresponding to each of these systems. Transfer 10 μl from each system to the center of the appropriate filter paper square. After they have been air-dried (about 10 min), drop the squares into 2 L of boiling 10% trichloroacetic acid (TCA). **CAUTION: Use care! TCA is highly corrosive! Wear your safety goggles!** After 10 min, remove the beaker containing the TCA from the hot plate. The filter paper squares will settle to the bottom. Decant the 10% TCA **(CARE!)** and add 1 L of 5% TCA to the beaker. Carefully stir the contents of the beaker with a long glass rod. After the squares have settled to the bottom, decant the 5% TCA and wash with another liter of 5% TCA. After again decanting the 5% TCA, wash the squares *twice* with 300-ml batches of acetone. Using long metal forceps, retrieve the filter paper squares from the wash beaker. Air-dry the squares for 15 min, then insert each square into a vial containing 10 ml of a scintillation counting cocktail. Count in scintillation counter.

Method C

1. Add 5 μl of 0.2 N KOH to each system in Parts II and III and incubate for 15 min at 30°C. (During the incubation, number a series of glass fiber filters with pencil so that they correspond to the incubation system series. Also

prepare 10% and 5% TCA solutions by diluting the 100% stock TCA solution.)

2. Spot 15-μl aliquots from each system on the appropriate glass fiber disc.
3. Place each disc in a small plastic cup containing 10 ml of cold 10% TCA.
4. Let stand for 10 min.
5. Transfer each disc in turn to a glass filter holder mounted on a 250-ml filter flask attached to a water aspirator.
6. Wash each filter three times with 10 ml 5% TCA for each wash.
7. Wash twice with 95% ethanol (about 10 ml).
8. Dry the filters in a vacuum oven.
9. Transfer filters to vials containing 10 ml of a toluene-based scintillation fluid and count in an LSC.

Note: *Complete the experimental procedure by disposing of all radioactive wastes according to directions provided.*

CALCULATIONS

All calculations should be expressed in terms of dpm/μl of [^{35}S]methionine in the original sample. All values should be corrected for decay of the [^{35}S]methionine, which has a half-life of 87.2 days. (You will be given the exact value of the initial specific radioactivity at the time of the experiment.)

REPORT

Your report should include a table, in three parts. The first part should summarize the effect of varying mRNA concentration. The second part should summarize the effect of added Na^+, and the third part should summarize the effect of incubation time on protein biosynthesis.

NOTES ON GLASSWARE PREPARATION

This experiment has been designed to minimize the need for glassware preparation, through the use of presterilized plastic ware whenever possible. Where extra glassware is needed, proceed as follows:

1. Wash each piece in an alkaline detergent such as Alconox®, then drain as completely as possible.
2. Wash each piece in 3 N HCl, then drain as above.

3. Rinse several times in the highly purified water supplied, and drain as completely as possible.
4. Wrap pipets (or put them into pipet cans) and cap beakers, flasks, etc. with aluminum foil.
5. Dry-sterilize all glassware items by baking in an oven at 200–210°C for 12 hours (or overnight). Keep all pieces in their protective wrappers until needed.
6. Sterile water may be made by autoclaving the highly purified water at 15 psi for not less than 30 min. Some workers recommend addition of 0.01 volume of diethyl pyrocarbonate to the water prior to autoclaving, but this is not always required.

NOTES ON REACTION COCKTAILS

The general composition of reaction cocktails employed by investigators in the field of in vitro protein biosynthesis does not vary greatly. In most instances, the differences relate to altered concentrations rather than to omissions or additions. Indeed, for the most careful work, the concentrations of K^+ and Mg^{2+} must be "fine-tuned" or optimized for a given mRNA. You will also note that the total volumes of the incubation mixtures in the examples cited are not the same, but this is largely a matter of convenience. Where reference numbers are given, they refer to citations at the end of this exercise.

Wheat Germ Translation Kit of Bethesda Research Laboratories

Hepes buffer (pH 7.5)	20 mM
Potassium acetate	30 mM
Magnesium acetate	0.1 mM
ATP	1.2 mM
GTP	0.1 mM
Creatine phosphate	5.5 mM
Creatine kinase	0.2 mg/ml
Spermidine phosphate	80 μM
19 amino acids	50 μM

Reticulocyte Lysate System of Schrader and O'Malley (1987)

Phosphate buffer (pH 7.5)	4 mM
Potassium chloride	86 mM
Magnesium chloride	3 mM
ATP	—
GTP	—
Creatine phosphate	6 mM
Creatine kinase	5 mg/ml

Hemin	15 μM
19 amino acids	80 μM

Reticulocyte Lysate System of London (1976)

Tris buffer (pH 7.6)	10 mM
Potassium chloride	76 mM
Magnesium acetate	2 mM
ATP	1 mM
GTP	0.2 mM
Creatine phosphate	5 mM
Creatine kinase	0.1 mg/ml
Hemin	10–30 μM
19 amino acids	6–170 μM

REFERENCES

Bethesda Research Laboratories, Bethesda, MD *Product Profile: Wheat-Germ In Vitro Translation System.*

Chirgwin, J. M., Przybyla, A. E., MacDonald, R. J., and Rutter, W. J. Isolation of biologically active ribonucleic acid from sources enriched in ribonuclease. *Biochemistry 18:*5294–5299 (1979).

Davis, L. G., Dibner, M. D., and Battey, J. F. *Basic Methods in Molecular Biology.* Elsevier New York (1986).

Lodish, H. Translational control of protein synthesis. *Annu. Rev. Biochem. 45:*39–72 (1976).

London, I. M. Role of hemin in the regulation of protein biosynthesis in erythroid cells. *Fed. Proc. 35:*2218–2222 (1976).

Nevins, J. R. Pathway of eukaryotic mRNA Formation. *Annu. Rev. Biochem. 52:*441–466 (1983).

Pelham, H.R.B., and Jackson, R. J. An efficient mRNA-dependent translation system from reticulocyte lysates. *Eur. J. Biochem. 67:*247–256 (1976).

Schrader, W. T., and O'Malley, B. W. *Laboratory Methods Manual for Hormone Action and Molecular Endocrinology,* Secs. 8-12 through 8-14. Houston Biological Assoc., Houston, TX (1987).

Restriction Enzyme Map of a Modified Plasmid*

This experiment provides an introduction to some of the methodology used in the study of DNA structure. The material to be examined is a plasmid preparation extracted from an *E. coli* culture. Here, we shall work with the plasmid pBR322, and the *E. coli* strain HB101 or LE392. (If a plasmid of the pUC series were selected, then the JM103 strain could be used.) High-speed centrifugation of the crude plasmid extract (from ruptured *E. coli* cells) in a cesium chloride (CsCl) gradient separates the major nucleic acid components; their location in the gradient is visualized by use of the intercalating agent, ethidium bromide (EtBr). The gradient region containing the plasmid is freed of CsCl and ethidium bromide and then is treated with a battery of restriction endonucleases; the product mixture is separated by agarose gel electrophoresis. A modified plasmid, the same as that to be purified but with an extraneous DNA insert, is supplied by the teaching staff. This modified plasmid is submitted to the same nuclease and electrophoretic protocol as the original purified plasmid. By comparison of the gel displays of the DNA fragments from the plasmid, modified plasmid, and a standard mixture of fragments of known size (kilo bases), the plasmid is mapped and the site of insertion in the modified plasmid is determined.

The two procedures presented below differ slightly in the method used to remove CsCl, protein, and RNA contaminants from plasmid fractions (column vs. dialysis).

The restriction endonucleases used in this experiment—EcoRI, PstI, and AVAI—are those for pBR312 fragmentation; other plasmids would be treated with a different complement of enzymes.

CAUTION: Ethidium bromide is reportedly mutagenic (see *Registry of Toxic Effects of Chemical Substances*), **and therefore gloves should always be used when handling either the solution or the solid.**

Several of the procedures to be followed are lengthy. Some, especially the centrifugal separation, are best conducted overnight. To avoid unwanted, time-dependent changes, it is important to complete the steps promptly.

*Preparatory reading: Section I, Chapters 2, 3, 6, 8, and 11.

MATERIALS AND REAGENTS

1. n-*Butanol:* For removal of excess ethidium bromide.
2. *Cesium chloride:* Finely powdered solid of highest purity; used to adjust the density of DNA solutions prior to centrifugation.
3. *Ethidium bromide (EtBr):* Dissolve in water at a concentration of 10 mg/ml (w/v).
4. *Cell suspension medium:* A 25% (w/v) solution of sucrose, made in 50 mM Tris-HCl buffer (pH 8.0).
5. *Cell wash medium:* Tris-HCl buffer, 10 mM, containing 1 mM EDTA, adjusted to pH 8.0.
6. *Lysozyme solution:* Crystalline lysozyme is dissolved in 0.25 mM Tris-HCl (pH 8.0) to give a concentration of 10 mg/ml.
7. *EDTA:* Adjust the pH of a 0.25 M solution to 8.0 with 0.1 N NaOH.
8. *Brij-58 detergent solution:* This contains 1% (w/v) of Brij-58 (a commercial detergent), 0.4% (w/v) sodium deoxycholate, and 62.5 mM EDTA, all dissolved in Tris-HCl, 50 mM, adjusted to pH 8.1. This mixed detergent solution is used to rupture bacterial cell membranes.
9. *Biogen A-15M (200–400 mesh):* An agarose-based exclusion gel with an upper exclusion limit of approximately 15×10^6 daltons. Prior to use, the gel should be equilibrated with a solution of 0.5 M NaCl containing 10 mM Tris + 1 mM EDTA, adjusted to pH 8.4. The equilibrated gel is then packed into columns, approximately 2.5×15 cm in size.
10. *95% ethanol:* This is used for cold precipitation of the isolated DNA.
11. *DNA suspension buffer (TE buffer):* Tris-HCl, 10 mM, with 0.1 mM EDTA, adjusted to pH 7.4. This is used to suspend the isolated DNA prior to digestion by the restriction enzyme(s).
12. *Digestion buffer:* Composed of 75 mM NaCl, 6 mM β-mercaptoethanol, 10 mM $MgCl_2$, 1 mg/ml (w/v) gelatin, all in Tris-HCl, 10 mM, adjusted to pH 7.4. This is used as the vehicle for the restriction enzyme(s), which is (are) ordinarily sold at a concentration of 1 U/ml in a mixture of salt, glycerol, EDTA, reducing agent, and albumin. The enzyme unit is defined as that activity which degrades 1.0 μg of λ phage DNA in 15 min at 37°C.
13. *Concentrated (5×) DNA electrophoresis buffer:* This contains 0.025% (w/v) bromphenol blue, 0.025% xylene cyanol, 50% (v/v) glycerol, 4.5 mM EDTA, 25 mM sodium acetate, and 100 mM Tris-HCl, all adjusted to pH 7.8. For use in casting the agarose gels, this must be diluted five fold. The 5× buffer is used *only* in preparing the samples for application to the gels.
14. *Agarose gels:* These are cast from a 1% (w/v) solution of powdered agarose in a special apparatus which provides a slab bearing 20 sample wells. Its use will be demonstrated.

15. *Nucleic acid molecular weight standards:* These are commercially prepared by treating DNA with HindIII restriction endonuclease to yield eight fragments, the sizes of which range from 125 bp to 23.1 kb.
16. *Ribonuclease A, heat-treated:* Prepare a 100 mg/ml solution of RNase A in TE buffer; heat for 10 min at 70°C to inactivate DNase.
17. *Ribonuclease T1 (RNase T1):* An endonuclease specific for Gp↓N, it is commercially available; the most concentrated preparation should be used.
18. *Proteinase K:* This is commercially available as a solid.
19. *5 M NaCl.*
20. *5 M ammonium acetate.*
21. *5% (w/v) SDS.*
22. *Quick-Seal® centrifuge tubes and heat-sealing apparatus.*
23. *Restriction endonucleases:* EcoRl, 20 units/μl; Pstl, 20 units/μl; Aval, 10 units/μl; all are in buffered 50% glycerol. Specifications may vary a bit, depending on commercial source.

The summary outlines given below are designed for students working in groups of two:

PROCEDURE A

Outline

1. Collect and wash the cells that contain the native plasmid DNA.
2. Solubilize the bacterial cell walls and spin out the cellular debris.
3. Mix the supernatant, containing chromosomal and plasmid DNA, with cesium chloride and ethidium bromide for overnight density-gradient centrifugal separation of the two DNA types. As will be demonstrated, the density adjustment must be made with considerable care.
4. Collect the plasmid DNA band, and remove the cesium chloride by exclusion gel chromatography.
5. Digest the plasmid DNA samples with the restriction enzymes.
6. Separate the products by electrophoresis on agarose slabs, along with known molecular weight markers to provide points of reference.
7. Observe the results under ultraviolet light, then photograph the gels under ultraviolet light to provide a permanent record.
8. From the photograph, or from the original gel, analyze the results in terms of sequence organization.

Because of the length of this experiment, it is useful to divide the total operations according to laboratory periods. The details of a suggested division are set forth in days, as indicated below.

Day 1

Pellet the cells contained in a 1-L culture by centrifugation for 10 min at ~ 4500 x *g* rpm and 4°C. Use metal or heavy glass centrifuge tubes.

Resuspend the cells in 200 ml of *cell wash medium;* then repeat the centrifugation procedure used for initial cell collection. **Note carefully:** *The remaining steps in the DNA isolation must be done at ice-bath temperature! Warmer temperatures will result in degradation of DNA!*

Resuspend the washed cells in 10 ml of *cell suspension medium,* then transfer the suspension to a 125-ml Erlenmeyer flask. Add 2.75 ml of lysozyme solution carefully to the center of the suspension, swirl, and place on an ice bath for 5 min. Add 4.1 ml of 0.25 M EDTA, swirl, and let stand for 5 min. Add 16.5 ml of the Brij-58 detergent solution and swirl for 10 min. The material in the flask should now be quite viscous, as the DNA is set free. *DNA may be fragmented by shearing forces; proceed carefully.*

Transfer the material to an equal number of tubes suitable for use with the Spinco SW-27 rotor, load the head into the centrifuge, and centrifuge for 30 min at 25,000 rpm and 4°C.

When the rotor has come to a stop, collect the tubes and combine the supernatants in a graduated cylinder. Read the A_{260} of this solution, diluting a small aliquot if necessary. Note the volume of the combined supernatants, then add 0.87 g of finely powdered cesium chloride per milliliter of solution. It is permissible to stir the solution *gently* with a fine glass rod, but it is probably better to allow the solid salt to dissolve undisturbed. When all of the cesium chloride has dissolved, note the volume again, then add 50 μl of ethidium bromide solution per milliliter of solution.

It is important to determine that the density of the solution has been properly adjusted. This will be done with an Abbé refractometer (to be demonstrated). The refractive index of your solution should be between 1.3960 and 1.3970. This is a fast, if indirect, assessment of the density of the solution.

Transfer the mixture to the Quick-Seal® ultracentrifuge tubes; fill to top and seal the tubes. This ends the exercise for day 1. The staff will centrifuge the tubes at 200,000 x *g* for 48 hours at 20°C in a 60 Ti rotor. They will then assist in the retrieval and storage of the samples after centrifugation.

Day 2

Mount your centrifuge tube vertically with a clamp on a ring stand. *Do not disturb the contents.* Snip off the sealed tip.

With a long-wavelength ultraviolet lamp, identify the two DNA bands, which should be visible in your sample. The upper, lighter band contains linear plasmid DNA plus linear chromosomal DNA. The lower, more dense band contains supercoiled plasmid DNA. RNA is pelleted at the bottom of the tube.

Use a 2- to 5-ml syringe with an 18-gauge needle to collect the plasmid band, by piercing the side of the tubes a few millimeters below the band and aspirating until the fluorescent band is removed. **CAUTION: Use protective safety glasses; ultraviolet light is harmful to your eyes!**

Transfer the supercoiled plasmid DNA band to a 15-ml Corex tube. Extract the DNA three times with equal volumes of *n*-butanol to remove excess ethidium bromide. If necessary, centrifuge the mixture for a few minutes to separate the phases. Use either the SS-34 rotor, fitted with rubber sleeves, or the SS-24 rotor directly. Discard the *n*-butanol extract.

Load the aqueous phase containing the plasmid DNA onto a prepared and preequilibrated Bio-Gel® A-15M column. Using the same buffer with which the gel was equilibrated, collect fractions, each containing 50 drops. The A_{260} of the fractions should be measured to locate the eluted DNA. It is perhaps quicker to first measure every other fraction, then check more closely where absorbance is significant. The concentration of DNA in μg/ml can be estimated as $10 \times A_{260}$ in a 1-cm light path cell.

Pool the DNA-containing fractions and measure their combined volumes. Measure the A_{260} of this solution, returning the aliquot measured to the pool. Precipitate the DNA by addition of 3 volumes of 95% ethanol; mix well by gentle stirring with a fine glass rod; allow precipitation to continue by storage overnight at −20°C. This ends the exercise for day 2.

Day 3

Pellet the precipitated DNA in the centrifuge at $27{,}000 \times g$ for 10–15 min. Discard the supernatant and allow the pellet to drain briefly by inverting the tube. Carefully resuspend the *dry* pellet in sufficient *DNA suspension buffer* to give a DNA concentration of 200 μg/ml.

You are now ready to digest *both* plasmid samples (the one you prepared *and* the modified plasmid given to you) with the enzymes EcoRI, PstI, and AvaI. Aliquots of DNA from each sample will be digested with a single enzyme only, with all possible combinations of the enzymes taken two at a time, and by all enzymes at once. Thus, you will have *seven different digests* for each DNA sample. Proceed as follows.

Place 5 μl of DNA solution into each of a series of seven 400-μl microcentrifuge tubes. Since you wish to digest two samples, you will require two series of tubes. Add to each tube 10 μl of digestion buffer and 2 μl of enzyme. (For those digests involving more than one enzyme, add 2 μl of each.) Close all of the tubes with the attached caps. To ensure good mixing of the contents of the tubes, centrifuge them briefly in the microcentrifuge. Finally, incubate the capped tubes for 60 min at 37°C in a thermostatted waterbath.

While enzyme digestion is taking place, prepare an additional 400-μl tube for each DNA sample by adding 5 μl of the undigested DNA and 10 μl of *digestion buffer.* Hold these tubes until the remaining digestions are complete.

At the end of the digestion period, add 5 μl of the *concentrated (5×) electrophoresis buffer* to all of the tubes. Ensure good mixing by a brief period of centrifugation in the microcentrifuge.

Obtain a standarde mixture of DNA size standards from the instructor. Add 0.2 volumes of loading buffer to ~2 μg of marker DNA. Heat to 65°C for ~5 min.

A 1% (w/v) agarose gel containing ethidium bromide (15 μl per 100 ml) will have been prepared by the staff; this will have 20 wells formed by inserting a comb into the gel while it was hardening. (Alternatively, the ethidium bromide may be added either to the samples or to the covering buffer; addition to the gel is our preference.)

Unlike the previous electrophoretic gels you have used, this agarose gel will be run in a horizontal plane. Note that the cast gel should be mounted with the wells numbered 1–20 to the left-hand end; we will assume that the numbers increase from the front of the gel to the rear as you face it.

You will be given further advice during the exercise concerning sample loading. Here it is sufficient to note that *you must keep careful record of which sample is loaded into which well.* Place 30 μl of the samples not treated with nuclease into wells 1 and 2, EcoRI-treated samples into wells 3 and 4, etc. Reserve the last two wells for the size standards (~30 μl of the heated standards).

Connect the power supply leads to the electrophoresis chamber so that the *anode (positive electrode)* is farthest away from the sample wells. Adjust the power supply to provide a current of about 75 mA. Continue the electrophoresis until the tracking dye (bromthymol blue) has migrated approximately 10 cm; the xylene cyanol dye will run approximately at a 400-kb equivalent distance.

Shut off the power, remove the electrical connections, and carefully remove the gel from the apparatus. Examine it under ultraviolet light.

Bring your gel to the instructor, who will provide you with Polaroid prints of the final gels. From migration data on the known markers, you will be able to estimate restriction fragment lengths and so to construct a restriction map of the original DNA species. This ends the exercise.

PROCEDURE B

Outline

1. Follow Procedure A, steps 1 and 2.
2. Treat cell extract with RNases and protease.
3. Mix preparation with cesium chloride and ethidium bromide, and centrifuge.
4. Collect the plasmid DNA band, extract with *n*-butanol, and dialyze.

5. Concentrate the dialyzed DNA preparation with *n*-butanol, and digest with endonucleases.

6. Follow Procedure A, steps 6 through 8.

Day 1

Add 4.0 ml of the lysozyme solution to 20 ml of cell suspension in 0.25 M Tris (pH 8.0). Incubate at room temperature for 5 min with gentle swirling on a rotary shaker. Add 14.4 ml of 0.25 M EDTA (pH 8.0), and swirl gently for an additional 5 min at room temperature. Add 12 ml of 5 M NaCl and mix carefully. Add 12 ml of 5% (w/v) SDS and swirl gently, to avoid foaming, for 10 min. The preparation should look uniformly translucent when swirling is complete.

Transfer the preparation to an even number of tubes for use with an SW-27 rotor. Chill on ice for a minimum of 1 hour to overnight; the contents of the tubes should not pour out when inverted. Centrifuge at 132,000 × *g* for 30 min at 4°C. Combine and save the supernatants; discard pellets.

Incubate the combined supernatants with 1 ml of heat-treated RNase A and 1 μl of RNase T1 (~1000 units) for 0.5–1.0 hour at 37°C.

Add 3 mg of proteinase K and incubate.

Adjust the incubated supernatant to the nearest volume, and add CsCl and EtBr according to the following table:

	No. of Tubes			
	2	**4**	**6**	**8**
Adjusted vol. (ml)	62	124	186	248
CsCl (g)	59.0	118	177	236
1% (w/v) EtBr (ml)	6.2	12.4	18.6	24.5

Mix carefully but thoroughly. The concentration of CsCl may be checked by refractometry as in Procedure A. Transfer the preparation to Quick-Seal® centrifuge tubes, seal, and centrifuge at 113,000 × *g* for 48 hours in a 60 Ti rotor at 20°C. The staff will assist in retrieving and storing the tubes.

Day 2

Mount the tube with a clamp on a ring stand; *be careful not to disturb the contents.* With a long-wavelength ultraviolet lamp, identify the fluorescent bands. **(CAUTION: Use protective eye goggles!)** Snip off the cap of the tube, and remove the upper band containing linear chromosomal DNA and linear plasmid (as described in Procedure A). Discard.

Collect the lower band containing plasmid DNA and transfer it to a 50-ml orange-capped plastic tube. Extract the DNA solution three times with equal volumes of *n*-butanol to remove EtBr.

Dialyze the aqueous phase against three changes of 3–5 L of 10 mM Tris (pH 7.5), 0.1 mM EDTA, overnight.

Day 3

Transfer the dialyzed DNA sample to an appropriately sized tube and add an equal volume of *n*-butanol; centrifuge in a SS-34 rotor at $\sim 500 \times g$ for 5 min, and discard the upper phase. Repeat the steps in this extraction until the volume of the aqueous phase has been reduced to ~ 300 μl. (**Note:** This is a volume reduction step in which the water, but not the DNA, is extracted into the *n*-butanol.) Transfer the DNA solution carefully to a 3-ml microcentrifuge tube. Add 18 μl of 5 M NH_4OAc (pH 5.5), and mix well. Add 600 μl of cold 95% ethanol, mix, and place tube on dry ice for 10 min. Spin down the DNA pellet for 10 min in a table-top microcentrifuge.

Decant the supernatant ethanol and dry the tube under vacuum. (A vacuum desiccator or vacuum oven at room temperature will serve.)

Add 100 μl of TE buffer to dissolve the DNA pellet. Take 5 μl, dilute with 995 μl of TE buffer, measure the A_{260} of the solution, and adjust the DNA concentration to 0.5 μg/μl. (The concentration of DNA in μg/μl will be about 10 times the A_{260} reading.)

From this point on, proceed with the restriction endonuclease digestions as described in Procedure A.

REPORT

Submit:

1. An elution profile of your own DNA preparation, if you used Procedure A. Plot A_{260} (ordinate) vs. fraction number from the Bio-Gel® column (abscissa). From this, calculate the total DNA yield.

2. The photograph of your agarose gel. Prepare from this a graph of the log of the number of base pairs (ordinate) vs. the mobility of the DNA standards (abscissa).

3. A table of the fragment sizes produced by each of the enzymatic digestions.

4. A restriction map for each of the plasmids.

REFERENCES

Bauer, W., and Vinograd, J. *J. Mol. Biol. 33:*141 (1968).

Burrell, C. J., MacKay, P., Greenaway, P. J., Hofschneider, P. H., and Murray, K. *Nature 279:*43 (1979).

Davis, L. G., Dibner, M. D., and Battey, J. E. *Basic Methods in Molecular Biology.* Elsevier, New York (1986).

Maniatis, T., Hardison, R. C., Lacy, E., Laurer, J., O'Connell, C., Quon, D., Sim, G. K., and Efstratiadis, A. *Cell 15:*687 (1978).

Registry of Toxic Effects of Chemical Substances (RTECS), Compound No. SF7950000.

Sutcliffe, J. G. *Cold Spring Harbor Symp. 43:*77 (1979).

Watson, J. D. *Molecular Biology of the Gene,* 3rd ed. W. A. Benjamin, Menlo Park, CA (1976).

SECTION III

APPENDIX

Self-Study Questions

Presented here are the kinds of questions you may be expected to answer; they are not comprehensive, but they afford you a means of examining your understanding of the material that has been presented.

pH, ACIDS, BASES, AND BUFFERS

1. Calculate the pH of a 0.1 M solution of lithium lactate.
2. The K_d for lactic acid is 1.38×10^{-4}. How many grams of lactic acid must be added to 2.0 g of NaOH to make 500 ml of a buffer, the pH of which is 4.00? If the specific gravity of lactic acid is 1.44, how many milliliters of lactic acid are required? What is the molarity of this buffer?
3. You need to make a buffer solution, pH = 5.7 and concentration = 0.075 M. You have a choice of three starting materials: (1) citric acid, (2) succinic acid, or (3) acetic acid (sp. gr. = 1.049).
   a. With which of these materials would you start, and how much of the selected acid would you weigh out in order to make 1 L of buffer?
   b. With what would you adjust the pH, and how much would you weigh out?
4. A polyprotonic buffer (0.06 M) was carefully made up and standardized against a glass electrode. The observed pH was noted. A sample of the buffer was then diluted 10 fold with distilled water and the pH was again measured. The second value was distinctly larger than the first. Explain this observation, and state if this is a property of all buffers to the same degree.
   Suppose you took a third sample of the same buffer and added to it NaCl (solid) until the Cl^- concentration = 0.5 M. What would the salt addition do to the observed pH? Relate this last observation to the general problem of protein isolation in biochemistry.
5. A native polypeptide, such as insulin, can be titrated with acid and base so as to measure the ability of the protein to bind or release H^+ over the pH range 2–10. What is the significance of a statement such as "insulin showed an acid-binding capacity of approximately 100 mol per 10^5 g of protein"? What is the significance of the further observation that the insulin precipitated in the pH interval, 4.1–6.7?
6. You are to set up an enzyme assay system in which the buffer is imidazole (pK = 6.95). For technical reasons, the pH of the system must remain constant to ± 0.1 pH units and there must not be less than 5 mmol/L of the deprotonated (salt) form of the

buffer at any time. You also know that during the assay, 50 μmol of inorganic phosphate is set free in the 1-ml cuvet. What is the *minimum* buffer concentration, expressed in mmol/L, with which you could begin the assay?

7. A 10-mg sample of a certain monocarboxylic acid was titrated with 0.01 M NaOH; the titration required 5.01 ml to reach the phenolphthalein end point. Give the gram molecular weight of the unknown acid (to the nearest 10 g).
8. 15.5 ml of a 0.15 M citric acid solution was mixed with 25.0 ml of 0.12 M NaOH and then diluted to 150 ml. The final pH was 4.38. Calculate the pK_{a2} of citric acid.
9. In a certain enzyme system (total volume 2.5 ml), conversion of the substrate to product is expected to generate a maximum of 0.14 meq of H^+. A pH of 8.6 must be maintained within 0.2 pH units. From the list supplied to you, select an appropriate buffer and calculate the minimum buffer concentration that must be used.
10. Phenolphthalein (P) is colorless in acid solution and pink to deep red in basic solutions. Its pK = 6.1, λ_{max} of the ionized form is 545 nm, and $\epsilon_{545}^{mM} = 16.64$. P was added to a sample of a colorless solution of unknown pH. The A_{545} of the mixture was 0.047, and the final concentration of P was 8.49 μM. Estimate the pH of the solution.

SPECTROPHOTOMETRY

1. You are told that the $\epsilon_{260nm}^{1\%}$ of a certain compound X is 3.48. If the molecular weight of compound X is 1247, what is the millimolar extinction coefficient of compound X?
2. The following absorbance data apply to compounds X and Y:

	260 nm	280 nm
Molar absorbance, X	14,700	3,100
Molar absorbance, Y	200	9,300
Unknown solution X + Y	0.732	0.361

What are the concentrations of X and Y in the unknown solution?

3. Substances A and B have the following spectrophotometric properties:

$\lambda_{max}^{A} = 480$ nm, $\epsilon_{\mu M}^{A} = 2.2$, $\epsilon_{\mu M}^{B} = 0.15$
$\lambda_{max}^{B} = 520$ nm, $\epsilon_{\mu M}^{A} = 1.4$, $\epsilon_{\mu M}^{B} = 0.25$

A mixture of A and B contained 0.25 μM A plus an unknown concentration of B. A sample of this mixture gave readings of $A_{480} = 1.10$ and $A_{520} = 1.27$. What is the concentration of B in this mixture?

4. A solution of hypoxanthine in 1.2 N HCl has a λ_{max} = 248 nm, where $\epsilon_M = 11.45 \times 10^3$. What is the highest and what is the lowest concentration of such a solution that could be measured by spectrophotometry using normally available instrumentation?
5. A plot of absorbance vs. [X] for a certain analysis did not pass through the origin. At [X] = 0, A was 0.32. What is the significance of this observation? Suppose, instead, that at $A = 0$, the line passed through a point representing [X] = 0.63 mg/ml. What would be the significance of this observation?
6. A plot of A vs. [X] for a certain analysis passed through the origin and remained linear up to a certain value of [X], above which it

curved to the right. Briefly explain possible causes for the observed curvature of the line.

7. For each of the commonly used analyses for total protein, review the structural characteristic of proteins upon which the analysis is based. Which protein assay would be recommended for determination of the concentration of (a) gelatin, (b) RNase A, and (c) a copolymer of tryptophan and glutamic acid?
8. What might you use to check the wavelength accuracy of a spectrophotometer? How might you check the accuracy of the absorbance scale?

RADIOCHEMISTRY AND SCINTILLATION COUNTING

1. Name three methods for determining efficiency of a scintillation counting system; briefly indicate the basis of each method.
2. Why does one usually add more than one fluor to a scintillation cocktail? How does addition of several fluors affect the problem of chemiluminescence?
3. In the statistics of counting (or of anything else), why does one ideally consider variance rather than standard deviation?
4. In a series of measurements, the dpm due to a ^{14}C-labeled metabolite gave values of $\bar{x}$ and s of 145,000 and 2,300, respectively. If you wish to report a value for μ, with 95% confidence limits, what would you write down? (**Note:** See Chapter 12, section on Calculation of the Mean, The Variance, and The Standard Deviation.)
5. Polynucleotide kinase from *E. coli* transfers a γ-P_i from ATP to a 5′-OH terminus of DNA (one transfer for each DNA strand). DNA normally contains 5′-phosphoryl termini, but these can be converted to -OH groups by pretreatment with alkaline phosphatase.

   You are given 4.7 μg of homogeneous (all strands the same length) DNA which has already been treated with phosphatase. This was then treated with [γ-^{32}P]ATP at a specific activity of 3 mCi/μmol in the presence of polynucleotide kinase. The DNA was then precipitated with acid, washed, and counted with an efficiency of 85%. The observed count was 1870 cpm. What was the molecular weight of the DNA? (**Note:** Assume that all molecules of DNA had an identical structure.)
6. For how many minutes would a sample with an *approximate* mean counting rate of 1600 cpm have to be counted to obtain a counting rate accurate to 1% (relative error)?
7. Two students independently collected $^{14}CO_2$ from experimental metabolic systems. They each counted triplicate samples with the results shown below. (The values given represent cpm for the samples.)

Student A	Student B
37,500	39,000
38,200	36,400
37,000	37,000

   Can you say, with 95% level of confidence, that there is no significant difference in the means of the results of student A and student B? Why?
8. Assume that bis(3,5-dibromosalicyl)fumarate (bDBSF) is uniformly labeled with ^{14}C in C-2 and C-3 of the fumarate moiety *and* in the carboxyl carbons of the

dibromosalicyl moieties. The specific activity of the reagent is 100 mCi/mmol. After the reaction with hemoglobin (Expt. 8), what would be the specific activities of the major protein bands observed in the isoelectric focusing gels? Identify the bands and express the specific activities as μCi/g. (MWs: Hb = 65,000 and bDBSF = 670).

9. In order to set up a quantitative immunoassay, it is necessary to label a certain protein (MW = 250×10^3) with ^{125}I. Iodination is presumed to introduce one iodine atom at one position ortho to the tyrosine phenolic groups. There is only one tyrosine per mole of protein. The K^{125} I used is carrier free (all of its iodide atoms are radioactive); its $t_{1/2} = 5.18 \times 10^6$ sec. What is the theoretical specific activity of the iodinated protein expressed as μCi/ug?
10. Three common isotopes important in biochemistry are ^{3}H, ^{14}C, and ^{32}P. They are all β^- emitters with maximum emission energies of 0.0186, 0.156, and 1.71 MeV, respectively. In calibrating a liquid scintillation spectrometer for use with the three isotopes taken in any mixture of two at a time, for which isotope would the gain setting be highest and for which lowest?
11. The decay constant for ^{32}P is 0.0488. From this, estimate the half-life of the isotope.
12. A sample of an algal culture was grown in a synthetic medium containing $^{35}SO_4^{2-}$ and $H^{32}PO_4^{2-}$ as the sole sources of S and P, respectively. The specific activity of the sulfate was 2.9×10^6 dpm/μmol, and of the phosphate 5.2×10^6 dpm/μmol. After several hours, the cells were harvested and washed, and a cell-free, deproteinized extract was prepared. When this was analyzed by paper chromatography, a spot containing ^{35}S and ^{32}P was identified. Material was eluted and examined in a dual-channel liquid scintillation spectrometer. Standards were simultaneously counted. The data given in the table below are corrected for background. Considering only the eluted material, calculate: (a) the corrected radioactivity of ^{35}S (in dpm); (b) the corrected radioactivity of ^{32}P (in dpm); (c) the amount of ^{35}S in the sample (μmoles); (d) the amount of ^{32}P in the sample; (μmoles); (e) the ratio of P/S in the unknown compound.

Added	Observed Counts	
	Channel 1	**Channel 2**
^{32}P standard (58,000 dpm)	39,000 cpm	14,000 cpm
^{35}S standard (110,000 dpm)	18,000 cpm	56,000 cpm
Eluted sample	74,600 cpm	38,500 cpm

CENTRIFUGATION

1. Define the term *sedimentation coefficient.* On what factors does it depend?
2. Cesium chloride is frequently employed to adjust the density of media in differential centrifugation studies. Why is CsCl used in place

of the much less expensive material, KCl?

3. In the preparation of subcellular fractions, directions commonly call for preparation of a 10% homogenate of the tissue, followed by one or more steps involving centrifugal speeds of 800–8000 rpm. Why should one not use a 25% homogenate, especially when tissue samples are small?
4. Ficoll® is a synthetic polymer frequently employed in place of sucrose in preparation of density gradients. Ficoll® is quite expensive. What advantages, relative to sucrose, makes researchers invest so much in this synthetic polymer?
5. State the formula by which one converts centrifugal speed to multiples of the gravitational force. Define the terms of your equation.
6. The ultracentrifuge was, until recently, a very important tool in the determination of molecular weights of macromolecules. It has largely been replaced in recent years by a different procedure. What is the replacement tool, why is it now so popular, and how does it compare in accuracy with the ultracentrifugal method it supplanted?
7. You are given a mixture of substances A and B; their respective sedimentation constants are 16S and 40S. They are to be separated in a swinging-bucket rotor having inner and outer radii of 3.8 and 8.9 cm. How long must the centrifuge run at 50,000 rpm to obtain the best separation of A from B? (Ignore acceleration and deceleration times.)

CHROMATOGRAPHY

1. Briefly (~150 words) outline the mechanism by which molecular species are separated in gel permeation (gel exclusion) chromatography.
2. You are given a mixture of four different solutes, A, B, C, and D, with molecular weights of 250, 2,500, 25,000, and 250,000, respectively. Using only chromatographic means, how would you *most efficiently* separate these solutes?
3. List *at least five* distinct factors that might degrade the performance of a gel exclusion column. Indicate how each might be minimized by correct techniques.
4. A fraction emerging as a very sharp peak from a Sephadex G-25 column was further examined with a DEAE–Sephadex column with a NaCl gradient. Two peaks were recovered from the ion-exchange medium. What possible explanation(s) can you construct to account for these phenomena?
5. Biotin carboxylase was passed through a Sephadex G-100 column (fractionation range, 4,000–150,000). It eluted in the 25th of a series of 5-ml fractions. A second sample was treated with the detergent, sodium dodecyl sulfate (SDS), then passed through the column in an SDS-containing medium. A single eluting protein peak was observed in a much later fraction. In a few sentences, explain these observations.
6. In a molecular exclusion chromatography experiment, a gel column with a diameter of 2.5 cm and a bed height of 20 cm was used. The distribution coefficient, K_d, for the material being studied was 0.65, and the void volume was 30 ml. The flow rate was set at 0.8 ml/min, and 2-ml fractions were collected. In which fraction(s) would the eluted material appear?

7. What effects would the following have on zone spreading in exclusion chromatography?
   a. Solvent velocity
   b. Column cross section (diameter)
   c. Column length
8. Why is the resolution of TLC frequently higher than that of separations performed in columns?
9. You are given a mixture known to contain at least five separate components. Try as you may, you are unable to resolve these into discrete fractions by any single chromatographic column at your disposal. After some careful thought, you apply the mixture to a single TLC plate and process it. Results indicate that all five components have been resolved. How was this accomplished?
10. Discuss the general problem of analyte detection in TLC. How does the problem differ from detection in column chromatography?
11. The particle size of a TLC medium is reported to be 40–50 μm. Suppose that the size had been reduced by grinding to 10–15 μm. What effect might this have on the separations, and why?
12. Is it possible to distinguish between cis and trans unsaturated fatty acid methyl esters by any form of TLC other than that used in Experiment 5?
13. Is there another way in which the separated spots in Experiment 5 might have been visualized (without going to the cost of radioactively tagged materials)?
14. What would be the effect of increasing cross-linking of a polystyrene ion-exchange resin, bearing weakly acidic groups ($-CH_2COO^-H^+$), on the recovery of amino acids from a protein hydrolysate? How would increased cross-linking affect separation of macromolecules on the same resin?
15. Several eponymic abbreviations have been applied to describe modified cellulosic ion-exchangers. For each of the eponyms listed, write the names of the modifying groups attached to the cellulose matrix: TEAE, CM, ECTEOLA. What would be their structures at pH 2, pH 7, and pH 12?
16. What effect, if any, would the following have on the efficiency of modified cellulose ion-exchangers?
   a. pH of the medium
   b. Operation of the column at 4°C
   c. Length–diameter ratio of the column
   d. Application of pressure to the top of the column
   e. Buffer concentration
   f. Addition of detergents to the medium
   g. Ionic strength of the medium
17. In discussion of gradient elution techniques with modified cellulose ion-exchangers, a useful formula applicable to *linear gradients* is:

$$c = \frac{(C_a - C_b)v}{V} + C_b$$

where:

c = concentration passing through the column at any instant

V = total volume of the gradient

v = volume of gradient solution actually collected

C_a = concentration of the unstirred solution

C_b = concentration of the stirred solution

State at least one unwritten assumption regarding the use of the above formula. For the highest resolution in separating a given mixture, how should the volume, *V*, compare with the volume, *v*? For a linear gradient, how should the volume of solution *a* compare with the volume of solution *b* in the gradient mixer?

18. Iodoacetate carboxymethylates the -SH groups of hemoglobin. Relative to HbA, how would $HbA(CH_2COOH)_n$ behave on (a) a G-25 Sephadex column? (b) an isoelectric focusing gel? (c) a DEAE-cellulose column?
19. Triethyloxonium fluoroborate $[C_2H_5)_3OBF_4^-]$ reacts with the carboxyl group of several proteins to yield the ethyl ester. Relative to the ummodified protein, how would the esterified protein behave on a column containing (all at pH 7.0)
    a. Sephadex G-25
    b. DEAE–cellulose
    c. An isolectric focusing gel
    d. Carboxymethyl (CM)-cellulose
20. Why is it preferable to use the salt of a strong acid and strong base to form an ionic gradient, rather than the salt of a weak acid or base? How does a salt gradient work to effect sharper separations in ion-exchange chromatography?
21. The application of a salt gradient to an ion-exchange column may also generate a shallow pH gradient. Explain.
22. Briefly (~150 words) explain how affinity chromatography differs in principle from ion-exchange chromatography.
23. In practical terms, what factors define the capacity, resolving power, and specificity of an affinity column?
24. Outline at least two different methods for affixing a nonprotein ligand to an inert support to make an affinity column packing.
25. The acronym, HETP, stands for "height equivalent to a theoretical plate." It is a concept derived from the theory of the distillation process. How does this concept relate to chromatography as a separation science?
26. Compare the relative advantages/disadvantages of GLC and HPLC with respect to the following:
    a. Speed
    b. Resolution
    c. Detection modes available
    e. Instrumental costs
    f. Solute recovery
    g. Separation of cis/trans isomers
27. A series of fatty acids were separated by GLC. Retention times were noted to be 2.3, 3.5, and 4.1 min. Assuming the MW of the first eluted peak to be 256, how specifically can you define the MWs for the later-eluting compounds? What precautions must you keep in mind in making these predictions? What pretreatment would you have applied to this sample?

ELECTROPHORESIS

1. Write the generalized equation for the movement of charged particles in an electric field and in an ionic medium. Define each of the terms in the equation.
2. You have separated a new enzyme from a biological system. After extensive purification, you still note the presence of two bands on a Coomassie-Blue, stained gel. You assume the minor band is a degradation product; you are not surprised to find no enzymic activity when the minor band is cut out of the gel, extracted and assayed. You are

surprised when the major band was similarly recovered and found inactive. Give a reasonable explanation and describe how you could test your hypothesis.

3. Many different media have been developed for use as electrophoretic supports, including strips of cellulose or cellulose acetate and gels of agarose, polyacrylamide, or starch. Comment on the general advantages or disadvantages of each of these media.
4. Other than the use of Coomassie Blue and similar dyestuffs, how many ways can you describe for detection of materials separated by electrophoresis?
5. In a laboratory exercise, several students made up their own batches of polyacrylamide gel. These were used to examine a single sample of a cellular extract provided for the class. Each student believed (s)he had followed directions exactly for gel preparation. However, student A got a very good separation of proteins from the extract; student B got a vast smear, with very little in the way of resolved bands; and student C got one very dark band located near one end of the gel. What explanations are possible?
6. By what means, other than SDS, can one cause subunit disaggregation of polymeric proteins? Can these means be employed in electrophoresis?
7. In a few sentences, explain why the resolution of polyacrylamide gel electrophoresis is higher than resolution in similar experiments performed with starch gels or on cellulose or cellulose acetate media.
8. Briefly describe two methods for determining the pH gradient along the length of an isoelectric focusing gel. Compare their advantages and disadvantages.
9. How could you detect protein in any of the gel tubes without any staining at all? Why is this method not in more general use?
10. Why are the native, but *not* the SDS, gels preelectrophoresed?
11. How can a tube gel electrophoresis system be modified for preparative use?
12. What special precautions would have to be taken to scale up isoelectric focusing gel systems for preparative use?
13. Briefly describe three methods for extracting a protein band from a gel.
14. In an experiment to determine the extent of phosphorylation of a receptor protein under varying conditions, what procedure would you suggest, and what information would you need, to evaluate the net charge on this protein.
15. State the principles that explain the observation that "disc" gel electrophoresis is capable of higher resolution than "zone" gel electrophoresis.

IMMUNOCHEMISTRY

1. Briefly describe the special attributes of immunoglobulins which qualify them for use in biochemical assays. What are their disadvantages?
2. Discuss the use of adjuvants in the production of antibodies as reagents; indicate what sorts of materials are contained in adjuvants and why they are added.
3. What particular differences exist between gels used for electrophoresis and immunodiffusion.
4. Does the absence of a precipitin band or arc in an immunodiffusion experiment always indicate

absence of an antigen? Explain your answer.

5. Briefly outline the principles of any radioimmunoassay; compare it to a similar nonradioactive immunoassay from the standpoint of detecting the end point or other result.
6. Define the following terms: hapten, epitope, antigenic determinants, polyvalent antibody.
7. Differentiate clearly between single-antibody and double-antibody immunoassays. Which of these two types might, intrinsically, be more sensitive?
8. State an unwritten assumption which should lead one to be quite critical of immunoassay results in general.
9. What is the generalized structure of an antibody?
10. Suppose that in Experiment 11 (Immunodiffusion section), 1 M NaCl had accidentally been added to the PBS. How would this have altered the experimental results?
11. What advantages might immunoelectrophoresis have over immunodiffusion?
12. What is the generalized structure of a precipitin? Illustrate with a sample sketch.
13. How does it happen that certain antibodies can react with more than one antigen?
14. Discuss the advantages and disadvantages of the combined use of immunochemistry and affinity chromatography.

LIGAND–PROTEIN BINDING (EXPTS. 4, 15, 16, AND 17)

1. What is the meaning of the terms *latent heat of fusion* and *latent heat of vaporization?* What are the values of these constants for water?
2. If polyvalent ions are better precipitants than monovalent ions, why not use $MgSO_4$, instead of $(NH_4)_2SO_4$, to fractionate proteins?
3. Compare the CBG binding system with the avidin–biotin binding system. What features do the two systems have in common? In what significant ways do the two systems differ?
4. Many binding systems show a considerable temperature dependence; that is, the binding affinities change quite markedly over a range of 10°C. Construct a plausible explanation for this observation.
5. How could you demonstrate that progesterone has an affinity for XAD-2 resin as well as for CBG?
6. In the disc-binding assay, a 50-μl aliquot was applied to each disc. Suppose you had applied a 200-μl aliquot, what might have been the consequences?
7. CBG is known to be a glycoprotein. Are any other glycoproteins to be found in human blood plasma or serum? Name five.
8. CBG contains sialic acid residues as part of the carbohydrate moiety covalently attached to the peptide structure. Sialic acid contributes to the relatively low pI value of CBG. Suggest a different component of carbohydrate moieties which might give to a protein a relatively higher pI value.
9. If [^{3}H]cortisol were not available, how else could one proceed to make measurements similar to those made here? From your reading and/or imagination, outline the details of such a radioisotope-free experimental system.
10. Why was it necessary to keep the wash buffer ice-cold to ensure a proper DEAE-disc assay?

11. What problems could have been introduced if the DEAE discs had been handled with bare hands?
12. If the equilibrium association constant for the binding of cortisol to serum albumin is given as about 10^5 or less, could the disc assay be used with such a system? Suggest an alternative method for measuring the binding parameters of such an albumin–cortisol system.
13. The molecular extinction coefficient for cortisol is 16,000 at 241 nm. The specific radioactivity of the [^{3}H]cortisol with which you worked (Expts. 16 and 17) is given in the laboratory notes. Taking into account the errors predictable in spectrophotometry and in liquid scintillation counting, how would you compare these methods as to sensitivity and dependability in determining the concentration of cortisol in a test solution?
14. Human blood contains another steroid binding protein, the sex steroid binding protein (hSBP), which binds dihydrotestosterone with an equilibrium constant of 10^9. How could this have affected your data (Expts. 15 and 17)? Suggest an assay for determining the concentration of hSBP.
15. Since CBG and albumin both bind cortisol to varying extents (see above and your laboratory notes) and since both proteins are present in human blood, what would be the distribution of bound cortisol between these two proteins in normal blood? Refer to your notes for total cortisol concentration in blood.
16. In Experiment 4, what would a plot of A_{500} vs. biotin added look like if the dissociation constant (K_d) for biotin were 5.8×10^6 and for HABA 1×10^{-15}?
17. Qualitatively, how would a plot of A_{500} vs, μg biotin added differ from the plots actually generated if the biotin additions had been made using 50-μl aliquots of a 0.040 mM biotin solution?
18. How would you determine if all four biotin binding sites on avidin are equivalent? What might your data look like if they were not?
19. Biotin also binds to streptavidin (SA) with the same affinity as to avidin (A). The isoelectric pH for SA is 7.4 and for A is 9.5. Under the incubation conditions of this experiment,
    a. How would the net electrical charge on SA compare to that of A?
    b. Why might there be more nonspecific binding of various nonbiotin molecules to A as compared to SA?
20. What is the total concentration, in moles per liter, of a protein, P, that possesses one binding site per molecule for a ligand, L, if analysis by equilibrium dialysis of a solution containing P + L gives the following data:

$$[L_{free}] = 10^{-5}\ M$$
$$[L_{bound}] = 5 \times 10^{-6}\ M$$
$$K_{assoc} = 10^5\ M^{-1}\ (S/mol)$$

21. What are the similarities and differences between the protonation of a protein and the binding of a small neutral molecule to a protein with respect to calculation of binding parameters?
22. A purified protein, SBP (sex steroid binding protein), was incubated with a ligand, testosterone, in an equilibrium dialysis system at 4°C until equilibrium was attained. The concentration of testosterone was varied in a series of incubations and the following data were accumulated:

$\bar{\nu} = \frac{\text{ng Testosterone/L}}{\text{ng SBP/L}} \times 10^3$	Free Testosterone Concentrations (ng/L)
0.20	14.30
0.75	57.70
1.30	130.0
1.75	219.0
2.50	500.0
3.25	162.5

Use an appropriate plot of the data to calculate a value for n and for K_a (the association constant) for the steroid–protein interaction (MWs: SBP = 74,000; testosterone = 288.4). Use proper units.

ENZYME CHEMISTRY (EXPTS. 9, 10, 12, 13, 14, AND 18)

1. What is the function of the phenylmethylsulfonyl fluoride added to the chicken muscle tissue during the homogenization? Name at least three other compounds that could perform the same general function. How do they differ in effectiveness?
2. Suppose you had a sample of Reactive Blue in free solution. What might have happened had you added some of this solution to any one of your assay tubes in Experiment 9?
3. Suggest two additional methods for following the course of the citrate synthase reaction, one a chemical elapsed time assay and the other a procedure for continuous real-time spectrophotometric monitoring of the reaction. Answer in two or three sentences, but include chemical equations.
4. Name two inorganic phosphatase inhibitors. How do they act?
5. Describe a method, other than the colorimetric method, which may be used to monitor phosphatase activity. What starting materials would be necessary to prepare?
6. What is the kinetic mechanism for the LDH reaction? How would the K_m (NADH) be determined?
7. In the enzyme substrates used in the acylase reaction (Expt. 14), the ^{14}C was randomly incorporated into the alanine carbons, not in the acetyl carbons. Could this situation have been reversed, that is, so that the substrates were synthesized from [U-^{14}C]acetic anhydride and unlabeled D- or L-alanine? Explain any advantages or disadvantages.
8. In *S. cerevisiae* the phosphorylation of ADP to ATP is linked to the formation of a "high-energy" intermediate. Name this intermediate and draw its structure.
9. Every cook (and certain others) understands that bakers' yeast will ferment sucrose as well as glucose. How do you explain this phenomenon?
10. Makers of Burp's Beer proudly tell us that their beverage is "naturally carbonated." How might this assist them in keeping the alcohol content of their product within legal limits?
11. Our friends at Burp's also tell us that they use only the finest

"mountain-grown" barley, which they allow to sprout. The sprouted barley is then dried, after which it is known as malt. The malted barley, along with other grain, is then fermented with brewers' yeast. In biochemical terms, state why the barley is made to sprout, and what the malting process has to do with expediting and promoting the fermentation process.

12. True champagne is a carbonated wine with an alcohol content that approaches 14% by volume. On the basis of your own experimental observations, how is it possible to reach such a high alcohol concentration in the presence of a simultaneously high concentration of CO_2?
13. Other than alcohol, name at least two kinds of foodstuffs the preparation of which involves a fermentation process.
14. Other than alcohol, name at least two nonfoodstuffs that are produced on an industrial scale by a fermentation process.
15. Name one feature common to *all* fermentation processes.
16. During the initial linear stage of an NAD-linked dehydrogenase assay, the ΔA_{340} was 0.12/min. The cuvet contained 1.25 mg of protein in a reaction volume of 2.5 ml. What was the specific activity of the enzyme, in international units?

MOLECULAR BIOLOGY (EXPTS. 20, 21, AND 22)

1. Explain, briefly, why the kinetics of RIF-inhibition differs from that for CAP or for glucose in β-galactosidase induction.
2. If, instead of measuring β-galactosidase activity, you had measured [^{3}H]leucine or [^{3}H]uridine incorporation, what results might have been obtained? What disadvantages might have been encountered, had the end point of this experiment been based on some such incorporation study rather than on an enzyme assay?
3. Is glucose an inhibitor of RNA synthesis or of protein synthesis, or neither? Suggest some experiments that would test your answer or that would clarify the role played by glucose in β-galactosidase induction.
4. How could the sensitivity of the β-galactosidase assay be improved?
5. Streptomycin might have been used to inhibit β-galactosidase induction. How is it presumed that streptomycin acts to inhibit protein biosynthesis?
6. a. How does induction of the tryptophan (Trp) operon differ from that of the lactose operon?
   b. *Outline* an experimental protocol to demonstrate Trp operon induction.
7. How could Experiment 20 be expanded to test more rigorously the level of induction as transcriptional or translational?
8. Name an example of steroid hormonal induction of a protein. Outline a current view concerning a mechanism of action. Could the level of inhibition be tested with a protocol similar to that for lactose induction? Outline such a procedure.
9. Other than induction, name three other physiologic mechanisms for the regulation of enzyme activity.
10. Unlike mammalian erythrocytes, avian erythrocytes are nucleated and they also contain mitochondria. Would avian erythrocytes be a suitable source for preparation of in vitro translation systems? Would they be a useful

source for isolation of an mRNA? Explain your opinions in a sentence or two.

11. Specifically, what are the advantages of [^{35}S]methionine, compared to [^{3}H]leucine, as a label for in vitro translation systems? Would added Na^+ affect each substrate equally?
12. Reference to the ancillary information concerning reaction cocktails (Expt. 21) shows that some contain hemin and others contain spermidine. What is the chemical nature of hemin and of spermidine? In a few sentences, state a mechanism by which each is assumed to enhance the activity of in vitro translation systems.
13. According to Abelson [*Annu. Rev. Biochem. 48:*1050 (1979)], the mRNA for ovalbumin has a 76-base leader sequence, an 1149-base coding sequence, and an untranslated, 634-base sequence at the 3′-end of the coding sequence followed by a polyA tail. What functions, if any, can be ascribed to (a) the leader sequence and (b) the untranslated, postcoding sequence of this message?

Elementary Enzyme Kinetics*

Before embarking on Experiment 9 (as well as subsequent enzyme-associated experiments), you may find a brief review of enzyme kinetics to be helpful. It is important to keep in mind the basic assumptions upon which rest the validity of the Michaelis–Menten equation as well as the derivative graphic plots.

ASSUMPTIONS OF THE ANALYSIS

1. We identify a single substrate, *S;* a single product, *P*; and an enzyme, *E,* which catalyzes conversion of *S* to *P.* We can write an equation which states that:

$$[E] + [S] \underset{k_2}{\overset{k_1}{\rightleftharpoons}} [ES] \underset{k_4}{\overset{k_3}{\rightleftharpoons}} [E] + [P] \qquad \text{(A2-1)}$$

   where [ES] is the enzyme–substrate complex.
2. The reaction is in a *steady state;* that is, [ES] is constant and [E] is in equilibrium with [ES].
3. [P]<<<[S], and [S] is substantially constant. This assumption can apply *only* under initial reaction conditions. The reaction velocity under these conditions is defined as the initial velocity, v_0.
4. As is required of a true catalyst, [E]<<<[S].
5. For any system that fulfills the above assumptions, there is a velocity, V_{max}, which is observed when [ES] = $[E_T]$; that is, when all of the enzyme is in the form of [ES]. At any other observed velocity, we can be sure that $v < V_{max}$ and that [ES] = $[E_T] - [E_{free}]$.

BASIS OF THE MICHAELIS–MENTEN EQUATION

From the law of mass action, we can write that

$$k_1[E][S] + k_4[E][P] \rightleftharpoons k_2[ES] + k_3[ES] \qquad \text{(A2-2)}$$

At *initial reaction conditions* (see assumption 3, above) we can write that

$$k_1[E][S] \rightleftharpoons (k_2 + k_3)[ES]$$

If we define the Michaelis–Menten constant, K_m, as $(k_2 + k_3)/k_1$, then

$$\frac{[E][S]}{[ES]} = \frac{k_2 + k_3}{k_1} = K_m \qquad \text{(A2-3)}$$

*Consult this appendix before performing Experiments 9, 10, 11, 12, and 13.

From assumption 5, above, we can write that

$$V_{max} = k_3[E_T] \qquad \text{(A2–4)}$$

Otherwise,

$$v = k_3[ES] \qquad \text{(A2–5)}$$

Since $[E_T] = [E] + [ES]$, it follows that

$$[E] = [E_T] - [ES] \qquad \text{(A2–6)}$$

Substituting values from Eqs. A2–4 and A2–5 into Eq. A2–6, we arrive at

$$[E] = \frac{V_{max}}{k_3} - \frac{v}{k_3} = \frac{(V_{max} - v)}{k_3} \qquad \text{(A2–7)}$$

The value of [E] derived from Eq. A2–7 can be substituted into Eq. A2–3, to give

$$\frac{[S]}{k_3\,[ES]\,(V_{max} - v)} = K_m \qquad \text{(A2–8)}$$

Since $v = k_3[ES]$, we can rewrite Eq. A2–8 in the form

$$\frac{[S]}{v}(V_{max} - v) = K_m \qquad \text{(A2–9)}$$

Equation A2–9 is more commonly written, after it has been solved for v, as:

$$\frac{[S]V_{max}}{[S] + K_m} = v \qquad \text{(A2–10)}$$

If we now look at the specific case where $v = V_{max}/2$, we find that

$$\frac{V_{max}}{2} = \frac{[S]V_{max}}{[S] + K_m} \qquad \text{(A2–11)}$$

so that

$$\frac{1}{2} = \frac{[S]}{[S] + K_m} \qquad \text{(A2–12)}$$

Finally, solution of Eq. A2–12 leads to

$$K_m = [S] \qquad \text{(A2–13)}$$

In other words, K_m is that value of [S] at which the reaction velocity is half-maximal, or $v = V_{max}/2$. Note that K_m is independent of [E].

GRAPHIC EVALUATION OF K_m AND V_{max}

Various means for evaluating these parameters of an enzyme have been developed, based on graphic analysis of measured reaction velocities and substrate concentrations. Several of these are displayed in Fig. A2–1. A variety of computer programs have been written to simplify the task of data reduction, although it is quite possible in simpler cases to estimate V_{max} and K_m "by hand."

PATTERNS OF INHIBITION

Competitive Inhibition

Imagine an inhibitor, *I*, which can compete with the substrate for a site on an enzyme, according to the *reversible* reaction:

$$[E] + [I] \rightleftharpoons [EI]$$

Because *I* is an inhibitor, the EI complex cannot undergo catalytic conversion. Formation of EI prevents some fraction of the enzyme from exerting its normal function; in other words, the *effective* concentration of enzyme is reduced. When *I* and *S* are present together, the observed reaction velocity is lower than when *S* is present

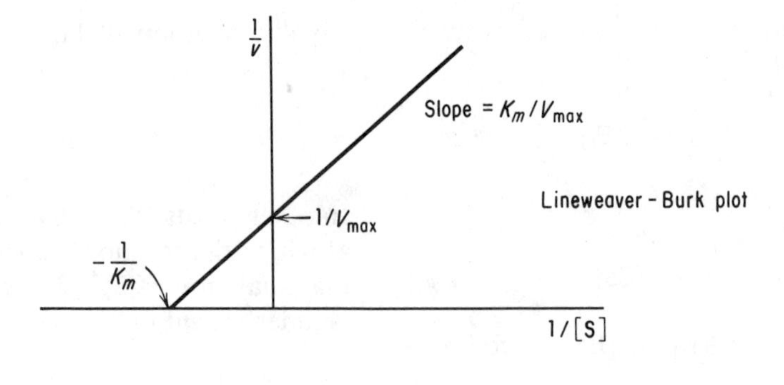

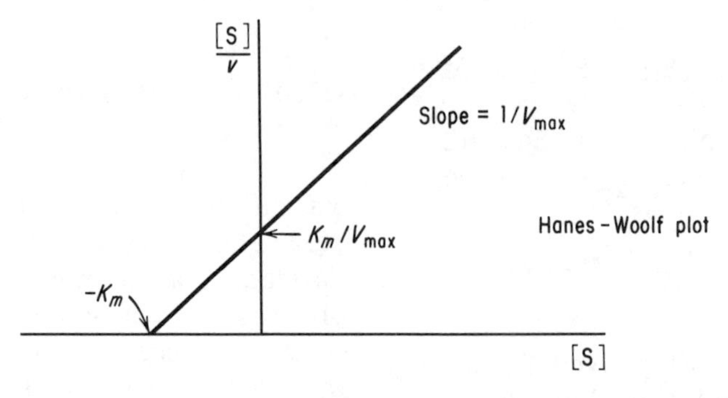

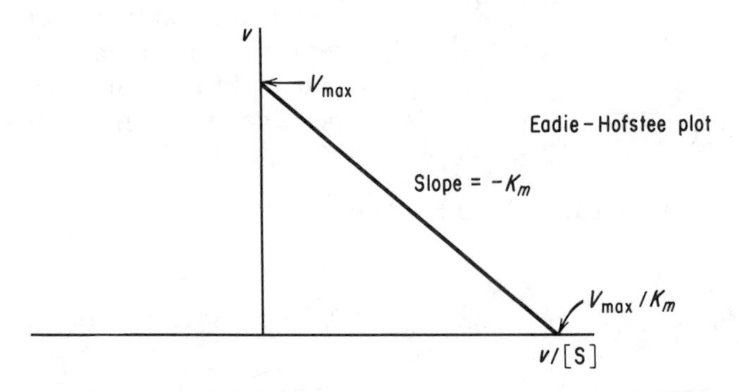

Figure A2–1. Three different kinetic plotting schemes.

alone. We can state that under initial reaction conditions, and for any value of [E], [S], and [I],

$$V = \frac{V_{\max}}{1 + \frac{K_m}{[\mathrm{S}]}\left(1 + \frac{[\mathrm{I}]}{K_I}\right)} \qquad \text{(A2–14)}$$

The inhibitor constant, K_I, is given by that concentration of the inhibitor that reduces v to $V_{max}/2$.

By taking reciprocals of both sides of Eq. A2–14, one obtains:

$$\frac{1}{v} = \frac{1}{V_{\max}} + \left[\frac{K_m}{V_{\max}[\mathrm{S}]}\left(1 + \frac{[\mathrm{I}]}{K_I}\right)\right] \qquad \text{(A2–15)}$$

A plot of $1/v$ versus $1/[S]$ again gives a straight line. As [I] is varied, one gets a family of straight lines, as is shown below. Certain useful expressions are also identified.

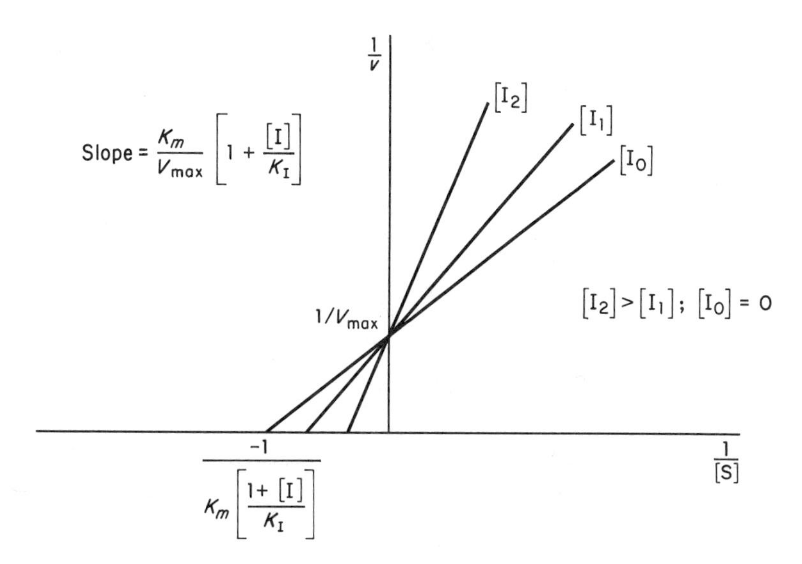

From such a family of curves it is possible to obtain a fairly good approximation of K_I by a secondary graphic solution, as follows: The slope of each line can be set equal to a new variable, y, so that

$$y = \frac{K_m}{V_{max}}\left[1 + \frac{[I]}{K_I}\right] \qquad \text{(A2-16)}$$

Equation A2–16 gives a straight line when y is plotted as a function of [I], as is shown below with useful terms identified.

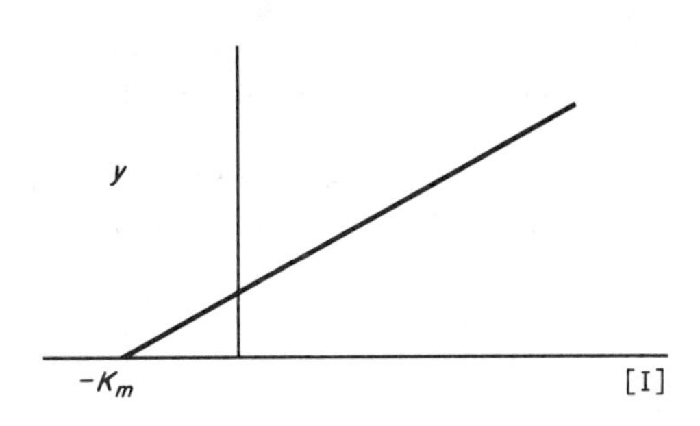

NONCOMPETITIVE INHIBITION

Imagine yet a different case, in which an inhibitor, I, reacts *either* with the free enzyme, E, *or* with the enzyme-substrate complex, ES. The two possible equilibria are

$$[E] + [I] \rightleftharpoons [EI], \text{ or } [ES] + [I] \rightleftharpoons [IES]$$

Imagine further that IES does *not* cleave to yield product; that is,

$$[IES] \not\rightleftharpoons [IE] + [P]$$

A general expression for v can be written as before,

$$v = \frac{V_{max}[S]}{K_m(1 + [I]/K_I) + [S](1 + [I]/K_I)} \qquad \text{(A2-17)}$$

Taking the reciprocals of both sides yields

$$\frac{1}{v} = \left[\frac{1}{[S]} \cdot \frac{K_m}{V_{max}}\left(1 + \frac{[I]}{K_I}\right)\right] + \frac{1}{V_{max}}\left(1 + \frac{[I]}{K_I}\right) \qquad \text{(A2-18)}$$

At [I] = 0, v = the uninhibited velocity. As [I] increases, a family of curves similar to those shown below is generated.

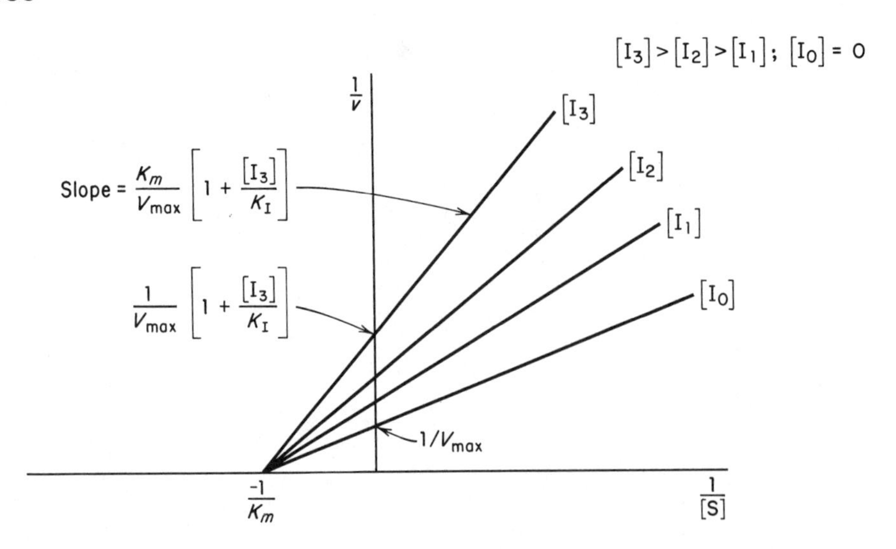

Other patterns of kinetic behavior—for example, sigmoidal patterns and uncompetitive patterns—are known and sometimes encountered. Even simple systems such as those described above sometimes give nonlinear results, frequently as a result of improper choice of the range of [S] or [I].

From a family of curves like the one shown above, it is possible to obtain a fairly good estimate of K_I by a secondary graphic solution. The slope of each line can be set equal to a new variable, y, so that

$$y = \frac{K_m}{V_{max}}\left(1 + \frac{[I]}{K_I}\right) \tag{A2–19}$$

When y is plotted as a function of [I], a single straight line is obtained, from which $-K_I$ may be estimated as the intercept on the [I] axis.

Although the above algebraic manipulations are based on use of the Lineweaver–Burk equation, it must be emphasized that this type of plot is not recommended. As pointed out in Chapter 12, the Lineweaver–Burk expression gives disproportionate statistical weight to small and large values of v. This problem is satisfactorily discussed by Cornish-Bowden (1979).

ASSAY OF ENZYME ACTIVITY: SPECIAL COMMENTS ON DETERMINATION OF INITIAL REACTION VELOCITIES

The reasons for running enzyme assays vary and may call for either semiquantitative or quantitative protocols. Semiquantitative methods are used simply to determine whether or not there is any catalytic activity present in a protein sample. For example during the purification of an enzyme by column chromatography, it may be sufficient to determine in which of the eluted protein fractions the enzyme activity resides; the procedure used should be rapid, sensitive, and convenient. For this purpose, a fixed-elapsed-time assay will suffice (see below). However, to compare accurately the activities of two different enzyme preparations, or to determine the values for K_m and V_{max} for an enzyme, it is preferable to monitor the course of the reaction either at several time points or, even better, continuously. The latter can usually be done by identifying an observable optical property of a substrate (or a product) and following its increase or decrease with time. Once the data are collected, they can be processed in

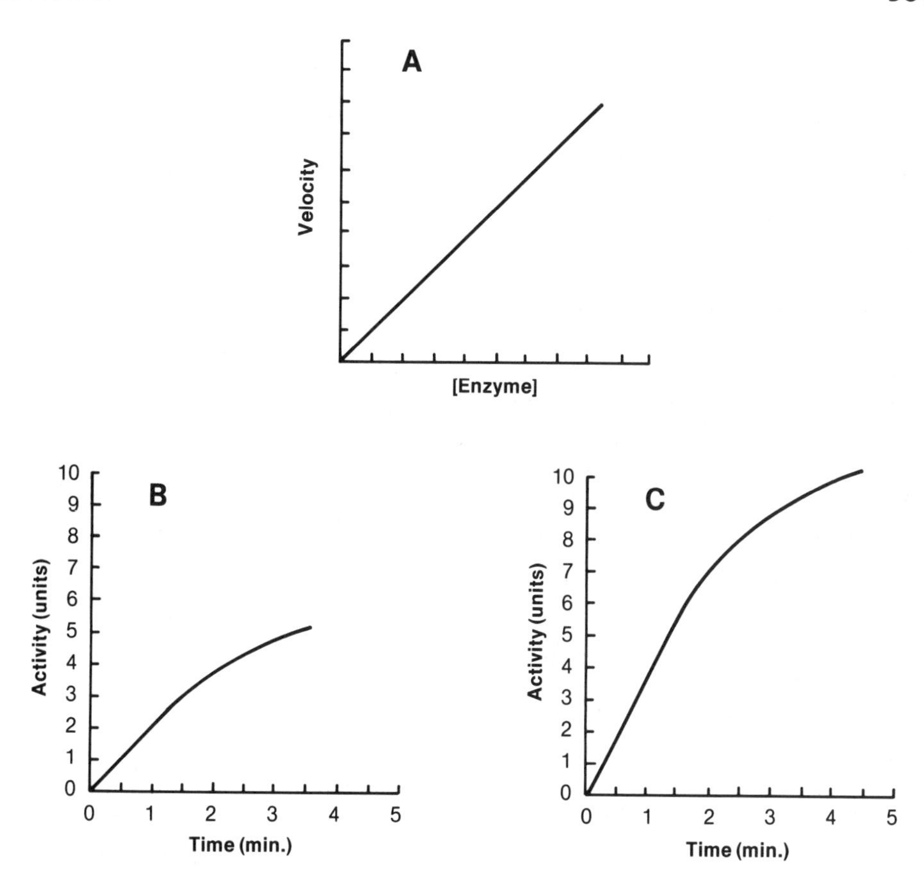

Figure A2–2. Examples of proper and improper plots of enzyme kinetic data.

several ways to obtain a value for the velocity; it is important to understand the conditions under which such data processing is valid.

As described in previous sections of this appendix, various forms of the Michaelis–Menten equation can be used to generate linear plots of the data obtained from an enzyme experiment—that is, the velocity, v, and substrate concentration, [S]. However, it is valid to use these equations only if the conditions specified in their derivation are fulfilled in the experimental protocol. In particular, it is necessary for the activity data to be collected under pseudo-zero-order conditions, that is, where $[S] \gg [E]$ so that $[E_T] = [ES]$. If this is not the case, then the calculated velocity will not be proportional to the total enzyme concentration—a necessary condition if the assay is to determine the enzyme concentration (Fig. A2–2A). In most properly devised experiments, this condition exists in the early phase of the experiment, that is, in the first minute or so. If one plot's activity vs. time (Fig. A2–2B,C), the curve will first have a linear phase (pseudo-zero order) which is followed by a second phase in which there is a falling off of activity as the substrate is utilized. Figure A2–2B represents data from a system in which the enzyme concentration is one-half that in the system represented in Fig. A2–2C. It can be seen in Fig. A2–2B that the velocity (i.e., activity/time), at 0.5 and 1.0 min is 2 units/min. Beyond that time the curve is no longer linear, and at 3 min the velocity is 1.6 units/min. In Fig. A2–2C, the calculated velocity should be twice that of Fig.

A2–2B; this is true at 0.5 and 1.0 min (v = 4 units/min) but beyond the linear portion of the curve at 3 min, the velocity is 3 units/min. If a 3-min elapsed-time assay had been used to determine enzyme concentrations in these two systems, then an erroneous conclusion would have been reached, that the enzyme concentration in system C was 1.8 times that in system B, instead of twice. If 0.5 or 1.0 min was the elapsed assay time, then the correct result of "twice" would have been obtained. The necessity for calculating velocities using data from the linear portion of an activity/time plot is therefore obvious; these velocities are the initial velocities or v_0. The concentrations of the substrates and enzyme in any assay system must be adjusted, therefore, so that the data collected, when plotted, will exhibit a linear phase.

REFERENCES

Cornish-Bowden, A. *Fundamentals of Enzyme Kinetics,* pp. 25–30. Butterworths, London (1979).

Marshall, A. G. *Biophysical Chemistry.* John Wiley & Sons, New York (1978).

Montgomery, R., and Swenson, C. A. *Quantitative Problems in the Biochemical Sciences,* 2nd ed. W. H. Freeman, San Francisco (1976).

Purich, D. L., ed. *Contemporary Enzyme Kinetics and Mechanism.* Academic Press, New York (1983).

Segal, I. H. *Biochemical Calculations,* 2nd ed. John Wiley & Sons, New York (1976).

Walsh, C. *Enzymatic Reaction Mechanisms.* W. H. Freeman, San Francisco (1979).

Zeffren, E., and Hall, P. L. *Study of Enzyme Mechanisms.* Wiley-Interscience, New York (1973).

Fractionation of Proteins with Ammonium Sulfate

Proteins may be differentially precipitated ("salted out") from aqueous solutions by additions of neutral salts. Although sodium chloride and sodium sulfate are sometimes used for this purpose, the most commonly employed salt is ammonium sulfate. At least three reasons dictate this choice. Polyvalent ions are better precipitants than monovalent ions; thus sulfates are more effective than chlorides. Ammonium sulfate is severalfold more soluble than sodium sulfate, allowing for greater salt concentrations in the protein-containing solution. Finally, the temperature dependence of the solubility of ammonium sulfate is small. Saturated ammonium sulfate solution is 4.1 M at 25°C and 3.9 M at 0°C. A minor disadvantage of ammonium sulfate is its effect on the pH of solutions to which it is added. The pH of a 0.1 M solution of ammonium sulfate is 5.5. Some methods call for neutralization of the salt solution, but others do not (one must pay careful attention to this point in each particular case).

Fractionation by ammonium sulfate (or sodium sulfate) depends largely on partial dehydration of dissolved proteins as bound water is replaced by salt ions. The salt may be added either as a solution (usually a saturated solution) or as a solid. A drawback to the use of such solutions is the large increase in volume of the sample, but this is offset by the relative ease of adding a liquid, compared to slow addition of a powder.

In either case, precipitation of certain proteins is quite temperature dependent. It is also very sensitive to localized excess concentration of the precipitant ions. For these reasons, the salt should be added *slowly,* with gentle but thorough stirring to be sure that local concentration gradients are avoided as much as possible. If a solution is added, it should be added dropwise from a separatory funnel or similar device. If solid salt is used, it must be finely powdered to promote rapid dissolution. When the stipulated quantity has been added, the final solution should be well stirred for some time to be sure that a true equilibrium has been reached. It is common to define the salt concentrations required to produce a given fractionation in terms of either final salt molarity or percent saturation of the solution.

Ammonium sulfate solutions are distinctly nonindeal: The addition of this salt to water is accompanied by a significant volume change. Thus, if one starts with 1 L of water, saturation requires addition of 767 g (=5.8 mol) of the salt, to give a final concentration of 4.1 mol/L. Obviously, when one adds smaller quantities of the salt, the error due to nonideal behavior is less. Most frequently, it is, in

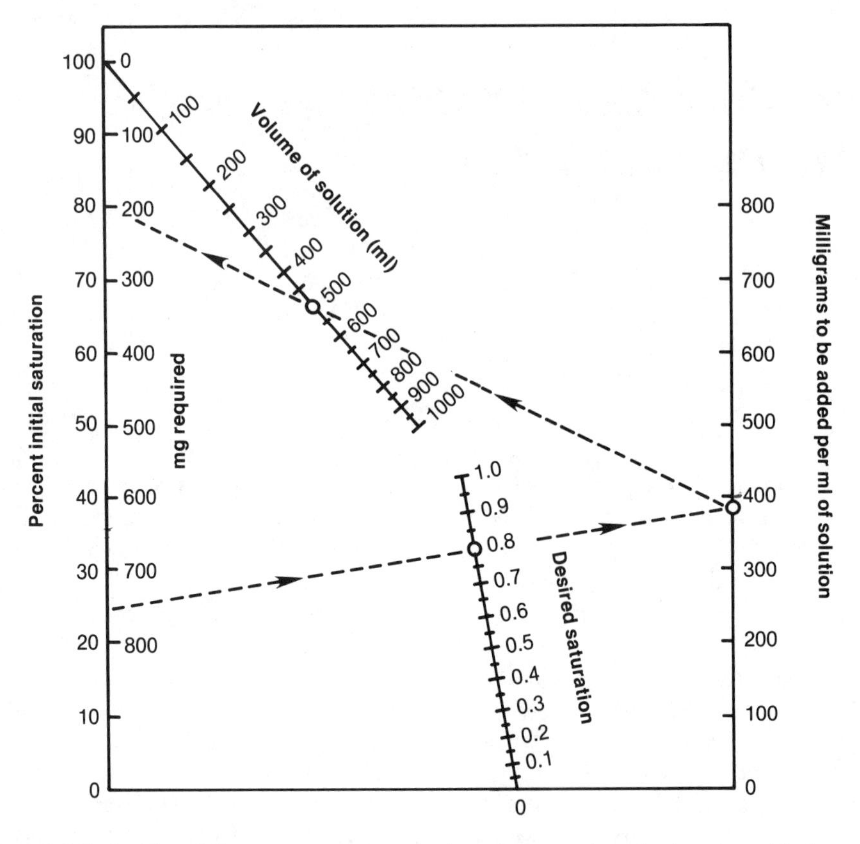

Figure A3–1. Use of the nomogram to calculate the quantity of salt to be added to make a solution of specified saturation. For full explanation, see text. (Adapted from M. Dixon, *Biochem. J.* 54:457, 1953, by permission of The Biochemical Society, London.)

fact, ignored, but one must be aware of the error thus incurred.

Certain other conventions that have become accepted must also be clearly understood. In describing a certain ammonium sulfate solution as being "at 30% saturation," one is implying that the solution contains 30% of 4.1 mol/L, *unless the method used specifies information to the contrary.*

Although there are valid objections to these ambiguities of usage, what matters is that a given method behaves reproducibly. The precision of separation is usually well within the boundaries of the errors mentioned.

Calculation of the quantity of salt to be added is aided by the use of tables or nomograms. A typical nomogram adapted from Dixon by Green and Hughes is given in Fig. A3–1. The left-hand axis gives the *initial percent saturation.* The solid diagonal line in the lower right indicates the *desired saturation.* If a straight-edge connecting these points is extended to the right-hand axis, the point of intersection indicates the *quantity of salt (mg/ml) required.* If a straight-edge through that point is allowed to intersect the angled line in the upper left (indicating the *volume of solution*), an extension of this line gives *the weight of salt* (on the inner side of the left-hand axis) *to be added.*

Equivalent or very similar information can also be presented in the form of tables; consult the references in the list below. Note, again, that these tables have been calculated for saturation at 25°C.

In working with saturated solutions, rather than with the solid salt, the procedure is simplified by use of the following formulae:

$$\frac{C_i V_i - C_f V_f}{C_f - C_s} = V_s$$

and

$$\frac{C_i V_i + V_s C_s}{V_i + V_s} = C_f$$

where:

C_i = initial concentration in the protein-containing solution, expressed as percent saturation
C_f = final concentration in the protein-containing solution, expressed as percent saturation
V_i = initial volume of the protein-containing solution
C_s = concentration of the salt solution, expressed as percent saturation
V_s = volume of the salt solution to be added

REFERENCES

Cooper, T. G. *Tools of Biochemistry.* John Wiley & Sons, New York (1977).

Dixon, M. *Biochem. J. 54:*457 (1953).

Dixon, M., cited by Green, A. A., and Hughes, W. L. *Methods Enzymol. 1:*76–77. Academic Press, New York (1955).

Scopes, R. K. Techniques for protein purification. In *Techniques in Protein and Enzyme Chemistry,* Part 1, *Biol:*7–18. Elsevier, Amsterdam (1978).

Index